LAYING the FOUNDATION

A Resource and Strategies Guide for Middle Grades Life and Earth Science

www.apstrategies.org

Acknowledgments

Funding for the *Laying the Foundation* series was provided through a grant from the O'Donnell Foundation.

Advanced Placement Strategies gratefully acknowledges the tireless efforts of the following educators to write and edit the *Laying the Foundation* series.

Project Directors

René McCormick
AP* Strategies, Inc.
Dallas, Texas

Lisa McGaw
AP* Strategies, Inc.
Dallas, Texas

Authors

Carol Brown
Saint Mary's Hall
San Antonio, Texas

Lynn Kirby
Kealing Junior High
Austin, Texas

Lisa McGaw
AP* Strategies, Inc.
Dallas, Texas

Hugh Henderson
Plano Senior High School
Plano, Texas

Carol Leibl
James Madison High School
San Antonio, Texas

Mary Payton
AP Strategies, Inc.
Dallas, Texas

Jason Hook
Kealing Junior High
Austin, Texas

René McCormick
AP Strategies, Inc.
Dallas, Texas

Debbie Richards
Bryan High School
Bryan, Texas

Editor

Mary Payton
Editor In Chief

Contributing Authors

Randy Baskin
Rider High School
Wichita Falls, Texas

Judy Cordell
Nolan High School
Fort Worth, Texas

Brian Kaestner
Saint Mary's Hall
San Antonio, Texas

Adrian Carrales
Kealing Junior High
Austin, Texas

Ron Esman
Plano Senior High School
Plano, Texas

René Moses
Carroll High School
Southlake, Texas

Andrew Cordell
Fort Worth Country Day
School
Fort Worth, Texas

Jeff Funkhouser
Northwest High School
Justin, Texas

Mary Anne Potter
Micron Systems
Dallas, Texas

Other Contributors and Reviewers

Pat Chriswell
Fort Bend Baptist Academy
High School
Sugar Land, Texas

Kristen Jones
A&M Consolidated High
School
College Station, Texas

Dennis Ruez, Jr.
UT Austin, Department of
Geology
Austin, Texas

Charlotte Taggart
Abilene High School
Abilene, Texas

Syllabi Contributors

Rhonda Alexander
Robert E. Lee High School
Tyler, Texas

Carol Brown
St. Mary's Hall
San Antonio, Texas

Tom Campbell
Keeling Junior High
Austin, Texas

Lynn Cook
Putnam City Schools
Oklahoma City, Oklahoma

Luiz DeCarvalho
Carroll High School
Southlake, Texas

Denise DeMartino
Westlake High School
Austin, Texas

Chuy Garcia
Hyde Park Baptist
Austin, Texas

Lawanna Jenkins
Hodges Bend Middle School
Houston, Texas

Sharon Hamilton
Fort Worth Country Day
Fort Worth, Texas

Nancy Nixon
Covington Middle School
Austin, Texas

Patti O'Conner
James Madison High School
San Antonio, Texas

Camie Fillpot
O'Henry Middle School
Austin, Texas

Nancy Ramos
Northside Health Careers
High School
San Antonio, Texas

Jackie Snow
Troy High School
Troy, Texas

Charlotte Taggart
Abilene High School
Abilene, Texas

Production

Sonya Pullen
AP* Strategies, Inc.
Dallas, Texas

Table of Contents

Assessment

Appendixes

Introduction to the Laying the Foundation Series

The Laying the Foundation Series in Science is designed to support classroom teachers in better preparing students for Advanced Placement science courses. We believe this goal is readily accomplished through well-designed science programs that begin in the middle grades. These guide books are also designed to assist a school or school district in building a strong science vertical team. Each guide is designed to provide the teacher with insight into the process skills, content skills, and assessment strategies that will better prepare students as they pursue Advanced Placement science courses and other advanced coursework.

Each guide begins with a set of Foundation Lessons. The Foundation Lessons target the process skills needed to provide a solid foundation for further scientific study. These lessons should serve as a spring board for establishing the basic expectations for a science vertical team and should be reinforced at each grade level.

Pre-AP teachers often say, "If I just knew what to teach, I would teach it!" Content skills specific to each course are outlined in the appropriate LTF guide. The Pre-AP course should have a greater depth and breadth of content. Therefore, a course scope along with sample syllabi from successful Pre-AP teachers is provided to aid teachers in designing their course. It is also our belief that connections between math and science should be demonstrated and emphasized to students from an early age. The middle grades content has been divided into two LTF guides, one for life and earth sciences and one for chemistry and physics. Some middle schools have found an integrated chemistry and physics course offered to students in Algebra I to be very successful. Middle grade teachers with an integrated science course covering all four disciplines should utilize both of the middle school guides. The remaining three guides in the series focus on first-year courses in biology, chemistry and physics.

Currently, there are no Pre-AP science textbooks. The lessons found in the LTF guides model how to add the fullness to each course that is currently lacking in most widely adopted textbooks. Teacher pages included with each lesson provide correlations to the TEKS and the National Science Standards as well as a connection to the relevant AP course outline. In addition, these pages provide helpful hints for setting up laboratory activities and insightful teaching strategies. Finally, the teacher pages offer content assistance for the lessons dealing with topics that are not typically found in the on-level science textbook.

We hope that you will use these guides to enhance instruction in your classroom, formulate strong horizontal and vertical teams, and better prepare students to succeed on AP science exams. It is also our hope that you use the lessons from the guides with all of your students. Ultimately, we believe student achievement depends upon the strong expectation that all of your students are preparing for Advanced Placement science classes or advanced course work.

— René McCormick and Lisa McGaw

Overview of Laying the Foundation in Middle Grades Life and Earth Science

Middle school science teachers have the opportunity to reach out to all students and inspire a passion for science. Through science, teachers can encourage students to become enthusiastic lifelong learners. The mastery of specific concepts and basic skills lays the foundation for future success in science. This process must begin in the 6th, 7th and 8th grades. Middle school teachers play an important role in encouraging students to pursue higher academic standards and goals, which include AP classes.

Middle school courses across the country vary widely in scope, sequence, and objectives. The Pre-AP strategies presented in this guide are intended to help teachers meet their course's basic objectives yet still encourage a depth of understanding and skills that prepares students for challenging high school courses. The labs and activities presented in this book engage students in learning beyond the textbook and beyond the course minimums.

The laboratory experience is an essential and critical part of learning for middle school students. This book emphasizes activities that require students to: identify a problem, develop and test a hypothesis, use a variety of equipment to collect data, communicate the data through charts and graphs, and form a conclusion based on analyzing data. It is important for middle school teachers to introduce complex problems that have multiple correct solutions. Teaching with an open-ended approach allows students to become comfortable with this mindset.

Teaching with an inquiry approach in middle school is an excellent way to encourage students to think for themselves. There are several different approaches to inquiry learning. In the purest of inquiry-based labs, the student is given a problem and allowed to brainstorm methods for finding a solution. Many traditional labs can be modified into *guided inquiry* activities. Guided inquiry lessons provide basic instructions, yet allow students to solve problems for themselves. Using inquiry methods will create an environment of active student learners. Even material that is presented in a lecture format can be enhanced with questioning techniques that encourage students to participate in an exchange of ideas.

In addition to content and skills, assessment is an important component in advanced academic courses. This guide includes several different types of sample assessments. Lab practicals, formal exams, and rubric-based projects are all appropriate assessments that provide students consistent verbal and written feedback on their performance.

Laying the Foundation in Middle Grades Life and Earth Science will provide teachers with concrete examples of advanced academic coursework and Pre-AP strategies for middle school science. Experienced teachers have supplemented each lesson with possible student answers and helpful teacher notes. Though not intended to be an all-inclusive curriculum guide, this book is a resource for teachers looking to go beyond a traditional textbook approach to middle school science. Using the suggested syllabi as a guide, teachers can arrange the labs and activities to meet the needs of students at their school. The effective use of a vertical team of teachers will also help ensure student success.

The College Board's motto, "Equity and Excellence", can only be realized through individual teachers having the highest expectations for all students. We hope this guide will help.

— Jason Hook and Lynn Kirby

Process Skills
Progression Chart

PROCESS SKILLS PROGRESSION CHART

	Factual Knowledge	Conceptual Understanding	Reasoning and Analysis
Acquire Data By Experimentation and Observation	Identify scientific equipment, instruments, and technology Know and observe safety precautions Follow a procedure	Choose appropriate equipment and technology	Work collaboratively to obtain scientific data
Record and Manipulate Data	Measure and record data in SI units Make and record observations	Determine variables to be measured Estimate and approximate quantities Solve mathematical equations using data	Design a data table or chart as appropriate Create appropriate graphical representations of data Analyze error Apply statistical analysis such as standard deviation, percent error, and chi square
Graph and Analyze Data	Plot data points Label the axes Title the graph	Translate graph into words Scale axes Calculate slope, area, and intercepts Construct line of best fit, curve fits, regression equations	Evaluate line of best fit, curve fit, and regression equation Detect patterns in data Interpret physical meaning of slope, area, and intercepts Interpolate, extrapolate and predict from a graph Transform data into linear form Recognize cause and effect relationships Draw appropriate conclusions Apply conclusions to new situations and further investigations

	Factual Knowledge	Conceptual Understanding	Reasoning and Analysis
Communicate and Share Results		Translate data into words Read and understand scientific articles	Defend results and conclusions in both written and oral format Relate concepts to unifying themes
Design Experiments	State the purpose Practice identifying variables	Design and use models to explain scientific concepts Understand the importance of controls Apply steps of scientific method to solve a problem Formulate a feasible and practical procedure	Formulate testable questions and hypotheses Critique experimental designs Predict outcomes Make environmentally friendly choices when designing experiments
Demonstrate Mathematical Problem-solving Skills	Identify relevant given information	Substitute values into an equation and solve Use dimensional analysis Estimate reasonable answers	
Use Technology	Recognize useful data tools such as graphing calculators, probes, data collection device and computers	Use data collection tools such as graphing calculators, probes, data collection device and computers	

Foundation
Lessons

The Scientific Method
Exploring Experimental Design

Unit Overview

OBJECTIVE
Students will identify and apply the steps of the scientific method.

LEVEL
All

NATIONAL STANDARDS
UCP.1, UCP.2, UCP.3, A.1, A.2, G.2

TEKS
6.1(A), 6.2(A), 6.2(B), 6.2(C), 6.2(D), 6.2(E), 6.3(A)
7.1(A), 7.2(A), 7.2(B), 7.2(C), 7.2(D), 7.2(E), 7.3(A)
8.1(A), 8.2 (A), 8.2(B), 8.2(C), 8.2(D), 8.2(E), 8.3(A)
IPC: 1(A), 2(A), 2(B), 2(C), 2(D), 3(A)
Biology: 1(A), 2(A), 2(B), 2(C), 2(D), 3(A)
Chemistry: 1(A), 2(A), 2(B), 2(C), 2(D), 2(E), 3(A)
Physics: 1(A), 2(A), 2(B), 2(C), 2(D), 2(E), 2(F), 3(A)

CONNECTIONS TO AP
AP Science courses all contain a laboratory component where the scientific method will be used.

TIME FRAME
Two 45 minute class periods

MATERIALS

Come Fly With Us student pages   *Scientific Method Practice 1* student pages
Penny Test Lab student pages   *Scientific Method Practice 2* student pages

TEACHER NOTES

Modern scientific inquiry or science (from *scientia*, Latin for knowledge) is generally attributed to the historical contributions of Galileo Galilei and Roger Bacon, though some historians believe that their practices were inspired by earlier Islamic tradition. In spite of the rich human tradition of scientific inquiry, there is, today, no single or universal method of performing science. According to the National Science Teachers Association, science is "characterized by the systematic gathering of information through various forms of direct and indirect observations and the testing of this information by methods including, but not limited to, experimentation." Although this definition is helpful to explain the process of science, it does not specify a list of experimental steps that one should logically progress through to perform an experiment. (An experiment can be defined as an organized series of steps used to test a probable solution to a problem, commonly called a hypothesis, or educated guess.) Despite the absence of a *standard* scientific method, there is a generally agreed upon model that describes how science operates.

Steps of the Scientific Method

1. State the problem: What is the problem? This is typically stated in a question format.
   - *EXAMPLE: Will taking one aspirin per day for 60 days decrease blood pressure in females ages 12-14?*

2. Research the problem: The researcher typically will gather information on the problem. They may read accounts and journals in the subject or be involved in communications with other scientists.
   - *EXAMPLE: Some people relate stories to doctors that they feel relief from high blood pressure after taking one aspirin per day. It is not scientific if the idea is untested or if one person reports this (called anecdotal evidence).*

3. Form a probable solution, or hypothesis, to your problem. Make an educated guess as to what will solve the problem. Ideally this should be written in an *if-then* format.
   - *EXAMPLE: If a female aged 12-14 takes one aspirin per day for 60 days, then it will decrease her blood pressure.*

4. Test your hypothesis: **Do an experiment**.
   - *EXAMPLE: Test 100 females, ages 12-14, to see if taking one aspirin a day for 60 days lowers blood pressure in those females.*

   Independent Variable (I.V.): The variable you change, on purpose, in the experiment. To help students remember it suggest the phrase "**I** change it" emphasizing the **I**ndependent variable.
   - *EXAMPLE: In this described experiment, taking an aspirin or not would be the independent variable. This is what the experimenter changes between his groups in the experiment.*

Dependent Variable (D.V.): The response to the I.V.
- *EXAMPLE: The blood pressure of the individuals in the experiment, which may change from the administration of aspirin.*

Control: The group, or experimental subject, which does not receive the I.V.
- *EXAMPLE: The group of females that does not get a dose of aspirin.*

Constants: Conditions that remain the same in the experiment.
- *EXAMPLE: In this scenario some probable constants would include: only females were used, only females at around the same age, the same dosage of aspirin was given to all the individuals in the experimental group for the same defined time interval—60 days, the same brand of aspirin was given, the same type of diet was ideally given to the members of the experimental group as well as the same activity level prescribed.*

5. Recording and analyzing the data: What sort of results did you get? Typically data is organized into data tables. The data is then graphed for ease of understanding and visual appeal.
   - *EXAMPLE: Out of 100 females, ages 12-14 yrs., 76 had lower blood pressure readings after taking one aspirin per day for 60 days.*

6. Stating a conclusion: What does all the data mean? Is your hypothesis correct?
   - *EXAMPLE: It appears that taking one aspirin per day for 60 days decreases blood pressure in 76% of the tested females ages 12-14, therefore the original hypothesis has been verified, that taking aspirin can decrease blood pressure.*

7. Repeating the work: Arguably, the most important part of scientific inquiry! When an experiment can be repeated and the same results obtained by different experimenters, that experiment is validated.

Included in this unit is a hands-on lab, the *Penny Test Lab*, that can be modified for use at any level (although it was initially designed for middle school use), to teach the steps of the scientific method. Students are given the simple task of determining the number of drops that can fit on the "Lincoln" side of a penny. As the lab is designed, the students quickly learn that even the most simple of experiments can contain many hidden variables that decrease the validity of the experiment.

Another student-centered activity has been included called *Come Fly with Us*. This activity makes a great first day activity to get kids warmed up to the scientific method. They will examine what happens to the spin direction of a paper helicopter when you fold the blades in different directions. Students construct a paper helicopter to test their hypothesis about how the helicopter will fly upon folding the blades in different directions.

Suggested Teaching Procedure

Day 1

1. Present notes on the steps of the scientific method as you see fit. Although this part is teacher-directed, ideally the steps should be presented as more of a discussion. Some questions to ask during your discussion are:
   - "What is the variable that the scientist changes?"
   - "What makes a valid experiment?"
   - "Why is it important to have detailed procedures for other scientists to repeat your experiment?"
   - "Why is the control such an important part of the experiment?"

   Also, students can be asked to imagine a scientific problem while the teacher asks students, "What is the independent variable in your problem?" and so on. An example of a scientific question was just presented in these teacher notes. You can use this example or make up another example to illustrate the prescribed steps. Students should record the steps in their notebooks.

2. After students take notes, pass out the student activity pages for *Come Fly With Us*.

3. Students should read the directions and perform the prescribed tasks in the procedure, applying new-found scientific method knowledge to the activity. Students should complete and turn in *Come Fly with Us* before leaving.

4. Assign *Scientific Method Practice 1* reading and questions for homework. Students are to return the completed questions the following class period.

Day 2

5. Use the answers that follow to review *Scientific Method Practice 1* after collecting the students' papers.

6. Pass out the student activity pages for the *Penny Test Lab*. Students should read the instructions and perform the lab during class.

7. After completing the lab, students should turn in a lab write-up at the end of the period.

8. Assign *Scientific Method Practice 2* reading and questions for homework. Students should answer questions and return the completed assignment the following class period.

Day 3

9. After collecting the students' papers, use the answers that follow to review *Scientific Method Practice 2* with students.

10. At this point you can begin an introduction to *Can Mosquitoes Transmit HIV Roleplay*, a complex student inquiry-based activity also found in this *Laying the Foundation* guide.

The Scientific Method
Exploring Experimental Design

Come Fly With Us

OBJECTIVE
Students will practice applying the steps of the scientific method to a problem.

LEVEL
All levels

NATIONAL STANDARDS
UCP.1, UCP.2, UCP.3, A.1, A.2, G.2

TEKS
6.1(A), 6.2(A), 6.2(B), 6.2(C), 6.2(D), 6.2(E), 6.3(A)
7.1(A), 7.2(A), 7.2(B), 7.2(C), 7.2(D), 7.2(E), 7.3(A)
8.1(A), 8.2 (A), 8.2(B), 8.2(C), 8.2(D), 8.2(E), 8.3(A)
IPC: 1(A), 2(A), 2(B), 2(C), 2(D), 3(A)
Biology: 1(A), 2(A), 2(B), 2(C), 2(D), 3(A)
Chemistry: 1(A), 2(A), 2(B), 2(C), 2(D), 2(E), 3(A)
Physics: 1(A), 2(A), 2(B), 2(C), 2(D), 2(E), 2(F), 3(A)

CONNECTIONS TO AP
Using the scientific method by acquiring data through experimentation and design of experiments are all fundamental skills needed for the AP Science courses.

TIME FRAME
45 minutes

MATERIALS

28 models of helicopter (provided)   28 pairs of scissors
28 pens or pencils

TEACHER NOTES
Come Fly With Us is an effective way for students to experimentally test a variable in a simple activity. This activity is designed to be the first activity that students do after learning the steps of the scientific method. The students can apply their newfound knowledge in a meaningful way.

Students cut out and fold a paper helicopter according to the instructions on the lab. After constructing the simple helicopter, students are instructed to fold the blades of the helicopter in opposing directions. Students generate a hypothesis as to how they think it will affect the direction of spin. The students then test their hypothesis and fly the helicopter after folding the blades in each direction. The students will discover that folding the blade one way will produce a clockwise spin of the helicopter. Folding the blades in the opposite direction will produce a counterclockwise spin of the helicopter. The students will ideally discover that applying the independent variable (folding the blades in opposing directions) causes the dependent variable to change (the direction of spin clockwise or counterclockwise). The students must also take into consideration the constants in their experiment: holding the helicopter at the same initial height, maintaining a stable wind environment, no other external forces acting on the helicopter, and holding the helicopter at the T each time. Although a control setup is not at first apparent, it is illustrated later.

The control is best illustrated in the second half of *Come Fly With Us*. A fictitious student, Bonita, believes that adding mass (paper clips) will stabilize her paper helicopter and increase the flight time. The independent variable is the presence or absence of the added mass. The control, by definition, does not receive the independent variable. Therefore, the control in Bonita's experiment is a paper helicopter with no paper clips added. The students can typically clearly envision the idea of a control. Further discussion of a control can describe how the experimental subject can only be truly tested when the results of the control setup are compared to the results of the experimental setup. Then the effect of the independent variable upon the dependent variable can clearly be seen. An extension of this lab could be to have the students actually try testing the extra weight and seeing how it affects the flight time.

POSSIBLE ANSWERS TO THE CONCLUSION QUESTIONS

1. In the helicopter experiment, what was the independent variable?
   - Folding the blades in different directions, with the black circle up and the white square down, or with the black circle down and the white square up.

2. What was the dependent variable?
   - The dependent variable is the direction of spin, clockwise or counterclockwise.

3. List three things you should try to keep constant each time you try this experiment.
   - There are many correct answers for this question. Possible answers include:
     - holding the helicopter in the same place (on the body versus the wing)
     - holding it at the same height
     - making sure there is no cross breeze each time
     - using the same helicopter
     - adding no extra force when letting it go each time

4. What is the problem question in Bonita's experiment?
   - Will adding extra mass in the form of paper clips to the helicopter stabilize it, making it stay in the air longer?

5. What is Bonita's hypothesis?
   - If additional paperclips are added to the helicopter, then the helicopter will be stabilized resulting in a longer flight time.

6. What is her independent variable?
   - Bonita's independent variable is the addition of paper clips (weight) to the helicopter.

7. What is her dependent variable?
   - Bonita's dependent variable is the amount of time the helicopter stays in the air.

8. What should her constants be?
   - Her constants should be the same as those listed in #3, plus: use the same size paper clips, attach the paper clips to the same place on the helicopter each time, etc...

9. What can she use for a control?
   - Her control is the same helicopter with no added mass.

10. Why should Bonita retest her experiment between 5-10 times?
    - Bonita should retest to make sure her results are reasonable and valid.

TEACHER PAGES

The Scientific Method
Exploring Experimental Design
Scientific Method Practice 1

POSSIBLE ANSWERS TO THE CONCLUSION QUESTIONS
NOTE: Problem, hypothesis and conclusion should all match in wording!

1. What was Erika's problem? [The problem should be stated as a question.]
   - Is the oven heating to the correct temperature? OR
   - Why didn't the cake rise?

2. What was Erika's hypothesis? [This is an answer to your problem question.]
   - No, the oven is not heating to the correct temperature. OR
   - The cake did not rise because the oven was not heating to the correct temperature.

3. What was Erika's conclusion? [This states whether your hypothesis was correct.]
   - The oven is heating to the correct temperature. OR
   - The oven was heating to the correct temperature and therefore could not have been the cause of the cake's failure to rise.

4. Which step in the scientific method do you think Erika should do next? Explain your reasoning.
   - Form a new hypothesis OR gather more information OR repeat the experiment.

5. List two other hypotheses which might explain why the cake did not rise.
   - Answers will vary

The Scientific Method
Exploring Experimental Design

Penny Test Lab

OBJECTIVE
Students will learn about controls and variables in an experiment. Additionally, they will learn what constitutes valid experimental procedure.

LEVEL
All levels

NATIONAL STANDARDS
UCP.1, UCP.2, UCP.3, A.1, A.2, G.2

TEKS
6.1(A), 6.2(A), 6.2(B), 6.2(C), 6.2(D), 6.2(E), 6.3(A)
7.1(A), 7.2(A), 7.2(B), 7.2(C), 7.2(D), 7.2(E), 7.3(A)
8.1(A), 8.2 (A), 8.2(B), 8.2(C), 8.2(D), 8.2(E), 8.3(A)
IPC: 1(A), 2(A), 2(B), 2(C), 2(D), 3(A)
Biology: 1(A), 2(A), 2(B), 2(C), 2(D), 3(A)
Chemistry: 1(A), 2(A), 2(B), 2(C), 2(D), 2(E), 3(A)
Physics: 1(A), 2(A), 2(B), 2(C), 2(D), 2(E), 2(F), 3(A)

CONNECTIONS TO AP
Using the scientific method by acquiring data through experimentation and design of experiments are all fundamental skills needed for the AP Science courses.

TIME FRAME
50 minutes

MATERIALS

28 pennies
28 eyedroppers
28 small beakers of water
28 paper towel
28 pencils or pens

28 calculators
28 pieces of graph paper (in appendix)
28 metric rulers
28 pieces of notebook paper

Laying the Foundation in Middle Grades Life and Earth Science

TEACHER NOTES

The *Penny Test Lab* takes a simple problem, how many drops of water will fit onto the "Lincoln" side of a penny, and expands it into an excellent lab that can be used to study the steps of the scientific method. To summarize, students use an eyedropper to determine how many drops of water will fit onto the penny before it spills over. This seemingly simple task generates many diverse results. By definition, a valid experiment is one that can be repeated by anyone else with the same results obtained. The lack of similar student results verifies that there are hidden variables that are unaccounted for in the procedure. This lab procedure is designed to show the students what an invalid experiment looks like.

Some of the hidden variables include pennies of different ages and conditions and different droppers (some plastic, some glass). Also, no exact procedure is given to the students about how to hold the dropper, how much pressure to put on the dropper, how to make a drop, how to drop it onto the penny, or from what height the drop should be released.

Before class, draw on the chalkboard or the overhead a data table with three rows and as many columns as there are students in the classroom. The first row should be labeled "student initials". The second row will be labeled "predicted" and the bottom row be labeled "observed".

One of the first steps of the procedure is for students to make a prediction about the number of drops they believe will fit on the "Lincoln" side of a penny. If students have never done a lab like this before or have no knowledge of the cohesive properties of water, they tend to underestimate the number of drops that will actually fit. Have students write their initials and their predicted number of drops on the chalkboard. Students then perform three trials, take an average from these three trials, and round it to the nearest whole number. Once they have completed this, they should write their average whole number of drops in the space provided on the class data table.

student initials	jh	hs	if	hg	sc	br
predicted #	12	8	13	20	26	11
actual #	23	16	35	12	22	13

A partial data table is shown here.

After everyone has completed their trials, students will analyze the class data by graphing the frequency of ranges of drops. Together as a class, you and the students will count the number of people that averaged 0-10 drops. Repeat this procedure for the ranges of 11-20, 21-30, 31-40, 41-50, etc. Instruct students to make a bar graph (technically a histogram since the intervals are equivalent) showing the number of people on the *y*-axis versus the range of drops on the *x*-axis.

Here is an example of a possible student histogram.

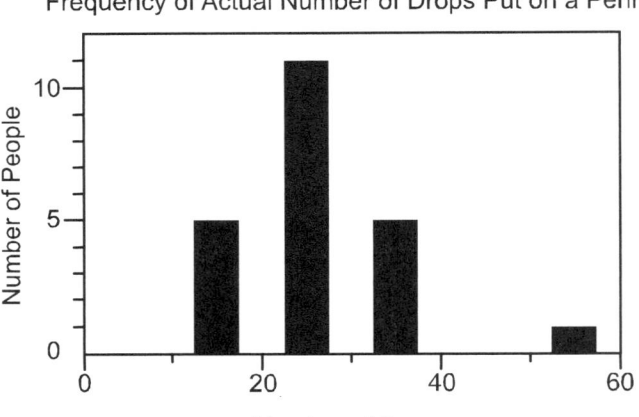

Frequency of Actual Number of Drops Put on a Penny

Before the students answer the questions, lead a discussion with students regarding what is considered a valid experiment versus an invalid experiment. Most students come to the conclusion that this is an invalid experiment due to the diverse results. Also, the discussion can include student ideas of hidden variables. These will include: different droppers, different pennies, no exact procedure for dropping, no definition of "drop," unstable table, etc. You can reveal to the students that this lab was purposefully designed to produce invalid results so that students could begin to understand that even in a simple task there can be many hidden variables.

POSSIBLE ANSWERS TO THE CONCLUSION QUESTIONS AND SAMPLE DATA

A typical data table for the student's 3 trials:

	Trial #1	Trial #2	Trial #3	Average
Number of Water Drops	_____	_____	_____	_____

1. Using your bar graph, determine if the average number of drops for each experimenter is about the same?
   - No, the results are not the same.

2. List four reasons why the actual number of drops for each experimenter was similar or dissimilar.
   - Student answers will vary
   - Four possible reasons:
     a. different types pennies
     b. no standard way of administering drops
     c. the table could be uneven
     d. each dropper is physically different and delivers drops of varying volume

3. Are the results of this experiment "valid"? Why or why not? Be sure to think about what makes an experiment valid.
   - Results are not valid in this experiment. To make an experiment valid, the results should be repeatable regardless of experimenter.

4. In this experiment, there were a limited number of constants. Name two of them.
   - Water (as opposed to alcohol or some other fluid)
   - Using pennies (as opposed to using nickels and pennies, etc…)

5. What was the independent variable in this experiment?
   - Student answers will vary. Some variation or factor that could affect the outcome (the dependent variable) could be accepted as the independent variable: height of dropper, size of dropper hole, pressure used when squeezing the dropper, size of drops, and so on.

6. What was the dependent variable in this experiment?
   - The dependent variable is the number of drops that fit on the head side of a penny.

7. Is it possible to state definitively how many drops of water will fit on the "Lincoln" side of a penny with this lab procedure? Why or why not?
   - Using this procedure it is not possible to state exactly how many drops fit onto a penny. This is not a valid experiment. There are too many hidden variables.

The Scientific Method
Exploring Experimental Design

Scientific Method Practice 2

POSSIBLE ANSWERS TO THE CONCLUSION QUESTIONS

1. State the problem in the form of a question.
   - What causes fresh water to freeze at a higher temperature than sea water?

2. Form a hypothesis to answer the problem question above based on the fact that fresh water does not contain salt.
   - The salt in sea water lowers the temperature at which water freezes.

3. According to the data table above, at what temperature did the experiment begin?
   - The experiment began at 25°C.

4. At what time intervals were the temperature measurements taken?
   - The time intervals were 5 minutes.

5. What conclusions can you draw from these graphs about the effect of salt on the freezing point of water?
   - Salt lowers the freezing point of water.

6. What can you say about the rate at which the temperature in the fresh water container dropped compared to the rate at which the temperature in the salt water container dropped?
   - The rate at which the temperature in the fresh water container dropped was the same as the rate at which the temperature in the salt water container dropped.

7. What was the independent variable in Stephanie and Amy's experiment?
   - The independent [manipulated] variable was the addition of salt.

8. What was the dependent variable?
   - The dependent [responding] variable was the temperature at which the water froze.

9. Explain why detailed, step-by-step written procedures are an essential part of any scientific experiment.
   - When a scientist writes a report on his or her experiment, it must be detailed enough so that scientists throughout the world can repeat the experiment for themselves. In many cases, it is only when an experiment has been repeated by scientists worldwide that it is considered to be accurate.

10. The following hypothesis is suggested to you: Water will heat up faster when placed under the direct rays of the sun than when placed under indirect, or angled, rays of the sun. Design an experiment to test this hypothesis. Be sure to number each step of your procedure. Identify your independent variable, dependent variable and control. Identify those things which will remain constant during your experiment.
    • Answers will vary.

TEACHER PAGES

The Scientific Method
Exploring Experimental Design

Overview

PURPOSE

Through this series of activities you will identify and apply the steps of the scientific method.

MATERIALS

Come Fly With Us   *Scientific Method Practice 1*
Penny Test Lab   *Scientific Method Practice 2*

PROCEDURE

Day 1

1. Take notes in your notebook from your teacher's discussion about the steps of the scientific method.

2. Do the *Come Fly With Us* activity in class. Turn in at the end of the period.

3. Do the *Scientific Method Practice 1* activity for homework. Be ready to turn in the write-up at the beginning of the next period.

Day 2

4. Turn in *Scientific Method Practice 1* to your teacher at the beginning of class. Your teacher will review the correct answers.

5. Begin *Penny Test Lab* after collecting your materials.

6. Turn in the *Penny Test Lab* write-up to your teacher after completing it.

7. Do the *Scientific Method Practice 2* activity for homework. Be ready to turn in the write-up at the beginning of the next period.

Day 3

8. Turn in *Scientific Method Practice 2* to your teacher at the beginning of class. Your teacher will review the correct answers.

The Scientific Method
Exploring Experimental Design

Come Fly With Us

This assignment is intended to be a quick and easy guide to the methods scientists use to solve problems. It should also give you information about how to "wing your way" through your own experiments. You are going to start by making a model helicopter with the attached instructions. You will be given a problem question, and it is your job to write a suitable hypothesis. Remember, your hypothesis should be a possible answer to the problem question and it should be based upon what you already know about a topic.

GLOSSARY OF WORDS USED IN CONDUCTING EXPERIMENTS

- **problem**: scientific question that can be answered by experimentation.
- **hypothesis**: an educated prediction about how the independent variable will affect the dependent variable stated in a way that is testable. This should be an "If…then…" statement.
- **variable**: a factor in an experiment that changes or could be changed
- **independent variable**: the variable that is changed on purpose.
- **dependent variable**: the variable that responds to the independent variable.
- **control**: the standard for comparison in an experiment; the independent variable is not applied to the control group.
- **constant**: a factor in an experiment that is kept the same in all trials.
- **repeated trials**: the number of times an experiment is repeated for each value of the independent variable.

PURPOSE
In this assignment you will practice applying the steps of the scientific method to a problem by experimentation.

MATERIALS

model of helicopter (provided)   scissors
pen

PROCEDURE

1. Find the section labeled Hypothesis on your student answer page. Read the problem question and respond with an appropriate hypothesis. Remember to use an "If...then..." format.

2. Once you have made your hypothesis, you should test it for accuracy. Stand on a chair and hold your helicopter by the "T" at shoulder level.

3. Drop the helicopter and note whether it spins clockwise or counterclockwise. Repeat this test several times.

4. Refold the blades so that the square on section Y shows when you look down on top of the helicopter.

5. Stand on a chair and hold your helicopter by the top of the "T" at shoulder level. Drop the helicopter and note whether it spins clockwise or counterclockwise.

6. Repeat this test several times.

Name _____

Period _____

The Scientific Method
Exploring Experimental Design

Come Fly With Us

PROBLEM

How will changing the direction that the paper helicopter blades are folded affect the "flight" of the helicopter?

HYPOTHESIS

ANALYSIS

You have just performed an experiment. Experiments involve changing something to see what happens. In this case, you refolded the helicopter blades. You made this change on purpose to learn about its effect on the flight of the helicopter. The parts of an experiment that change are called *variables*.

When designing an experiment, you should choose one variable that you will purposely change. You will measure the effect of this *independent variable* on another variable that you think will respond to the change. The responding variable is called the *dependent variable*.

If you kept every variable except the folds the same in each test, you were making it a fair test. Why? Only the variable you changed could be causing the dependent variable to change because everything else was kept constant.

To have a fair test, you also need a *control*, or a standard for comparison. A control for the helicopter experiment would be an "unchanged" helicopter against which you could compare the results. You could make another helicopter as your standard for comparison and not refold its blades.

It is important to note that in some experiments, it is impossible to have a control that is completely unchanged. For example, let's say you are trying to determine the effect of light from different light sources on plant growth. The control plant needs some kind of light in order to live through the experiment. So, you have to choose one light source — any one say, normal sunlight — to be the standard of comparison.

After you refolded the blades of the helicopter, you dropped the helicopter several times and observed the results. These repeated trials enable you to be more confident of your results. If you conducted your experiment only once, the results could be due to an error or a chance event, such as a draft. But, when you repeat your experiment many times and each time achieve similar results, you can be more confident that your findings are not due to an error or chance.

Read the following paragraph and then answer the conclusion questions that follow using complete sentences:

> Bonita wanted to know if adding mass to her paper helicopter would affect how long it would stay in the air. She predicted that adding some mass would help to stabilize the helicopter and keep it in the air longer than a helicopter without extra mass. She experimented with different numbers of paper clips attached to her helicopter.

CONCLUSION QUESTIONS

1. In the helicopter experiment, what was the independent variable?

2. What was the dependent variable?

3. List three things you should try to keep constant each time you try this experiment.
   a.
   b.
   c.

4. What is the problem question in Bonita's experiment?

5. What is Bonita's hypothesis?

6. What is her independent variable?

7. What is her dependent variable?

8. What should her constants be?

9. What can she use for a control?

10. Why should Bonita retest her experiment between 5-10 times?

PAPER HELICOPTER MODEL

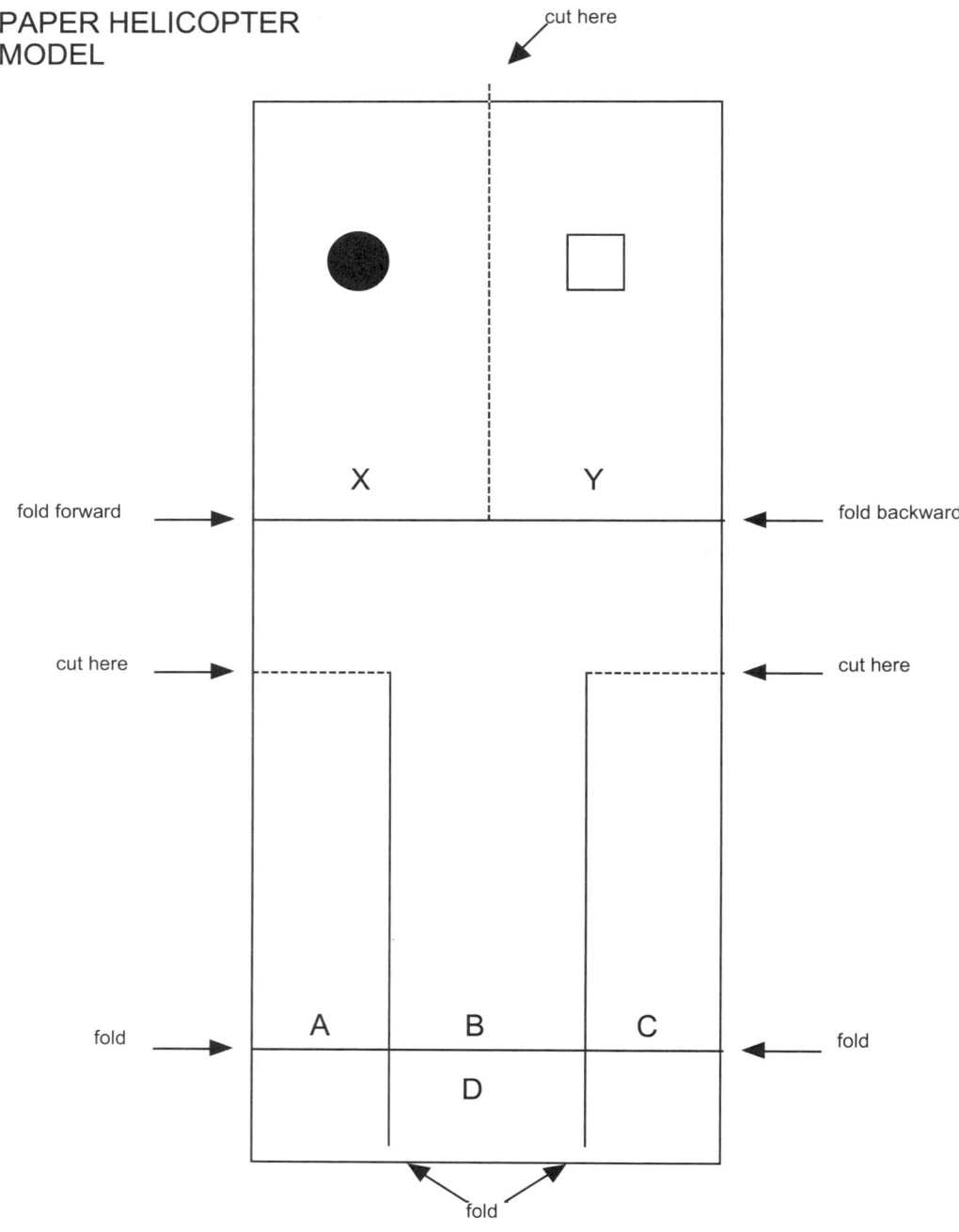

INSTRUCTIONS:
1. Cut out the rectangular helicopter (above).
2. Now cut along dotted lines.
3. Fold along the solid lines: section C behind section B, section A behind section B, and section D behind section B.
4. Complete the helicopter by folding blade X with the dot up and blade Y in the opposite direction with the square down.

Name _____

Period _____

The Scientific Method
Exploring Experimental Design

Scientific Method Practice 1

DIRECTIONS: *Read the following paragraphs and then answer the questions that follow on a separate sheet of paper. Use complete sentences to answer all questions. Be certain to restate the question in your answer.*

Science differs from other subject areas in the way it seeks to answer questions. This approach to problem solving is called the scientific method. The scientific method is a systematic approach to problem solving. The basic steps, in no particular order, of the scientific method are:

- Stating the problem
- Gathering information on the problem
- Forming a hypothesis
- Performing experiments to test the hypothesis
- Recording and analyzing data
- Stating a conclusion
- Repeating the work

Erika baked a cake for her mother's birthday. When the cake was taken from the oven, Erika noticed that the cake had not risen. She guessed that the oven had not heated to the correct temperature. She set up the following experiment to test her hypothesis.

First, Erika put a thermometer in the oven. She then turned the oven dial to 375°F. She noticed that the preheating light came on when she turned the oven on. She waited until the preheating light went out, indicating that the oven was up to temperature. Erika then read the thermometer within the oven. It read 400°F. Erika concluded that the oven was heating properly.

CONCLUSION QUESTIONS

1. What was Erika's problem? [The problem should be stated as a question.]

2. What was Erika's hypothesis? [This is an answer to your problem question.]

3. What was Erika's conclusion? [This states whether your hypothesis was correct.]

4. Which step in the scientific method do you think Erika should do next? Explain your reasoning.

5. List two other hypotheses which might explain why the cake did not rise.

Name _____

Period _____

The Scientific Method
Exploring Experimental Design

Penny Test Lab

PURPOSE
In this activity you will learn about controls and variables in an experiment. You will also learn what constitutes valid experimental procedure.

MATERIALS

penny	calculator (optional)
eyedropper	graph paper
water	ruler
paper towel	pen

DEFINITIONS
- **variable**: things in an experiment that change or could be changed.
- **independent variable**: variable that is changed on purpose.
- **dependent variable**: variable that responds to the independent variable.
- **constant**: things in an experiment that are kept the same in all trials.

PROCEDURE
1. Answer each of the following questions using complete sentences. For fill-in-the blank statements, copy the entire sentence.

2. Copy the lab purpose onto your paper.

3. Your task is to guess how many drops of water will fit on the "Lincoln" side of a penny.

   [Copy the following statement.] PROBLEM: How many drops of water will fit onto the "Lincoln" side of a penny?

4. [Copy the following statement and make a prediction by filling in the blank.] HYPOTHESIS: I predict that _____ drops of water will fit on the head side of a penny.

5. After you have made your hypothesis and have written it down on your lab paper, you will write it on the chalkboard under the heading "Predicted Number of Drops."

6. Copy the following chart onto your paper. Be neat and use a ruler!

TEST RESULTS:

	Trial #1	Trial #2	Trial #3	Average
Number of Water Drops	_____	_____	_____	_____

7. Test to see if your hypothesis is correct. Place your penny on a paper towel and, using the eyedropper, add water to the "Lincoln" side of the penny, one drop at a time, counting each drop until the water spills over. Do not count the drop that causes the water to spill over. Write the number of drops you counted under Trial #1 on your chart. Repeat this procedure two more times, filling in the number of drops you counted for each trial under the appropriate heading on your Test Results chart.

8. Find the average of your three trials, round your answer to a whole number (no decimals), and write the average number of drops on your Test Results chart. Then write your average on the chalkboard under the heading "Actual Average".

9. Write a sentence that will serve as your conclusion for this experiment. Remember that your conclusion should state whether your hypothesis was correct.

10. Make a bar graph of the class test results using the data from "Actual Average" on the chalkboard. The *x*-axis (horizontal line) should be titled "Average Number of Drops" and the *y*-axis (vertical line) should be titled "Number of Tests." Before graphing, you will need to organize the class data into ranges — make a chart that shows how many people got averages between 0-10, 11-20, 21-30, etc. When you have finished your bar graph, give it an appropriate title.

Answer the conclusion questions on your paper. Be sure to use complete sentences.

CONCLUSION QUESTIONS

1. Using your bar graph, determine if the average number of drops for each experimenter is about the same.

2. List four reasons why the actual number of drops for each experimenter was similar or dissimilar.

3. Are the results of this experiment "valid"? Why or why not? Be sure to think about what makes an experiment valid.

4. In this experiment, there were a limited number of constants. Name two of them.

5. What was the independent variable in this experiment?

6. What was the dependent variable in this experiment?

7. Is it possible to state definitively how many drops of water will fit on the "Lincoln" side of a penny with this lab procedure? Why or why not?

Name _____

Period _____

The Scientific Method
Exploring Experimental Design

Scientific Method Practice 2

DIRECTIONS: *Answer these questions on a separate sheet of paper using a black or dark blue ink pen. Use complete sentences to answer all questions. Be certain to restate the question in your answer!*

Stephanie and Amy were vacationing in Canada. Bundled up in warm clothing, they walked along the beach. Glistening strips of ice hung from the roofs of the beach houses. Only yesterday, Stephanie commented, these beautiful icicles had been a mass of melting snow. Throughout the night, the melted snow had continued to drip, freezing into lovely shapes. Near the ocean's edge, Amy spied a small pool of sea water. Surprisingly, she observed, it was not frozen as were the icicles on the roofs. What could be the reason, they wondered?

A scientist might begin to solve the problem by gathering information. The scientist would first find out how the sea water in the pool differs from the fresh water on the roof. This information might include the following facts: The pool of sea water rests on sand, while the fresh water drips along a tar roof. The sea water is exposed to the cold air for less time than the fresh water. The sea water is saltier than the fresh water.

Using all of the information that has been gathered, the scientist might be prepared to suggest a possible solution to the problem. A proposed solution to a scientific problem is called a hypothesis. A hypothesis almost always follows the gathering of information about a problem. Sometimes, however, a hypothesis is a sudden idea that springs from a new and original way of looking at a problem.

A scientist (or a science student) does not stop once a hypothesis has been suggested. In science, evidence that either supports a hypothesis or does not support it must be found. This means that a hypothesis must be tested to show whether it is correct. Such testing is usually done by performing experiments.

Experiments are performed according to specific rules. By following these rules, scientists can be confident that the evidence they uncover will clearly support or not support a hypothesis. For the problem of the sea water and freshwater, a scientist would have to design an experiment that ruled out every factor but salt as the cause of the different freezing temperatures. Stephanie and Amy, being excellent science students, set up their experiment in the following as follows.

First, they put equal amounts of fresh water into two identical containers. Then Stephanie added salt to only one of the containers. [The salt is the variable. In any experiment, only one variable should be tested at a time. In this way, scientists can be fairly certain that the results of the experiment are caused by one and only one factor — in this case the variable of salt.] To eliminate the possibility of hidden or unknown variables, Stephanie and Amy conducted a control experiment. A control experiment is set up exactly like the one that contains the variable. The only difference is that the control experiment does not contain the variable. Scientists compare the results of an experiment to a control experiment.

In the control experiment, Stephanie and Amy used two containers of the same size with equal amounts of water. The water in both containers was at the same starting temperature. The containers were placed side by side in the freezing compartment of a refrigerator and checked every five minutes. But only one container had salt in it. In this way, they could be fairly sure that any differences that occurred in the two containers were due to the single variable of salt. In such experiments, the part of the experiment with the variable is called the experimental setup. The part of the experiment with the control is called the control setup.

Stephanie and Amy collected the following data: the time intervals at which the containers were observed, the temperatures of the water at each interval, and whether the water in either container was frozen or not. They recorded the data in the tables below and then graphed their results.

WATER (control setup)

Time (min)	0	5	10	15	20	25	30	
Temperature (°C)	25	20	15	10	5	0*	-10	

   *Asterisk means liquid has frozen

WATER WITH SALT (experimental setup)

Time (min)	0	5	10	15	20	25	30	
Temperature (°C)	25	20	15	10	5	0	-10*	

Stephanie and Amy might be satisfied with their conclusion after just one test. For a scientist, however, the results from a single experiment are not enough to reach a conclusion. A scientist would want to repeat the experiment many times to be sure the data were reproducible. So, a scientific experiment must be able to be repeated. And before the conclusion of a scientist can be accepted by the scientific community, other scientists must repeat the experiment and check the results. Consequently, when a scientist writes a report on his or her experiment, that report must be detailed enough so that scientists throughout the world can repeat the experiment for themselves. In most cases, it is only when an experiment has been repeated by scientists worldwide that it is considered to be accurate and worthy of being included in new scientific research.

By now it might seem as if science is a fairly predictable way of studying the world. After all, you state a problem, gather information, form a hypothesis, run an experiment, and determine a conclusion. Well, sometimes it isn't so neat and tidy.

In practice, scientists do not always follow all the steps in the scientific method. Nor do the steps always follow the same order. For example, while doing an experiment a scientist might observe something unusual or unexpected. That unexpected event might cause the scientist to discard the original hypothesis and suggest a new one. In this case, the hypothesis actually followed the experiment.

As you already learned, a good rule to follow is that all experiments should have only one variable. Sometimes, however, scientists run experiments with several variables. Naturally, the data in such experiments are much more difficult to analyze. For example, suppose scientists want to study lions in their natural environment in Africa. It is not likely they will be able to eliminate all the variables in the environment and concentrate on just a single lion. So, although a single variable is a good rule and you will follow this rule in almost all of the experiments you design or perform, it is not always practical in the real world.

There is yet another step in the scientific method that cannot always be followed. Believe it or not, many scientists search for the truths of nature without ever performing experiments. Sometimes the best they can rely on are observations and natural curiosity. Here is an example. Charles Darwin is considered the father of the theory of evolution, how living things change over time. Much of what we know about evolution is based on Darwin's work. Yet Darwin did not perform a single controlled evolutionary experiment! He based his hypotheses and theories on his observations of the natural world. Certainly it would have been better had Darwin performed experiments to prove his theory of evolution. But as the process of evolution generally takes thousands, even millions of years, performing an experiment would be a bit too time consuming.

CONCLUSION QUESTIONS

1. State Stephanie and Amy's problem in the form of a question.

2. Form a hypothesis (to answer the problem question above) based on the fact that fresh water does not contain salt.

3. According to the data table above, at what temperature did the experiment begin?

4. At what time intervals were the temperature measurements taken?

5. What conclusions can you draw from these graphs about the effect of salt on the freezing point of water?

6. What can you say about the rate at which the temperature in the fresh water container dropped compared to the rate at which the temperature in the salt water container dropped?

7. What was the independent variable in Stephanie and Amy's experiment?

8. What was the dependent variable?

9. Explain why detailed, step-by-step written procedures are an essential part of any scientific experiment.

10. The following hypothesis is suggested to you: Water will heat up faster when placed under the direct rays of the sun than when placed under indirect, or angled, rays of the sun. Design an experiment to test this hypothesis. Be sure to number each step of your procedure. Identify your independent variable, dependent variable and control. Identify those things which will remain constant during your experiment.

Numbers in Science
Exploring Measurements, Significant Digits, and Dimensional Analysis

OBJECTIVE
Students will be introduced to correct measurement techniques, correct use of significant digits, and dimensional analysis.

LEVEL
All

NATIONAL STANDARDS
UCP.1, UCP.3, A.1, G.2

TEKS
6.1(A), 6.2(A), 6.2(B), 6.2(C), 6.2(D), 6.4(A)
7.1(A), 7.2(A), 7.2(B), 7.2(C), 7.2(D), 7.4(A)
8.1(A), 8.2 (A), 8.2(B), 8.2(C), 8.2(D), 8.4(A)
IPC: 1(A), 2(A), 2(B), 2(C), 2(D)
Biology: 1(A), 2(A), 2(B), 2(C), 2(D)
Chemistry: 1(A), 2(A), 2(B), 2(C), 2(E)
Physics: 1(A), 2(A), 2(B), 2(C), 2(D), 2(F)

CONNECTIONS TO AP
Students are expected to report measurements and perform calculations with the correct number of significant digits.

TIME FRAME
180 minutes, depending on level

MATERIALS
small cube
metric ruler
200 mL beaker
large graduated cylinder
spherical object
tweezers
flexible tape measure
balance

TEACHER NOTES
Small wooden alphabet blocks or dice should be inexpensive and easy to obtain. Cubic shaped ice could also be used. Be sure to find a cube/graduated cylinder combination that ensures total submersion of the cube since its volume will be determined by water displacement.

Spherical objects might be a large marble or small rubber ball. Again, be sure to check the sphere/cylinder size to ensure that total submersion of the sphere is possible.

For the flexible tape measure, photocopy the metric side of a tape measure and have students cut it out on paper. Or, provide students with a length of string and metric ruler. The string can be wrapped around the sphere, marked, and then removed and measured.

This lesson is designed to introduce or reinforce accurate measurement techniques, correct usage of significant digits, and dimensional analysis. Dimensional analysis is also called the Factor-Label method or Unit-Label method and is a technique for setting up problems based on unit cancellations. Lecture as well as guided and independent practice of these topics should precede this activity. Students should be provided with reference tables containing metric and English conversion factors.

The purpose of significant digits is to communicate the accuracy of a measurement as well as the measuring capacity of the instrument used. Remind students repeatedly to take measurements including an estimated digit and to perform their calculations with the correct number of significant digits. Emphasize that points will be deducted for answers containing too many or too few significant digits. The correct number of significant digits to be reported by your students will depend entirely upon your equipment.

POSSIBLE ANSWERS TO CONCLUSION QUESTIONS AND SAMPLE DATA

Introduction
Left: _____ 5.75 mL _____

Middle: _____ 3.0 mL _____

Right: _____ 0.33 mL _____

DATA AND OBSERVATIONS

Data Table
Cube

Mass:	15.05 g (4sd)		
Dimensions	length: 3.68 cm (3sd)	width: 3.65 cm (3 sd)	height: 3.67 cm (3 sd)
Volume	Beaker initial volume: 100 mL (1sd)		Beaker Final volume: 150 mL (2 sd)
	Graduated cylinder initial volume: 175.0 mL (4 sd)		Graduated cylinder final volume: 225.1 mL (4 sd)

	Sphere		
Mass:	19.38 g (4sd)		
Dimensions	Circumference: 7.62 cm (3sd)		
Volume	Beaker initial volume: 100 mL (1sd)		Beaker final volume: 110 mL (2sd)
	Graduated cylinder initial volume: 175.0 mL (4sd)		Graduated cylinder final volume: 182.3 mL (4sd)

Formula for calculating the volume of a cube:	V = length x width x height
Formula for calculating the circumference of a circle:	C = πd
Formula for calculating the diameter of a circle:	d = 2r
Formula for calculating the volume of a sphere:	$V = \dfrac{4}{3}\pi r^3$

ANALYSIS

- Remember to follow the rules for reporting all data and calculated answers with the correct number of significant digits.
- You may need tables of metric and English conversion factors to work some of these problems.

1. For each of the measurements you recorded above, go back and indicate the number of significant digits in parentheses after the measurement. Ex: 15.7 cm (3sd)
    - The number of significant digits will be determined by the equipment you are using.

2. Use dimensional analysis to convert the mass of the cube to (a) mg and (b) ounces.

    (a) $15.05 \, \cancel{g} \times \dfrac{1000 \, mg}{1 \, \cancel{g}} = 15{,}050 \, mg$

    (b) $15.05 \, \cancel{g} \times \dfrac{1 \, \cancel{lb}}{454 \, \cancel{g}} \times \dfrac{16 \, oz}{1 \, \cancel{lb}} = 0.5304 \, oz$

3. Calculate the volume of the cube in cm^3.
    - V=l×w×h

        $V = 3.68 \, cm \times 3.65 \, cm \times 3.67 \, cm = 49.3 \, cm^3$

4. Use dimensional analysis to convert the volume of the cube from cm^3 to m^3.

- $49.3 \, cm^3 \times \dfrac{1 \, m}{100 \, cm} \times \dfrac{1 \, m}{100 \, cm} \times \dfrac{1 \, m}{100 \, cm} = 4.93 \times 10^{-5} \, m^3$

5. Calculate the volume of the cube in mL as measured in the beaker. Convert to cm^3 knowing that $1 \, cm^3 = 1 \, mL$.

- $V = V_{final} - V_{initial}$

 $V = 150 \, mL - 100 \, mL$

 $V = 50 \, mL = 50 \, cm^3$

6. Calculate the volume of the cube in mL as measured in the graduated cylinder. Convert to cm^3 knowing that $1 \, cm^3 = 1 \, mL$.

- $V = V_{final} - V_{initial}$

 $V = 225.1 \, mL - 175.0 \, mL$

 $V = 50.1 \, mL = 50.1 \, cm^3$

7. Using the density formula $D = \dfrac{mass}{volume}$, calculate the density of the cube as determined by the (a) ruler (b) beaker (c) graduated cylinder.

(a) $D = \dfrac{15.05 \, g}{49.3 \, cm^3}$

 $D = 0.305 \, \dfrac{g}{cm^3}$

(c) $D = \dfrac{15.05 \, g}{50.1 \, cm^3}$

 $D = 0.300 \, \dfrac{g}{cm^3}$

(b) $D = \dfrac{15.05 \, g}{50 \, cm^3}$

 $D = 0.3 \, \dfrac{g}{cm^3}$

8. Use dimensional analysis to convert these three densities into kg/m^3.

(a) $0.305 \, \dfrac{g}{cm^3} \times \dfrac{1 \, kg}{1000 \, g} \times \dfrac{100 \, cm}{1 \, m} \times \dfrac{100 \, cm}{1 \, m} \times \dfrac{100 \, cm}{1 \, m} = 305 \, \dfrac{kg}{m^3}$

(b) $0.3 \, \dfrac{g}{cm^3} \times \dfrac{1 \, kg}{1000 \, g} \times \dfrac{100 \, cm}{1 \, m} \times \dfrac{100 \, cm}{1 \, m} \times \dfrac{100 \, cm}{1 \, m} = 300 \, \dfrac{kg}{m^3}$

(c) $0.300 \, \dfrac{g}{cm^3} \times \dfrac{1 \, kg}{1000 \, g} \times \dfrac{100 \, cm}{1 \, m} \times \dfrac{100 \, cm}{1 \, m} \times \dfrac{100 \, cm}{1 \, m} = 300 \, \dfrac{kg}{m^3}$

T E A C H E R   P A G E S

- The bar above the last zero of the number 300 communicates it is a significant zero transforming the recorded answer from one significant digit to three. It is equally appropriate to teach your students to use scientific notation to effectively communicate three significant digits. The number would be correctly written as 3.00×10^2. Another way to communicate a number accurate to the ones position is to use a decimal at the end of the number. The number could be written as 300. representing that this measurement is accurate to the last digit.

9. Convert the mass of the sphere to (a) kg and (b) lbs.

    (a) $19.38\,g \times \dfrac{1\,kg}{1000\,g} = 0.01938\,kg$

    (b) $19.38\,g \times \dfrac{1\,lb}{454\,g} = 0.04269\,lbs$

10. Using the measured circumference, calculate the diameter of the sphere.
    - $C = \pi d$

    $d = \dfrac{C}{\pi}$

    $d = \dfrac{7.62\,cm}{3.14} = 2.43\,cm$

11. Calculate the radius of the sphere.
    - $d = 2r$

    $r = \dfrac{d}{2}$

    $r = \dfrac{2.43\,cm}{2} = 1.22\,cm$

12. Calculate the volume of the sphere from its radius.
    - $V = \dfrac{4}{3}\pi r^3$

    $V = \dfrac{4}{3}(\pi)(1.22)^3$

    $V = 7.61\,cm^3$

13. Calculate the volume of the sphere in mL as measured in the beaker. Convert to cm^3 knowing that 1 cm^3 = 1 mL.
    - $V = V_{final} - V_{initial}$

    $V = 110\,mL - 100\,mL$

    $V = 10\,mL = 10\,cm^3$

14. Calculate the volume of the sphere in mL as measured in the graduated cylinder. Convert to cm^3 knowing that $1\ cm^3 = 1\ mL$.

- $V = V_{final} - V_{initial}$

  $V = 182.3\ mL - 175.0\ mL$

  $V = 7.3\ mL = 7.3\ cm^3$

15. Using the density formula $D = \dfrac{mass}{volume}$, calculate the density of the sphere as determined by the (a) tape measure (b) beaker (c) graduated cylinder.

(a) $D = \dfrac{19.38\ g}{7.60\ cm^3}$

   $D = 2.55\ \dfrac{g}{cm^3}$

(c) $D = \dfrac{19.38\ g}{7.3\ cm^3}$

   $D = 2.7\ \dfrac{g}{cm^3}$

(b) $D = \dfrac{19.38\ g}{10\ cm^3}$

   $D = 2\ \dfrac{g}{cm^3}$

16. Use dimensional analysis to convert the densities into lbs/ft^3.

(a) $\dfrac{2.55\ g}{cm^3} \times \dfrac{1\ lb}{454\ g} \times \left(\dfrac{2.54\ cm}{1\ in}\right)^3 \times \left(\dfrac{12\ in}{1\ ft}\right)^3 = 159\ \dfrac{lbs}{ft^3}$

(b) $\dfrac{2\ g}{cm^3} \times \dfrac{1\ lb}{454\ g} \times \left(\dfrac{2.54\ cm}{1\ in}\right)^3 \times \left(\dfrac{12\ in}{1\ ft}\right)^3 = 100\ \dfrac{lbs}{ft^3}$

(c) $\dfrac{2.7\ g}{cm^3} \times \dfrac{1\ lb}{454\ g} \times \left(\dfrac{2.54\ cm}{1\ in}\right)^3 \times \left(\dfrac{12\ in}{1\ ft}\right)^3 = 170\ \dfrac{lbs}{ft^3}$

TEACHER PAGES

CONCLUSION QUESTIONS

1. Compare the densities of the cube when the volume is measured by a ruler, beaker and graduated cylinder. Which of the instruments gave the most accurate density value? Use the concept of significant digits to explain your answer.
   - The density of the cube had 3 significant digits when measured with the ruler. After subtracting to find the difference between the initial and final water levels in the graduated cylinder and beaker there are 2 significant digits when measured with the graduated cylinder but only 1 significant digit when measured with the beaker.
   - The ruler is the more accurate measure of the volume when compared to the volume obtained by water displacement using the graduated cylinder. The tweezers used to submerge the cube will contribute a small amount to the volume recorded since they contribute to the TOTAL amount of water displaced. See if you can get your students to come up with this concept!
   - Student answers may vary in significant digits depending on equipment used.

2. A student first measures the volume of the cube by water displacement using the graduated cylinder. Next, the student measures the mass of the cube before drying it. How will this error affect the calculated density of the cube? Your answer should state clearly whether the calculated density will increase, decrease or remain the same and must be justified.
   - The calculated density of the cube would increase.
   - Measuring a wet block will make the mass appear greater. Since mass is in the numerator of the equation $D = \dfrac{mass}{volume}$, the density value reported will be too large.

3. A student measures the circumference of a sphere at a point slightly higher than the middle of the sphere. How will this error affect the calculated density of the sphere? Your answer should state clearly whether the calculated density will increase, decrease, or remain the same and must be justified.
   - The density of the sphere would increase.
   - If the student measured the circumference at any point other than the center, the circumference would be reported as too low.
   - If the circumference is too small then the diameter will be too small.
   - If $\dfrac{C_{\downarrow}}{\pi} = d \ \therefore d_{\downarrow}$ the diameter is reported as too small, the radius will also be reported as too small.
   - If $\dfrac{d_{\downarrow}}{2} = r \ \therefore r_{\downarrow}$ the radius is reported as too small then the volume will be reported as too small.
   - If $V = \dfrac{4}{3}\pi r_{\downarrow}^{3} \ \therefore V_{\downarrow}$ the volume is reported as too small the density will be reported as too large.

   $$Density = \frac{m}{V_{\downarrow}} \ \therefore Density_{\uparrow}$$

Numbers in Science
Exploring Measurements, Significant Digits, and Dimensional Analysis

TAKING MEASUREMENTS

The accuracy of a measurement depends on two factors: the skill of the individual taking the measurement and the capacity of the measuring instrument. When making measurements, you should always read to the smallest mark on the instrument and then estimate another digit beyond that.

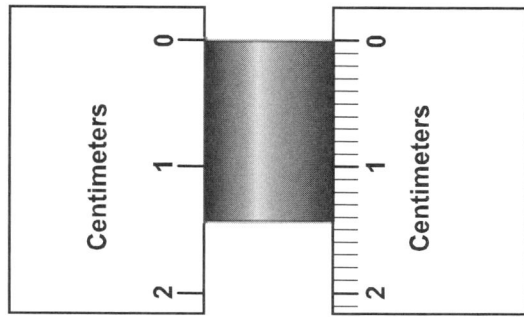

For example, if you are reading the length of the steel pellet pictured above using only the ruler shown to the left of the pellet, you can confidently say that the measurement is between 1 and 2 centimeters. However, you MUST also include one additional digit estimating the distance between the 1 and 2 centimeter marks. The correct measurement for this ruler should be reported as 1.5 centimeters. It would be incorrect to report this measurement as 1 centimeter or even 1.50 centimeters given the scale of this ruler.

What if you are using the ruler shown on the right of the pellet? What is the correct measurement of the steel pellet using this ruler? 1.4 centimeters? 1.5 centimeters? 1.40 centimeters? 1.45 centimeters? The correct answer would be 1.45 centimeters. Since the smallest markings on this ruler are in the tenths place we must carry our measurement out to the hundredths place.

If the measured value falls exactly on a scale marking, the estimated digit should be zero.

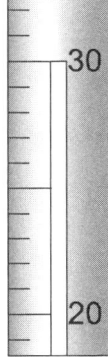

The temperature on this thermometer should read 30.0°C. A value of 30°C would imply this measurement had been taken on a thermometer with markings that were 10° apart, not 1° apart.

When using instruments with digital readouts you should record all the digits shown. The instrument has done the estimating for you.

When measuring liquids in narrow glass graduated cylinders, most liquids form a slight dip in the middle. This dip is called a *meniscus*. Your measurement should be read from the bottom of the meniscus. Plastic graduated cylinders do not usually have a meniscus. In this case you should read the cylinder from the top of the liquid surface. Practice reading the volume contained in the 3 cylinders below. Record your values in the space provided.

Left:_____

Middle: _____

Right:_____

SIGNIFICANT DIGITS

There are two kinds of numbers you will encounter in science, exact numbers and measured numbers. *Exact numbers* are known to be absolutely correct and are obtained by counting or by definition. Counting a stack of 12 pennies is an exact number. Defining 1 day as 24 hours are exact numbers. Exact numbers have an infinite number of significant digits.

Measured numbers, as we've seen above, involve some estimation. Significant digits are digits believed to be correct by the person making and recording a measurement. We assume that the person is competent in his or her use of the measuring device. To count the number of significant digits represented in a measurement we follow 2 basic rules:

1. If the digit is NOT a zero, it is significant.

2. If the digit IS a zero, it is significant if
   a. It is a sandwiched zero
      OR
   b. It terminates a number containing a decimal place

Examples:
> 3.57 mL has 3 significant digits (Rule 1)
> 288 mL has 3 significant digits (Rule 1)
> 20.8 mL has 3 significant digits (Rule 1 and 2a)
> 20.80 mL has 4 significant digits (Rules 1, 2a and 2b)
> 0.01 mL has only 1 significant digit (Rule 1)
> 0.010 mL has 2 significant digits (Rule 1 and 2b)
> 0.0100 mL has 3 significant digits (Rule 1 and 2b)
> 3.20×10^4 kg has 3 significant digits (Rule 1 and 2b)

SIGNIFICANT DIGITS IN CALCULATIONS

A calculated number can never contain more significant digits than the measurements used to calculate it.

Calculation rules fall into two categories:

1. <u>Addition and Subtraction</u>: answers must be rounded to match the measurement with the least number of decimal places.
   > 37.24 mL + 10.3 mL = 47.54 (calculator value), report as 47.5 mL

2. <u>Multiplication and Division</u>: answers must be rounded to match the measurement with the least number of significant digits.
   > 1.23 cm x 12.34 cm = 15.1782 (calculator value), report as 15.2 cm^2

DIMENSIONAL ANALYSIS

Throughout your study of science it is important that a unit accompanies all measurements. Keeping track of the units in problem can help you convert one measured quantity into its equivalent quantity of a different unit or set up a calculation without the need for a formula.

In conversion problems, equality statements such as 1 ft. = 12 inches, are made into fractions and then strung together in such a way that all units except the desired one are canceled out of the problem. Remember that defined numbers, such as the 1 and 12 above, are exact numbers and thus will not affect the number of significant digits in your answer. This method is also known as the Factor-Label method or the Unit-Label method.

To set up a conversion problem follow these steps.

1. Think about and write down all the "=" statements you know that will help you get from your current unit to the new unit.

2. Make fractions out of your "=" statements (there could be 2 fractions for each "="). They will be reciprocals of each other.

3. Begin solving the problem by writing the given amount with units on the left side of your paper and then choose the fractions that will let a numerator unit be canceled with a denominator unit and vice versa.

4. Using your calculator, read from left to right and enter the numerator and denominator numbers in order. Precede each numerator number with a multiplication sign and each denominator number with a division sign. Alternatively, you could enter all of the numerators, separated by multiplication signs, and then all of the denominators, each separated by a division sign.

5. Round your calculator's answer to the same number of significant digits that your original number had.

Example:
How many inches are in 1.25 miles?

Solution:

$$1\,ft = 12\,in \qquad \frac{1\,ft}{12\,in} \text{ OR } \frac{12\,in}{1\,ft}$$

$$5280\,ft. = 1\,mile \qquad \frac{5280\,ft.}{1\,mile} \text{ OR } \frac{1\,mile}{5280\,ft.}$$

$$1.25\,miles \times \frac{5280\,ft.}{1\,mile} \times \frac{12\,in.}{1\,ft.} = 79,200\,in.$$

As problems get more complex the measurements may contain fractional units or exponential units. To handle these problems treat each unit independently. Structure your conversion factors to ensure that all the given units cancel out with a numerator or denominator as appropriate and that your answer ends with the appropriate unit. Sometimes information given in the problem is an equality that will be used as a conversion factor.

Example: Suppose your automobile tank holds 23 gal and the price of gasoline is 33.5¢ per L. How many dollars will it cost you to fill your tank?

Solution: From a reference table we will find,
    1 L = 1.06 qt
    4 qt = 1 gal

We should recognize from the problem that the price is also an equality, 33.5¢ = 1 L and we should know that 100¢ = 1 dollar

Setting up the factors we find,

$$23\,gal \times \frac{4\,qt}{1\,gal} \times \frac{1\,L}{1.06\,qt} \times \frac{33.5¢}{1\,L} \times \frac{\$1}{100¢} = \$29$$

In your calculator you should enter $23 \times 4 \div 1.06 \times 33.5 \div 100$ and get 29.0754717. However, since the given value of 23 gal has only 2 significant digits, your answer must be rounded to \$29.

Squared and cubed units are potentially tricky. Remember that a cm^2 is really cm x cm. So, if we need to convert cm^2 to mm^2 we need to use the conversion factor 1 cm = 10 mm twice so that both centimeter units cancel out.

Example: One liter is exactly 1000 cm^3. How many cubic inches are there in 1.0 L?

Solution:
We should know that

$$1000 \text{ cm}^3 = 1 \text{ L}$$

From a reference table we find,

$$1 \text{ in.} = 2.54 \text{ cm}$$

Setting up the factors we find,

$$1.0 \; \cancel{L} \times \frac{1000 \, \cancel{cm} \times \cancel{cm} \times \cancel{cm}}{1 \, \cancel{L}} \times \frac{1 \, in}{2.54 \, \cancel{cm}} \times \frac{1 \, in}{2.54 \, \cancel{cm}} \times \frac{1 \, in}{2.54 \, \cancel{cm}} = 61 \, in^3$$

(The answer has 2 significant digits since our given 1.0 L contained two significant digits.)

As you become more comfortable with the concept of unit cancellation you will find that it is a very handy tool for solving problems. By knowing the units of your given measurements, and by focusing on the units of the desired answer you can derive a formula and correctly calculate an answer. This is especially useful when you've forgotten, or never knew, the formula!

Example: Even though you may not know the exact formula for solving this problem, you should be able to match the units up in such a way that only your desired unit does not cancel out.

What is the volume in liters of 1.5 moles of gas at 293 K and 1.10 atm of pressure?

The ideal gas constant is $\dfrac{0.0821 \; \text{L} \cdot \text{atm}}{\text{mol} \cdot \text{K}}$

Solution: It is not necessary to know the formula for the ideal gas law to solve this problem correctly. Working from the constant, since it sets the units, we need to cancel out every unit except L. Doing this shows us that moles and kelvins need to be in the numerator and atmospheres in the denominator.

$$\frac{0.0821 \; \text{L} \cdot \cancel{atm}}{\cancel{mol} \cdot \cancel{K}} \times \frac{1.5 \, \cancel{mol}}{} \times \frac{293 \, \cancel{K}}{} \times \frac{1}{1.10 \, \cancel{atm}} = 33 \, \text{L}$$

(2 significant digits since our least accurate measuement has only 2 sig.digs.)

NOTE: NEVER consider the number of significant digits in a constant to determine the number of significant digits for reporting your calculated answer. Consider ONLY the number of significant digits in given or measured quantities.

PURPOSE

In this activity you will review some important aspects of numbers in science and then apply those number handling skills to your own measurements and calculations.

MATERIALS

small cube   spherical object
metric ruler   tweezers
200 mL beaker   flexible tape measure
large graduated cylinder   balance

PROCEDURE

*Remember when taking measurements it is your responsibility to estimate a digit between the two smallest marks on the instrument.

1. Mass the small cube on a balance and record your measurement in the data table on your student page.

2. Measure dimensions (the length, width and height) of the small cube in centimeters, being careful to use the full measuring capacity of your ruler. Record the lengths in your data table.

3. Fill the 200 mL beaker with water to the 100 mL line. Carefully place the cube in the beaker and use the tweezers to gently submerge the cube. The cube should be just barely covered with water. Record the new, final volume of water.

4. Fill the large graduated cylinder ¾ of the way full with water. Record this initial water volume. Again, use the tweezers to gently submerge the cube and record the final water volume.

5. Mass the spherical object on a balance and record your measurement in the data table.

6. Use the flexible tape measure to measure the widest circumference of the sphere in centimeters. Be careful to use the full measuring capacity of the tape measure.

7. Fill the 200 mL beaker with water to the 100 mL line. Carefully place the spherical object in the beaker and, if needed, use the tweezers to gently submerge the sphere. Record the final volume of water from the beaker.

8. Fill the large graduated cylinder ¾ of the way full with water. Record this initial water volume. If needed, use the tweezers to gently submerge the sphere and record the new water volume.

9. Dry the cube and sphere and clean up your lab area as instructed by your teacher.

Name _____

Period _____

Numbers in Science
Exploring Measurements, Significant Digits, and Dimensional Analysis

DATA AND OBSERVATIONS

Data Table			
CUBE DATA			
Mass:			
Dimensions	length:	width:	height:
Volume	Beaker initial volume: 100 mL	Beaker final volume:	
	Graduated cylinder initial volume:	Graduated cylinder final volume:	
SPHERE DATA			
Mass:			
Dimensions	Circumference:		
Volume	Beaker initial volume: 100 mL	Beaker final volume:	
	Graduated cylinder initial volume:	Graduated cylinder final volume:	
Formula for calculating the volume of a cube:			
Formula for calculating the circumference of a circle:			
Formula for calculating the diameter of a circle:			
Formula for calculating the volume of a sphere:			

ANALYSIS

- Remember to follow the rules for reporting all data and calculated answers with the correct number of significant digits.
- You may need tables of metric and English conversion factors to work some of these problems.

1. For each of the measurements you recorded above, go back and indicate the number of significant digits in parentheses after the measurement. Ex: 15.7 cm (3sd)

2. Use dimensional analysis to convert the mass of the cube to
    a. mg

    b. ounces

3. Calculate the volume of the cube in cm^3.

4. Use dimensional analysis to convert the volume of the cube from cm^3 to m^3.

5. Calculate the volume of the cube in mL as measured in the beaker. Convert the volume to cm^3 knowing that $1 \ cm^3 = 1$ mL.

6. Calculate the volume of the cube in mL as measured in the graduated cylinder. Convert to cm^3 knowing that $1 \ cm^3 = 1$ mL.

7. Using the density formula $D = \dfrac{mass}{volume}$, calculate the density of the cube as determined by the

   a. ruler

   b. beaker

   c. graduated cylinder

8. Use dimensional analysis to convert these three densities into kg/m^3.

9. Convert the mass of the sphere to
   a. kg

   b. lbs.

10. Using the measured circumference, calculate the diameter of the sphere.

11. Calculate the radius of the sphere.

12. Calculate the volume of the sphere from its radius.

13. Calculate the volume of the sphere in mL as measured in the beaker. Convert to cm^3 knowing that 1 cm^3 = 1 mL.

14. Calculate the volume of the sphere in mL as measured in the graduated cylinder. Convert to cm^3 knowing that 1 cm^3 = 1 mL.

15. Using the density formula $D = \dfrac{mass}{volume}$, calculate the density of the sphere as determined by the

    a. tape measure

    b. beaker

    c. graduated cylinder

16. Use dimensional analysis to convert these three densities into lbs/ft^3.

CONCLUSION QUESTIONS

1. Compare the densities of the cube when the volume is measured by a ruler, beaker and graduated cylinder. Which of the instruments gave the most accurate density value? Use the concept of significant digits to explain your answer.

2. A student first measures the volume of the cube by water displacement using the graduated cylinder. Next, the student measures the mass of the cube before drying it. How will this error affect the calculated density of the cube? Your answer should state clearly whether the calculated density will increase, decrease or remain the same and must be justified.

3. A student measures the circumference of a sphere at a point slightly higher than the middle of the sphere. How will this error affect the calculated density of the cube? Your answer should state clearly whether the calculated density will increase, decrease, or remain the same and must be justified.

Literal Equations
Manipulating Variables and Constants

OBJECTIVE
Students will review how to solve literal equations for a particular variable.

LEVEL
Middle Grades: Chemistry/Physics, Chemistry I, Physics I

NATIONAL STANDARDS
UCP.1, UCP.2, G.2

TEKS
IPC: 4(A)
Chemistry: 2(C)
Physics: 3(B)

CONNECTIONS TO AP
In AP science courses, particularly physics and chemistry, the student is often given an equation and asked to solve it for a particular variable.

TIME FRAME
45 minutes

TEACHER NOTES
Throughout chemistry and physics courses, and at times in IPC, the students will need to be able to solve an equation for a particular variable to see how that variable depends on other variables and constants. Manipulation of variables without the substitution of numbers is an important skill in helping students understand that the variables depend on each other in a certain way regardless of any particular numbers which may be substituted into the equation. For example, in the equation $F_{net} = ma$ (Newton's second law), the acceleration a is always proportional to the net force F_{net} regardless of the value of the mass. In the equation $P_1V_1 = P_2V_2$ (Boyle's law), pressure P and volume V are always inversely proportional to each other.

The following examples and exercises illustrate the manipulation of many of the literal equations that commonly appear in physics and chemistry courses. Although the students may not understand the meaning of many of the equations during this practice exercise, practicing solving the equations will sharpen their algebra skills. When they ultimately learn the meaning of the equations, they will be more likely to feel comfortable with the conceptual understanding behind the equations rather than losing the meaning of the relationships among the variables in the algebraic manipulation.

A *literal equation* is one which is expressed in terms of variable symbols (such as d, v, and a) and constants (such as R, g, and π). Often in science and mathematics the students are given an equation and asked to solve it for a particular variable symbol or letter called the *unknown*.

The symbols, which are not the particular variable we are interested in solving for, are called *literals*, and may represent variables or constants. Literal equations are solved by isolating the unknown variable on one side of the equation, and all of the remaining literal variables on the other side of the equation. Sometimes the unknown variable is part of another term. A *term* is a combination of symbols such as the products ma or πr^2. In this case the unknown (such as r in πr^2) must be factored out of the term before we can isolate it.

The following rules, examples, and exercises will help you review and practice solving literal equations from physics and chemistry.

Suggested Teaching Procedure:

1. Review the procedure section with students. Emphasize keeping their equations neat and orderly.

2. Choose several of the listed examples to work with students on the overhead or chalkboard.

3. Instruct students to complete the remaining exercises in the space provided on their student answer pages.

ANSWERS TO EXERCISES

Directions: For each of the following equations, solve for the variable in **bold** print.

1. $v = \mathbf{a}t$

   - $\dfrac{v}{t} = \mathbf{a}$

2. $P = \dfrac{F}{\mathbf{A}}$

   - $P\mathbf{A} = F$

     $\mathbf{A} = \dfrac{F}{P}$

3. $\lambda = \dfrac{\mathbf{h}}{p}$

   - $\lambda p = \mathbf{h}$

4. $F(\mathbf{\Delta t}) = m\Delta v$

   - $\mathbf{\Delta t} = \dfrac{m\Delta v}{F}$

5. $U = \dfrac{G\mathbf{m_1}m_2}{r}$

   $Ur = G\mathbf{m_1}m_2$

   - $\dfrac{Ur}{Gm_2} = \mathbf{m_1}$

6. $C = \dfrac{5}{9}\left(\mathbf{F} - 32\right)$

   • $\dfrac{9}{5}C = \mathbf{F} - 32$

   $\mathbf{F} = \dfrac{9}{5}C + 32$

7. $v^2 = v_0{}^2 + 2\mathbf{a}\Delta x$

   $v^2 - v_0{}^2 = 2\mathbf{a}\Delta x$

   • $\dfrac{v^2 - v_0{}^2}{2\Delta x} = \mathbf{a}$

8. $K_{avg} = \dfrac{3}{2}k_B\mathbf{T}$

   $\dfrac{2}{3}K_{avg} = k_B\mathbf{T}$

   • $\dfrac{2K_{avg}}{3k_B} = \mathbf{T}$

9. $K = \dfrac{1}{2}m\mathbf{v}^2$

   $2K = m\mathbf{v}^2$

   • $\dfrac{2K}{m} = \mathbf{v}^2$

   $\mathbf{v} = \sqrt{\dfrac{2K}{m}}$

10. $v_{rms} = \sqrt{\dfrac{3RT}{\mathbf{M}}}$

   $v^2{}_{rms} = \dfrac{3RT}{\mathbf{M}}$

   • $\mathbf{M}v^2{}_{rms} = 3RT$

   $\mathbf{M} = \dfrac{3RT}{v^2{}_{rms}}$

11. $v_{rms} = \sqrt{\dfrac{3\mathbf{k}_B T}{\mu}}$

$v_{rms}{}^2 = \dfrac{3\mathbf{k}_B T}{\mu}$

- $\mu v_{rms}{}^2 = 3\mathbf{k}_B T$

$\dfrac{\mu v_{rms}{}^2}{3T} = \mathbf{k}_B$

12. $F = \dfrac{1}{4\pi\varepsilon_0}\dfrac{Kq_1 q_2}{\mathbf{r}^2}$

$4\pi\varepsilon_o \mathbf{r}^2 F = Kq_1 q_2$

- $\mathbf{r}^2 = \dfrac{Kq_1 q_2}{4\pi\varepsilon_o F}$

$\mathbf{r} = \sqrt{\dfrac{Kq_1 q_2}{4\pi\varepsilon_o F}}$

13. $\dfrac{1}{s_i} + \dfrac{1}{s_o} = \dfrac{1}{\mathbf{f}}$

- $\mathbf{f} = \dfrac{1}{\dfrac{1}{s_o} + \dfrac{1}{s_i}}$

14. $\dfrac{1}{C_{EQ}} = \dfrac{1}{\mathbf{C}_1} + \dfrac{1}{C_2}$

$\dfrac{1}{C_{EQ}} - \dfrac{1}{C_2} = \dfrac{1}{\mathbf{C}_1}$

- $\mathbf{C}_1 = \dfrac{1}{\dfrac{1}{C_{EQ}} - \dfrac{1}{C_2}}$

15. $V = \dfrac{4}{3}\pi \mathbf{r}^3$

    $\dfrac{3}{4}V = \pi \mathbf{r}^3$

- $\dfrac{3V}{4\pi} = \mathbf{r}^3$

    $\mathbf{r} = \sqrt[3]{\dfrac{3V}{4\pi}}$

16. $P + \mathbf{D}gy + \dfrac{1}{2}\mathbf{D}v^2 = C$

    $P + \mathbf{D}\left(gy + \dfrac{1}{2}v^2\right) = C$

- $C - P = \mathbf{D}\left(gy + \dfrac{1}{2}v^2\right)$

    $\mathbf{D} = \dfrac{C - P}{\left(gy + \dfrac{1}{2}v^2\right)}$

17. $P + Dgy + \dfrac{1}{2}D\mathbf{v}^2 = C$

    $\dfrac{1}{2}D\mathbf{v}^2 = C - P - Dgy$

- $D\mathbf{v}^2 = 2\left(C - P - Dgy\right)$

    $\mathbf{v}^2 = 2D\left(C - P - Dgy\right)$

    $\mathbf{v} = \sqrt{2D\left(C - \mathbf{P} - Dgy\right)}$

18. $x = x_0 + v_0 t + \dfrac{1}{2}\mathbf{a}t^2$

    $x - v_0 t = \dfrac{1}{2}\mathbf{a}t^2$

- $2\left(x - v_0 t\right) = \mathbf{a}t^2$

    $\mathbf{a} = \dfrac{2\left(x - v_0 t\right)}{t^2}$

19. $n_1 \sin \theta_1 = n_2 \sin \theta_2$

$$\frac{n_1 \sin \theta_1}{n_2} = \sin \theta_2$$

- $$\theta_2 = \sin^{-1} \left[\frac{n_1 \sin \theta_1}{n_2} \right]$$

20. $mg \sin \theta = \mu mg \cos \theta \left(\dfrac{M+m}{m} \right)$

$$\frac{mg \sin \theta}{mg \cos \theta} = \mu \left(\frac{M+m}{m} \right)$$

$$\frac{\sin \theta}{\cos \theta} = \mu \left(\frac{M+m}{m} \right)$$

- $$\tan \theta = \mu \left(\frac{M+m}{m} \right)$$

$$\theta = \tan^{-1} \left[\mu \left(\frac{M+m}{m} \right) \right]$$

Literal Equations
Manipulating Variables and Constants

A *literal equation* is one which is expressed in terms of variable symbols (such as d, v, and a) and constants (such as R, g, and π). Often in science and mathematics you are given an equation and asked to solve it for a particular variable symbol or letter called the *unknown*.

The symbols which are not the particular variable we are interested in solving for are called *literals*, and may represent variables or constants. Literal equations are solved by isolating the unknown variable on one side of the equation, and all of the remaining literal variables on the other side of the equation. Sometimes the unknown variable is part of another term. A *term* is a combination of symbols such as the products ma or πr^2. In this case the unknown (such as r in πr^2) must factored out of the term before we can isolate it.

The following rules, examples, and exercises will help you review and practice solving literal equations from physics and chemistry.

PROCEDURE
In general, we solve a literal equation for a particular variable by following the basic procedure below.

1. Recall the conventional order of operations, that is, the order in which we perform the operations of multiplication, division, addition, subtraction, etc.:
   a. Parenthesis
   b. Exponents
   c. Multiplication and Division
   d. Addition and Subtraction

   This means that you should do what is possible within parentheses first, then exponents, then multiplication and division from left to right, then addition and subtraction from left to right. If some parentheses are enclosed within other parentheses, work from the inside out.

2. If the unknown is a part of a grouped expression (such as a sum inside parentheses), use the distributive property to expand the expression.

3. By adding, subtracting, multiplying, or dividing appropriately,

   (a) move all terms containing the unknown variable to one side of the equation, and

   (b) move all other variables and constants to the other side of the equation. Combine like terms when possible.

4. Factor the unknown variable out of its term by appropriately multiplying or dividing both sides of the equation by the other literals in the term.

5. If the unknown variable is raised to an exponent (such as 2, 3, or ½), perform the appropriate operation to raise the unknown variable to the first power, that is, so that it has an exponent of one.

EXAMPLES

1. $F = m\mathbf{a}$. Solve for **a**.

   $F = ma$

   Divide both sides by m:

   $$\frac{F}{m} = \mathbf{a}$$

   Since the unknown variable (in this case a) is usually placed on the left side of the equation, we can switch the two sides:

   $$\mathbf{a} = \frac{F}{m}$$

2. $P_1 V_1 = P_2 \mathbf{V}_2$. Solve for $\mathbf{V}_2$.

   $P_1 V_1 = P_2 \mathbf{V}_2$

   Divide both sides by P_2:

   $$\frac{P_1 V_1}{P_2} = \mathbf{V}_2$$

   $$\mathbf{V}_2 = \frac{P_1 V_1}{P_2}$$

3. $v = \dfrac{d}{\mathbf{t}}$. Solve for **t**.

   Multiply each side by **t**:

   $$\mathbf{t} v = d$$

   Divide both sides by v:

   $$\mathbf{t} = \frac{d}{v}$$

4. $PV = n\mathbf{R}T$. Solve for $\mathbf{R}$.

   $PV = n\mathbf{R}T$

   Divide both sides by n:

   $$\frac{PV}{n} = \mathbf{R}T$$

   Divide both sides by $\mathbf{T}$:

   $$\frac{PV}{nT} = \mathbf{R}$$

   $$\mathbf{R} = \frac{PV}{nT}$$

5. $R = \dfrac{\rho\mathbf{L}}{A}$. Solve for $\mathbf{L}$.

   $$R = \frac{\rho\mathbf{L}}{A}$$

   Multiply both sides by A:

   $$RA = \rho\mathbf{L}$$

   Divide both sides by ρ:

   $$\frac{RA}{\rho} = \mathbf{L}$$

   $$\mathbf{L} = \frac{RA}{\rho}$$

6. $A = h(a + \mathbf{b})$. Solve for $\mathbf{b}$.

   Distribute the h:

   $$A = ha + h\mathbf{b}$$

   Subtract ha from both sides:

   $$A - ha = h\mathbf{b}$$

Divide both sides by h:

$$\frac{A - ha}{h} = \mathbf{b}$$

$$\mathbf{b} = \frac{A - ha}{h}$$

7. $P = P_0 + \rho\mathbf{g}h$. Solve for $\mathbf{g}$.

Subtract P_0 from both sides:

$$P - P_0 = \rho\mathbf{g}h$$

Divide both sides by ρh:

$$\frac{P - P_0}{\rho h} = \mathbf{g}$$

$$\mathbf{g} = \frac{P - P_0}{\rho h}$$

8. $U = \frac{1}{2}\mathbf{Q}V$. Solve for $\mathbf{Q}$.

Multiply both sides by 2:

$$2U = \mathbf{Q}V$$

Divide both sides by V:

$$\frac{2U}{V} = \mathbf{Q}$$

$$\mathbf{Q} = \frac{2U}{V}$$

9. $U = \frac{1}{2}k\mathbf{x}^2$. Solve for $\mathbf{x}$.

Multiply both sides by 2:

$$2U = k\mathbf{x}^2$$

Divide both sides by k:

$$\frac{2U}{k} = \mathbf{x}^2$$

Take the square root of both sides:

$$\sqrt{\frac{2U}{k}} = \mathbf{x}$$

$$\mathbf{x} = \sqrt{\frac{2U}{k}}$$

10. $T = 2\pi\sqrt{\dfrac{\mathbf{L}}{g}}$. Solve for $\mathbf{L}$.

Divide both sides by 2π:

$$\frac{T}{2\pi} = \sqrt{\frac{\mathbf{L}}{g}}$$

Square both sides:

$$\frac{T^2}{4\pi^2} = \frac{\mathbf{L}}{g}$$

Multiply both sides by g:

$$\frac{gT^2}{4\pi^2} = \mathbf{L}$$

$$\mathbf{L} = \frac{gT^2}{4\pi^2}$$

11. $F = \dfrac{Gm_1m_2}{\mathbf{r}^2}$. Solve for $\mathbf{r}$.

Multiply both sides by $\mathbf{r}^2$:

$$F\mathbf{r}^2 = Gm_1m_2$$

Divide both sides by F:

$$\mathbf{r}^2 = \frac{Gm_1m_2}{F}$$

Take the square root of both sides:

$$\mathbf{r} = \sqrt{\frac{Gm_1m_2}{F}}$$

12. $\dfrac{h_i}{h_o} = -\dfrac{s_i}{\mathbf{s}_o}$. Solve for $\mathbf{s}_o$.

Cross-multiply:

$$h_i \mathbf{s}_o = -h_o s_i$$

Divide both sides by h_i:

$$\mathbf{s}_0 = -\dfrac{h_o s_i}{h_i}$$

13. $\dfrac{1}{R_{EQ}} = \dfrac{1}{R_1} + \dfrac{1}{R_2} + \dfrac{1}{\mathbf{R}_3}$. Solve for $\mathbf{R}_3$.

Subtract $\dfrac{1}{R_1} + \dfrac{1}{R_2}$ from both sides:

$$\dfrac{1}{R_{EQ}} - \dfrac{1}{R_1} - \dfrac{1}{R_2} = \dfrac{1}{\mathbf{R}_3}$$

Take the reciprocal of both sides:

$$\dfrac{1}{\dfrac{1}{R_{EQ}} - \dfrac{1}{R_1} - \dfrac{1}{R_2}} = \mathbf{R}_3$$

$$\mathbf{R}_3 = \dfrac{1}{\dfrac{1}{R_{EQ}} - \dfrac{1}{R_1} - \dfrac{1}{R_2}}$$

This equation could be solved further with several more algebraic steps.

14. $F = qvB \sin\theta$. Solve for $\boldsymbol{\theta}$.

Divide both sides by qvB:

$$\dfrac{F}{qvB} = \sin\theta$$

Take the inverse sine of both sides:

$$\theta = \sin^{-1}\left[\dfrac{F}{qvB}\right]$$

15. $\mu mg \cos\theta = mg \sin\theta$. Solve for μ.
    Divide both sides by $mg\cos\theta$:

$$\mu = \frac{mg \sin\theta}{mg \cos\theta} = \frac{\sin\theta}{\cos\theta} = \tan\theta$$

Name _____

Period _____

Literal Equations
Manipulating Variables and Constants

EXERCISES

Directions: For each of the following equations, solve for the variable in **bold** print. Be sure to show each step you take to solve the equation for the **bold** variable.

1. $v = \mathbf{a}t$

2. $P = \dfrac{F}{\mathbf{A}}$

3. $\lambda = \dfrac{\mathbf{h}}{p}$

4. $F(\mathbf{\Delta t}) = m\Delta v$

5. $U = \dfrac{G\mathbf{m_1}m_2}{r}$

6. $C = \dfrac{5}{9}(\mathbf{F} - 32)$

7. $v^2 = v_0{}^2 + 2\mathbf{a}\Delta x$

8. $K_{avg} = \dfrac{3}{2}k_B\mathbf{T}$

9. $K = \dfrac{1}{2}m\mathbf{v}^2$

10. $v_{rms} = \sqrt{\dfrac{3RT}{\mathbf{M}}}$

11. $v_{rms} = \sqrt{\dfrac{3\mathbf{k}_B T}{\mu}}$

12. $F = \dfrac{1}{4\pi\varepsilon_0} \dfrac{Kq_1 q_2}{\mathbf{r}^2}$

13. $\dfrac{1}{s_i} + \dfrac{1}{s_o} = \dfrac{1}{\mathbf{f}}$

14. $\dfrac{1}{C_{EQ}} = \dfrac{1}{\mathbf{C}_1} + \dfrac{1}{C_2}$

15. $V = \dfrac{4}{3}\pi\mathbf{r}^3$

16. $P + \mathbf{D}gy + \dfrac{1}{2}\mathbf{D}v^2 = C$

17. $P + Dgy + \dfrac{1}{2}D\mathbf{v}^2 = C$

18. $x = x_0 + v_0 t + \dfrac{1}{2}\mathbf{a}t^2$

19. $n_1 \sin\theta_1 = n_2 \sin\boldsymbol{\theta_2}$

20. $mg \sin\boldsymbol{\theta} = \mu mg \cos\boldsymbol{\theta}\left(\dfrac{M+m}{m}\right)$

TEACHER PAGES

Graphing Skills
Reading, Constructing and Analyzing Graphs

Unit Overview

OBJECTIVE

The purpose of this lesson is to provide the teacher with basic graphing skill lessons to be used throughout the science course of study. There are many different kinds of graphs and each has a fairly specific use. By teaching graphing at all grade levels students should be able to choose the best type of graph to represent their data. Students should also become familiar with ways to analyze their graphed information realizing that there is meaning to their graph.

LEVEL

All

NATIONAL STANDARDS

UCP.1, UCP.2, G.2

TEKS

6.2(E)
7.2(E)
8.2(E)
IPC: 2(C)
Biology: 2(C)
Chemistry: 2(D)
Physics: 2(C), 2(E)

CONNECTIONS TO AP

Each of the AP Science courses requires that students are able to read, construct and analyze graphs.

TIME FRAME

45 minutes

MATERIALS

    graph paper
    pencil
    data

TEACHER NOTES

Graphing is a skill that should be introduced at each grade level and reinforced throughout the year whenever data is available. The analysis level of the graph will vary depending on the mathematical ability of your students. The lessons presented here may be used as stand alone lessons or may be combined as a general review of graphing skills.

Graphing is an essential tool in science. Graphs enable us to visually communicate information. The lessons that follow will focus on bar graphs, pie charts and line graphs. Goals for this series of lessons on graphing include:

- Choosing an appropriate display for data (which type of graph to construct)
- Identifying data to be displayed on the x and y axes
- Scaling a graph properly
- Labeling a graph with axes labels, title, units, and legend or key if necessary
- Extrapolating and interpolating data points
- Understanding relevant relationships such as slope and area under the curve

Graphing Skills
Reading, Constructing and Analyzing Graphs

Bar Graphs and Histograms

OBJECTIVE

Students will become familiar with basic bar graphing skills to be used throughout their science course of study.

LEVEL

All

NATIONAL STANDARDS

UCP.1, UCP.2, G.2

TEKS

6.2(E)
7.2(E)
8.2(E)
IPC 2(C)
Biology 2(C)
Chemistry 2(D)
Physics 2(C), 2(E)

CONNECTIONS TO AP

Each of the AP Science courses requires that students are able to read, construct and analyze graphs.

TIME FRAME

45 minutes

MATERIALS

(For each student working individually)

> 4 sheets of quadrille graph paper
> pencil
> data

TEACHER PAGES

TEACHER NOTES

Graphing is a skill that should be introduced at each grade level and reinforced throughout the year whenever data is available. Before allowing students to begin this exercise, you should do the following:

- Distinguish between the four types of graphs represented in the student introduction. Key points to be made are as follows:
  - *Simple Bar Graph*: The width of bars must be the same.
  - *Grouped Bar Graph*: The width of bars must be the same. Each bar within a group needs some distinguishing mark—different colors, different markings, etc. The student must provide a legend so that the graph may easily be interpreted.
  - *Composite Bar Graph*: The width of bars must be the same. Each different component of the bar must have some distinguishing mark—different colors, different markings, etc.
  - *Histogram:* The width of bars must be the same. Clearly make the distinction here for the students that in a histogram the bar itself represents a range of independent variables rather than a single value.
- Encourage students to decide which sets of data belong on the *x*-axis and the *y*-axis. The key here is to place the independent variable on the *x*-axis and the dependent variable on the *y*-axis.
- Show the students how to properly scale a graph. The scale represents the range of frequency values shown on the graph. Visually show students how to properly accomplish this task by making an overhead transparency of a piece of the graph paper. Write down the range, count the squares on the graph paper and decide the scale by spacing appropriately. Emphasize the importance of using the entire length and width of the paper when creating the axes.
- Demonstrate proper labeling of a graph with a title, axes labels and units, and keys if necessary. Titles are usually given at the top of each graph and provide an overview of the information that is given in the graph. The axes labels should provide specific information as to what is represented. It is customary to label a graph in the format *y* vs *x*. For example if a graph is described as "Temperature vs Time" then temperature should be on the *y* axis and time should be on the *x*-axis. Depending on the level of your students you may wish to require this format.
- In analyzing a graph, show the students how to read the graph and interpret their graph by using interpolation and extrapolation. Point out that it is difficult to identify specific interpolation (between) points on bar graphs but that rough comparisons are easy. Extrapolation (beyond) points is also difficult (if not impossible) to determine on bar graphs.
- As students work on this activity monitor them closely to ensure that they have correct labels and scales on the graphs they are constructing.

SAMPLE DATA
Student graphs should look similar to the samples found below.

PART I: SIMPLE BAR GRAPH

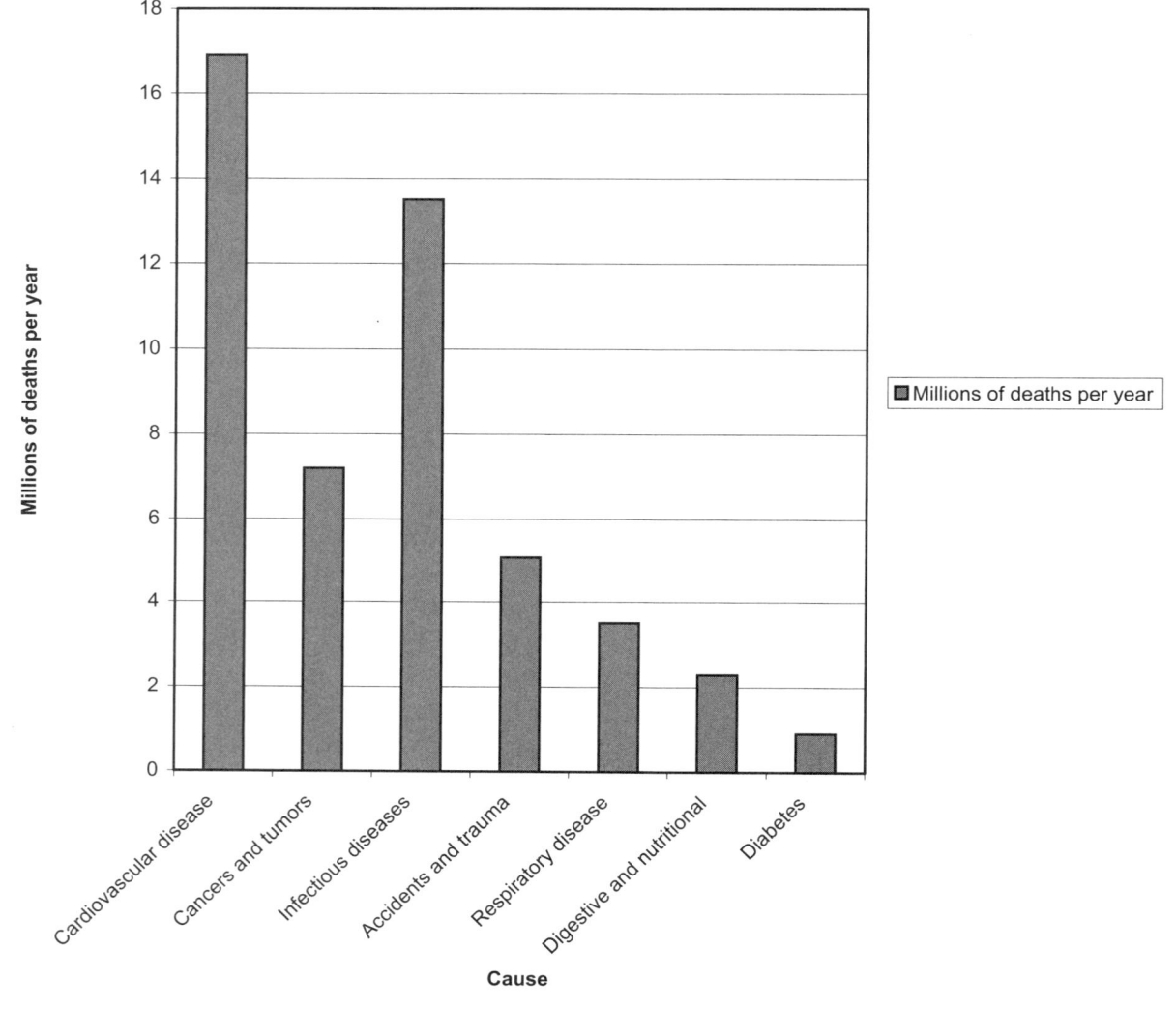

PART II: GROUPED BAR GRAPH

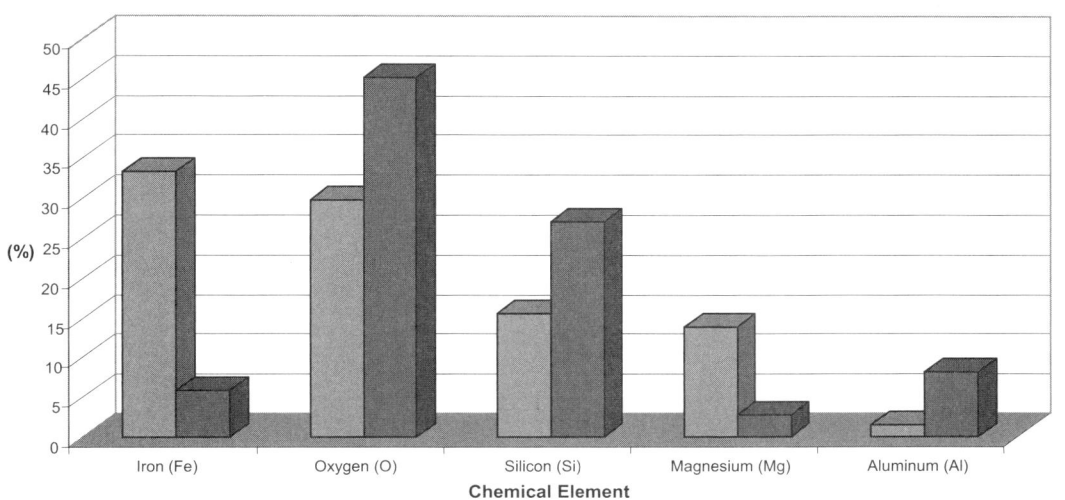

Some Common Elements of the Earth

PART III: COMPOSITE BAR GRAPH

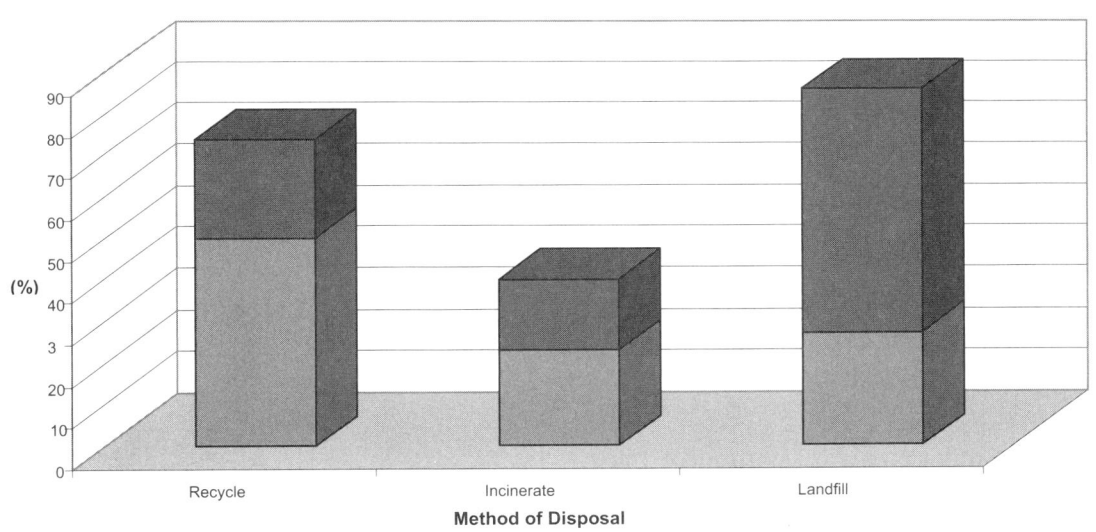

Solid Waste Disposal in US and Japan

PART IV: HISTOGRAM

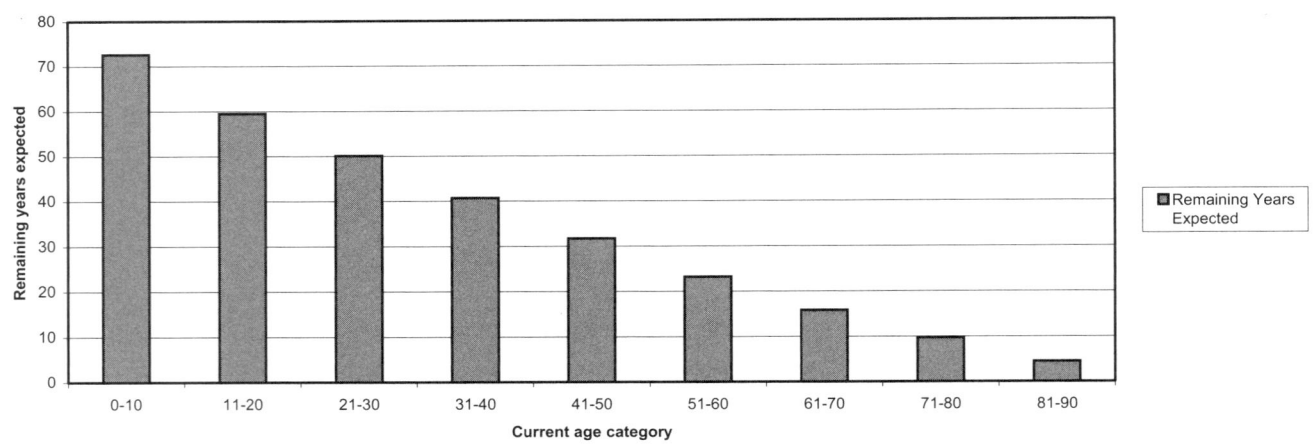

Life Expectancies in the US

POSSIBLE ANSWERS TO THE CONCLUSION QUESTIONS

Using the graphs that you constructed, answer the following questions.

PART I: SIMPLE BAR GRAPH
1. How many deaths due to accidents and trauma occur per year?
   - 5.1 million deaths

2. Can you predict the number of deaths due to cancers and tumors for the next ten years? Explain.
   - No. This type of extrapolation is a weakness of bar graphs.

PART II: GROUPED BAR GRAPH
1. What element is the most abundant in the Earth's crust?
   - Oxygen — about 45%.

2. What element is the most abundant within the Earth?
   - Iron — about 33%.

TEACHER PAGES

PART III: COMPOSITE BAR GRAPH

1. What is the favored method of waste disposal for the Japanese population? Cite possible reasons for this.
   - Recycling — about 50% of their waste is disposed of this way.
   - Possible reasons might include: minimum available undeveloped land; overcrowded population.

2. What is the favored method of waste disposal for the US population? Cite possible reasons for this.
   - Landfill — about 60% of waste is disposed of this way.
   - Possible reasons might include: plenty of undeveloped land; expense associated with recycling.

PART IV: HISTOGRAM

1. Make a prediction about the remaining years of life that would be expected for someone in the current age category of 91-100.
   - Some students will attempt to find a mathematical trend in the data by looking at how the years change over each interval. Students should answer between 0 and 2 years if using this type of pattern.

2. Is the answer to question 7 an accurate number? Why or why not? Cite specific reasons.
   - No, the answer in 7 is not an accurate number.
   - This is very difficult to predict from a graph of this type.

3. What type of data is easily represented by a bar graph?
   - Data that compares amounts or frequency of occurrence.

4. Why is a legend (or key) necessary in the grouped and composite bar graphs?
   - Legends are necessary to distinguish between the bars.

5. Explain why it is difficult to make direct comparisons between recycling in Japan and the US using the composite bar graph that you drew.
   - It is difficult for comparison since the bars are merged together. In order to compare accurately, you would have to accurately measure the length of each piece separately.

6. What is the importance of scaling?
   - Scaling is important for making accurate comparisons.

7. Distinguish between the dependent and the independent variable for each of the graphs that were constructed. On which axis should the independent variable be placed?

	Dependent Variable	**Independent Variable**
Simple Bar Graph	Millions of deaths per year	Cause of death
Grouped Bar Graph	Percentages in Earth and crust	Chemical elements
Composite Bar Graph	Percentage of waste	Method of disposal
Histogram	Remaining years expected	Current age

- The *x*-axis is usually used for the independent variable.

TEACHER PAGES

Graphing Skills
Reading, Constructing and Analyzing Graphs

Pie Charts

OBJECTIVE
Students will become familiar with basic pie chart graphing skills to be used throughout their science course of study.

LEVEL
All

NATIONAL STANDARDS
UCP.1, UCP.2, G.2

TEKS
6.2(E)
7.2(E)
8.2(E)
IPC 2(C)
Biology 2(C)
Chemistry 2(D)
Physics 2(C), 2(E)

CONNECTIONS TO AP
Each of the AP Science courses requires that students are able to read, construct and analyze graphs.

TIME FRAME
45 minutes

MATERIALS
(For each student working individually)

2 sheets of blank paper	data
pencil	protractor
compass	colored pencils

TEACHER NOTES

Graphing is a skill that should be introduced at each grade level and reinforced throughout the year whenever data is available. Before allowing students to begin this exercise, you should do the following:

- Emphasize the usefulness of pie charts for showing percentages.
- Remind students that each pie chart should always total 100%. They will be asked to calculate the percentage for the second set of data. To find the percentage:

$$\% = \frac{\text{specific sample of data}}{\text{total data collected}} \times 100$$

- Point out that a legend or key will be necessary for this type of display.
- Reinforce that labels for each wedge and a title are always necessary.
- Students may need to be instructed on proper use of a compass and protractor. An easy way to construct a pie chart is to have students draw a circle with the compass. Using the protractor, make four marks on the outside of the circle in 90° intervals. Next, have the students divide each quadrant into five equal sections. Each section will represent 5%. This makes estimation of points quite simple.

SAMPLE DATA

Student graphs should look similar to the samples found below.

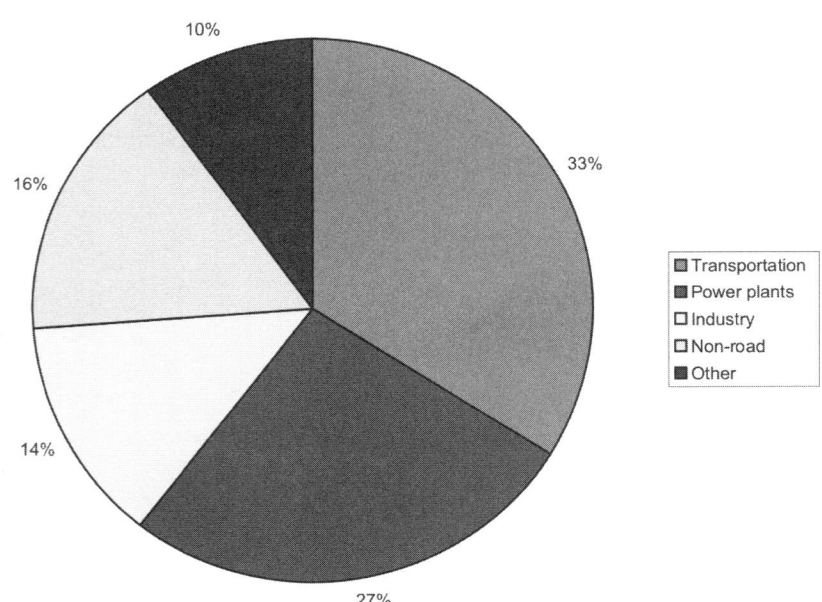

Sources of Nitrogen Oxide Pollutants

Laying the Foundation in Middle Grades Life and Earth Science

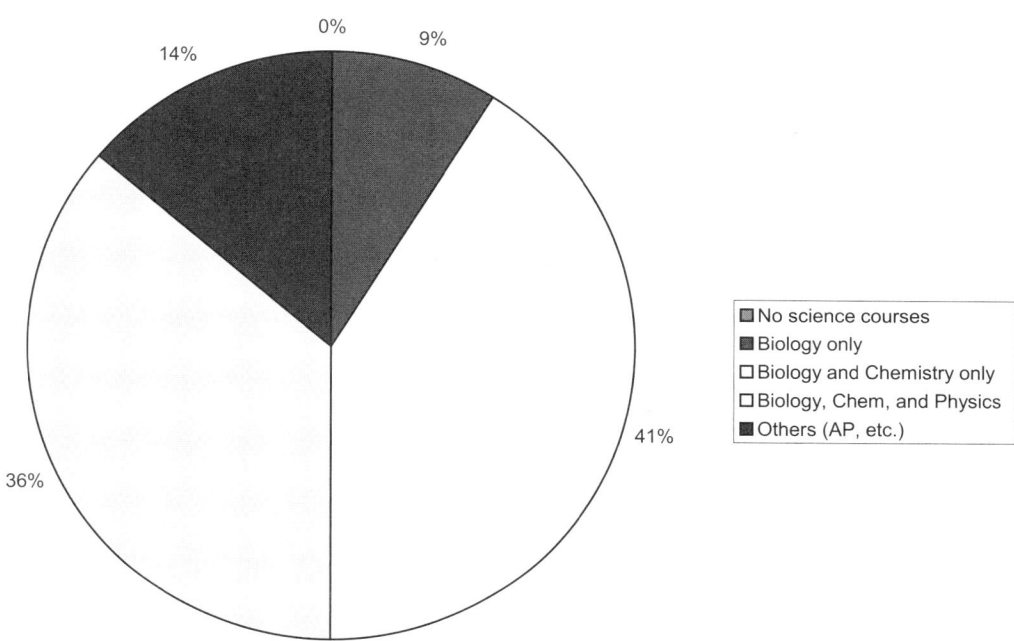

Science Courses Completed by Adults Ages 25-30

- No science courses
- Biology only
- Biology and Chemistry only
- Biology, Chem, and Physics
- Others (AP, etc.)

POSSIBLE ANSWERS TO THE CONCLUSION QUESTIONS

Using the graphs that you constructed, answer the following questions:

Sample Data Set 1: Sources of Nitrogen Oxide Air Pollution

1. How much nitrogen oxide air pollution is due to transportation?
   - 33%

2. Not taking into account the set of data labeled "other", what category contributes the least nitrogen oxide air pollutants into our environment?
   - Industry

3. Can you predict from a graph of this type the amount of nitrogen oxide air pollution that will be contributed by industry in the next ten years? Explain your reasoning.
   - No
   - The pie graph has no time period associated with it.

Sample Data Set 2:

1. Why must a percentage calculation be performed on the data before making the graph?
   - Pie graphs show parts of a whole, or 100%. Since there were 500 people polled, the data had to be calculated as if the sample size were 100.

2. Describe the trend displayed by this pie chart. Be specific.
   - Everyone polled had taken at least one science course. Most of those polled had taken at least two courses and many had taken three courses. The "other" category represents a small population of students who continued their science course of study or took a different science elective that was not listed.

3. Most adults polled between the ages of 25-30 years of age completed which science courses during their high school career?
   - Biology and chemistry

4. Describe the type of data that can be displayed using pie charts. List three specific places where you might see pie charts printed.
   - Pie charts visually display data that can easily be categorized into percentages.
   - Newspapers, science textbooks, and consumer labeling.

Graphing Skills
Reading, Constructing and Analyzing Graphs

Line Graphs

OBJECTIVE
Students will become familiar with basic line graphing skills to be used throughout their science course of study.

LEVEL
All

NATIONAL STANDARDS
UCP.1, UCP.2, G.2

TEKS
6.2(E)
7.2(E)
8.2(E)
IPC 2(C)
Biology 2(C)
Chemistry 2(D)
Physics 2(C), 2(E)

CONNECTIONS TO AP
Each of the AP Science courses requires that students are able to read, construct and analyze graphs.

TIME FRAME
45 minutes

MATERIALS
(For each student working individually)

| 4 sheets of quadrille graph paper | data |
| pencil | ruler |

TEACHER PAGES

TEACHER NOTES

Graphing is a skill that should be introduced at each grade level and reinforced throughout the year whenever data is available. In this exercise students will construct line graphs using paper and pencil. Line graphs will be constructed and analyzed more than any of the other types of graphs throughout the science course of study so you should make sure that all students are comfortable with this activity.

Upon completion of the graphing exercise, analysis of the graph follows. Students should understand basic paper/pencil construction methods of graphing before proceeding to the Foundation Lessons which use Microsoft Excel and Graphical Analysis. Before allowing students to begin this exercise, you should have students make a practice graph at their desks while you make one at the overhead or chalkboard. In your example discuss the following points:

- Ask students if they can identify the *x* and *y*- axes. Label these on your sample.
- Ask students about the terms independent and dependent variable. Do they know where to place each on their graph? Below the *x*-axis write the label "independent" and beside the *y*-axis write the label "dependent". Remind the students that each of the variables should have some type of unit associated with them. Write the word "unit" in parentheses after the word "independent" and "dependent" variable on your example.
- Ask students to identify the last missing component of the graph. [Title] Write the term "title" at the top of the graph. Point out that it is common to title a graph using the dependent variable (*y*-axis vs. the independent variable (*x*-axis) format. Encourage students to give descriptive titles to their graph and not just re-name the axes. The graphs that the students construct should have "good" titles. Titles that tell exactly what information the author is trying to represent with the graph. The title should be concise, clear and complete.
- Using any set of generic data points, illustrate to students how to determine the range for each variable and how to determine the scale for each axis. Point out that graph paper must always be used to construct a graph. Each square along an axis must represent the same increment. Encourage students to use as much space as available to construct their graph. Label each axis on the graph with the proper numbers according to data given.
- Illustrate how to plot points on the graph. Emphasize that each point represents both an *x* and a *y* component. Remind students that the plotting of points in science is the same as plotting an ordered pair in the math class. (in math the points are always given as (*x,y*). Encourage the use of pencil to plot the data set.
- Discuss with students the importance of using a graph to understand relationships. In the science classroom graphs are not generally connect-the-dot graphs. It is common practice to draw the best smooth curve or the line of best fit that relates the data. Illustrate how this is done with the data previously plotted. If more than one set of data is to be displayed on any one graph, remind students that a legend or key will be necessary to identify each line.
- Point out to students that as their mathematical and scientific skills increase, the usefulness and meaning of their graphs will also increase.
- Illustrate the following analysis techniques:
  - Interpolation of data — find a value that lies on the smooth curve or line *between* two actual data points.

  - Extrapolation of data — find a value that lies on the smooth curve or line beyond the actual plotted points. Data can be extrapolated both on the front and back end of the line/curve.

o   Linear regression of data — If the plotted relationship generates a straight line, have students write the equation for the straight line in slope-intercept form [y = mx + b]. To calculate the slope of the line use the equation:

$$slope = \frac{rise}{run} = \frac{\Delta y}{\Delta x} = \frac{y_2 - y_1}{x_2 - x_1}$$

The y-intercept can be found by extending the line of best fit backwards until it crosses the y-axis

POSSIBLE ANSWERS TO THE CONCLUSION QUESTIONS AND SAMPLE DATA

Sample Data Set 1: The following set of data was collected while experimenting with position and time of a miniature motorized car traveling on a straight track.

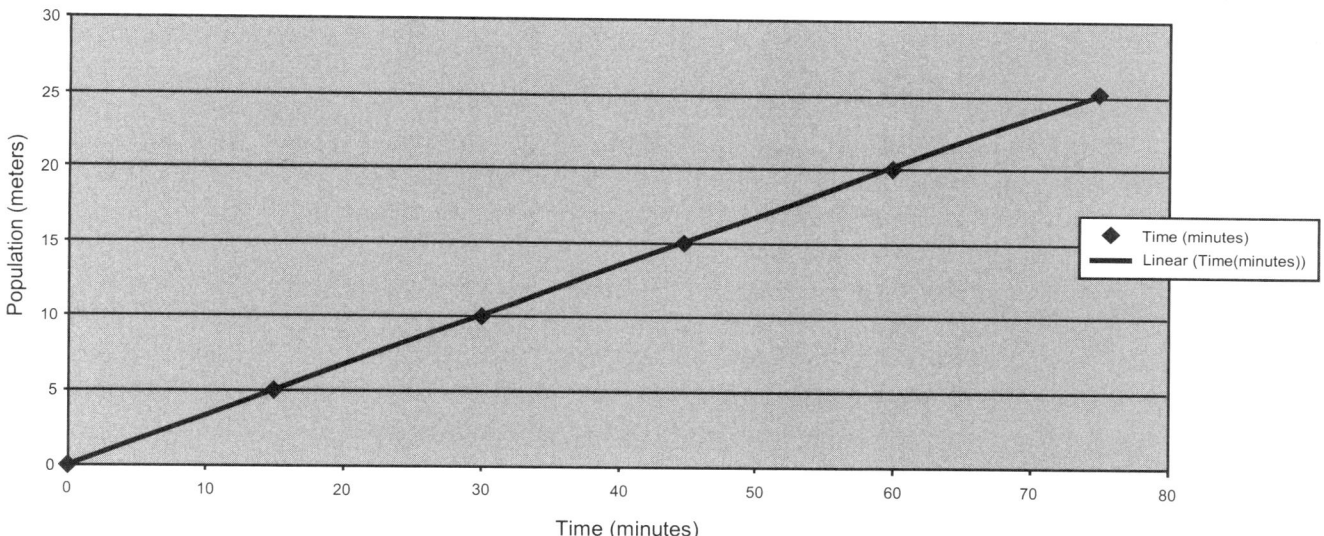

1. What is the independent variable for this graph? Explain.
   - Time is the independent variable. It is the property that is controlled by the experimenter. It is plotted on the x-axis.

2. What would be the position of the car after 25 minutes?
   - About 7.5 to 8.0 meters. Arriving at this answer requires that the student use knowledge of interpolation.

3. If the experiment were carried out for 80 minutes, what would be the position of the car?
   - About 28 meters. Arriving at this answer requires that the student use knowledge of extrapolation.

4. Calculate the slope of the line drawn. What does the slope of this line represent? Explain.

- $slope = \dfrac{rise}{run} = \dfrac{\Delta y}{\Delta x} = \dfrac{y_2 - y_1}{x_2 - x_1}$

- Students may choose any set of points to solve. Answers may vary slightly. Possible solution might be: $slope = \dfrac{rise}{run} = \dfrac{\Delta y}{\Delta x} = \dfrac{10m - 5m}{30min - 15min} = \dfrac{5m}{15min} = .33 \, m/min$

- The slope represents velocity — distance per time.

5. Write the equation for a straight line including the value that was determined for slope.

- y   =   m   x   + b

  position (m)   = (.33 m/min) (time) + 0

Sample Data Set 2: The following set of data was collected during an experiment to find the density for an unknown metal.

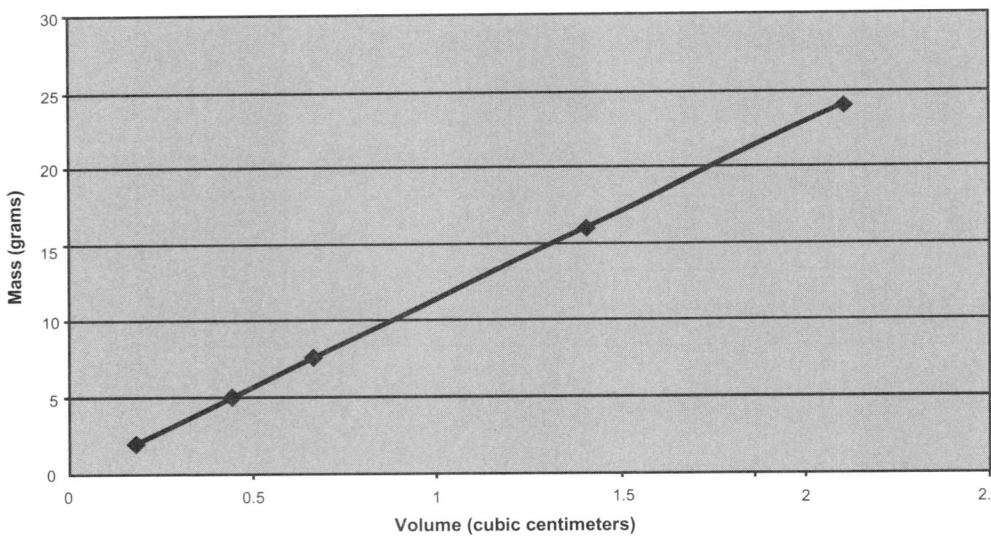

Density: Mass vs. Volume

1. What values were considered when creating the scale for each axis in this experiment?
   - Mass: the range of the numbers was from 2.00 to 24.00 grams. The scale was chosen so that each line represents 5 grams. (student answers may vary as chosen scales may vary)
   - Volume: the range was from 0.18 to 2.11 cm^3. The scale was chosen so that each mark represented 0.50 cm^3.

2. What does a data point on this graph actually represent?
   - Each data point represents a mass and a volume measurement. The relationship between the two is the density which is the slope of the best-fit line.

3. What volume would a 20.00 gram sample of this substance occupy?
   - About 1.75 cm^3. Arriving at this answer requires that the student use knowledge of interpolation.

4. Calculate the density of the substance. (HINT: calculate the slope of the line.)
   - $slope = \dfrac{rise}{run} = \dfrac{\Delta y}{\Delta x} = \dfrac{y_2 - y_1}{x_2 - x_1}$ Students may choose any set of points to solve. Answers may vary

   slightly. Possible solution might be: $slope = \dfrac{rise}{run} = \dfrac{\Delta y}{\Delta x} = \dfrac{15g - 5g}{1.44cm^3 - .44cm^3} = \dfrac{10}{1.00} = 10.0\,g/cm^3$

   - The slope represents density — mass per unit volume.

5. Write the equation for a straight line including the value that was determined for slope.

   - $\quad$ y $\quad$ = $\quad$ m $\quad\quad$ x $\quad$ + $\quad$ b

     mass (g) $\quad$ = $\quad$ (10.0 g/cm^3 $\quad$ (volume(cm^3)) $\quad$ + $\quad$ 0

6. Use the equation and find the mass when the volume is 5.00 cm^3.

   - $\quad$ y $\quad$ = $\quad$ m $\quad\quad$ x $\quad$ + $\quad$ b

     mass (g) $\quad$ = $\quad$ (10.0 g/cm^3) $\quad$ (volume (cm^3)) $\quad$ + $\quad$ 0

     mass (g) $\quad$ = $\quad$ (10.0 g/cm^3) $\quad$ (5.00 cm^3) $\quad$ + $\quad$ 0

     mass (g) $\quad$ = $\quad$ 50.0 grams

TEACHER PAGES

Sample Data Set 3: The following set of data was collected during an experiment studying the effect of light intensity on rate of photosynthesis.

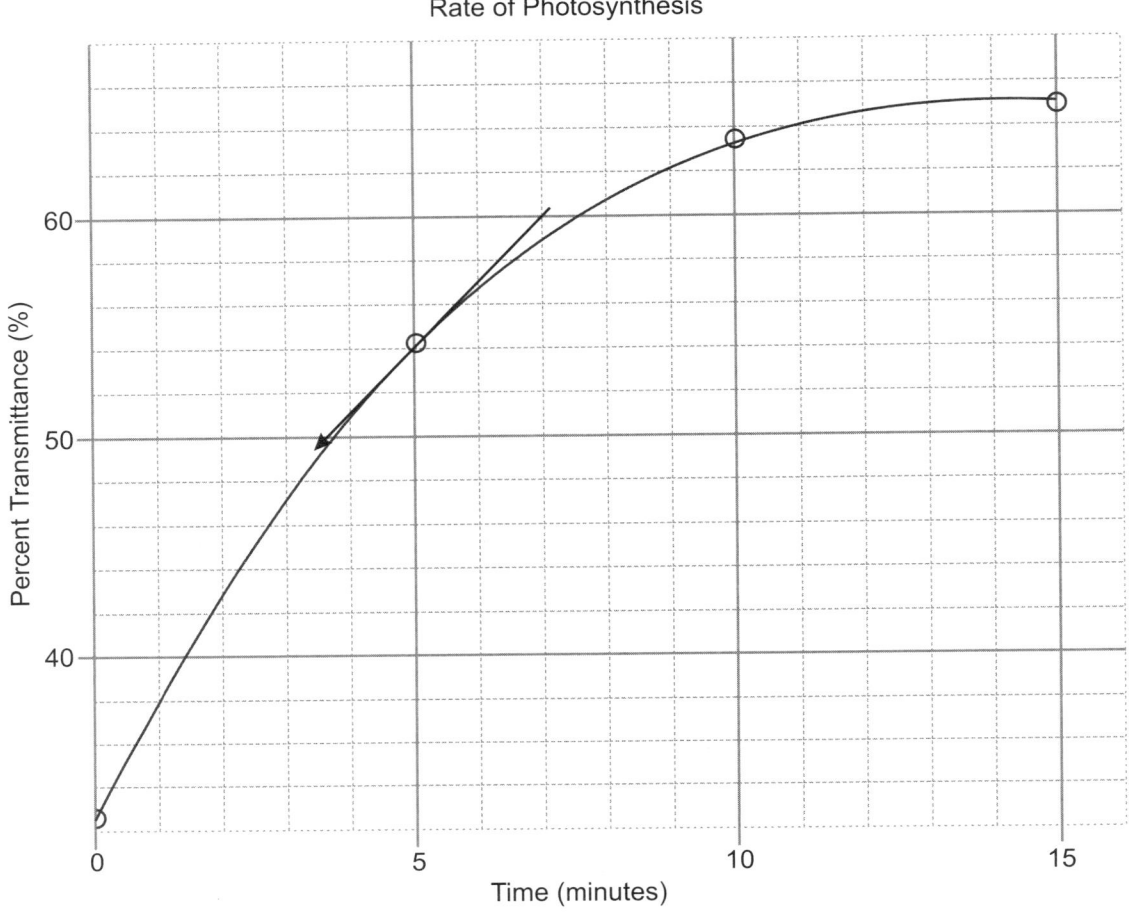

1. Does this graph represent a linear relationship? Why or why not?
   - This graph is not a linear relationship. The slope of the line changes with time.

2. What is the dependent variable in this graph? Explain.
   - The dependent variable is the percent transmittance. Time is the independent variable, the one that the person performing the experiment has control over. At each time interval, the transmittance is measured.

3. If the experiment were continued for 30 minutes, what trend in percent transmittance could be expected?
   - The graph begins to level somewhat between 10 and 15 minutes. It seems logical that the percent transmittance would soon level off or begin to fall.

Note: Middle school teachers may wish to omit the remainder of this lesson.

4. Calculate the slope of the line at 5 minutes. What does this represent?
   - A tangent line must be drawn at 5 minutes.
   - $\text{slope} = \dfrac{\text{rise}}{\text{run}} = \dfrac{\Delta y}{\Delta x} = \dfrac{60\% - 50\%}{7\,\text{min} - 4.5\,\text{min}} = \dfrac{10}{2.5} = 4.09\%/\text{min}$
   - The slope represents the instantaneous rate. Notice that the slope of the line is different at 10 minutes.

Sample Data Set 4: The following set of data was collected during a titration experiment of a diprotic acid and sodium hydroxide.

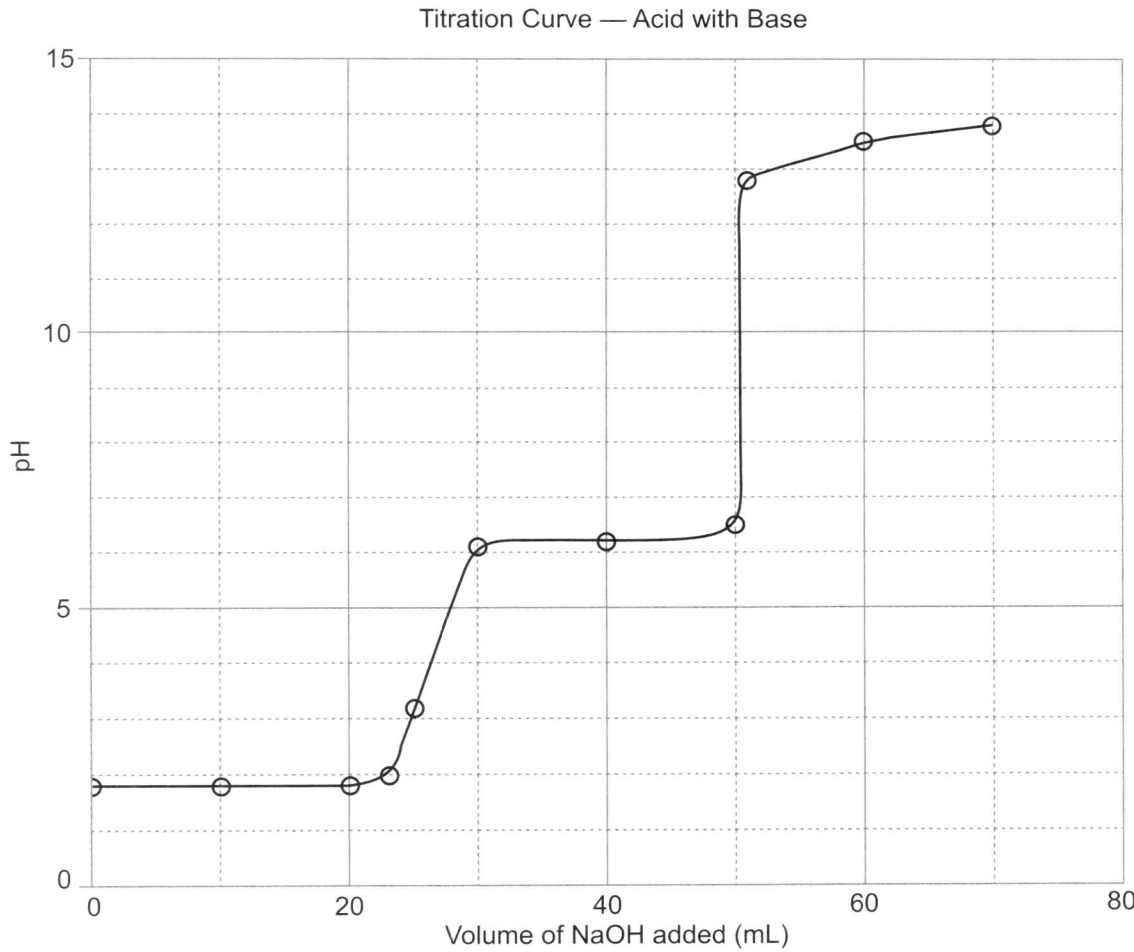

Titration Curve — Acid with Base

1. What is the pH of the solution after 20.0 mL of NaOH are added? After 30.0 mL are added? Would it have been easy to predict this answer?
   - The pH at 20.0 mL = 1.80; the pH at 30.0 mL = 6.10
   - The graph is not a direct relationship and thus, would not have made this easy to predict such a large jump in pH.

2. Graphs often help us to understand the progress of a chemical reaction. In the titration graph for this set of data, there are two relatively sharp, upward curves. The middle of these steep rising portions represent equivalence points (point at which the moles of acid and base are equal). Identify the volume of NaOH needed to reach each of the equivalence points.
   - The first equivalence point occurs around 28 mL. (the midpoint of the sharp, upward curve must be taken)
   - The second equivalence point occurs around 50 mL.

3. What is the pH at 65 mL. What is the pH expected to do beyond this point with greater additions of the base NaOH? Explain.
   - The pH at 65 mL is approximately 13.50.
   - Beyond 65 mL the pH will gradually rise and level off. Since NaOH is a strong base it will eventually reach close to a pH of 14.00.
   - The curve begins leveling off after the second equivalence point, therefore, the pH will not drastically change with more NaOH added. The pH after the equivalence point is only due to the amount of NaOH that is in excess of the acid.

Graphing Skills
Reading, Constructing and Analyzing Graphs

Bar Graphs and Histograms

Bar graphs are very common types of graphs. They are found in almost all science books, magazines, and newspapers. They can be useful tools in scientific study by allowing us to visually compare amounts or frequency of occurrences between different data sets. Bar graphs can be used to show how something changes over time or to compare items with one another. When reading or constructing this type of graph you should pay close attention to the title, the label on the axes, the unit or scale of the axes, and the bars.

In a simple bar graph the specific group or experimental subject is assigned the *x*-axis (horizontal) and the *y*-axis (vertical) is known as the frequency axis. In general, the *x*-axis will be divided into time periods or measurements while the *y*-axis is designated for the frequency of occurrences. When data is grouped, the *x*-axis always represents the grouped data while the y-axis shows the frequency data. A composite bar graph is often useful when displaying the sum of various dependent variables when the values are a fraction of the whole. Histograms are very similar to simple bar graphs with one exception — the bar represents a range of values rather than one single value and the intervals must all be of equal magnitude. Study the sample graphs below before completing this exercise.

Simple Bar Graph

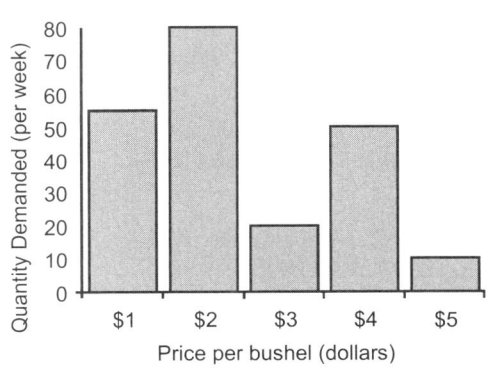

Price of Wheat vs. Quantity Demanded

Grouped Bar Graph

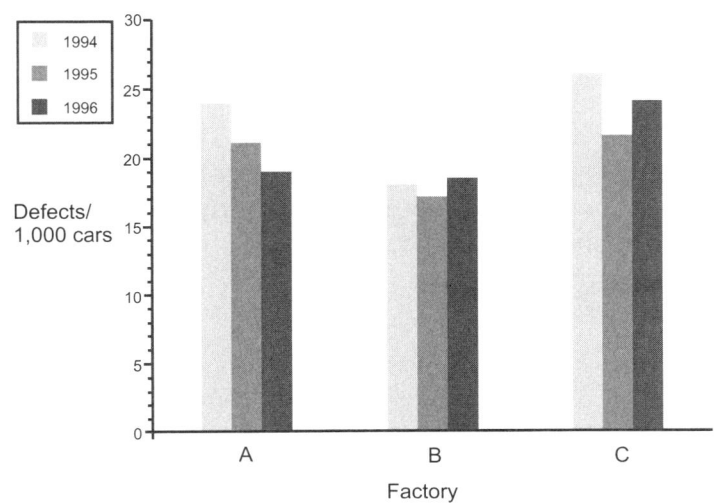

Car Defects vs. Different Factories Over Time

Composite Bar Graph

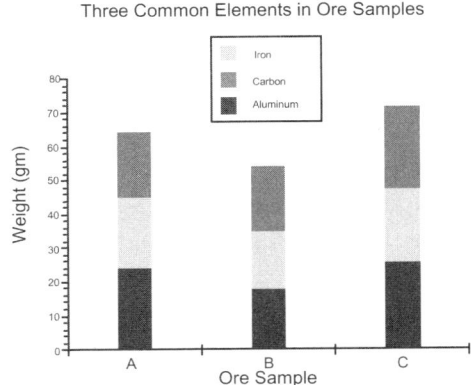

Three Common Elements in Ore Samples

Histogram

Population Distribution

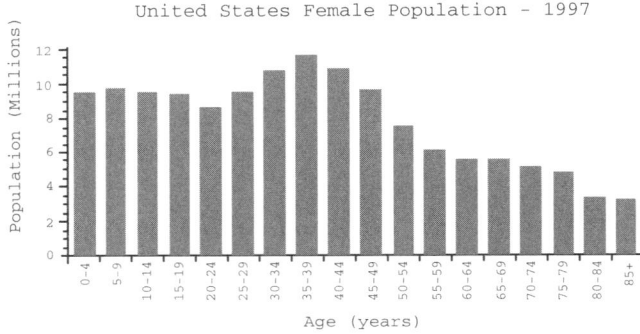

PURPOSE

In this exercise you will create simple bar graphs, grouped bar graphs, composite bar graphs and histograms. You will be expected to properly label each of your graphs and analyze each one by making statements about trends in the data.

MATERIALS

4 sheets of graph paper   data
pencils   straight edge

PROCEDURE

PART I: SIMPLE BAR GRAPH

1. Obtain one piece of graph paper and a pencil.

2. Study the data table below.

Leading Causes Of Death Worldwide	
Cause	Deaths Per Year (millions)
Cardiovascular disease	16.9
Cancers and tumors	7.2
Infectious diseases (includes AIDS, malaria, etc.)	13.5
Accidents and trauma	5.1
Respiratory disease	3.5
Digestive and nutritional	2.3
Diabetes	0.9

3. Choose the data to be graphed on the *x*-axis and the *y*-axis.

4. Survey the data and determine an appropriate scale for each axis. Be sure to utilize as much of the graph paper as possible o display your data. Use your pencil to lightly mark the scale of your *x* and *y*-axes. Have your teacher check your scale before proceeding any further. When making a bar graph, the individual bars should be constructed with the same width. You may decide the width of your bars.

5. When your teacher approves, construct your simple bar graph. Be sure to label each axis with units and give your graph a title.

PART II: GROUPED BAR GRAPH

1. Study the following data and follow the same procedure as Part I with a clean sheet of graph paper. This data should be graphed as a grouped bar graph and include a legend or key to indicate what each bar represents.

Some of the Most Common Chemical Elements of the Earth		
Chemical Element	Whole Earth (%)	Crust (%)
Iron (Fe)	33.3	5.8
Oxygen (O)	29.8	45.2
Silicon (Si)	15.6	27.2
Magnesium (Mg)	13.9	2.8
Aluminum (Al)	1.5	8.2

PART III: COMPOSITE BAR GRAPH

1. Study the following data and follow the same graphing procedure with a clean sheet of graph paper. This data should be graphed as a composite bar graph. You will need to include a legend and be sure to place the waste disposal methods in the same order for each bar drawn.

Solid Waste Recycled, Incerated, and Landfilled in US and Japan		
	Japan	US
Recycle (%)	50	24
Incinerate (%)	23	17
Landfill (%)	27	59

PART IV: HISTOGRAM

1. Study the following data and follow the same graphing procedure. This data should be graphed as a histogram. It is important that histograms have the same interval and width for each bar. For example, each bar might represent 10 years in the data table below.

Life Expectancies in the US	
Current Age	Remaining Years Expected
0-10	72.6
11-20	59.5
21-30	50.1
31-40	40.7
41-50	31.7
51-60	23.2
61-70	15.8
71-80	9.7
81-90	4.5

Graphing Skills
Reading, Constructing and Analyzing Graphs

Bar Graphs and Histograms

CONCLUSION QUESTIONS

Using the graphs that you constructed, answer the following questions.

PART I: SIMPLE BAR GRAPH

1. How many deaths due to accidents and trauma occur per year?

2. Can you predict the number of deaths due to cancers and tumors for the next ten years? Explain.

PART II: GROUPED BAR GRAPH

1. What element is the most abundant in the earth's crust?

2. What element is the most abundant within the earth?

PART III: COMPOSITE BAR GRAPH

1. What is the favored method of waste disposal for the Japanese population? Cite possible reasons for this.

2. What is the favored method of waste disposal for the US population? Cite possible reasons for this.

PART IV: HISTOGRAM

1. Make a prediction about the remaining years of life that would be expected for someone in the current age category of 91-100.

2. Is the answer to question 7 an accurate number? Why or why not? Cite specific reasons.

3. What type of data is easily represented by a bar graph?

4. Why is a legend (or key) necessary in the grouped and composite bar graphs?

5. Explain why it is difficult to make direct comparisons between recycling in Japan and the US using the composite bar graph that you drew.

6. What is the importance of scaling?

7. Distinguish between the dependent and the independent variable for each of the graphs that were constructed. On which axis should the independent variable be placed?

	Dependent Variable	Independent Variable
Simple Bar Graph		
Grouped Bar Graph		
Composite Bar Graph		
Histogram		

Graphing Skills
Reading, Constructing and Analyzing Graphs

Pie Charts

Pie charts are very commonly found in newspapers, magazines and textbooks. A pie chart is a very good way to represent percentages.

PURPOSE
In this activity you will practice constructing and analyzing basic pie charts.

MATERIALS

blank paper	data
pencil	protractor
compass	colored pencils

Safety Alert
1. The sharp point on the compass should only be placed on paper.

PROCEDURE
1. After observing your teacher demonstrate the use of the compass and protractor, use Sample Data Set 1 to construct a pie chart on a piece of blank paper. Use different colors to represent different sections of your graph.

2. Be sure to label your chart with an appropriate title and be sure to provide a legend or key that distinguishes each component.

Sample Data Set 1: Sources of Nitrogen Oxide Air Pollution

Sources of Nitrogen Oxides	Percentages (%)
Power plants	53%
Transportation	68%
Industry	27%
Non-road	32%
Other	20%

3. Use Sample Data Set 2 to construct a second pie chart on another blank piece of paper. Be sure to label appropriately. *Note*: Before beginning construction of this graph, you must calculate component percentages. Show your work on the student answer page.

Sample Data Set 2: 500 adults between the ages of 25-30 were polled as to which science courses they completed in their high school years. The following data was collected.

Science Courses Completed	Number of Adults
No science courses	0
Biology only	45
Biology and Chemistry only	205
Biology, Chemistry and Physics only	180
Other classes not listed (AP, etc.)	70

Name _____

Period _____

Graphing Skills
Reading, Constructing and Analyzing Graphs

Pie Charts

ANALYSIS

1. Staple your two graphs behind this answer page.

2. Show your work here for Sample Data 2 - percentage calculations.

CONCLUSION QUESTIONS

Using the graphs that you constructed, answer the following questions:

Sample Data Set 1: Sources of Nitrogen Oxide Air Pollution

1. How much nitrogen oxide air pollution is due to transportation?

2. Not taking into account the set of data labeled "other", what category contributes the least nitrogen oxide air pollutants into our environment?

3. Can you predict from a graph of this type the amount of nitrogen oxide air pollution that will be contributed by industry in the next ten years? Explain your reasoning.

Sample Data Set 2:

1. Why must a percentage calculation be performed on the data before making the graph?

2. Describe the trend displayed by this pie chart. Be specific.

3. Most adults polled between the ages of 25-30 years of age completed which science courses during their high school career?

4. Describe the type of data that can be displayed using pie charts. List three specific places where you might see pie charts printed.

Name _____

Period _____

Graphing Skills
Reading, Constructing and Analyzing Graphs

Line Graphs

There are all kinds of charts and graphs used in the science classroom. Graphs are useful tools in science. Trends in data are easy to visualize when represented graphically. A line graph is beneficial in the classroom for many different types of data. Line graphs are probably the most widely used scientific graph. They can be used to show how something changes over time, the relationship of two quantities, and can be readily used to *interpolate* (predict between measured points on the graph) and *extrapolate* (predict beyond the measured points along the same slope) data points that were not actually measured in the lab setting. The analysis of these graphs provides very valuable information.

PURPOSE
In this activity you will learn the basic procedure for constructing and analyzing line graphs.

MATERIALS

4 sheets of graph paper	data
pencil	ruler

PROCEDURE

1. Follow along with your teacher as a sample line graph is constructed. Label a blank piece of graph paper as your teacher explains the important components of a line graph.

2. Use the sample sets of data below to construct line graphs. Place only one graph on each sheet of graph paper and use as much of the graph as possible to display your points. ***Do not connect the dots!*** Draw the best smooth curve or line of best fit as your teacher demonstrated.

3. Following the steps below will help ensure that all components of the graph are correctly displayed.
   a. **Identify the variables**. Independent on the x-axis and dependent on the y-axis.
   b. **Determine the range**. Subtract the lowest value data point from the highest value data point—for each axis separately.
   c. **Select the scale units**. Divide each axis uniformly into appropriate units using the maximum amount of space available. (Remember that the axes may be divided differently but each square along the same axis must represent the same interval.)
   d. **Number and label each axis**. Be sure to include units where appropriate as part of the axis label.
   e. **Plot the data points as ordered pairs**. (x,y)

    f. **Draw the best straight line or best smooth curve**. Use a straight edge to draw your line in such a way that equal numbers of points lie above and below the line.

    g. **Title the graph**. The title should clearly describe the information contained in the graph. It is common to mention the dependent variable first followed by the independent variable.

4. After creating graphs for the 4 data sets below, use the graphs to answer the conclusion questions on your student answer page.

Sample Data Set 1: The following set of data was collected while experimenting with position and time of a miniature motorized car traveling on a straight track.

Position (meters)	Time (minutes)
0	0
15	5
30	10
45	15
60	20
75	25

Sample Data Set 2: The following set of data was collected during an experiment to find the density for an unknown metal.

Mass (g)	Volume (cm^3)
2.00	0.18
5.00	0.44
7.50	0.66
16.00	1.41
24.00	2.11

Sample Data Set 3: The following set of data was collected during an experiment studying the effect of light intensity on rate of photosynthesis.

Percent Transmittance (%)	Time (minutes)
32.5	0
54.3	5
63.5	10
65.0	15

Sample Data Set 4: The following set of data was collected during an acid-base titration experiment.

pH	Volume of NaOH (mL)
1.80	0.00
1.80	10.00
1.82	20.00
2.00	23.00
3.20	25.00
6.10	30.00
6.20	40.00
6.50	50.00
12.80	51.00
13.50	60.00
13.80	70.00

Name _____

Period _____

Graphing Skills
Reading, Constructing and Analyzing Graphs

Line Graphs

DATA AND OBSERVATIONS

Staple your completed graphs behind this answer page.

CONCLUSION QUESTIONS

Using the graphs that you constructed, answer the following questions:

Sample Data Set 1:

1. What is the independent variable for this graph? Explain.

2. What would be the position of the car after 25 minutes?

3. If the experiment were carried out for 80 minutes, what would be the position of the car?

4. Calculate the slope of the line drawn. What does the slope of this line represent? Explain.

5. Write the equation for a straight line including the value that was determined for slope.

Sample Data Set 2:

1. What values were considered when creating the scale for each axis in this experiment?

2. What does a data point on this graph actually represent?

3. What volume would a 20.00 gram sample of this substance occupy?

4. Calculate the density of the substance. (HINT: calculate the slope of the line.)

5. Write the equation for a straight line including the value that was determined for slope.

6. Use the equation and find the mass when the volume is 5.00 cm^3.

Sample Data Set 3:

1. Does this graph represent a linear relationship? Why or why not?

2. What is the dependent variable in this graph? Explain.

3. If the experiment were continued for 30 minutes, what trend in percent transmittance could be expected?

4. Calculate the slope of the line at 5 minutes. What does this represent?

Sample Data Set 4:

1. What is the pH of the solution after 20.0 mL of NaOH are added? After 30.0 mL are added? Would it have been easy to predict this answer?

2. Graphs often help us to understand the progress of a chemical reaction. In the titration graph for this set of data, there are two relatively sharp, upward curves. The middle of these steep rising portions represent equivalence points (point at which the moles of acid and base are equal). Identify the volume of NaOH needed to reach each of the equivalence points.

3. What is the pH at 65 mL. What is the pH expected to do beyond this point with greater additions of the base NaOH? Explain.

Microsoft Excel
Using Excel in the Science Classroom

OBJECTIVE
Students will take data and use an Excel spreadsheet to manipulate the information. This will include creating graphs, manipulating data, finding averages and calculating standard deviation.

LEVEL
All

NATIONAL STANDARDS
UCP.1, UCP.2, A.1, A.2, E.1, E.2, G.2

TEKS
6.2 (E), 6.4(A)
7.2(E), 7.4(A)
8.2 (E), 8.4(A), 8.4(B)
IPC: 2(C)
Biology: 2(C)
Chemistry: 2(D)
Physics: 2(C), 2(E)

CONNECTIONS TO AP
Graphing skills, data management, using technology

TIME FRAME
30 minutes (for each lesson)

MATERIALS
Computers with Microsoft Excel software

TEACHER NOTES
This foundation lesson contains four sub-lessons: bar graphs, line graphs, scatter plots with linear regressions, and data management. You may want to teach each lesson as a stand alone, or as they are relevant to a current lab. The graphing and data lessons can be completely independent of one another.

Sample data has been provided for you to use if you would like to teach these as a stand-alone lesson. It is probably best used as a follow up to a data collection lab so that students can use real data.

Microsoft Excel
Using Excel in the Science Classroom

Part I: How to Make a Bar Graph

PURPOSE

To use the software program Microsoft Excel to generate a bar graph.

MATERIALS

data from this handout
computer
Microsoft Excel software

PROCEDURE

In science class you have collected data to see how much the density of water changes as you add grams of salt. Your teacher wants you to take the data and produce and a bar graph using Excel. The data is as follows:

Grams of Salt	Density
0	1.00
5	1.03
10	1.07
15	1.11
20	1.14

1. Open the Excel program on your computer. A blank workbook will appear. Notice that the columns are identified with letters and the rows are identified by numbers.

2. In the box "A1", type Grams of Salt.

3. In the box "B1", type Density. If you need to make a box larger, take your cursor to the top of the column and place it between two boxes. A double arrow should appear and you can stretch the column to the size you need.

4. Enter the data in the boxes below each section. Be careful to enter the coordinating data in the correct row.

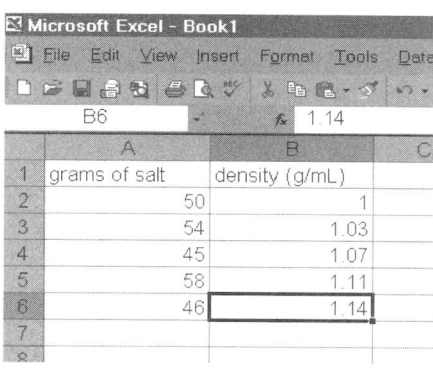

5. On your toolbar there is a very small, colorful bar graph icon. This is called the Chart Wizard. Click on the Chart Wizard icon.

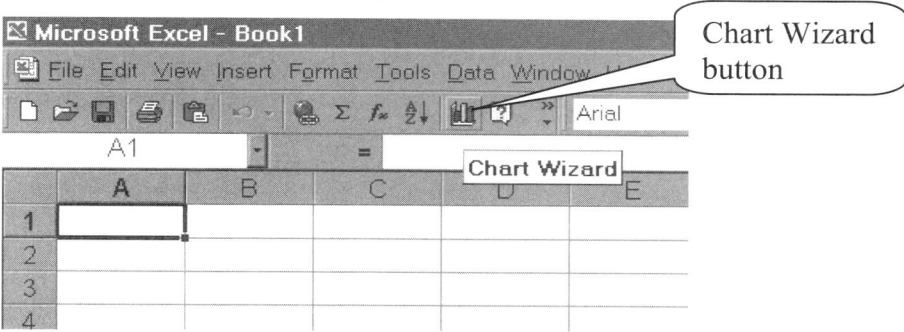

6. After clicking the Chart Wizard button, the first window that opens identifies chart type. Choose "Column" on the left-hand side, and under the chart sub-type on the right side click on the first choice available.

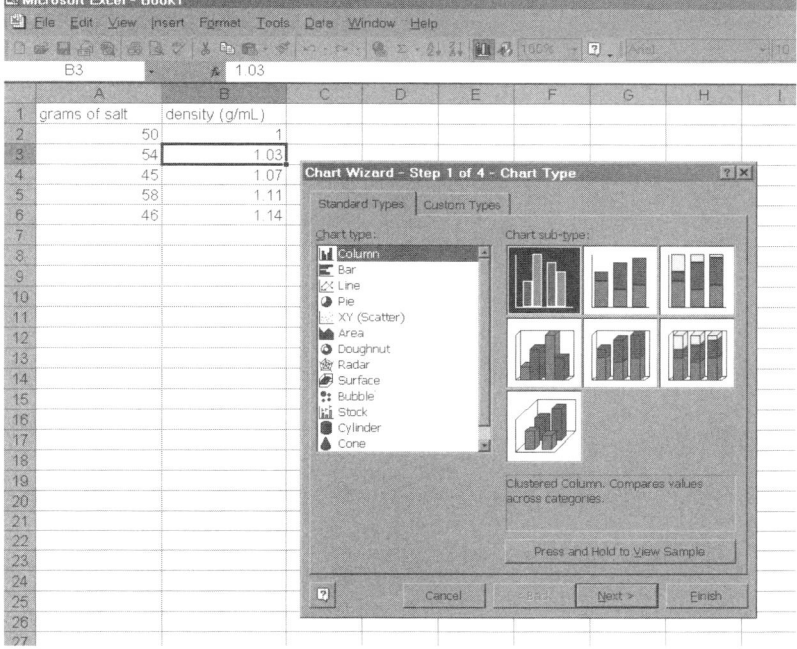

7. Click Next.

8. The next window that appears has two tabs: Data Range and Series. Click on Series and **remove all existing data sets from the series box**.

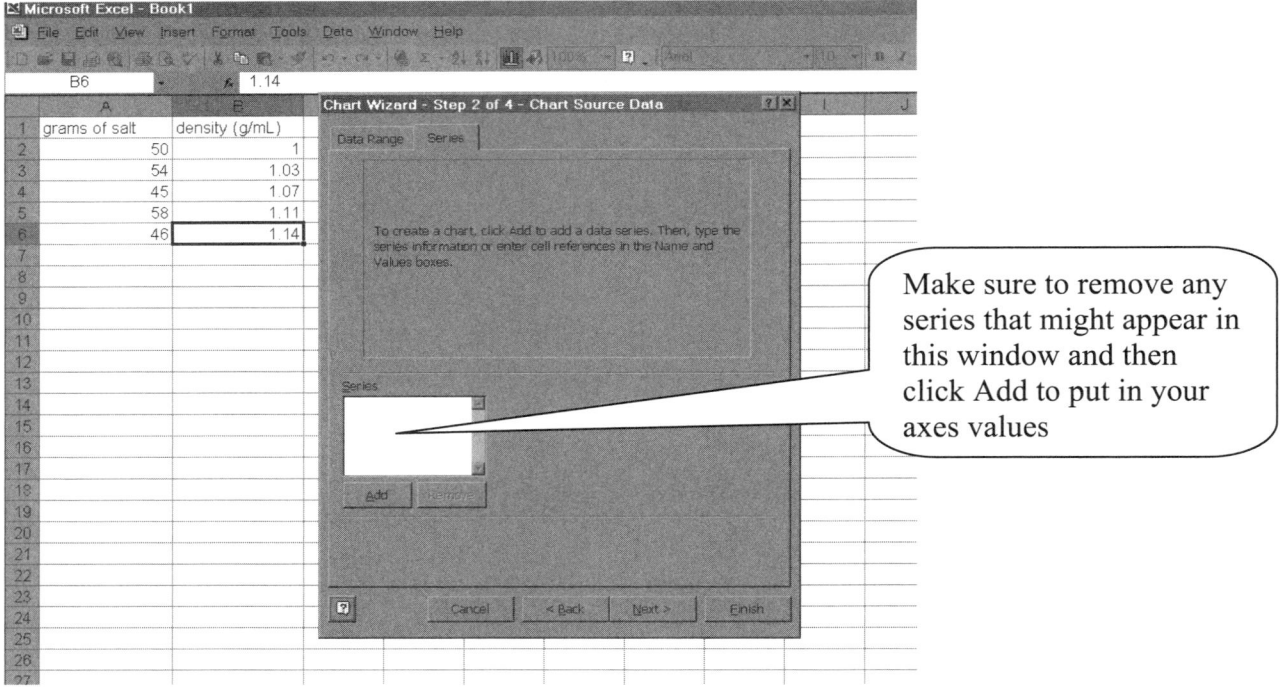

Make sure to remove any series that might appear in this window and then click Add to put in your axes values

9. Now click **Add** to add your data series. On the bottom of the window is "Category (X) Axis Labels". In the right corner of the Category (X) axis labels is a small button with a tiny graph containing a red arrow. Click on this button.

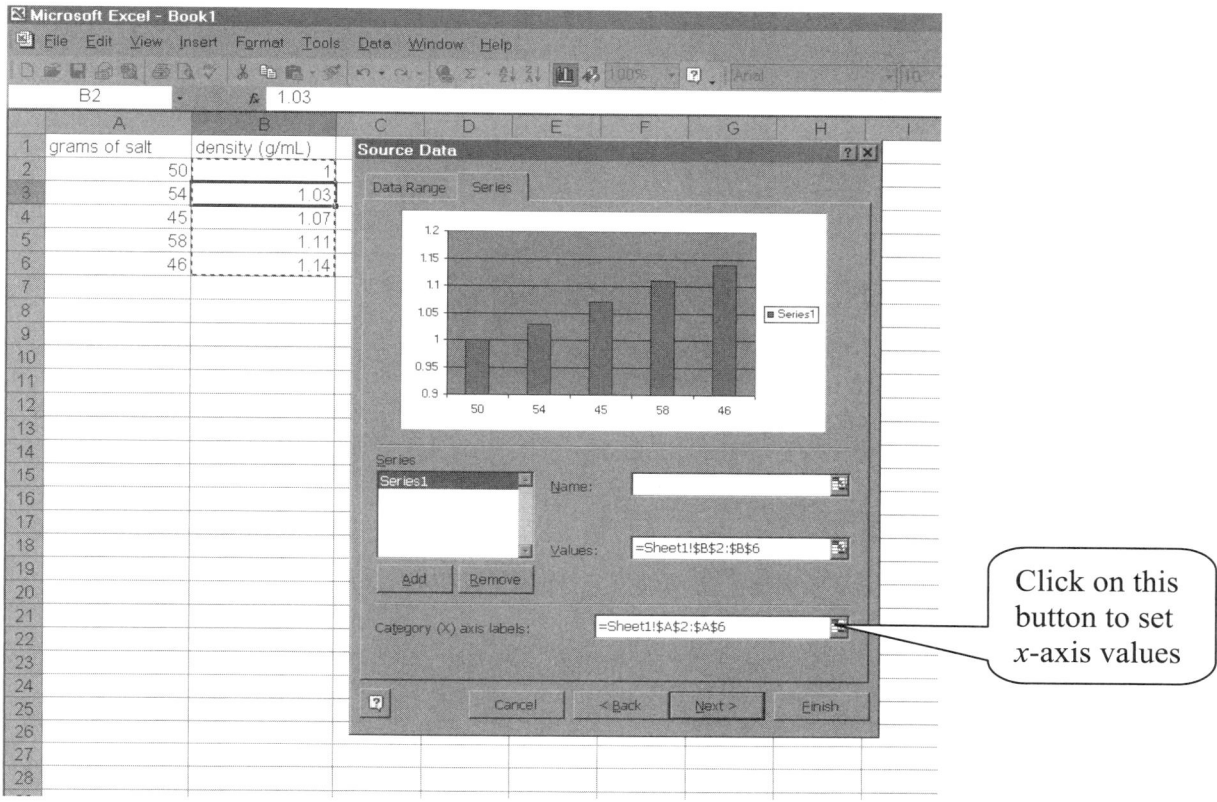

10. Clicking on the button takes you back to your spreadsheet of data. With your mouse, highlight the data you want on the *x*-axis, in this case, the grams of salt, or boxes B1-F1. Press Enter after highlighting.

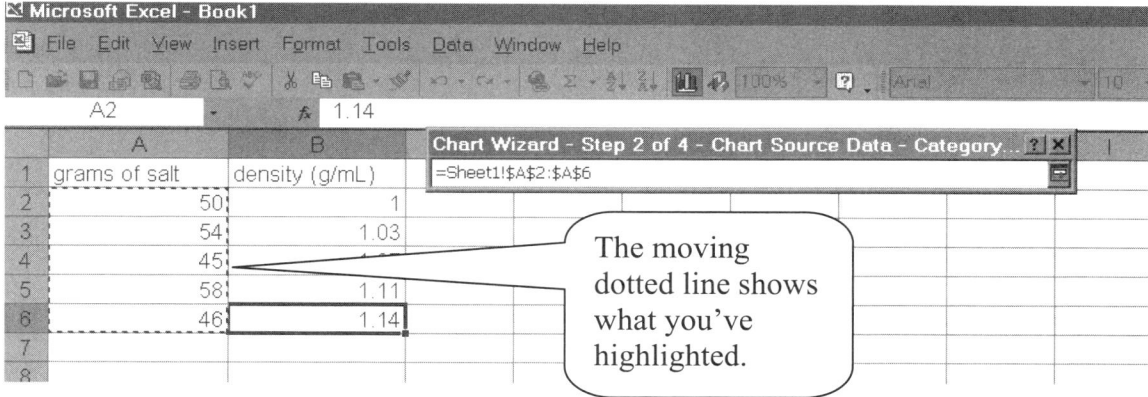

11. The chart wizard screen should now reappear. Click on the small graph button next to the spot labeled Values. This is your *y*-axis label.

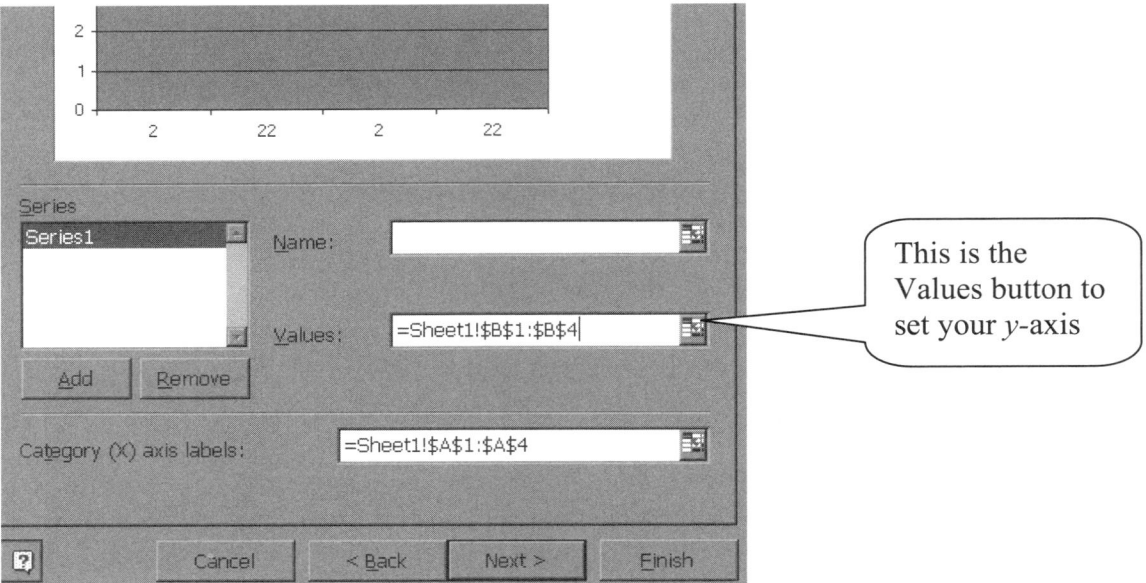

12. After clicking on the Values button, the computer takes you back to your spreadsheet and now you want to highlight your *y*-axis values, in this case density, B2-F2. Press Enter after highlighting.

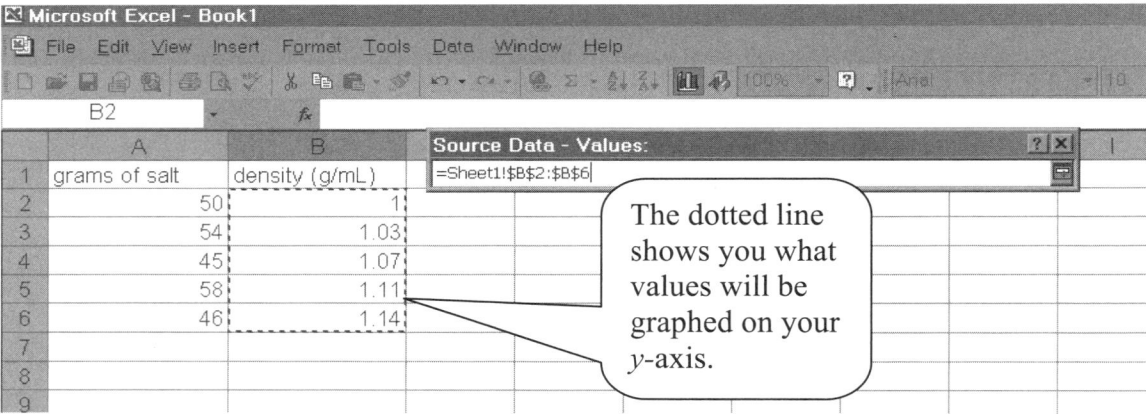

13. The Chart Wizard screen will reappear. Click next on the bottom of the screen.

14. The new screen allows you to name your graph and label your axis. Fill in the blanks with the appropriate information and click Finish.

15. You now have finished your bar graph and Excel will ask you if you want the graph to appear on your spreadsheet, or on a separate page. Choose whichever you need. Below is a copy of the graph inserted into the spreadsheet page.

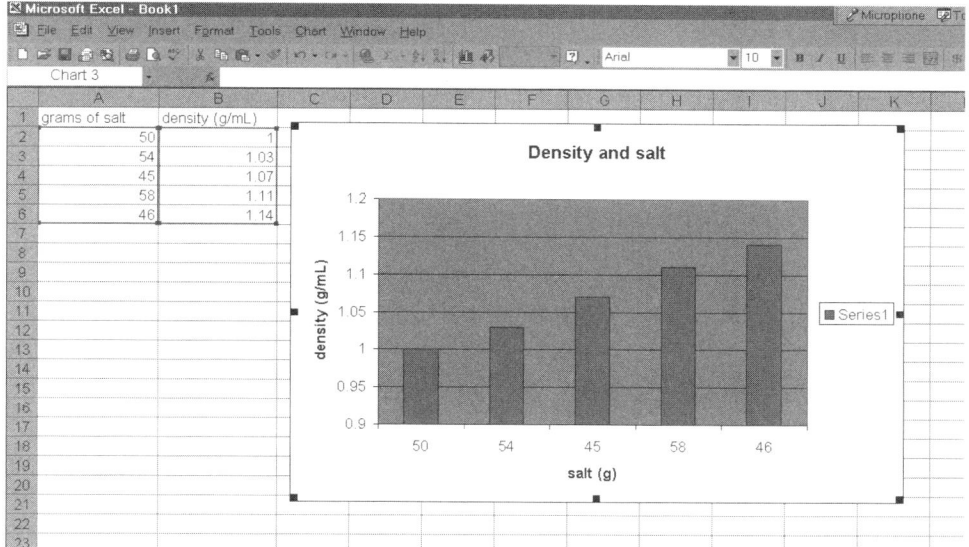

16. You may now print your completed graph by selecting print from the file menu on the task bar.

Microsoft Excel
Using Excel in the Science Classroom

Part II: How to Make a Line Graph

PURPOSE

To use the software program Microsoft Excel to create a line graph.

MATERIALS

data from this handout
computer
Microsoft Excel software

PROCEDURE

In science class you have collected data to see how much the density of water changes as you add grams of salt. Your teacher wants you to take the data and produce and a line graph using Excel. The data is as follows:

Grams of Salt	Density
0	1.00
5	1.03
10	1.07
15	1.11
20	1.14

1. Open an Excel Workbook. Notice that the columns are identified with letters and the rows are identified by numbers.

2. In the box "A1", type Grams of Salt.

3. In the box "B1", type Density. If you need to make a box larger, take your cursor to the top of the column and place it between two boxes until a double arrow appears. Now stretch the column to the size you need.

4. Enter the data in the boxes below each section. Be careful to enter the coordinating data in the correct row.

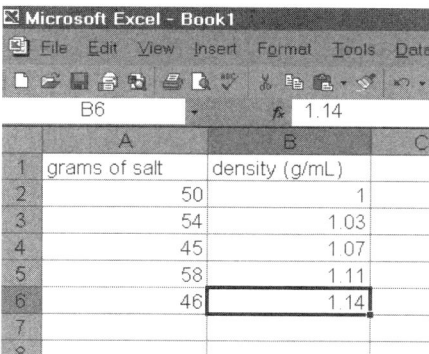

5. On your toolbar there is a very small, colorful bar graph icon. This is called the Chart Wizard. Click on the Chart Wizard icon.

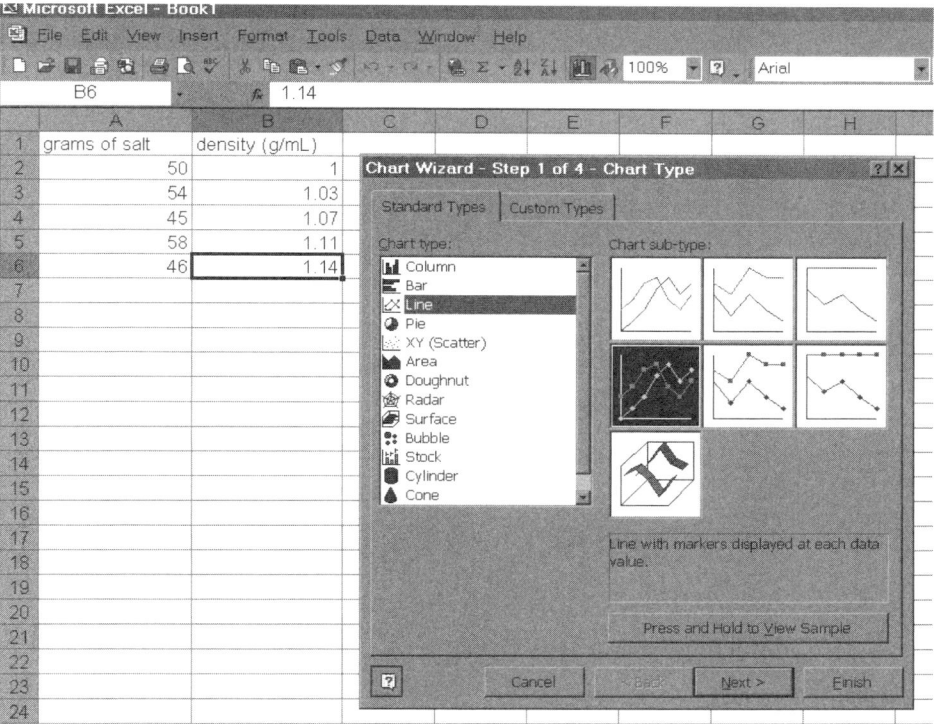

6. The first window to open is to identify chart type. Choose "Line" on the left hand side, and under the chart sub-type on the right side click on the first choice on the second line. Click Next.

7. The next window that appears has two tabs: Data Range and Series. Click on Series and **remove all existing data sets from the series box**.

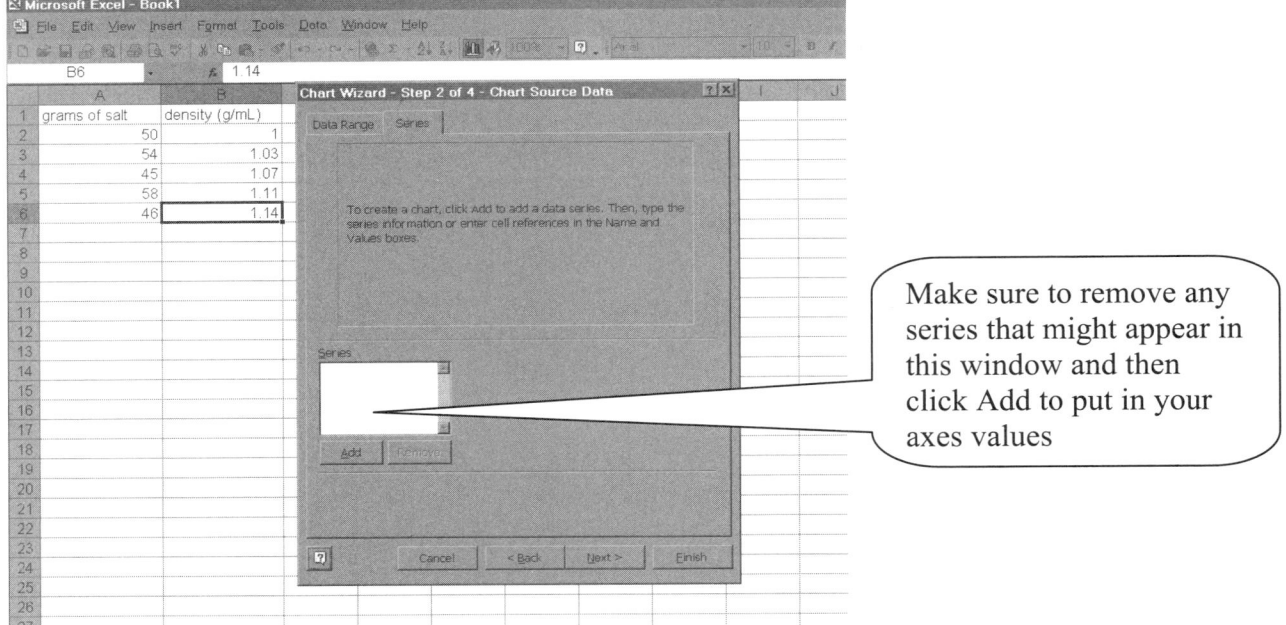

Make sure to remove any series that might appear in this window and then click Add to put in your axes values

8. Now click **Add** to add your data series. On the bottom of the window is "Category (X) Axis Labels". In the right corner of the Category (X) axis labels is a small button with a tiny graph containing a red arrow. Click on this button.

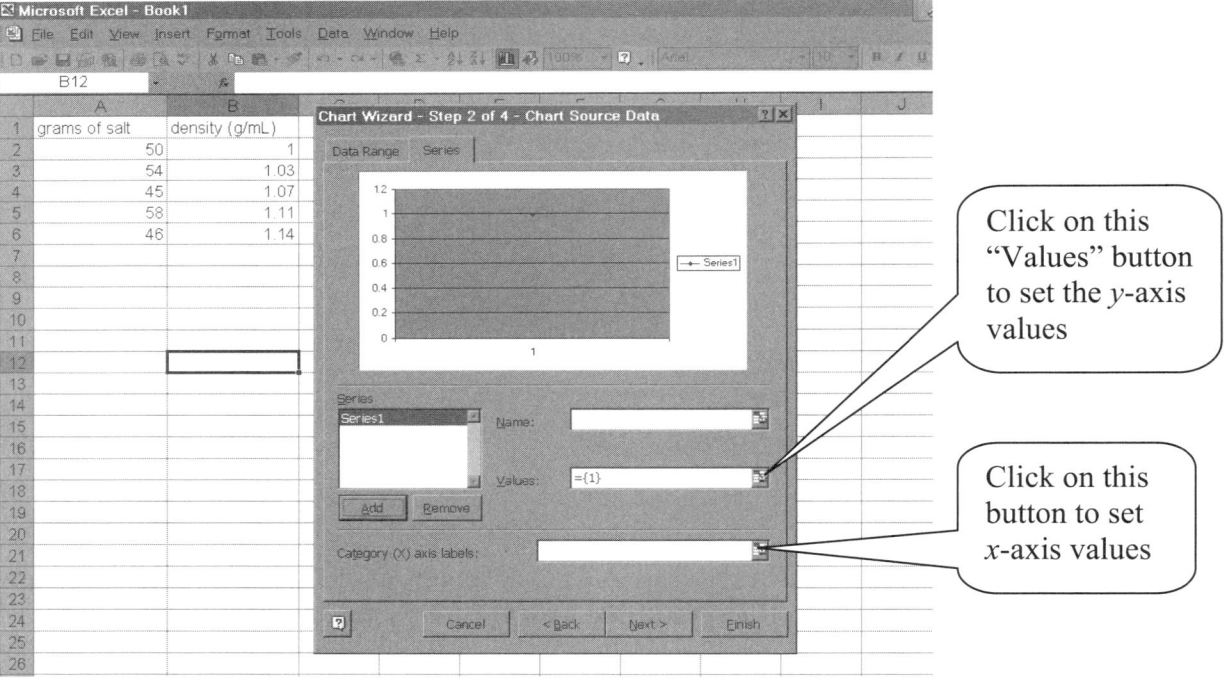

Click on this "Values" button to set the *y*-axis values

Click on this button to set *x*-axis values

9. Clicking on the button takes you back to your spreadsheet of data. With your mouse, highlight the data you want on the *x*-axis, in this case the grams of salt, or boxes B1-F1. Press Enter after highlighting.

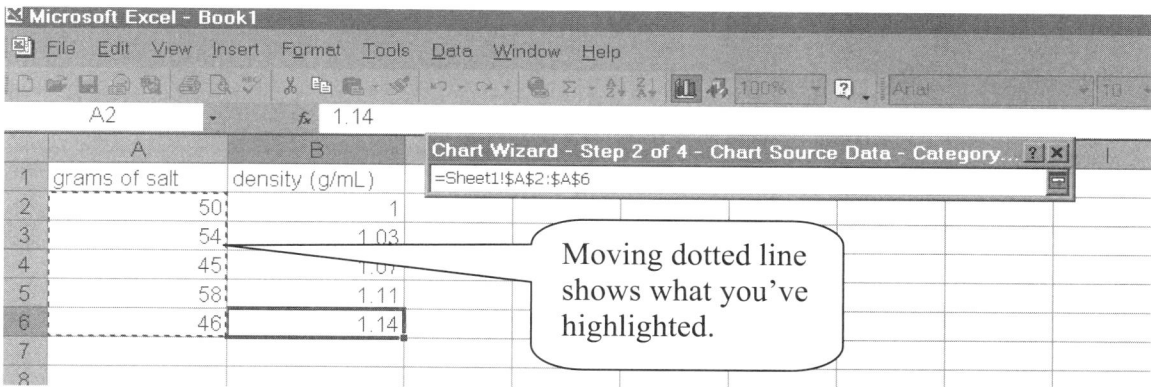

Moving dotted line shows what you've highlighted.

10. The Chart Wizard screen should now reappear. Click on the small graph button next to the spot labeled Values. This is how you add data to your *y*-axis.

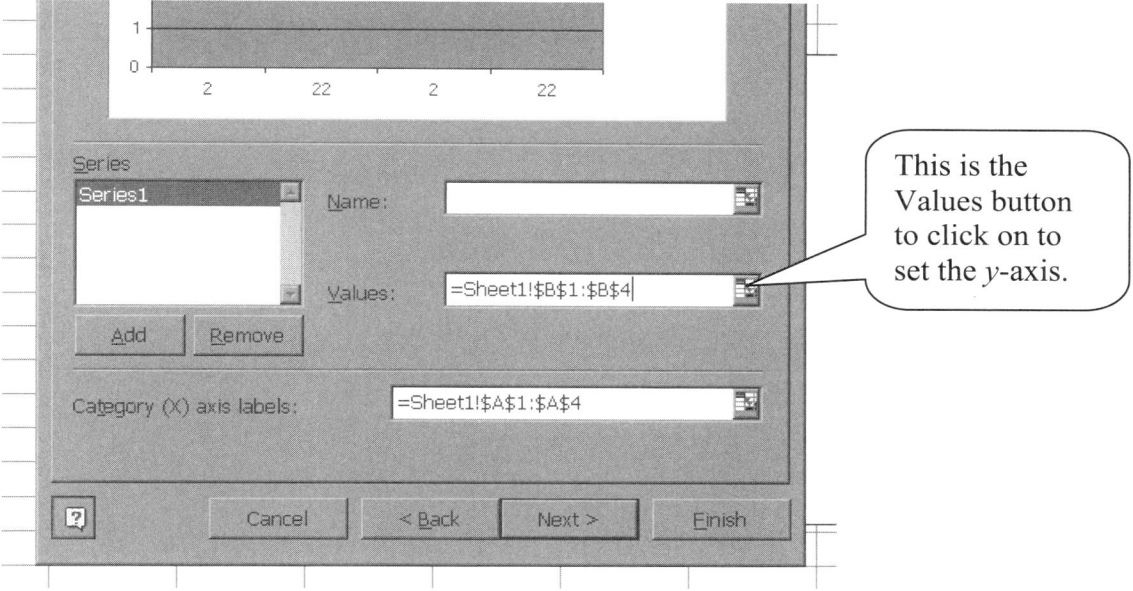

This is the Values button to click on to set the *y*-axis.

11. Clicking on the button takes you back to your spreadsheet and you now want to highlight your *y*-axis values, in this case density, B2-F2. Press Enter after highlighting.

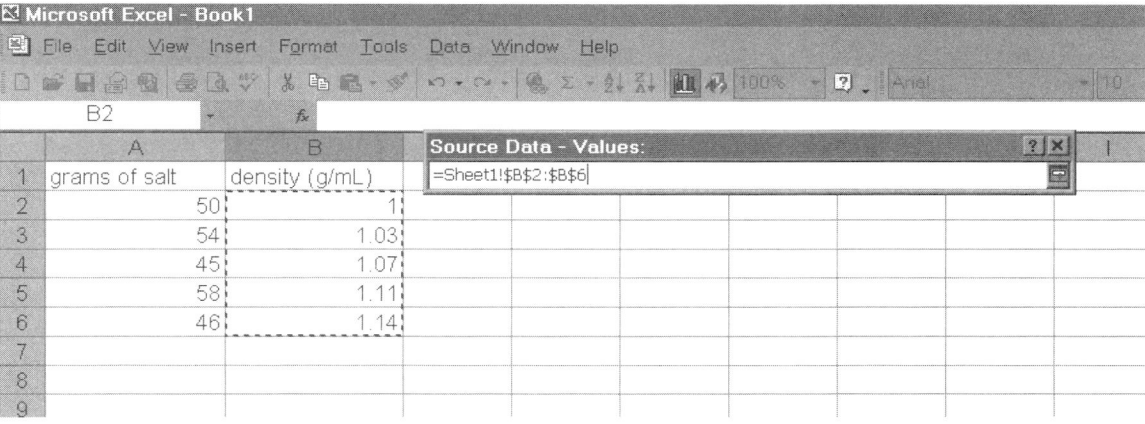

12. The Chart Wizard screen should reappear. Click Next on the bottom of the screen.

13. The new screen allows you to name your graph and label your axis. Fill in the blanks with the appropriate information and click Finish.

14. You now have finished your graph and Excel will ask you if you want the graph to appear on your spreadsheet, or on a separate page. Choose the one you need. Below is a copy of the graph inserted into the spreadsheet page.

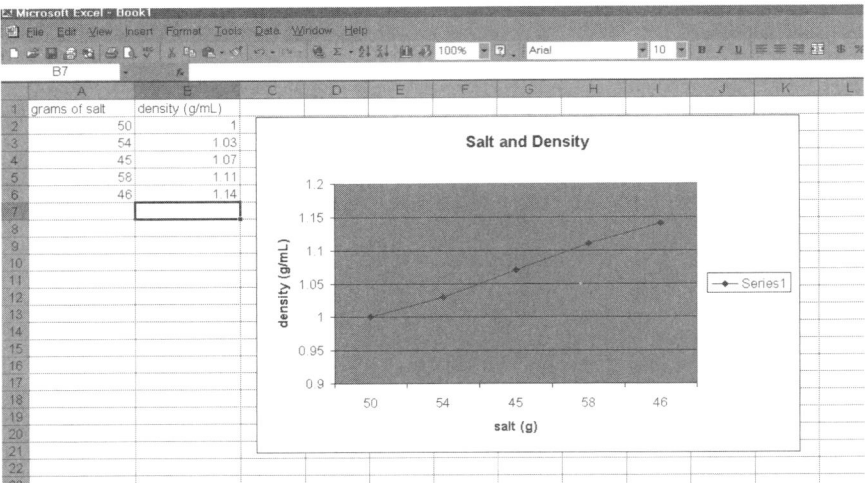

15. You may now print your completed graph by selecting print from the file menu on the task bar.

Microsoft Excel
Using Excel in the Science Classroom

Part III: How to Create a Scatter Plot and Linear Regression Equation

PURPOSE
To use the software program Microsoft Excel to create a line graph.

MATERIALS

> data from this handout
> computer
> Microsoft Excel software

PROCEDURE
In science class you have collected data to see how much the density of water changes as you add grams of salt. Your teacher wants you to take the data and produce and a line graph using Excel. The data is as follows:

Grams of Salt	Density
0	1.00
5	1.03
10	1.07
15	1.11
20	1.14

1. Open an Excel Workbook. Notice that the columns are identified with letters and the rows are identified by numbers.

2. In the box "A1", type Grams of Salt.

3. In the box "B1", type Density. If you need to make a box larger, take your cursor to the top of the column and place it between two boxes until a double arrow appears. Now stretch the column to the size you need.

4. Enter the data in the boxes below each section. Be careful to enter the coordinating data in the correct row.

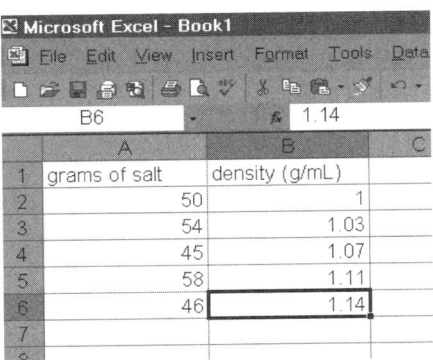

5. To create a scatter plot graph, click on the small, colorful bar graph icon. This is called the Chart Wizard.

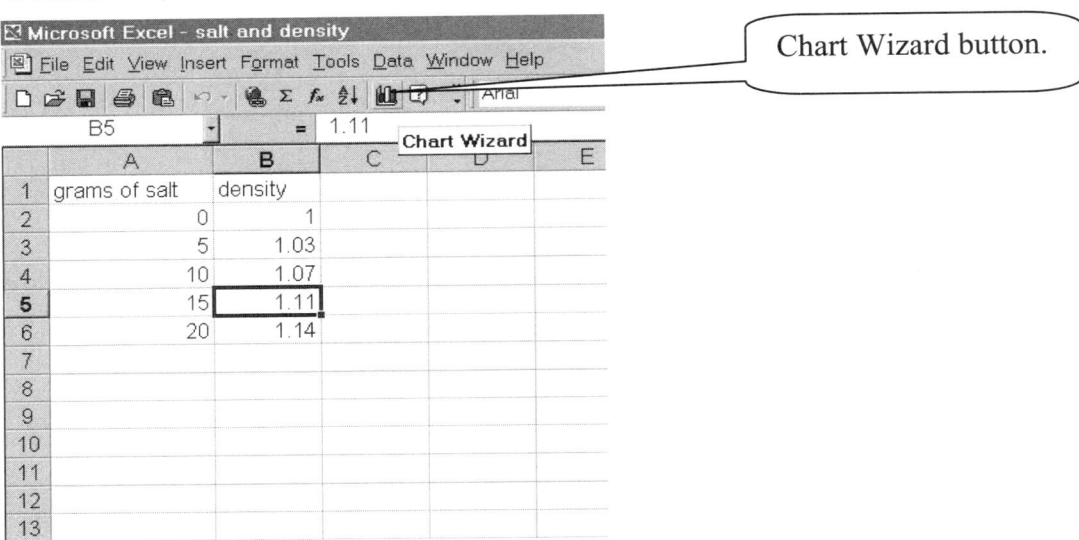

6. After clicking on the chart wizard button a dialogue screen will appear that allows you to choose Chart type. Choose XY SCATTER as the type of chart. *Do not choose a subtype with any lines connecting the dots*. Click Next.

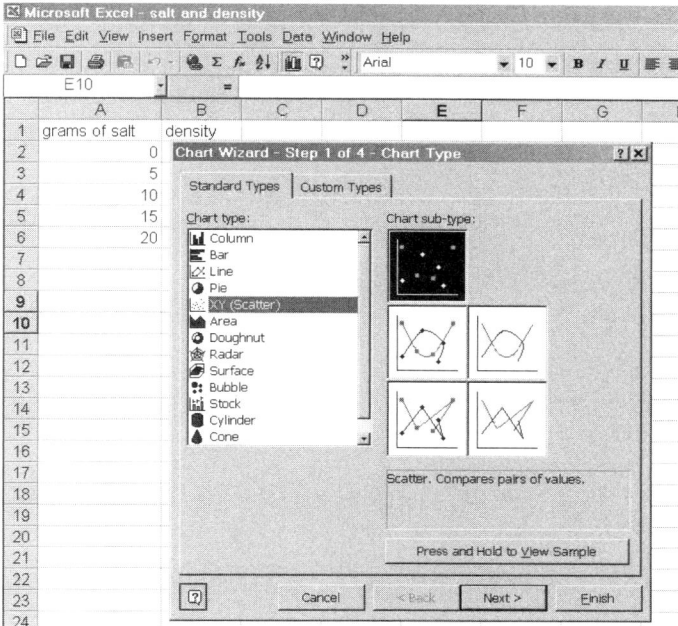

7. At the next dialogue box you will see a preview of your graph. Click the Series tab at the top of the box.

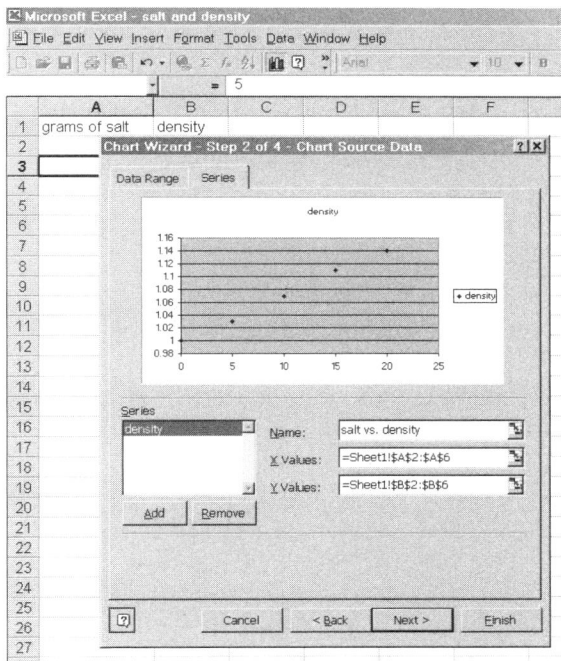

8. On the bottom left of this box it lists the series of data being plotted. To the right there is a blank cell where you can name this series. Name the series in terms of what variables are being graphed in the *y* vs. *x* format (i.e. density vs. salt).

9. Under the Name cell there are two cells that are labeled <u>X</u> values and <u>Y</u> values. The letters in these boxes correspond to the columns in the worksheet. Make sure the data is plotted on the correct axis. If they are not where you want them, click on the small button next to the <u>X</u> values button and it will take you back to your data table. Highlight the column you want to be plotted on your *x*-axis. Do the same for the *y*-axis.

10. When you are satisfied that the correct columns are being plotted and you have named your series, click Next. The next dialogue box, Chart Options, gives you the opportunity to label your axes (include units!). Click Next when you are finished.

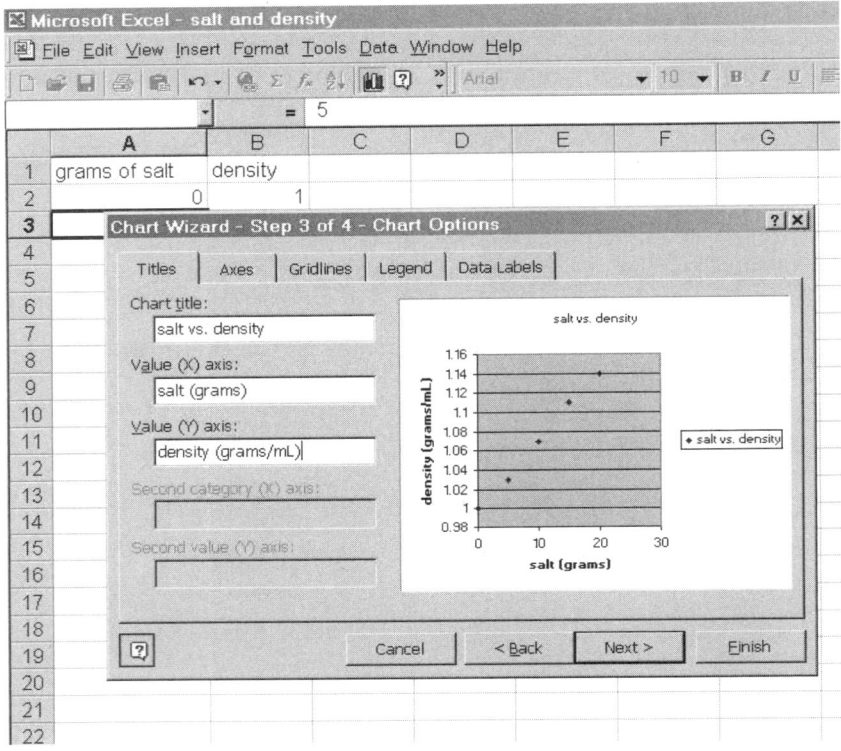

11. The final dialogue box will ask you if you want the graph to appear on your spreadsheet, or on a separate page. Click Finish when you are done.

12. To add a mathematically calculated regression line or best fit curve, choose Add Trendline from the Chart pull-down menu on your toolbar.

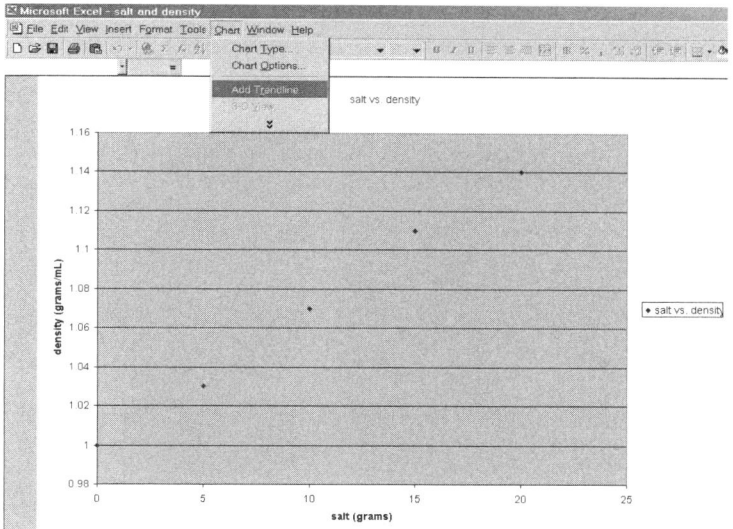

13. The next dialogue box allows you to choose the type of regression you desire.

14. The Options tab allows you to see the mathematical equation and correlation constant (R^2) if the boxes are checked for these options.

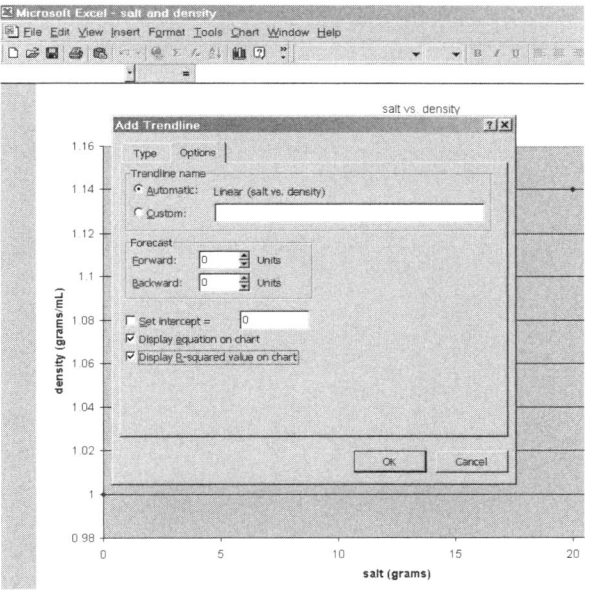

15. If you need to extrapolate data beyond the range of data you have calculated, increase the numbers in the forecast box.

16. If the preset *y*-intercept of 0 causes your graph axis and area to shift too much, set your *y*-intercept more within your data range.

17. You can also title this regression line something that tells about its origin (i.e. linear regression, power regression, etc.)

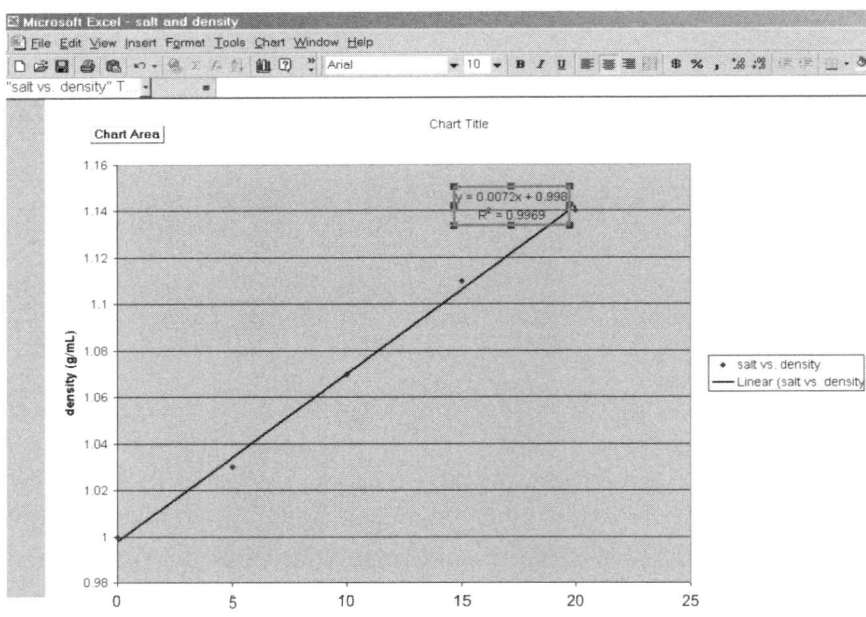

Microsoft Excel
Using Excel in the Science Classroom

Part IV: How Excel Can Manage Data

PURPOSE

To use the software program Microsoft Excel for manipulating data and determining statistical information.

MATERIALS

data from this handout
computer
Microsoft Excel software

PROCEDURE

The table below contains data collected to see how the circumference of the human head relates to the length of the face. For 5 students the data is as follows:

Circumference of Head (cm)	Length of face (cm)
50	11
54	13
45	10
58	14
46	9

1. Open an Excel Workbook. Notice that the columns are identified with letters and the rows are identified by numbers.

2. In the box "A1", type Circumference of Head (cm).

3. In the box "B1", type Length of Face (cm). If you need to make a box larger, take your cursor to the top of the column and place it between two boxes until the double arrows appear. Click and stretch the column to the size you need.

4. Enter the data in the boxes below each section. Be careful to enter the coordinating data in the same row.

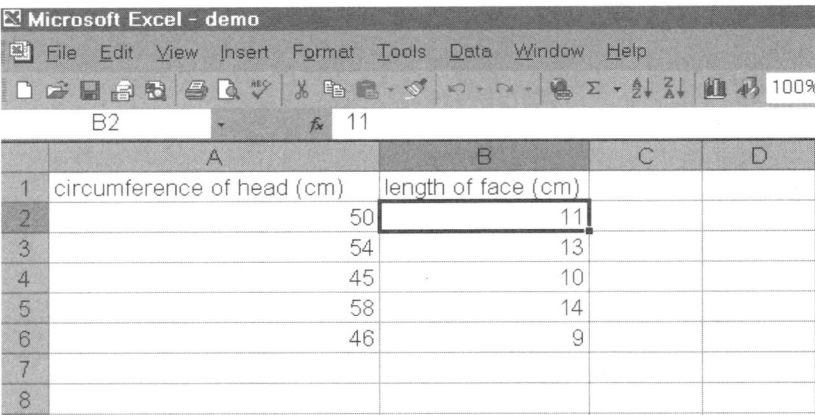

5. You are now going to have the computer calculate an index value for each person by dividing the length of the face by the circumference of the head. Label the new column in C1, skull index.

6. Click in box C2. Notice on the lower tool bar there is an empty box next to a small *fx*. Put your cursor in the box and type an equal sign (=).

7. Following the equal sign enter B2/A2. Press Enter.

8. Your spreadsheet will now reappear and you will see an index number in box C2. Right click your mouse on C2 and choose copy and then drag your mouse down column C for as far as there is data. This will apply the same formula to all of these cells.

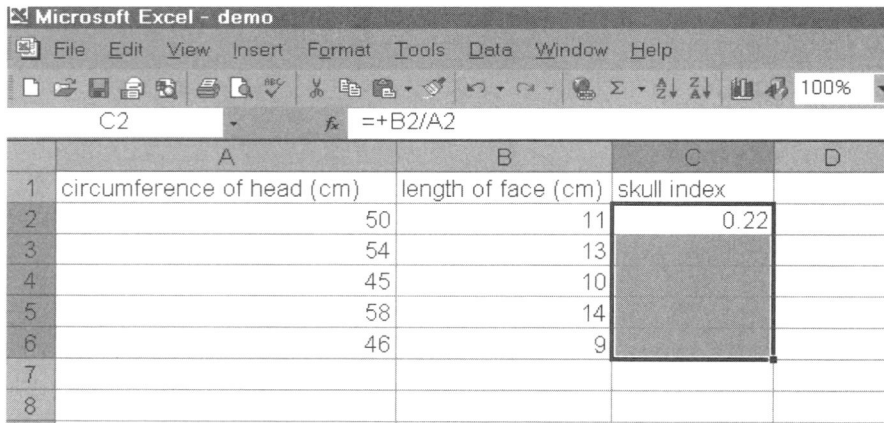

9. Press enter. Excel will calculate and fill in all the indices.

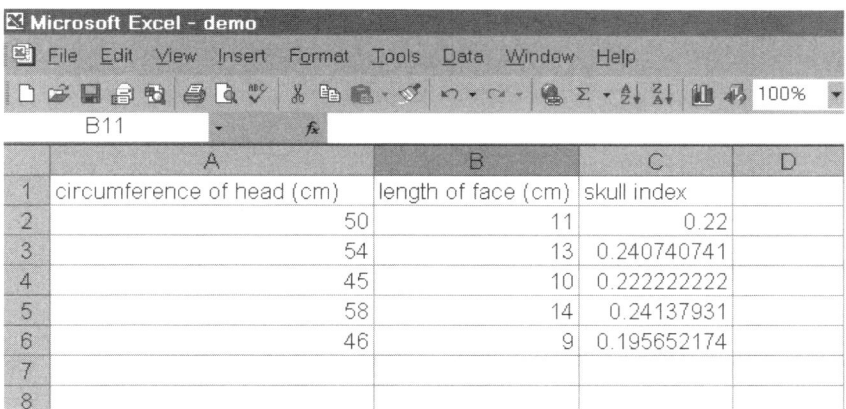

10. To reduce the numbers to two significant figures, right click on the number in cell C3 and select format cell. Click in number and then choose 2 decimal places.

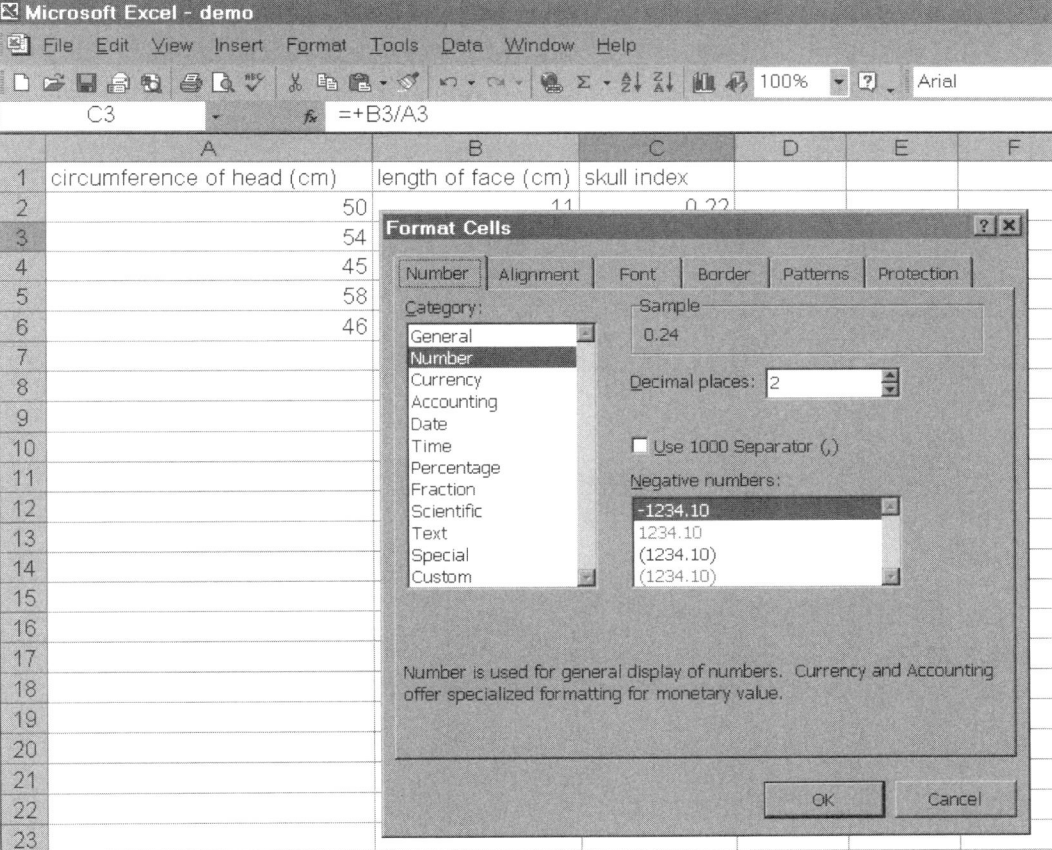

11. Click OK to exit this dialogue box and notice that one cell has changed to two decimal points. Right click on cell C3, select copy, and drag down the rest of your column. Press Enter and all numbers should change to two decimals.

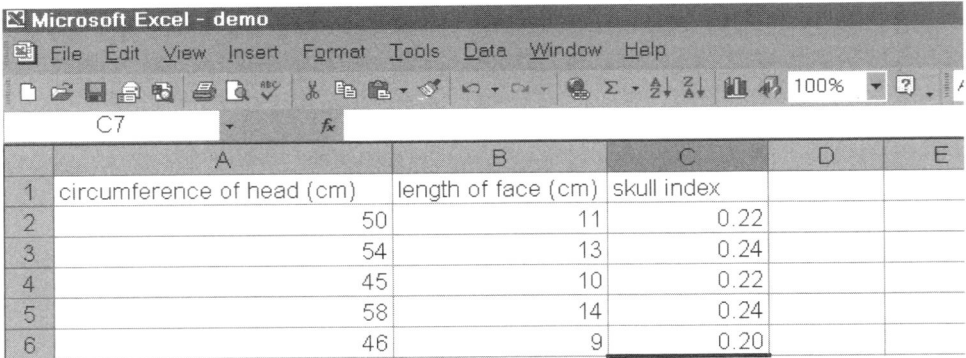

12. To calculate the average of the skull index, click in box C7, below your last index value.

13. Click on *fx* and choose average from the select a function box.

14. The next dialogue box asks you to identify what you want averaged. Highlight the five index values and then press Enter.

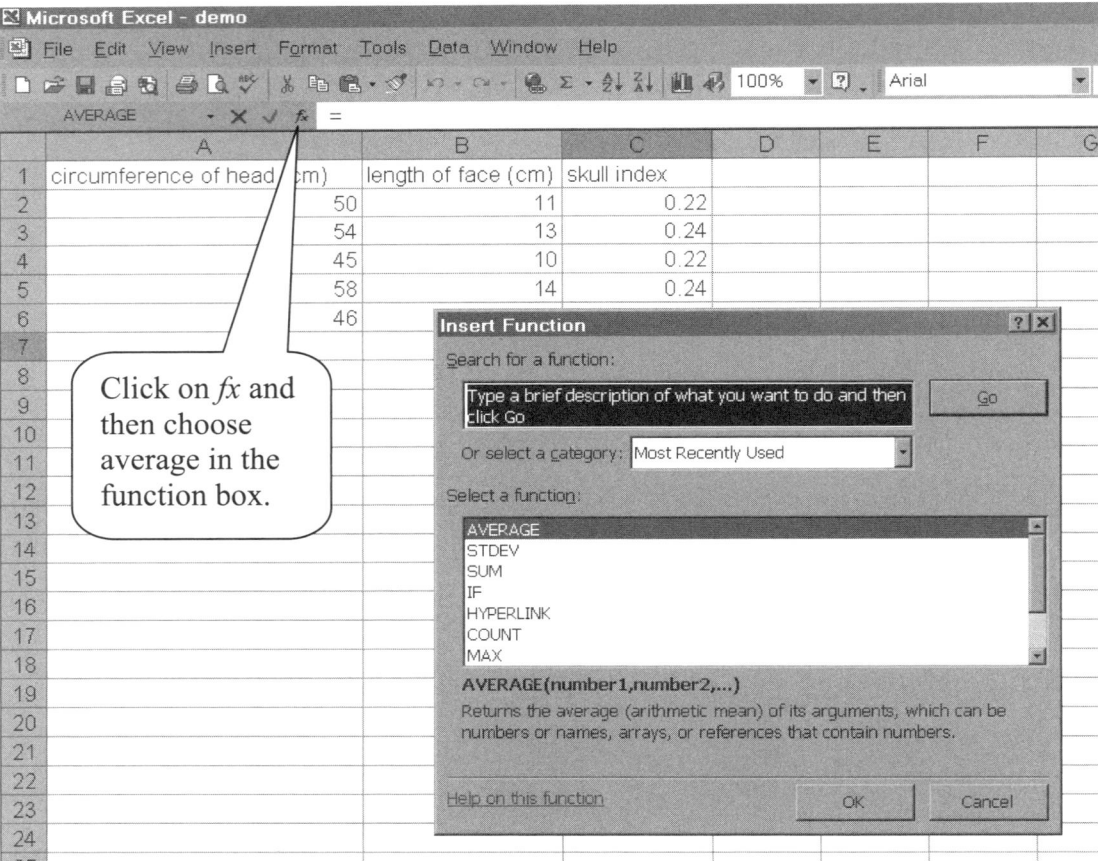

Click on *fx* and then choose average in the function box.

15. The next dialogue box asks you to identify what you want averaged. Highlight the five index values and then press enter. The average value will appear in cell C7.

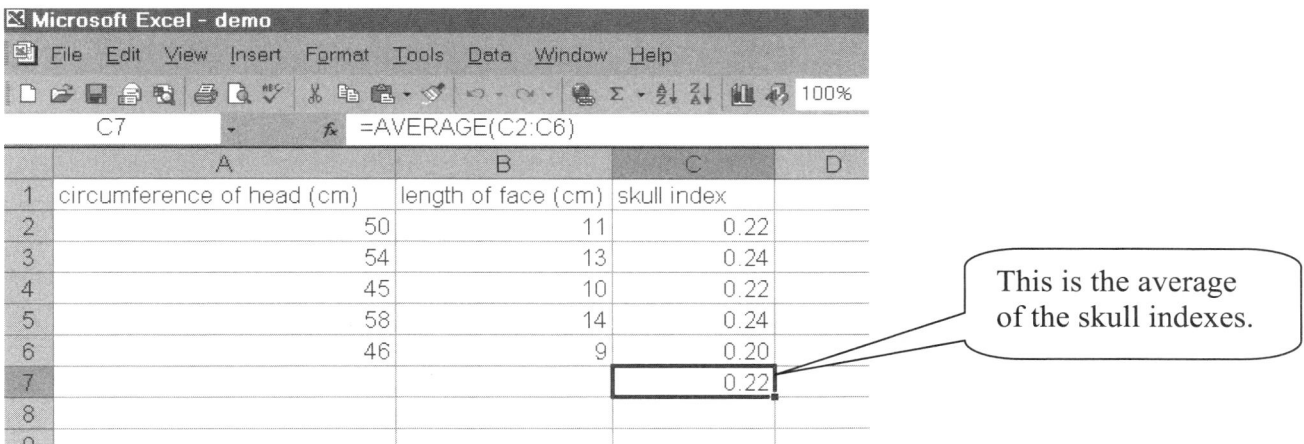

16. To calculate the sum, standard deviation, maximum or minimum you would follow the same procedure except in step #13 choose the appropriate function.

17. To graph your data follow the procedure outlined in Microsoft Excel Part I or II.

Graphing Calculator
Using the TI-83+ in the Science Classroom

OBJECTIVE

Students will follow the teacher's oral instructions through a series of calculator menus and settings. Students will then enter sets of data to generate graphs, regression equations, and interpolations.

LEVEL

All

NATIONAL STANDARDS

UCP.1, UCP.2, A.1, A.2, E.1, E.2, G.2

TEKS

6.2 (E), 6.4(A)
7.2(E), 7.4(A)
8.2 (E), 8.4(A), 8.4(B)
IPC: 2(C)
Biology: 2(C)
Chemistry: 2(D)
Physics: 2(C), 2(E)

CONNECTIONS TO AP

All AP science courses use graphing skills to analyze data. The calculator is a tool to assist in manipulating data, seeing relationships, and drawing conclusions.

TIME FRAME

90 minutes or two class periods

MATERIALS

graphing calculator   link cords
teaching calculator   view screen and overhead projector
                                         OR TI Presenter

TEACHER NOTES

During the past decade, the graphing calculator, along with its companion data collection devices and probes, have swept the educational field in both mathematics and science. More than any other technology, it has changed the way science is taught. Because we are able to obtain excellent data very rapidly, we may now spend our time in analysis and synthesis of the data. This tool has been found to be extremely effective helping students at all levels understand complex scientific systems.

Unfortunately, many teachers were in school in a time BC (before calculator) and the graphing calculator may seem intimidating. These notes are intended to be a quick guide to navigation of the graphing calculator. The notes are based on a TI-83+, however, other calculators have similar functions.

This document is meant to be a guide to start students on the graphing calculator. It is also designed for you to use as a reference. You should take the students through the basic steps of using the calculator. If a teaching calculator and view screen is available, it facilitates the presentation.

Suggested Teaching Procedure:

TIP #1: You cannot hurt your calculator.

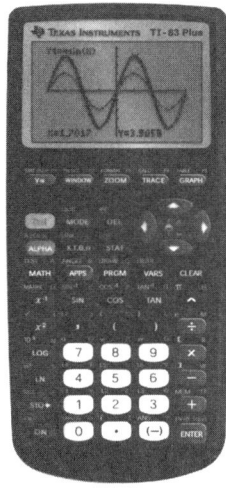

1. *Layout and Important buttons:*
   There are several important keys. Use the diagram to locate each of them.

   | ENTER | Used to enter data and execute functions. |

   ENTER Used to enter data and execute functions.

   2nd Accesses the yellow function above each of the keys

   ALPHA Accesses the green letters above the keys and allows for typing in text.

   ◄ ▲ ► ▼ Operate like a mouse or computer arrows.

   APPS Store applications which drive data collection devices or other study aids.

   Y= WINDOW ZOOM TRACE Graphing keys located across top that allow for graphing. They also serve as function keys for some applications.

2. *Adjusting the brightness of the screen*

Turn your calculator on. Use 2nd ▲ or 2nd ▼ arrow to increase or decrease the darkness of the printing on the screen. A number at the top right indicates your battery setting. If you are using rechargeable batteries, recharge the batteries when the setting level is at 6. You will get a warning if your batteries are running low.

3. *Setting mode*

Select MODE to see a selection menu, which looks like the one below. Use ◄ ▲ ► ▼ to move around on the screen. We will move through each line of this menu to see which of these functions you will most likely be using. Because many of the programs that drive the probes set the decimal at 2 or 3 decimal places, it is often necessary to re-set this function back to Float.

4. *Using the CATALOG*

The CATALOG has all of the calculator's functions listed in it. You will find the CATALOG by pressing 2nd 0. We will use the catalog to turn on statistical diagnostics. When you turn on diagnostics, you will enable your calculator to provide you with a correlation of regression when you attempt to curve fit. The closer the value of the "r" is to 1.000 or –1.000, the better the data fit the function which is analyzed. You will notice when you open the catalog it is locked into ALPHA mode. You can tell this by the ▣ in the middle of the curser. You can go to any letter in the alphabet by hitting that key. Press "D" which is above the x^{-1} button. Arrow down, ▼, to **Diagnostics On**. Select it by pressing ENTER. Execute it by pressing ENTER again. It will say, "DONE".

Tip #1: If you are selecting a function from a menu of functions, pressing ENTER once selects the function. Pressing ENTER a second time actually executes the function.

Tip #2: If you are having difficulty in navigation, read the entire screen!

5. *Numbers in Scientific Notation*

The EE function must be used. This is the 2nd , . The , is located above the 7 key. It is very important that you not use the x 10^ method of putting numbers in scientific notation. Operations that are commutative will work regardless of how the number is entered. Non-commutative operations will not be calculated properly.

- **Exercise:** Note how entering the number incorrectly affects the value of the answer: The calculator interprets 2.5E-2 as a single number and divides the numerator by that number. In the second, and incorrect example, the calculator interprets 2.5 as one factor and 10^{-2} as another factor, so the numerator is divided by 2.5 and multiplied by 10^{-2}.

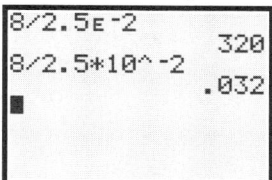

6. *Managing memory in TI-83+*

   Enter [2nd][+] to get the following screen. This menu allows you to do several important operations. The operations you will be using most include the following:

   - **1: About** will tell you about your operating system. Because the 83+ has a flash memory, it is possible to constantly update your operating system as new advances come out. To do this, you will need a GraphLink cable and a program called TI Connect. You can download this program *free* from the Internet.

   - **2: Mem Mgmt/Del** allows you to delete programs or applications you may not be using.

   - 4: **ClrAllLists** is the easiest way to remove data from multiple lists.

   - **7: Reset** will reset all of your memory. At this point you will have the choice of resetting your RAM or just your defaults.

   One of the advantages to the TI-83+ is that it allows you to place programs into an archived file to free up memory that is needed to do other tasks. Programs that are in archived files have an * in front of them. To take them out of archives, arrow to 2:Mem Mgmt/Del. Arrow to 7:Prgm. Scroll to the program you wish to take out of archives and press [ENTER]. This feature toggles on and off.

 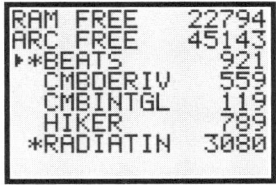

7. *Linking calculators and transferring programs*

   Firmly press the link cord into both calculators. Press [2nd] [X,T,Θ,n] for the calculator receiving information.

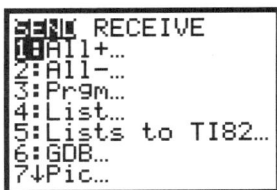

Arrow ▶ to receive and press ENTER. The screen should look like the one below.

Press ENTER. The receiving calculator should now display "Waiting".

Now set up the transmitting calculator by pressing 2nd X,T,Θ,*n*. This window displays the various information or items available for transfer. Use the down arrow, ▼, to select the desired information or item to be transferred. Press ENTER. A new window appears displaying more specific items. Use the down arrow, ▼, to select the item to be transferred and press ENTER. Additional items may be selected for transfer using this same method.

When you have finished selecting your items, arrow, ▶, to TRANSMIT and press ENTER. If a selected item is already present on the receiving calculator, such as a List, a dialog box will appear allowing the user to overwrite that particular item.

8. ***Entering numbers in a list***

Numbers to be analyzed or graphed are stored in the calculator's lists. The lists are handled by the statistical function. Press the STAT key.

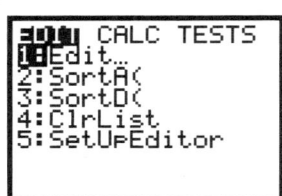

The operations you can perform from this menu are:

 1: Edit: This is where you will go to enter or change numbers in the list

 2: Sort A(: This will sort the numbers from smallest to largest

 3: Sort D(: This will sort the numbers from largest to smallest

 4: ClrList: This will allow you to clear a specific list, however you must enter an argument specifying which lists to clear.

 5: SetUpEditor: Run this option if you need to reestablish lists that have been deleted or altered

9. ***Managing Lists***

Press STAT then select EDIT, press ENTER. You have 6 lists and may enter up to 999 data points in each list. You may also create additional lists with the LIST function. However, that is rarely needed for ordinary classroom work.

Clearing lists: There are three major ways to clear lists.

(a) Use the arrow key to move to the top of the list where it says L1. Press clear, press ENTER. That specific list will be cleared.

(b) Go to 2nd + to get to (mem)ory. Go to 4:ClearAllLists. Press ENTER ENTER.

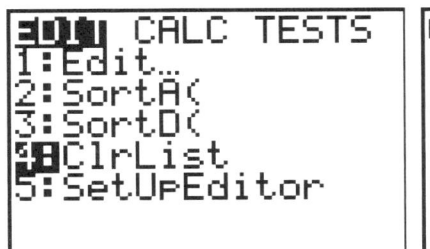

(c) Go to STAT. Arrow to 4:ClrList. Press ENTER. Clear Lists 1 and 2 by pressing 2nd 1 , 2nd 2). This method will only clear the list(s) that you tell it to clear. If you want to clear multiple lists, separate them by commas.

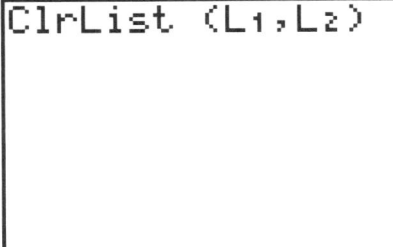

10. ***Working with sample data:***

This data is similar to the Middle Grades Chemistry lesson titled *"What's That Liquid?"*. It involves a simple density problem.

A Physical Science class took the following data. Students poured a liquid into a graduated cylinder, and took the mass of several pre-determined volumes.

Enter the volume in L_1. Enter the mass in L_2.

Volume (mL)	Mass (g)
2	52.4
5	56.0
8	59.6
15	68.0
20	74.0

11. *Clearing ⌨Y= functions*

Make sure that all equations from any previous user are removed from ⌨Y=. To do this, press ⌨Y= and position your cursor on the = sign and press ⌨CLEAR.

12. *Setting up STAT PLOT*

Go to ⌨2nd ⌨Y= to get to STAT PLOT.

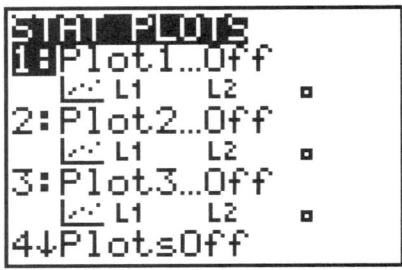

Press ⌨ENTER to get to Plot 1. Use the down arrow, ⌨▾, and ⌨ENTER to turn on Plot 1. Move down a row. Select the first choice, a scatterplot. Be sure that XList is L_1 and YList is L_2. Arrow to the next row and choose a point protector style.

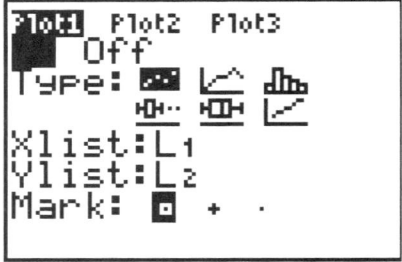

13. *Graphing Statistical Data*

Statistical data are most easily graphed by using the ZOOMSTAT. Press ⌨ZOOM and arrow down to ZOOMSTAT. Since this is also the 9th choice, it can be accessed by ⌨ZOOM ⌨9 as well. Your graph should now be shown on the screen.

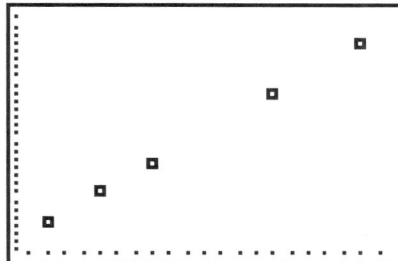

Tip #3: If you THINK you should be seeing a graph but you are not, try using $\boxed{\text{ZOOM}}$ $\boxed{9}$ to re-set your window to the range that includes your data.

14. ***Regressions***

If the data in your graph looks as though it might be linear, you should run a linear regression to find out the function. If students have done a considerable amount of graphing, they should understand the meaning of "line of best fit". A linear regression is simply a mathematical "line of best fit". Press $\boxed{\text{STAT}}$ and side arrow $\boxed{\blacktriangleright}$ over to CALC. Arrow down to 4:LinReg (ax + b). You should see the following screen.

```
EDIT CALC TESTS
1:1-Var Stats
2:2-Var Stats
3:Med-Med
4:LinReg(ax+b)
5:QuadReg
6:CubicReg
7↓QuartReg
```

Press $\boxed{\text{ENTER}}$ $\boxed{\text{ENTER}}$ to get to the screen shown below.

```
LinReg
  y=ax+b
  a=1.2
  b=50
  r²=1
  r=1
```

This will give a linear equation in the form of $y = \mathbf{a}x + \mathbf{b}$. In this case, the function is $y = 1.2x + 50$. The correlation coefficient or correlation of regression is given by "r" and in this case is 1. That would indicate that this is a very nice linear function. In a real-life laboratory situation, the correlation coefficient would not usually be 1.00, but rather more like 0.999 or 0.998. Again, the closer the correlation coefficient is to 1.000 or -1.000, the more likely the data fit the function chosen.

15. ***Pasting the function into*** Y=

This is one of the more difficult sets of keystrokes to make. Open Y= Press VARS. Since we are dealing with statistical data, arrow down ▼ to Statistics. ENTER. Arrow over ▶ to EQ and ENTER. The equation for the line in the form of $y = ax + b$ will be pasted into your Y= menu. Now press ZOOM 9 to see both the scatterplot and the regression line.

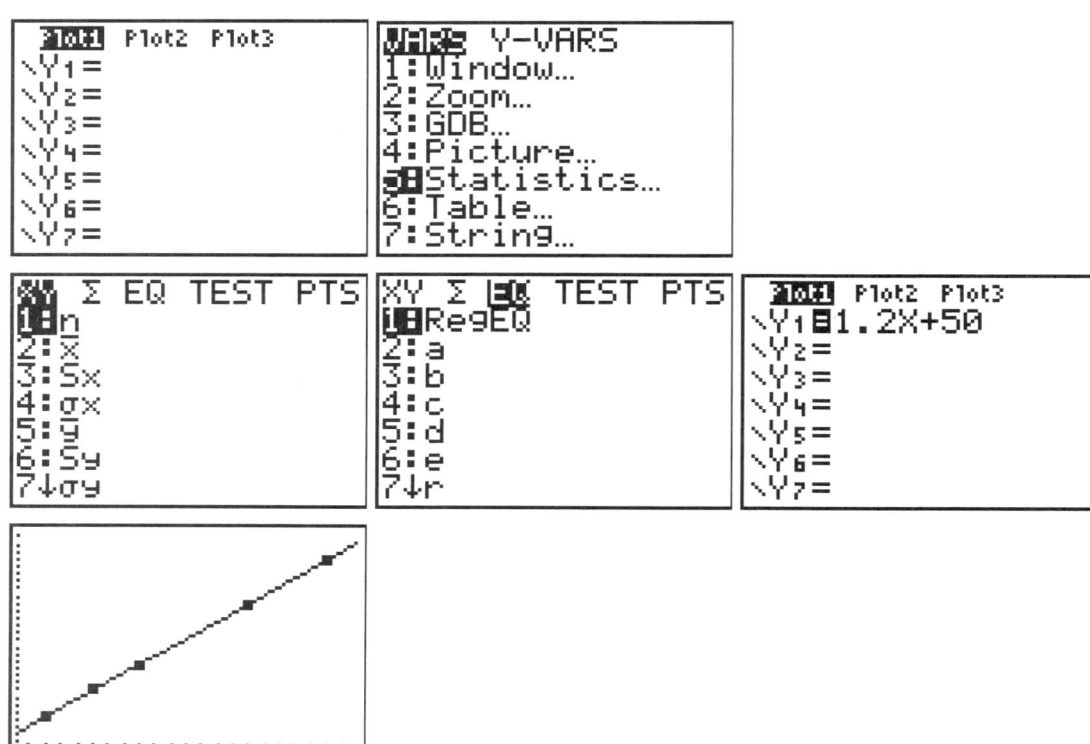

16. ***A second method to paste a function into*** Y=

If you are reasonably certain of the function that the data expresses, it is possible to run the regression and paste it into Y= in one step. Follow the screens to perform this set of operations. Start with STAT Calc 4:LinReg ENTER. Then go to VARS and side arrow ▶ to Y-VARS. Press ENTER ENTER ENTER. You can see the regression equation and it will be pasted into Y= at the same time. View your graph by pressing ZOOM 9.

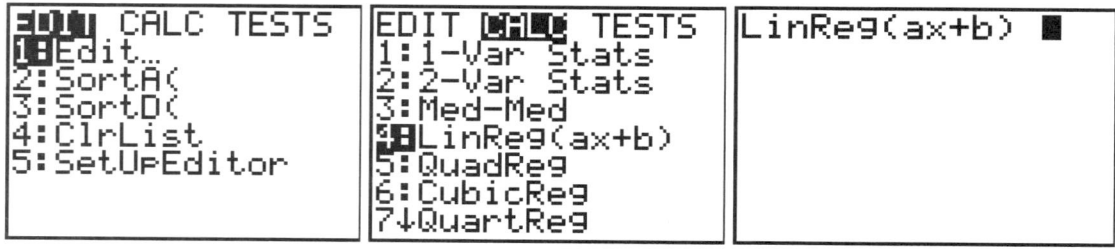

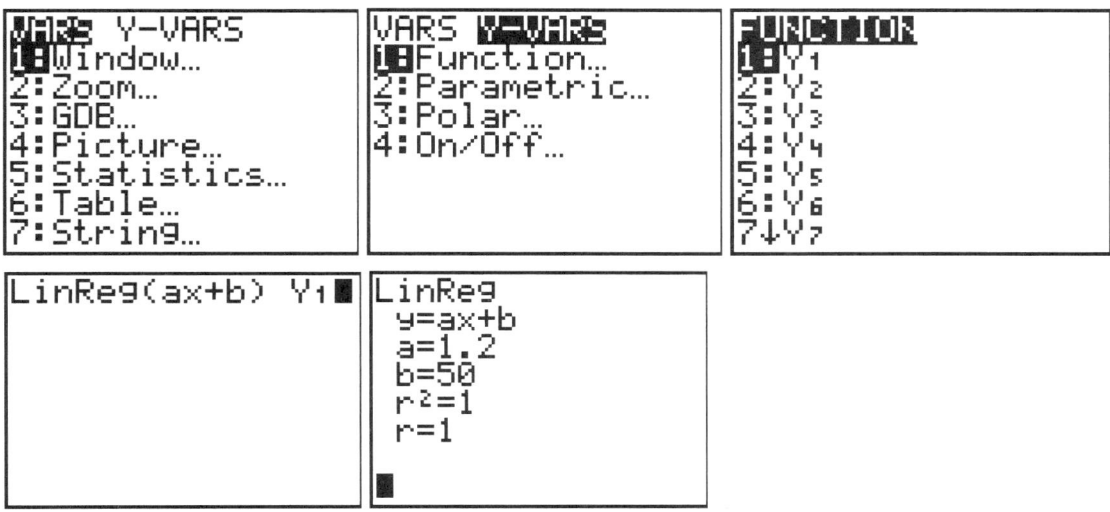

17. **Tracing on the graph**

Go to TRACE. In the upper left corner you should see a P1:L_1,L_2. This means that you are in a statistical plot of your data. Use the sideways arrows to move from data point to data point. The X and Y values are shown at the bottom of the screen. Now press the up arrow, ▲. Notice that the designation in the upper right hand corner has changed, and you should have your regression equation on the screen. Your cursor is now on the regression line so that you may interpolate values. Again the X and Y values are shown at the bottom of the screen.

18. **Helping students interpret the graph**

This is a very good time to help students realize that a graph is simply a mathematical picture of a real world situation. You can talk about independent and dependent variables, rewrite the equation in terms of words, and discuss the meaning of the value for b. For the example that we were using, Y is the mass of the liquid and the graduated cylinder; X is the volume of the liquid. The slope of the line, rise over run, is mass/volume and thus is the density. The y-intercept is the mass of the empty graduated cylinder. The equation can then be expressed in words: Total Mass = Density x Volume + Mass of graduated cylinder.

19. **Calculating a value**

We would like to be able to use this equation that has been developed. Students can be asked to apply their algebraic knowledge to solve certain problems about the situation. They can also use the graph itself to interpret information and make predictions. Let us say that we would like to know the total mass when there is 17-mL of liquid in the graduated cylinder. Press 2nd TRACE to get to CALC. 1: VALUE. Press ENTER You will see a prompt at the bottom of your graph that says X=. Enter 17 and press ENTER. The mass of the cylinder and liquid will be displayed at the bottom.

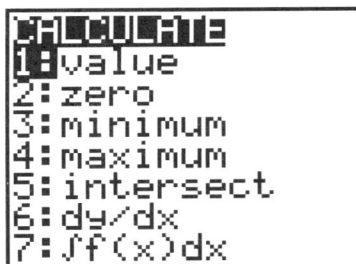 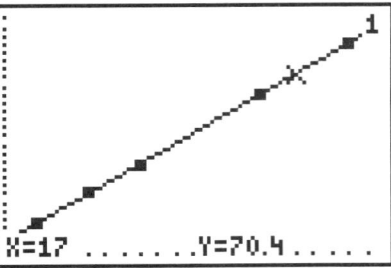

POSSIBLE ANSWERS TO THE CONCLUSION QUESTIONS

1. Enter the following data in your calculator and answer the questions.

   Pressure of a gas at various temperatures

L₁ Temperature (Kelvins)	L₂ Pressure (torr)
200	600
250	750
300	900
350	1050
400	1200

   a. What is the equation for this function?

      - $y=3x$

   b. What is the pressure when the temperature is 0 Kelvin?

      - 0 torr

2. Enter this as Pressure vs. Distance below the surface of water. Remember that the title of a graph is always the dependent variable (*y*-axis) vs. independent variable (*x*-axis). It is easiest if the *x* value is placed in L₁ and the y value is placed in L₂.

   Water pressure is measured at various depths in the ocean.

Distance below surface of water (feet)	Pressure (lb/in^2)
5	16.93
20	23.61
33	29.40
50	36.97

   a. What is the pressure at the surface of the water?

      - 14.7 lb/in^2

   b. What does this represent?

      - The atmospheric pressure at the surface.

3. On a distant planet, far, far away, the atmosphere is different from that of Earth. The speed of sound is not the same. The following data were collected:

Independent variable Temperature (Degrees Celsius)	Dependent variable Speed of sound (m/s)
0	276
10	289
20	302
30	315
50	341

a. What is the slope of this line?

- 1.3 m/s $^\circ$C

b. What is the temperature when the speed of sound is 295?

- 14.6 $^\circ$C

4. A small toy car is given a push to start it rolling down an inclined plane. The distance verses time is plotted as follows.

Time (seconds)	Distance (meters)
1	15.00
2	50.00
3	105.0
4	180.0
5	275.0
10	1050

a. Does this look like a linear function?

- No, it is curved.

b. Can you describe the motion of the car in words?

- The car is going faster and faster.

Challenge: What is the mathematical function that describes this data?

- $y = 10x^2 + 5x + 0$

What does the "b" in this particular equation represent?

- The initial speed before the car started down the inclined plane.

TEACHER PAGES

Graphing Calculator
Using the TI-83+ in the Science Classroom

The graphing calculator is an extremely valuable and powerful tool in both math and science classes. You will be using it in many of your classes during your educational experience. The document provided is intended to be a quick reference for many of the skills you will be using in science. Your teacher will walk you through the basic maneuvers and skills you are expected to know.

PURPOSE
In this activity you will practice using the graphing calculator to analyze data.

> graphing calculator
> pencil

PROCEDURE
Your teacher will take you through the basic functions found on your graphing calculator. You should follow along and do each of the steps on your own calculator. When you get home, go through the steps again on your own, using this guide.

1. *Layout and important buttons*:

    There are several important keys. Use the diagram to locate each of them.

    ENTER Used to enter data and execute functions.

    2nd   Accesses the yellow function above each of the keys

    ALPHA Accesses the green letters above the keys and allows for typing in text.

    ◄ ▲ ► ▼   Operate like a mouse or computer arrows.

    APPS Store applications which drive data collection devices or other study aids.

    Y= WINDOW ZOOM TRACE   Graphing keys located across top that allow for graphing. They also serve as function keys for some applications.

2. *Adjusting the brightness of the screen*

    Turn your calculator on. Use 2nd ▲ or 2nd ▼ arrow to increase or decrease the darkness of the printing on the screen. A number at the top right indicates your battery setting. If you are using rechargeable batteries, recharge the batteries when the setting level is at 6. You will get a warning if your batteries are running low.

3. ***Setting mode***

Select MODE to see a selection menu, which looks like the one below. Use the ◄ ▲ ▼ ► arrows to move around on the screen. We will move through each line of this menu to see which of these functions you will most likely be using. Because many of the programs that drive the probes set the decimal at 2 or 3 decimal places, it is often necessary to re-set this function back to Float.

4. ***Using the CATALOG***

The CATALOG has all of the calculator's functions listed it. You will find the CATALOG by pressing 2nd 0. We will use the catalog to turn on statistical diagnostics. When you turn on diagnostics, you will enable your calculator to provide you with a correlation of regression when you attempt to curve fit. The closer the value of the "r" is to 1.000 or −1.000, the better the data fit the function which is being analyzed. You will notice when you open the catalog it is locked into ALPHA mode. You can tell this by the ⓐ in the middle of the curser. You can go to any letter in the alphabet by hitting that key. Press "D" which is above the x^{-1} button. Arrow down, ▼, to Diagnostics On. Select it by pressing ENTER. Execute it by pressing ENTER again. It will say, "DONE".

Tip #1: If you are selecting a function from a menu of functions, pressing ENTER once selects the function. Pressing ENTER a second time actually executes the function.

Tip #2: If you are having difficulty in navigation, read the entire screen!

5. *Numbers in Scientific Notation*

The EE function must be used. This is the ⟨2nd⟩ ⟨,⟩. The ⟨,⟩ is located above the 7. It is very important that you not use the x 10⟨^⟩ method of putting numbers in scientific notation. Operations that are commutative will work regardless of how the number is entered. Non-commutative operations will not be calculated properly.

- **Exercise:** Note how entering the number incorrectly affects the value of the answer: The calculator interprets 2.5E-2 as a single number and divides the numerator by that number. In the second, and incorrect example, the calculator interprets 2.5 as one factor and 10^{-2} as another factor, so the numerator is divided by 2.5 and multiplied by 10^{-2}.

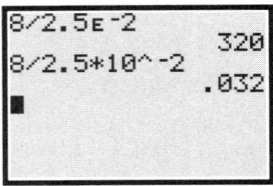

6. *Managing memory in TI-83+*

Enter ⟨2nd⟩⟨+⟩ to get the following screen. This menu allows you to do several important operations. The operations you will be using most include the following:

- **1: About** will tell you about your operating system. Because the 83+ has a flash memory, it is possible to constantly update your operating system as new advances come out. To do this, you will need a GraphLink cable and a program called TI Connect. You can download this free program from the Internet.

- **2: Mem Mgmt/Del** allows you to delete programs or applications you may not be using.

- **4: ClrAllLists** is the easiest way to remove data from multiple lists.

- **7: Reset** will reset all of your memory or your defaults.

One of the advantages to the TI-83+ is that it allows you to place programs into an archived file to free up memory that is needed to do other tasks. Programs that are in archived files have an * in front of them. To take them out of archives, arrow to 2:Mem Mgmt/Del. Arrow to 7:Prgm. Scroll to the program you wish to take out of archives and press ⟨ENTER⟩. This feature toggles on and off.

 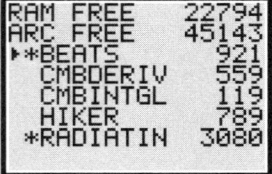

7. ***Linking calculators and transferring programs***

Firmly press in the link cord into both calculators. Press [2nd] [X,T,Θ,*n*] for the calculator receiving information.

Arrow [▶] to receive and press [ENTER]. The screen should look like the one below.

Press [ENTER]. The receiving calculator should now display "Waiting".

Now set up the transmitting calculator by pressing [2nd] [X,T,Θ,*n*]. This window displays the various information or items available for transfer. Use the down arrow, [▼], to select the desired information or item to be transferred. Press [ENTER]. A new window appears displaying more specific items. Use the down arrow, [▼], to select the item to be transferred and press [ENTER]. Additional items may be selected for transfer using this same method.

When you have finished selecting your items, arrow, [▶], to TRANSMIT and press [ENTER]. If a selected item is already present on the receiving calculator, such as a List, a dialog box will appear allowing the user to overwrite that particular item.

8. ***Entering numbers in a list***

Numbers to be analyzed or graphed are stored in the calculator's lists. The lists are handled by the statistical function. Press the [STAT] key.

The operations you can perform from this menu are:

1: Edit: This is where you will go to enter or change numbers in the list

2: Sort A(: This will sort the numbers from smallest to largest

3: Sort D(: This will sort the numbers from largest to smallest

4: ClrList: This will allow you to clear a specific list, however you must enter an argument specifying which list(s) to clear. If you want to clear multiple lists, separate them by commas.

5: SetUpEditor: Run this option if you need to reestablish lists that have been deleted or altered

9. *Managing Lists*

From the STAT EDIT screen, press ENTER. You have 6 lists and may enter up to 999 data points in each list. You may also create additional lists with the LIST function. However, that is rarely needed for ordinary classroom work.

Clearing lists: There are three major ways to clear lists.

(a) Use the arrow key to move to the top of the list where it says L1. Press clear, press ENTER. That specific list will be cleared.

(b) Go to 2nd + to get to (mem)ory. Go to 4:ClearAllLists. Press ENTER ENTER.

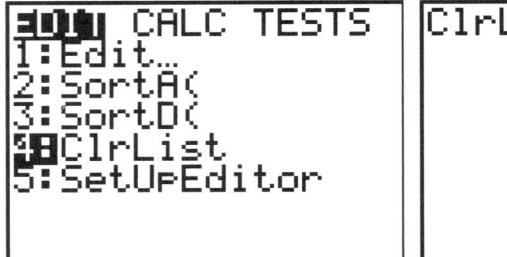

(c) Go to STAT. Arrow to 4:ClrList. Press ENTER. Clear List 1 and 2 by pressing 2nd 1 , 2nd 2 and pressing ENTER. This method will only clear the lists that you tell it to clear.

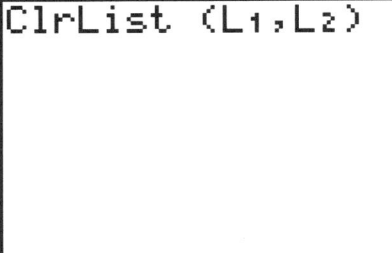

10. ***Working with sample data:***

Sample Data: Physical Science students poured liquid into a graduated cylinder and measured the mass of several pre-determined volumes.

Enter the volume in L_1. Enter the mass in L_2.

Volume (mL)	Mass (g)
2	52.4
5	56.0
8	59.6
15	68.0
20	74.0

11. ***Clearing [Y=] functions***

Make sure that all equations from any previous user are removed from [Y=]. To do this, press [Y=] and position your cursor on the = sign and press [CLEAR].

12. ***Setting up STAT PLOT***

Go to [2nd] [Y=] to get to STAT PLOT.

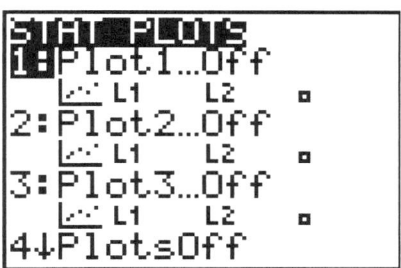

Press [ENTER] to get to Plot 1. Use the down arrow, [▾], and [ENTER] to turn on Plot 1. Move down a row. Select the first choice, a scatterplot. Be sure that XList is L_1 and YList is L_2. Arrow to the next row and choose a point protector style.

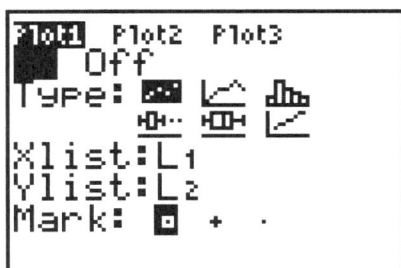

13. **Graphing Statistical Data**

Statistical data are most easily graphed by using the ZOOMSTAT. Press ZOOM and arrow down to ZOOMSTAT. Since this is also the 9th choice, it can be accessed by ZOOM 9 as well. Your graph should now be shown on the screen.

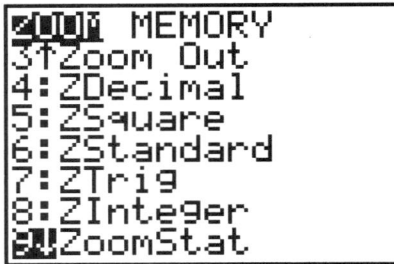 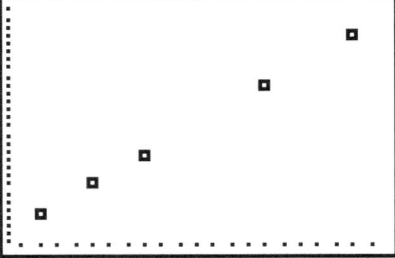

Tip #3: If you think you should be seeing a graph but you are not, try using ZOOM 9 to re-set your window to the range that includes your data.

14. **Regressions**

If the data in your graph looks as though it might be linear, you should run a linear regression to find out the function. If students have done a considerable amount of graphing, they should understand the meaning of "line of best fit". A linear regression is simply a mathematical "line of best fit". Press STAT and side arrow ▶ over to CALC. Arrow down to 4:LinReg (ax + b). You should see the following screen.

```
EDIT CALC TESTS
1:1-Var Stats
2:2-Var Stats
3:Med-Med
4:LinReg(ax+b)
5:QuadReg
6:CubicReg
7↓QuartReg
```

Press ENTER ENTER to get to the screen shown below.

```
LinReg
 y=ax+b
 a=1.2
 b=50
 r²=1
 r=1
```

This will give a linear equation in the form of **y = ax + b**. In this case, the function is $y = 1.2x + 50$. The correlation coefficient or correlation of regression is given by "r" and in this case is 1. That would indicate that this is a very nice linear function. In a real-life laboratory situation, the correlation coefficient would not usually be 1.00, but rather more like 0.999 or 0.998. Again, the closer the correlation coefficient is to 1.000 or −1.000, the more likely the data fit the function chosen.

15. ***Pasting the function into*** Y=

This is one of the more difficult sets of keystrokes to make. Open Y= Press VARS. Since we are dealing with statistical data, arrow down ▼ to Statistics. Press ENTER. Arrow over ▶ to EQ and press ENTER. The equation for the line in the form of *y*= a*x* + b will be pasted into your Y= menu. Now press ZOOM 9 to see both the scatterplot and the regression line.

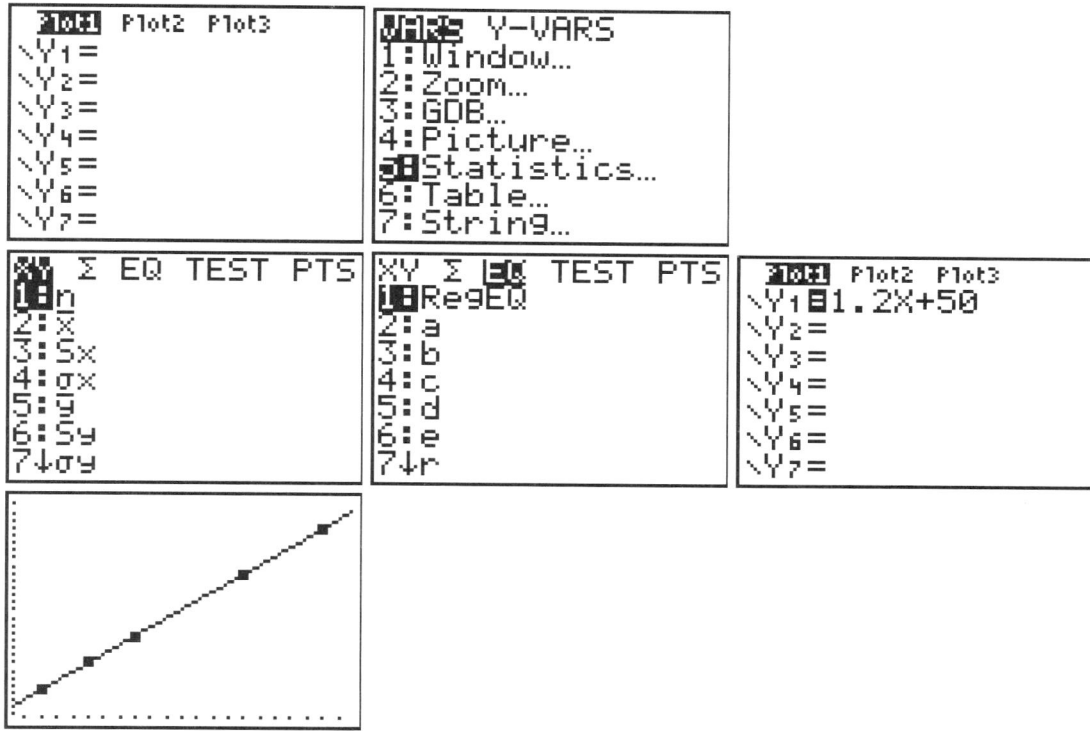

16. *A second method to paste a function into* Y=

If you are reasonably certain of the function that the data expresses, it is possible to run the regression and paste it into Y= in one step. Follow the screens to perform this set of operations. Start with STAT Calc 4:LinReg ENTER. Then go to VARS and side arrow ▶ to Y-VARS. Press ENTER ENTER ENTER. You can see the regression equation and it will be pasted into Y= at the same time. View your graph by pressing ZOOM 9.

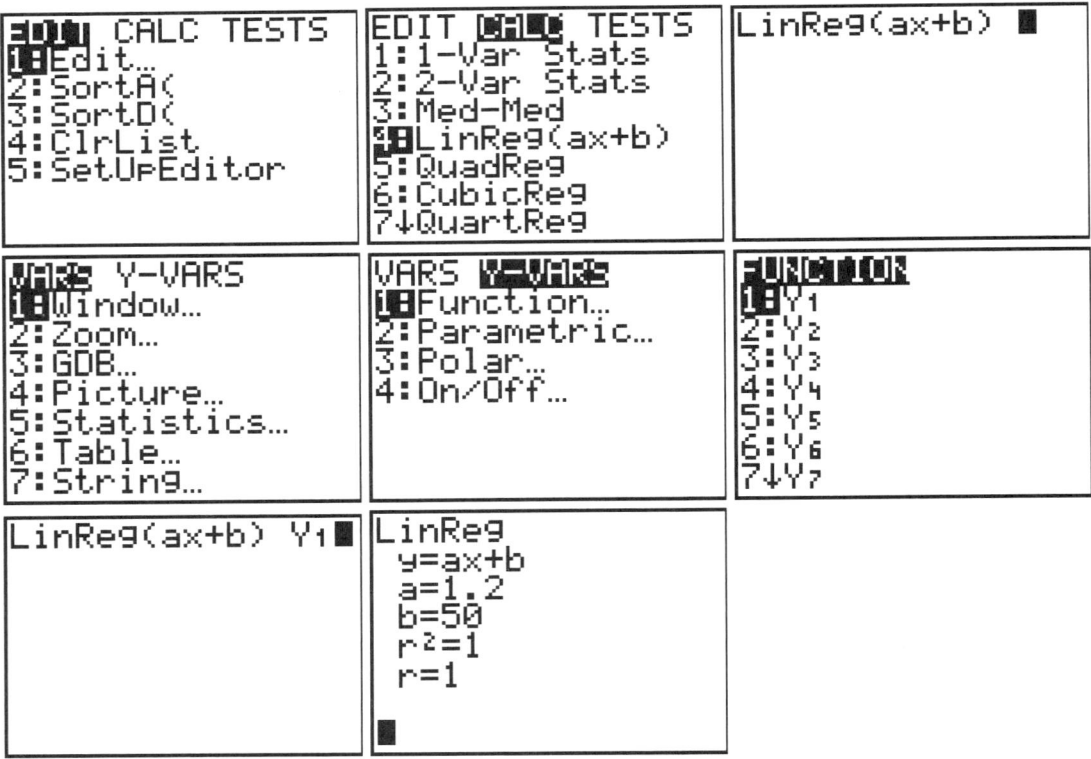

17. *Tracing on the graph*

Press TRACE. In the upper left corner you should see a P1:L₁,L₂. This means that you are in a statistical plot of your data. Use the sideways arrows to move from data point to data point. The X and Y values are shown at the bottom of the screen. Now press the up arrow, ▲. Notice that the designation in the upper right hand corner has changed, and you should have your regression equation on the screen. Your cursor is now on the regression line so that you may interpolate values. Again the X and Y values are shown at the bottom of the screen.

18. ***Calculating a value***

We would like to be able to use this equation that has been developed. Let us say that we would like to know the total mass when there is 17 mL of liquid in the graduated cylinder. Press 2nd TRACE to get to CALC. 1: VALUE. Press ENTER. You will see a prompt at the bottom of your graph that says X=. Enter 17 ENTER. The mass of the cylinder and liquid will be displayed at the bottom.

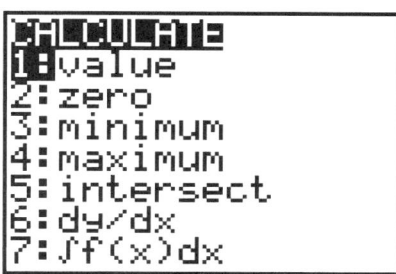
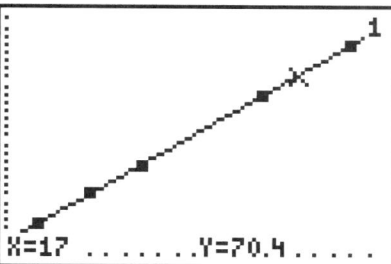

Name _____

Period _____

Using the Graphing Calculator
Analysis of Data Sets

EXERCISES

1. Enter the following data in your calculator and answer the questions.

   Pressure of a gas at various temperatures

L_1 Temperature (Kelvins)	L_2 Pressure (torr)
200	600
250	750
300	900
350	1050
400	1200

   a. What is the equation for this function?

   b. What is the pressure when the temperature is 0 Kelvin?

2. Enter this as Pressure vs. Distance below the surface of water. Remember that the title of a graph is always the dependent variable (y-axis) vs. independent variable (x-axis). Analysis is easiest if the x value is placed in L_1 and the y value is placed in L_2.

   Water pressure is measured at various depths in the ocean.

Distance below surface of water (feet)	Pressure (lb/in^2)
5	16.93
20	23.61
33	29.40
50	36.97

a. What is the pressure at the surface of the water?

b. What does this represent?

3. On a distant planet, far, far away, the atmosphere is different from that of Earth. The speed of sound is not the same. The following data were collected:

Independent variable Temperature (Degrees Celsius)	Dependent variable Speed of sound (m/s)
0	276
10	289
20	302
30	315
50	341

a. What is the slope of this line?

b. What is the temperature when the speed of sound is 295?

4. A small toy car is given a push to start it rolling down an inclined plane. The distance verses time is collected as follows.

Time (seconds)	Distance (meters)
1	15.00
2	50.00
3	105.0
4	180.0
5	275.0
10	1050

a. Does this look like a linear function?

b. Can you describe the motion of the car in words?

Challenge: What is the mathematical function that describes this data?

What does the "b" in this particular equation represent?

Data Collection Devices
Determining the Amount of Energy Found in Food

OBJECTIVE
Students will determine the amount of energy in a sample of peanut and walnut. The students will calculate the percent yield and percent error in the experiment and evaluate the sources of error.

LEVEL
All

NATIONAL STANDARDS
UCP.2, UCP.3, A.1, B.5, C.5, E.1, E.2, G.2

TEKS
6.1(A), 6.2(A), 6.2(B), 6.2(C), 6.2(D), 6.2(E), 6.4(A), 6.9(A)
7.1(A), 7.2(A), 7.2(B), 7.2(C), 7.2(D), 7.2(E), 7.4(A)
8.1(A), 8.2(A), 8.2(B), 8.2(C), 8.2(D), 8.2(E), 8.4(A), 8.4(B), 8.10(A), 8.10(C)
IPC: 1(A), 2(A), 2(B), 2(C), 2(D), 8(A), 8(B)
Biology: 1(A), 2(A), 2(B), 2(C), 2(D), 9(A)
Chemistry: 1(A), 2(A), 2(B), 2(C), 2(D), 2(E), 5(A), 5(B), 5(C)
Physics: 1(A), 2(A), 2(B), 2(C), 2(D), 2(E), 2(F), 7(A), 7(B)

CONNECTIONS TO AP
Use of probeware enhances the AP student to accurately and quickly obtain data for a variety of experiments. Energy is a common theme studied in all AP science courses.

TIME FRAME
50 minutes

MATERIALS
(For a class of 28 working in pairs)

14 TI-83 or TI-83 + graphing calculators (used for graphing calculator lab)	14 Lab Pro's or CBL's (used for graphing calculator lab)
14 computers (used for computer lab)	14 Lab Pro's or Serial Interface Device (used for computer lab)
14 10 cc syringes (used for graphing calculator lab)	14 10 mL graduated cylinder (used for computer lab)
14 sheets heavy aluminum foil	14 packages of matches
14 temperature probes	14 calorimeters
14 30 mL test tubes	1 balance
1 package of peanuts	1 package of walnuts
28 goggles	14 test tube clamps
28 aprons	14 corks with inserted needle

TEACHER NOTES

The calorimeter can be made by using a tin vegetable can and drilling a center hole large enough to fit a 30 mL test tube and many small holes to vent. See diagram in step 11 of student procedures.

1. Calculator Information

Students will need a graphing calculator and either the CBL 2 (Calculator-Based Laboratory System) or LabPro® interface. TI 83, 83+, 85, 86, 89 or 92 calculators may be used. The temperature probe that is needed for this lab comes from Vernier. The calculators and interfaces can also be purchased from Vernier. The address is:

> Vernier Software Company
> 13979 SW Millikan Way
> Beaverton, OR 97005-2886
> (503)-277-2299
> www.vernier.com

2. Calculator Software Information

This activity uses DATAMATE as the data collection program. This program must be updated periodically and can be downloaded free from Vernier. A TI-GraphLink cable is necessary for connecting the CBL 2 or LabPro to a computer for download or upload processes. Once updated, it can be installed on the graphing calculators directly from the CBL 2 or LabPro.

3. Computer Hardware Information

Macintosh System Requirements
Power Macintosh or newer computer
At least 32 MB of RAM
Operating System 7.5 or newer

Windows System Requirements
Windows 98/NT/ME/2000/XP
At least 32 MB of RAM
Pentium processor-based or compatible PC

4. Computer Software Information

The software that is used in conjunction with the temperature probe is **Logger Pro 3® Software**. It can be purchased for the PC or Mac. Start the program. Under file there will be a number of folders, open "Probes & Sensors", then open "Temperature Probes", then open, "Stainless Steel Temp Probe.MBL" or "Direct Connect Temp Probe.MBL" for the Direct Connect Temperature probe.

One nice feature of Logger Pro Software is that there are templates for the probes already installed. They can be modified to fit your own experiment and then saved under your own title for the students. The following steps work nicely for this experiment:

- It is important that you try this lab and software settings before working with students so that you can modify the templates or instructions to meet the particular needs of your own experiment.

- Be sure you have the experimental apparatus set up for this experiment by connecting the interface to the computer through the modem port and the temperature probe to Port 1 or Ch 1on the interface.

- Pull down the **Set Up Menu** and highlight **Data Collection**. Then click on **Mode**. Make sure that **Real Time Collect** is selected. Next, click on **Sampling** and select **minutes** and key in **5**. Below that, move the slide bar so that 10 pts/minute is selected. Press **OK**.

- The axes of the graph are modified by first highlighting the last number on the *x*-axis, Time (minutes). Now type in the amount of time that you wish to run the experiment. Type in 5. This means that the experiment will run for five minutes. The *y*-axis should have a range of 15-105. If it does not have this range, highlight the top number, type in 105 and then highlight the lower number and type in 15. The temperature of the water should increase 20-30 degrees Celsius.

- Once you have accomplished this, you can save your experiment so that you do not have to set up the software again. Pull down the **File Menu** and highlight **Save Experiment As**. A dialog box will appear. Type in the name you would like to use for your lab. Decide where you want to save this on the hard drive or a disk. Highlight **Save**. Another dialog box will appear and ask you if you would like to save the calibration for this experiment. Click **Yes**. Now the experiment can be used by your students without having to set up the software. As long as you use that same probe with the same computer, the calibration should be good for a long while.

ANSWERS TO PRE-LAB QUESTIONS

1. After reading the background information, define the terms autotroph and heterotroph.
   - An autotroph can make its own energy rich organic compounds from inorganic compounds. Heterotrophs can not make their own energy rich organic compounds and must obtain them from the environment.

2. What are two purposes of food?
   - Food is providing energy to the organism and comprises the building blocks for synthesizing their own polymers.

3. What is free energy?
   - Free energy is the amount of energy available to do work.

4. What is entropy?
   - Entropy is a measurement of disorder in the system.

5. What is a calorie, a Calorie and a kilocalorie?
   - A calorie is the amount energy it takes to raise the temperature of one gram of water, one degree Celsius. One thousand calories is equal to one Calorie or one kilocalorie.

6. Write the equations to be used to determine the number of calories, percent yield, and percent error.

   calories = (Final Temperature – Beginning Temperature) * 10 g

   $$\text{Kilocalories or Calories} = \frac{\text{calories}}{1000}$$

   $$\text{calories} = 10 \text{ g H}_2\text{O} \times (\text{final temperature} - \text{initial temperature}) \times 1\frac{\text{calorie}}{\text{g} \cdot {}^\circ\text{C}}$$

   $$\text{Kilocalories or Calories} = \frac{\text{calories}}{1000}$$

   $$\frac{\text{kcal}}{\text{g of food}} = \frac{\text{kcal}}{\text{mass of sample}}$$

   $$\text{Percent Error} = \frac{(\text{Experimental Value} - \text{Actual Value})}{\text{Actual Value}} \times 100$$

   $$\text{Percent Yield} = \frac{\text{Experimental Value}}{\text{Actual Value}} \times 100$$

   $$\text{Percent Error} = \frac{(\text{Experimental Value} - \text{Actual Value}) * 100}{\text{Actual Value}}$$

POSSIBLE ANSWERS TO THE CONCLUSION QUESTIONS AND SAMPLE DATA

Data Table						
Experiment	Mass of Sample	Temperature Before Burning	Temperature After Burning	# of calories	# of kilocalories	# of kilocalories/gram
Peanut # 1	0.23	22	62	400	0.4	1.74
Peanut # 2	0.2	21.5	65	435	0.435	2.18
Peanut # 3	0.24	22	61	390	0.39	1.63
Average						1.85
Walnut # 1	0.2	22	67	450	0.45	2.25
Walnut # 2	0.18	22	62	400	0.4	2.22
Walnut # 3	0.21	21.5	39	475	0.475	2.26
Average						2.24

1. Which nut contained the greatest number of kilocalories per gram?
   - The walnut has the greatest number of kilocalories per gram 2.24 kc/g

2. What was the shape of the graph displayed on your screen? What relationship did it establish?
   - The relationship is not linear. At first it increases and then slows down.

3. The nutritional labels show that peanuts are 6.00 kcal/gram and walnuts are 7.02 kcal/gram. What was your percent yield and was it greater than or less than the actual value?
   - Walnut = (2.24 / 7.02) * 100 = 31.9%
   - Peanut = (1.85 / 6.00) * 100 = 30.1%

4. Give two reasons why your answer varied from the actual values.
   a. One reason the percent yield is less is that some of the nut was not burned.
   b. A second is a large percent of the heat escaped into the environment and did not go into the test tube.

5. What is your percent error? Do you think that it is acceptable? Why or why not?
   - The percent error is 68% and 69% respectively

6. Nuts are high in lipids. What foods might you investigate to determine their calorie content for carbohydrates? Proteins?
   - Answers will vary.

7. A student fails to replace the 10 grams of water in the test tube between the second and third trials for the walnut. How will this error affect the calculated energy result? Your answer should clearly state that the calculated value will increase, decrease or remain the same. Mathematically justify your answer.

- The calculated energy value will decrease. The water was extremely hot as a result of trial 2. There is not sufficient time between trials for the very hot water to cool back down to room temperature. This will cause the change in temperature (ΔT) to be reported as too low.

  Since $\quad \text{Energy} = \text{mass} \times \Delta T \times 1 \dfrac{\text{calorie}}{\text{g} \cdot {}^\circ\text{C}}$ the quantity of energy will be reported as too low.

REFERENCE

Milani, Jean P., Revision Coordinator. *Biological Science, A Molecular Approach*. Dubuque: Kendall Hunt Publishing Company, 1990. pp. 628-630

Data Collection Devices
Determining the Amount of Energy Found in Food
Using a Graphing Calculator

All living systems require energy. In the case of autotrophs such as plants and related organisms, they have the ability to take inorganic compounds from the environment and make high energy compounds required for their survival. Heterotrophs, however, are unable to synthesize high energy, organic nutrients from inorganic compounds and therefore must obtain these high-energy nutrients from the environment. From a human point of view, their nutrition translates into our food. All food contains energy. The amount it contains depends on the type of food it is and the organic compounds it contains.

Food is a source of both energy and the basic building blocks needed for the synthesis of macromolecules. Living systems follow the 1st and 2nd Laws of Thermodynamics. The first law states that energy cannot be created or destroyed but only converted from one form to another. This energy conversion is never 100%. Energy is always lost in the form of heat. For living systems, this is especially true. The energy content in food is converted from chemical energy to mechanical energy as muscles contract. In fireflies, the chemical energy found in food is converted to electromagnetic energy as they light up the summer skies with their bioluminescence. It is also possible to convert the chemical energy in a candy bar into another form of chemical energy. The sugar in the candy converts to fat. The fat will be later stored in our adipose tissue. As stated earlier, heat energy always accompanies any energy transfer. This type of energy is never found by itself but is always associated with one of the other forms of energy. In each of the previous examples, as energy converts from one form to another a certain amount of heat energy was released.

Free energy is the amount of energy available to do work. Heat energy that is released during an energy conversion is considered to be wasted energy and is not available to do work. This explains why energy conversions are not 100% efficient. Suppose that a candy bar had 180 kilocalories of chemical energy. Suppose you ate the candy bar in order to have mechanical energy available to play a volleyball game. At best, your body would only use 72 kilocalories of the 180 available and the rest would be released as heat energy. That's over 108 kilocalories lost!

180 kcal   180 kcal
Total Energy   Total Energy

Candy Bar   ——— ENERGY CONVERSION ➝ Muscle Movement + Heat Energy
180 kcal   ————————————————————➤ 72 kcal + 108 kcal
Free Energy   ————————————————————➤ Free Energy + Wasted Energy

This brings us to the Second Law of Thermodynamics. Suppose there is a closed system which means energy can not enter or leave the system. This would mean that its *total* energy content could not increase or decease, it must therefore remain constant. As time passes, energy conversions occur, and more and more of the free energy available to do work converts to unusable heat energy. The unusable heat energy is unavailable to do work. If this process continued indefinitely, then eventually there would be no free energy and the system would decease in its order, becoming more chaotic. The Second

Law of Thermodynamics states that in a closed system, one in which energy can not enter or leave, there will be an increase in the system's disorder. Entropy is a measurement of the disorder of a system.

In biology, energy is often measured in calories. A calorie is the amount of heat energy needed to raise the temperature of one gram of water one degree Celsius. The amount of energy found in food is measured in kilocalories (symbolized by C or called Calories). So if a candy bar says it has 180 Calories, it really means that it has 180,000 calories. A calorimeter is used to determine the amount of calories found in food.

In this investigation you will burn a sample of food with known mass. Above the burning food sample is a test tube of water that contains 10 mL of water or 10 g of water. The temperature of the water is recorded before and after the sample is burned. The change in temperature, ΔT is calculated.

By using the equation $\text{Energy} = \text{mass} \times \Delta T \times 1 \dfrac{\text{calorie}}{\text{g} \cdot {}^\circ \text{C}}$ the change in temperature, ΔT, is multiplied by the number of grams of water used (10 g) and the specific heat of the water. This allows you to calculate the amount of heat energy absorbed by the water. Since the energy absorbed by the water was the result of the burning the food, it also represents the amount of energy contained in the food sample. Since the amount of energy you calculated is in calories, it needs to be converted to kilocalories using the relationship, 1000 calories equals one kilocalories or Calorie. To make the energy relative to the particular sample you used, the energy value should be divided by the mass of your sample. This will allow the energy per gram of your sample to be compared to the energy per gram of other samples.

$$\text{calories} = 10 \text{ g H}_2\text{O} \times (\text{final temperature - initial temperature}) \times 1 \dfrac{\text{calorie}}{\text{g} \cdot {}^\circ \text{C}}$$

$$\text{Kilocalories or Calories} = \dfrac{\text{calories}}{1000}$$

$$\dfrac{\text{kcal}}{\text{g of food}} = \dfrac{\text{kcal}}{\text{mass of sample}}$$

This lab will be repeated three times to obtain an average energy value and increase the validity of your experiment. Your teacher will give you the actual number of the Calories found in the food sample used. You are to determine the percent yield.

$$\text{Percent Yield} = \dfrac{\text{Experimental Value}}{\text{Actual Value}} \times 100$$

You can also determine the percent error by using the formula below:

$$\text{Percent Error} = \dfrac{(\text{Experimental Value - Actual Value})}{\text{Actual Value}} \times 100$$

PURPOSE

In this activity you will determine the amount of energy found in a walnut and a peanut by using a calorimeter and compare it to the actual value.

MATERIALS

temperature probe	graphing calculator
interface	aprons
matches	test tube
test tube clamps	heavy aluminum foil
calorimeter	goggles
balance	samples of walnut and peanut
10 cc syringe	

Safety Alert
1. CAUTION: Needles are sharp. Exercise care.
2. CAUTION: Be sure to wear goggles anytime a flame is present in the room.
3. CAUTION: If you burn yourself, immediately place the burned area under cold running water and notify your teacher.

PROCEDURE

1. Answer the pre-lab questions on your student answer page.

2. In the space marked HYPOTHESIS on your student answer page, identify whether the peanut or walnut contains the most energy.

3. Measure out three pieces of peanut and three individual pieces of walnut, approximately 0.2 g each. Determine the exact mass of each piece and record this information on your data table. Do not get the pieces mixed. Be able to identify which piece has what mass.

4. Slide the calculator and CBL2 or LabPro Interface in the bottom part of the cradle and it will click into place. Snap the calculator into the top portion of the cradle.

5. Plug the short black link cable into the link port on the bottom of the TI Graphing Calculator and the interface. Follow the instructions below to activate the **DATAMATE** program for data collection.

 a. TI-83/TI-73

 • Press [PRGM], then press the number key that precedes the **DATAMATE** program. Press [ENTER], then press [CLEAR] when you reach the Main screen.

 b. TI-83+

 • Press [APPS], and then press the calculator key for the number that precedes the **DATAMATE** program. Press [ENTER], then press [CLEAR] when you reach the Main screen.

 c. TI-86

 • Press [PRGM], then [F1] to select <**NAMES**>, and press a menu key to select <**DATAM**> (usually F1). Press [ENTER] and then press [CLEAR] when you reach the Main screen.

 d. TI-89/TI-92

 • Press [2nd], [–] [VAR LINK]. Use [▼] or cursor pad to scroll down to "Datamate", then press [ENTER]. Press [ )] to complete the open parenthesis that follows "Datamate" on the entry line and press [ENTER]. When you reach the Main screen, press [CLEAR].

6. Plug the temperature probe into channel 1 of the interface. At this time the interface should have automatically identified your temperature probe and the correct temperature should be displayed in the upper right hand corner. If the correct temperature is not displayed, do the following:

 • Select **SETUP** from the MENU by pressing [1]

 • Select "CH1" from the MENU and press [ENTER].

 • Select TEMPERATURE from the SELECT SENSOR MENU and press [ENTER].

 • Select the type of temperature probe that is connected to the interface.

 • Press [1] to indicate o.k.

7. Select **SET UP** from the MENU by pressing [1].

 • Select MODE by pressing the [▲] key and then [ENTER].

 • Select TIME GRAPH by pressing [2] and then [2] again to change the time settings.

8. Set up the calculator and the interface for data collection.

 • Enter "5" as the time between samples, in seconds. Press [ENTER].

 • Enter "60" as the number of samples (the interfaced will collect data for 5 minutes), press [ENTER].

9. Another window will appear with the summary of the probes and the length of the experiment. Press ⬜ 1 ⬜ to indicate OK. Press ⬜ 1 ⬜ again to return to the main menu.

10. A new window will appear, and the calculator is now ready to start the experiment. DO NOT press ⬜ 2 ⬜ until you are ready to run the experiment.

11. Obtain a calorimeter. Place one piece of the walnut on the needle anchored in the cork.
    - CAUTION: Needles are sharp. Exercise care when doing this step.
    - CAUTION: Be sure to wear goggles anytime a flame is lit in the room.

12. Place the nut, needle, and cork setup on a folded piece of heavy-duty aluminum foil and set the calorimeter over the setup. Put a rolled piece of masking tape around the test tube to prevent it from sliding through the hole. Slide the test tube into the hole in the top of the calorimeter. Adjust the test tube so that it is about 2 cm above the nut. See the diagram below:

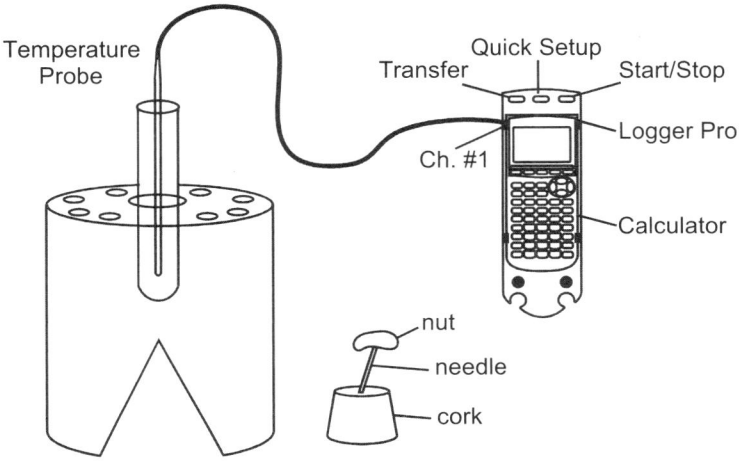

Calorimeter set-up with probe

13. Remove the calorimeter with the test tube in it from over the nut, being careful not to change the position of the test tube. Measure 10 cc of water and pour it in the test tube.

14. Put the temperature probe into the test tube. After about 15 seconds, the temperature should stabilize. Press ⬜ 2 ⬜ to begin data collection. The experiment will run for 5 minutes. There should be 4 short beeps and the quick setup light will flash. You should notice that the temperature is being graphed as the data is being collected.

15. When the experiment is complete, four short beeps will sound and the quick set up light will flash. Now a labeled, fitted graph will be displayed.

16. Using a match, set fire to the nut. Quickly and carefully, position the calorimeter over the burning nut.
    • CAUTION: If you burn yourself, immediately, place the burned area under cold running water. Notify your teacher.

17. You should notice that the temperature is being graphed as the data is being collected. It will continue taking data for five minutes. When the experiment is complete, 4 short beeps will sound and the quick setup light will flash. Upon completion of the data collection a labeled, fitted graph will be displayed.

18. Burn the nut completely. If your nut burns out before it is completely burned, quickly relight it.

    Use the ⬅ and ➡ keys to move the cursor. View the data points displayed at the bottom of the graph. Use these keys to determine the minimum and maximum temperature values of the water and record these in your data table on your student answer page.

19. Press ENTER on the calculator and you will return to the main menu. The parameters that you set for this experiment are still in the calculator. Another run of the experiment can be done without resetting the calculator.

20. Repeat this experiment two more times with the other pieces of walnut and three more times with the pieces of peanut. It is important not to mix the pieces of nuts so that correct energy per gram values will be obtained.
    • Caution: Be sure to get 10 cc of new water and take care in the handling of the test tube, as it is extremely hot.

21. Clean up your area and return your equipment to its original condition.

Data Collection Devices
Determining the Amount of Energy Found in Food Using a Computer

All living systems require energy. In the case of autotrophs such as plants and related organisms, they have the ability to take inorganic compounds from the environment and make high energy compounds required for their survival. Heterotrophs, however, are unable to synthesize high energy, organic nutrients from inorganic compounds and therefore must obtain these high-energy nutrients from the environment. From a human point of view, their nutrition translates into our food. All food contains energy. The amount it contains depends on the type of food it is and the organic compounds it contains.

Food is a source of both energy and the basic building blocks needed for the synthesis of macromolecules. Living systems follow the 1st and 2nd Laws of Thermodynamics. The first law states that energy cannot be created or destroyed but only converted from one form to another. This energy conversion is never 100%. Energy is always lost in the form of heat. For living systems, this is especially true. The energy content in food is converted from chemical energy to mechanical energy as muscles contract. In fireflies, the chemical energy found in food is converted to electromagnetic energy as they light up the summer skies with their bioluminescence. It is also possible to convert the chemical energy in a candy bar into another form of chemical energy. The sugar in the candy converts to fat. The fat will be later stored in our adipose tissue. As stated earlier, heat energy always accompanies any energy transfer. This type of energy is never found by itself but is always associated with one of the other forms of energy. In each of the previous examples, as energy converts from one form to another a certain amount of heat energy was released.

Free energy is the amount of energy available to do work. Heat energy that is released during an energy conversion is considered to be wasted energy and is not available to do work. This explains why energy conversions are not 100% efficient. Suppose that a candy bar had 180 kilocalories of chemical energy. Suppose you ate the candy bar in order to have mechanical energy available to play a volleyball game. At best, your body would only use 72 kilocalories of the 180 available and the rest would be released as heat energy. That's over 108 kilocalories lost!

| 180 kcal | 180 kcal |
| Total Energy | Total Energy |

Candy Bar —— ENERGY CONVERSION ⟶ Muscle Movement + Heat Energy
180 kcal ⟶ 72 kcal + 108 kcal
Free Energy ⟶ Free Energy + Wasted Energy

This brings us to the Second Law of Thermodynamics. Suppose there is a closed system which means energy can not enter or leave the system. This would mean that its *total* energy content could not increase or decease, it must therefore remain constant. As time passes, energy conversions occur, and more and more of the free energy available to do work converts to unusable heat energy. The unusable heat energy is unavailable to do work. If this process continued indefinitely, then eventually there would be no free energy and the system would decease in its order, becoming more chaotic. The Second

Law of Thermodynamics states that in a closed system, one in which energy can not enter or leave, there will be an increase in the system's disorder. Entropy is a measurement of the disorder of a system.

In biology, energy is often measured in calories. A calorie is the amount of heat energy needed to raise the temperature of one gram of water one degree Celsius. The amount of energy found in food is measured in kilocalories (symbolized by C or called Calories). So if a candy bar says it has 180 Calories, it really means that it has 180,000 calories. A calorimeter is used to determine the amount of calories found in food.

In this investigation you will burn a sample of food with known mass. Above the burning food sample is a test tube of water that contains 10 mL of water or 10 g of water. The temperature of the water is recorded before and after the sample is burned. The change in temperature, ΔT is calculated.

By using the equation $Energy = mass \times \Delta T \times 1 \dfrac{calorie}{g \bullet °C}$ the change in temperature, ΔT, is multiplied by the number of grams of water used (10 g) and the specific heat of the water. This allows you to calculate the amount of heat energy absorbed by the water. Since the energy absorbed by the water was the result of the burning the food, it also represents the amount of energy contained in the food sample. Since the amount of energy you calculated is in calories, it needs to be converted to kilocalories using the relationship, 1000 calories equals one kilocalories or Calorie. To make the energy relative to the particular sample you used, the energy value should be divided by the mass of your sample. This will allow the energy per gram of your sample to be compared to the energy per gram of other samples.

$$calories = 10 \text{ g } H_2O \times (final\ temperature - initial\ temperature) \times 1 \dfrac{calorie}{g \bullet °C}$$

$$Kilocalories\ or\ Calories = \dfrac{calories}{1000}$$

$$\dfrac{kcal}{g\ of\ food} = \dfrac{kcal}{mass\ of\ sample}$$

This lab will be repeated three times to obtain an average energy value and increase the validity of your experiment. Your teacher will give you the actual number of the Calories found in the food sample used. You are to determine the percent yield.

$$Percent\ Yield = \dfrac{Experimental\ Value}{Actual\ Value} \times 100$$

You can also determine the percent error by using the formula below:

$$Percent\ Error = \dfrac{(Experimental\ Value - Actual\ Value)}{Actual\ Value} \times 100$$

PURPOSE

In this activity you will determine the amount of energy found in a walnut and a peanut by using a calorimeter and compare it to the actual value.

MATERIALS

temperature probe

matches

test tube clamps

calorimeter

balance

10 mL graduated cylinder

computer with Logger Pro® installed

aprons

test tube

heavy aluminum foil

goggles

samples of walnut and peanut

Lab Pro® interface box

Safety Alert

1. CAUTION: Needles are sharp. Exercise care.
2. CAUTION: Be sure to wear goggles anytime a flame is present in the room.
3. CAUTION: Matches are flammable. If you burn yourself, immediately place the burned area under cold running water. Notify your teacher.

PROCEDURE

1. Answer the pre-lab questions on your student answer page.

2. In the space marked HYPOTHESIS on your student answer page, identify whether the peanut or walnut contains the most energy.

3. Plug the temperature probe into Port 1 or CH 1 of the interface box.

4. Mass three individual pieces of peanut and three individual pieces of walnut, approximately 0.2 g each. Record the exact mass of each piece in your data table. Store each nut piece on a small square of paper labeled with its mass. It is very important not to mix up the pieces.

5. Turn on the computer. Click on the folder called **Experiment Templates**, and open **Calorimeterss** (for stainless steel probe) or **Calorimeterdc** (for direct connect temperature probe). Logger Pro® should open.
   - CAUTION: Electricity is being used. Take care not to spill any liquids on any of the computer equipment or electrical outlets.

6. Obtain a calorimeter.

7. Place one of the walnuts on the needle anchored in the cork.
   - CAUTION: Needles are sharp. Exercise care.

8. Place nut, needle, and cork setup on a folded piece of heavy-duty aluminum foil and set the colorimeter over the setup. Put a rolled piece of masking tape around the test tube to prevent it from sliding through the hole. Slide the test tube into the hole in the top of the calorimeter. Adjust the test tube so that it is about 2 cm above the nut. See the diagram below:

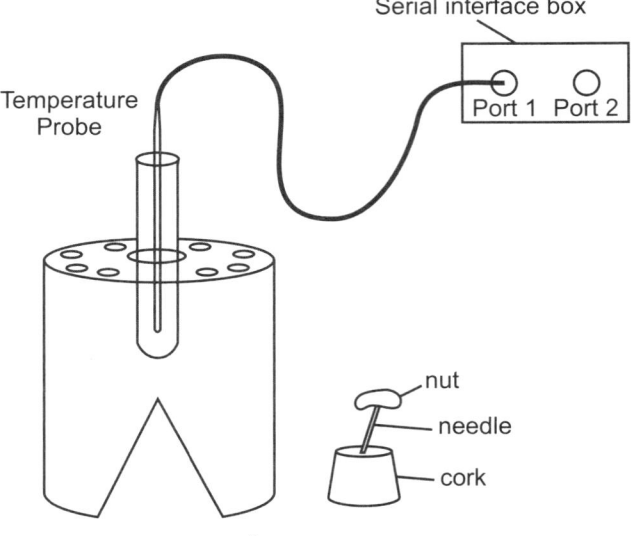

Calorimeter set-up with probe

9. Remove the calorimeter with the test tube in it from over the nut, being careful not to change the adjust position of the test tube. Measure 10 mL of water and pour it in the test tube.

10. Put the temperature probe into the test tube.

11. Using a match, set fire to the nut. Quickly and carefully position the calorimeter over the burning nut.
    - CAUTION: If you burn yourself, immediately, place the burned area under cold running water. Notify your teacher.

12. Click **Collect** and the temperature probe should start taking temperature readings immediately. Record the starting temperature.

13. Burn the nut completely. If your nut burns out before it is completely burned out, quickly relight it. When it burns out, look at the computer screen and record the final temperature. Note the shape of the graph and its axes.

14. Pull down the **Analyze Menu** and highlight **Statistics**. A statistics box will appear. It will list the maximum and minimum temperature. Record this information in your data table on the student answer page. Click the very small, gray square in the corner of the statistics window to close it.

15. Repeat this experiment two more times with the other pieces of walnut and three more times with the pieces of peanut. If you would like to keep the data that is being collected so that you can compare the various runs of the experiment, pull down the **Data Menu** and click **Store Latest Run**. This will cause the line on the graph to become a lighter red. If you do not wish to save your data you can **Delete the Latest Run** when you pull down the **Data Menu**. If you do neither of these, the Logger Pro® software will ask you if you would like to erase the previous data, you must click **Yes** to continue. It is important not to mix the pieces of nuts so that correct energy per gram values will be obtained.

16. If your teacher would like you to print a graph for this lab, pull down the **File Menu** and highlight **Print**. A dialog box will appear. Type in the number of desired copies and click **OK** or press Enter on the keyboard.

17. After you have finished close all windows on the computer. Clean up your area and return your equipment to its original condition.

Name _____

Period _____

Data Collection Devices
Determining the Amount of Energy Found in Food

HYPOTHESIS _____

DATA AND OBSERVATIONS _____

Data Table						
Experiment	**Mass of Sample**	**Temperature Before Burning**	**Temperature After Burning**	**# of calories**	**# of kilocalories**	**# of kilocalories/gram**
Peanut # 1						
Peanut # 2						
Peanut # 3						
Average						
Walnut # 1						
Walnut # 2						
Walnut # 3						
Average						

PRE-LAB QUESTIONS

1. After reading the background information, define the terms autotroph and heterotroph.

2. What are two purposes of food?

3. What is free energy?

4. What is entropy?

5. What is a calorie, a Calorie and a kilocalorie?

6. Write the equations to be used to determine the number of calories, percent yield, and percent error.

CONCLUSION QUESTIONS

1. Which nut contained the greatest number of kilocalories per gram?

2. What was the shape of the graph that screen displayed for you (if using a graphing calculator) or that Logger Pro® displayed for you (if using a computer) and what relationship did it establish?

3. The nutritional labels show that peanuts are 6.00 kcal/gram and walnuts are 7.02 kcal/gram. What was your percent yield and was it greater than or less than the actual value?

4. Give two reasons why your answer varied from the actual values.

5. What is your percent error? Do you think that it is acceptable? Why or why not?

6. Nuts are high in lipids. What foods might you investigate to determine their calorie content for carbohydrates? Proteins?

7. A student fails to replace the 10 grams of water in the test tube between the second and third trials for the walnut. How will this error affect the calculated energy result? You answer should clearly state that the calculated value will increase, decrease or remain the same. Mathematically justify your answer.

Computer Graphing Software
Using Graphical Analysis® 3 or Logger Pro® 3

OBJECTIVE

Students will learn to use Graphical Analysis® 3 or Logger Pro® 3 computer graphing software to graph and analyze data as well as generate a lab report by importing their data and graphs into a word processing document.

LEVEL

All

NATIONAL STANDARDS

UCP.1, UCP.2, A.1, A.2, E.1, E.2, G.2

TEKS

6.2 (E), 6.4(A)
7.2(E), 7.4(A)
8.2 (E), 8.4(A), 8.4(B)
IPC: 2(C)
Biology: 2(C)
Chemistry: 2(D)
Physics: 2(C), 2(E)

CONNECTIONS TO AP

All AP science exams ask students to read and interpret graphs.

TIME FRAME

90+ minutes

MATERIALS

    either Graphical Analysis® 3 software or Logger Pro® 3 software
    computer

TEACHER NOTES

Ideally, this lesson would follow a lesson on graphing calculators and the use of data collection devices. The data from the calculator can be imported into either of these pieces of software for further analysis. It is important to load the TI Connect software *first*, followed by the graphing software you have chosen. *Foundation Lesson VII: Data Collection Devices*, addresses the collection of data using Logger Pro. It is the intent of this lesson to simply use these computer graphing programs to analyze a set of sample data to learn the features of the program and then to generate a lab report by cutting and pasting the data and graphs into a word processing document. The students are directed to skip two of the more mathematically advanced tutorials and told they may do them count for extra credit.

Both of these graphing software programs are available for purchase from Vernier Software & Technology, 13979 SW Millikan Way, Beaverton, Oregon 97005-2886. You can also purchase on-line at www.vernier.com. At the time of this printing, Graphical Analysis® 3 can be purchased at a cost of $80 and Logger Pro® 3 at a cost of $100. This gives your campus a site license that allows you to load the program on any and every computer on campus and give a copy of the program to each of your students.

What is the difference between these two programs? Graphical Analysis® 3, GA3, is simply a user friendly graphing software program that can be used alone or with TI graphing calculator, CBL 2, or LabPro compatibility. Data can be imported from a TI graphing calculator, CBL 2 or LabPro into GA3 in a matter of seconds with the use of a TI-Graph Link cable. Logger Pro® 3, LP3, has GA3 embedded within it, but is also capable of communicating directly with a LabPro and probes to collect data bypassing the need for a calculator. Logger Pro® 3 houses all of the experiment files found in the TI and Vernier Lab Manuals and has more capability than GA3. Logger Pro also contains some sample movies that graph data as they play so that the data can be analyzed. You may also add your own movies and synchronize them with data collection for further analysis. Logger Pro® 3 is the software to purchase if you have access to several computers for student use in your laboratory.

The complete user's manual to either piece of software is available at www.vernier.com.

**IF you are using a USB TI Graph-Link, you need to first load TI Connect from the TI website, http://education.ti.com/us/product/accessory/connectivity/down/download.html.

THEN load the graphing software. TI Connect contains the driver for the USB Graph-Link. Neither Graphical Analysis nor Logger Pro contains this driver. **You will not be able to import data from the calculator if you load the graphing software first. If you are using the older gray or black Graph-links, you do not need TI Connect software at all.

Graphical Analysis® 3 Computer Requirements

Windows requirements:

- Windows 95, Windows 98, Windows 2000, Windows NT 4.x, Windows ME, and Windows XP.
- 133 MHz Pentium processor or better.
- 16 MB physical RAM plus free hard disk space (for virtual memory).
- Color monitor (>=256 colors)

Macintosh requirements:

- Mac OS 8.x, MacOS 9.x, MacOS X.
- 66 MHz PowerPC processor or better.
- 16 MB machine RAM, 8MB for the application partition.

Logger Pro® 3 Computer Requirements

- Windows 98®, 2000, ME, NT, or XP on a Pentium Processor or equivalent, 133 MHz, 32 MB RAM, 25 MB of hard disk space, for a minimum installation.
- Mac OS® 9.2, or Mac OS X (10.1 or newer), with 25 MB of hard disk space for a minimum installation.
- Using the movie feature of Logger Pro will require a faster processor and an additional 100 MB of hard disk space.

Note: Logger Pro cannot be used with the ULI or Serial Box interface.

Loading Graphical Analysis or Logger Pro onto Your Hard Drive

(Remember to load the TI Connect software first, before the graphing software)

To install Logger Pro or Graphical Analysis on a computer running Windows 98/2000/ME/NT/XP, follow these steps: [note that GA3 will run on Windows 95, but LP3 will NOT]

1. Place the software CD in the CD-ROM drive of your computer.

2. If you have Autorun enabled, the installation will launch automatically; otherwise choose Settings→Control Panel from the Start menu. Double click on Add/Remove Programs. Click on the Install button in the resulting dialog box.

3. The software installer will launch, and a series of dialog boxes will step you through the installation. You will be given the opportunity to either accept the default directory [recommended] or enter a different directory.

To install Graphical Analysis or Logger Pro on a computer running MacOS 8.x, MacOS 9.x, MacOS X, follow these steps:

1. Place the Graphical Analysis CD in the CD-ROM drive of your computer.

2. Double-click on the Install Graphical Analysis icon and follow the directions.

Setting Preferences Within the Software

There is one aspect of both programs that needs to be changed. The setting in the start-up file for both programs has the "Connect Lines" or "Connected Points" feature as the default. One of your primary goals is to have Pre-AP students mathematically model data rather than draw a dot-to-dot picture of the data. You will spend a great deal of time encouraging students to break this habit. You want students to mathematically model their data using curve-fitting so that they can use the equation of the curve to interpolate and extrapolate. This setting is easy to change, but must be done on each computer or on the master *copy* of the software that you issue to students. Follow these steps:

1. Start the program and double click on the blank graph. This dialogue box appears:

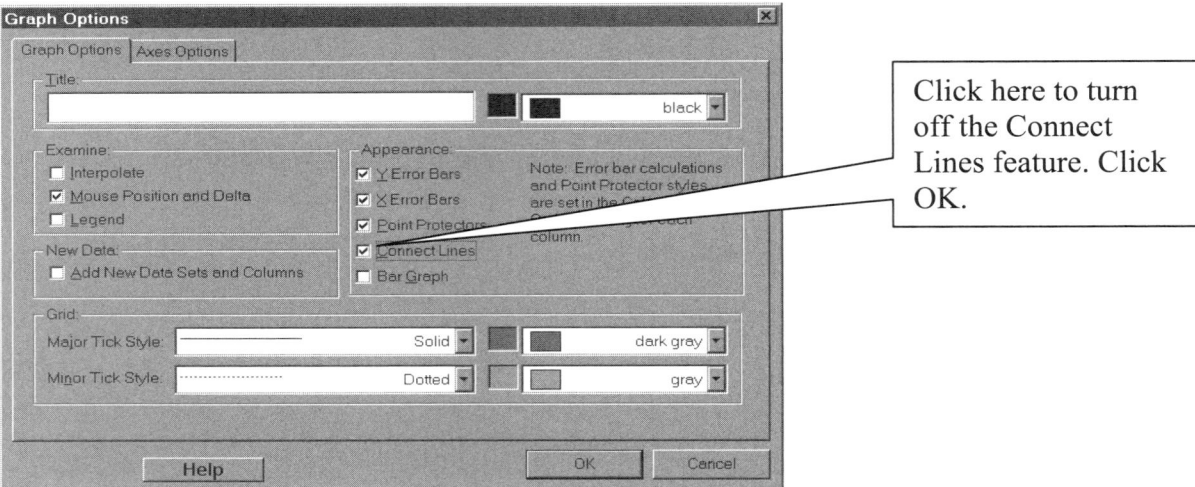

Click here to turn off the Connect Lines feature. Click OK.

2. Go to File and select preferences and make sure the Start up file box is checked. You can also change other features such as removing the automatic curve fitting option for students if you wish.

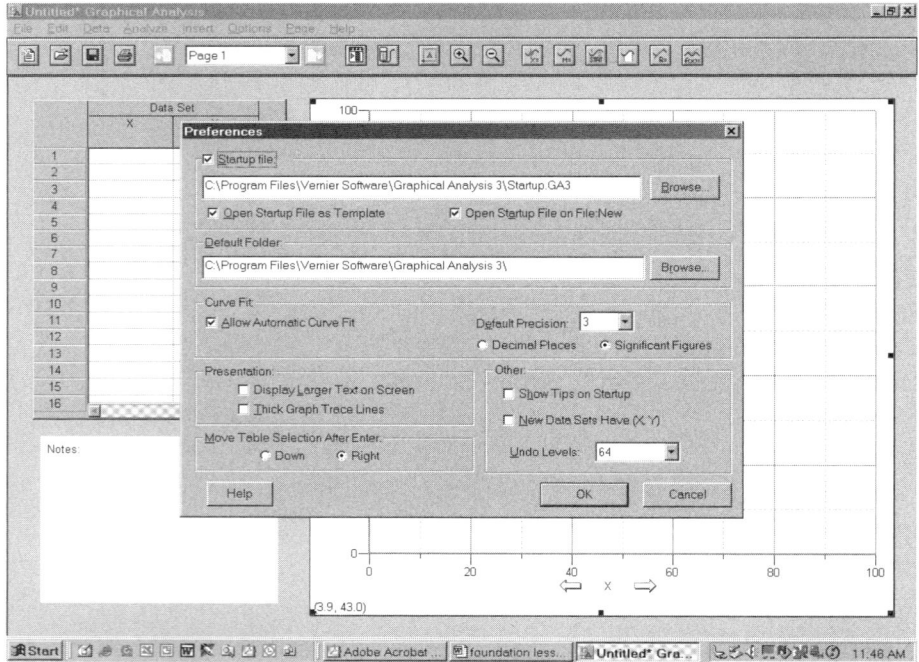

3. Go to File and select Save As. If using GA3, select the file titled startup.ga3. If using LP3, select the file titled startup.xmbl.

4. Right click on the file and scroll to the bottom of the pop up box and select Properties. The Properties dialogue box will appear. Click to remove the Read Only status of the startup file. Click OK. Click SAVE. The program will inform you that the file already exists and ask you if you wish to replace it, click YES.

5. Failure to disable the read only status will result in this error message:

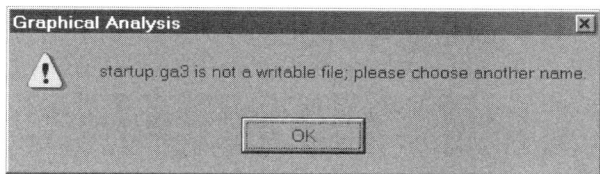

Click OK and simply repeat step 4.

6. If you did not get the error message, you were successful. However, you must restore your new and improved startup file to Read Only status to protect it from corruption. Go to File, Save As and select the startup file again. Right click and select properties at the bottom of the pop up box. Click to replace the Read Only checkmark. Click OK. Click Cancel.

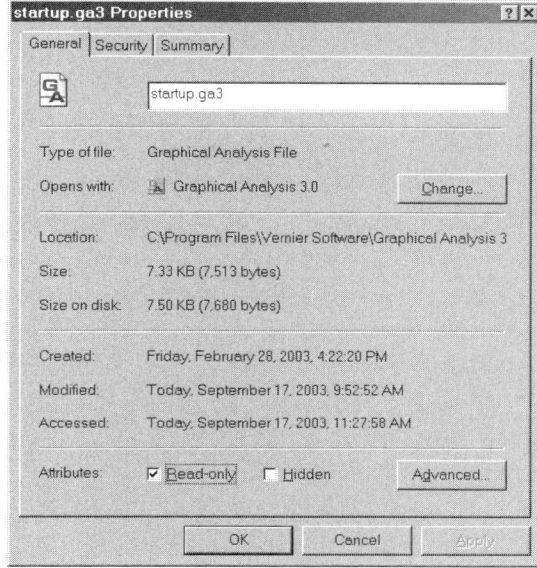

7. Repeat these steps on each of your student computers.

8. Since you may make copies of the CD for distribution to students you may wish to do these steps **once** on your own computer. When you are setting up the files to burn your master *copy* for distribution, simply delete the factory startup file before you burn the disk and replace it with the one from the hard drive on the computer you are working from once you have completed steps 1-7. The startup file is found in C:\Program Files\Vernier Software\Graphical Analysis 3 if using GA3. The startup file is found in C:\Program Files\Vernier Software\Logger Pro 3 if using LP3. Performing this step *before* burning your master *copy* CD for student distribution will save you a great deal of time and make the student's home computer begin the program just as it does at school.

NOTE: Since the tutorials in this lesson do NOT use the startup file, do not be surprised if this change is not apparent. It will take effect anytime a new file is created.

STUDENT SAMPLE

This sample can serve as your grading key. Since all students are following the same set of tutorials, their reports should be nearly identical and thus easy to grade.

Learning to Use Graphical Analysis [or Logger Pro]
Ima Student
8th Grade Science, Period 4

Basic Operations Tutorial Results

How much garbage was generated per person in 1994?

• 2511 lb/person/yr

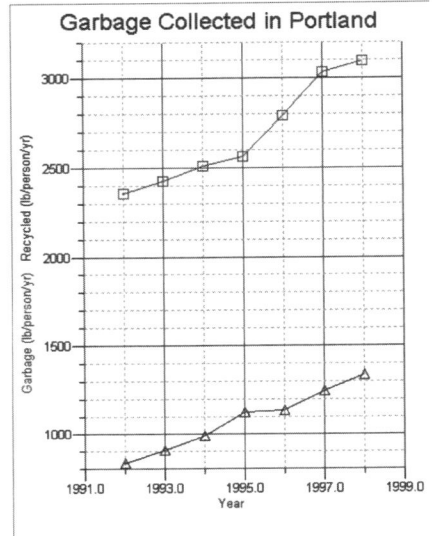

Customization Tutorial Results

	Data Set	
	X	Y
1	1.00	2.00
2	2.00	4.00
3	3.00	5.80
4	4.00	6.70
5	5.00	8.20
6	6.00	10.0
7	7.00	12.0
8	8.00	14.3
9	9.00	15.6
10	10.0	17.0
11		
12		
13		
14		
15		
16		
17		
18		

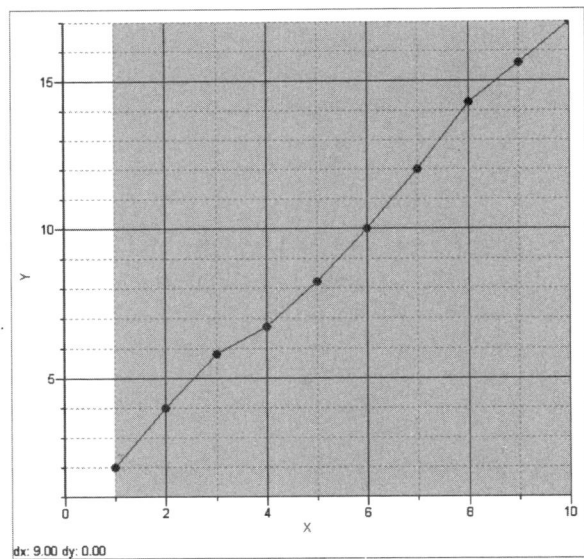

Viewing Graphs Tutorial Results

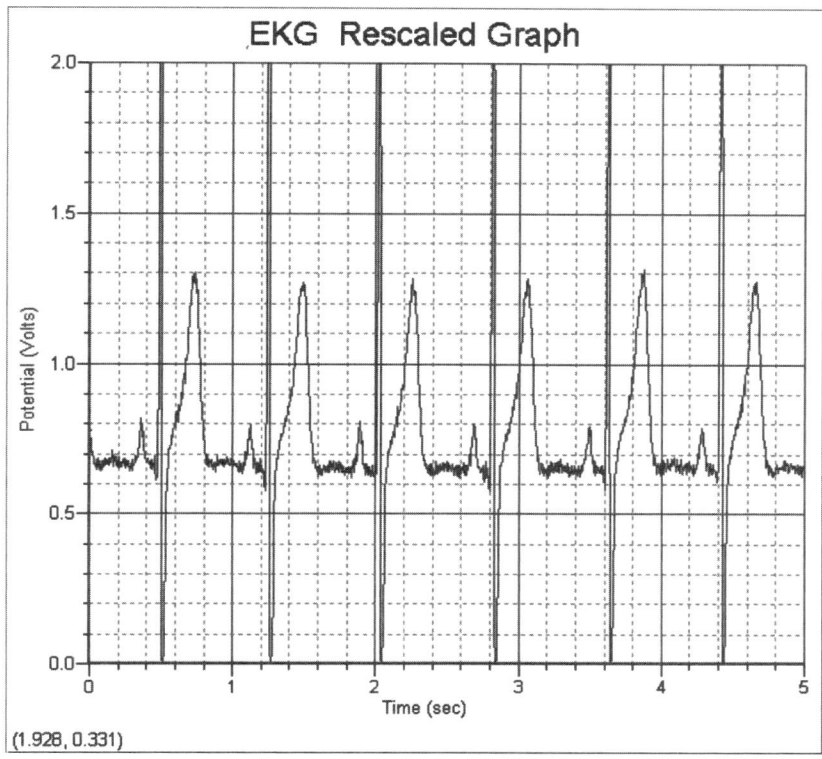

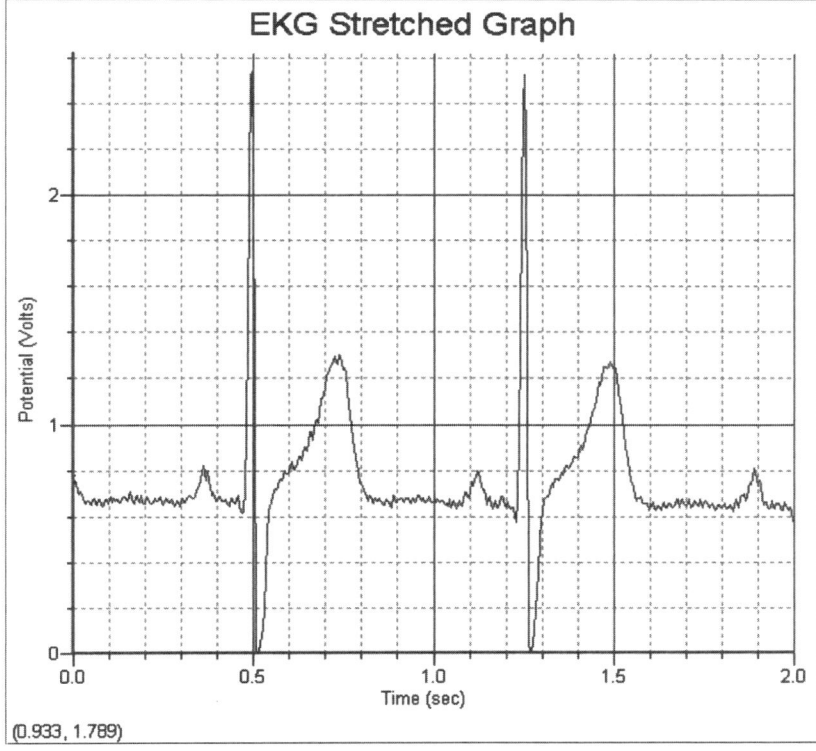

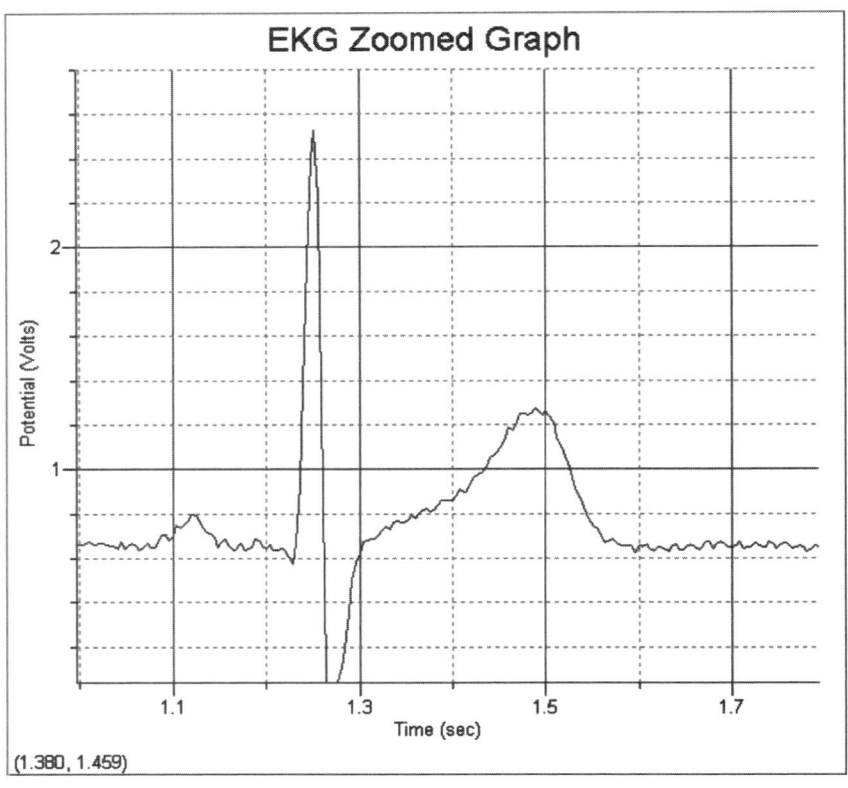

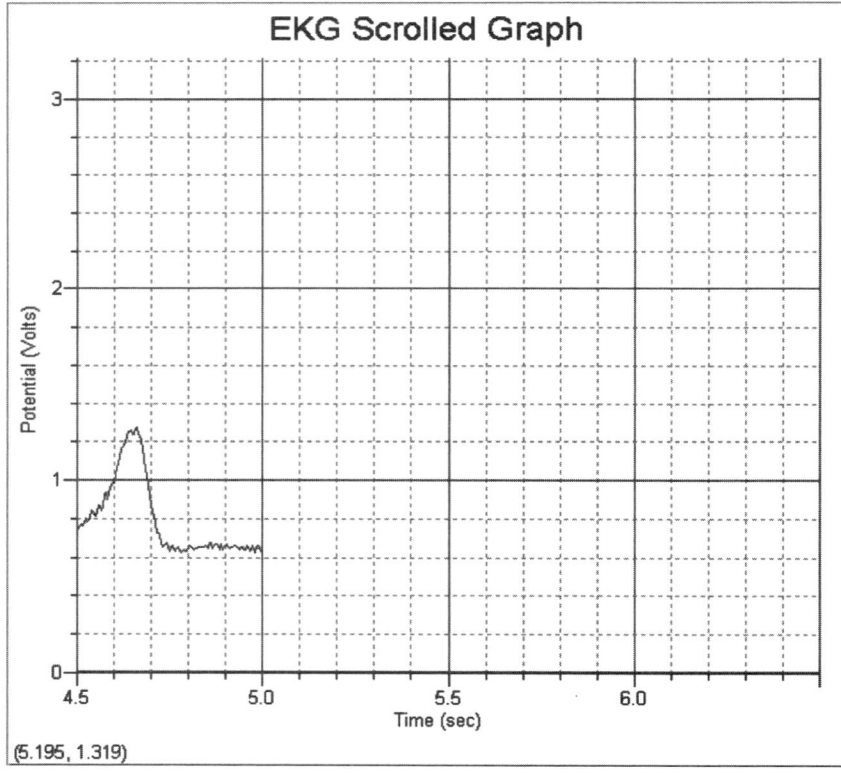

Linearization Part 1 Tutorial Results

Is Population proportional to Time Interval ^2?

- Yes since the graph of Population vs. Time Interval squared is linear.

The top graph on the right shows a different kind of relationship. What is it?

- It is an inverse relationship since y decreases as x increases.

Linearization Part 2 Tutorial Results

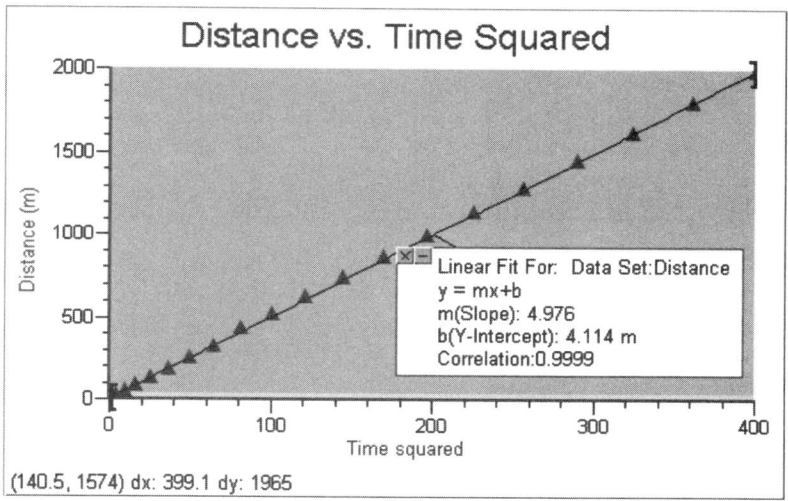

Is the graph now a straight line? If so, then you've found the relationship.

- Yes it is. The equation for the straght line is $y = 4.976x + 4.11$ with a correlation of 0.999. This means that Distance = 4.976 (time^2) + 4.11 m.

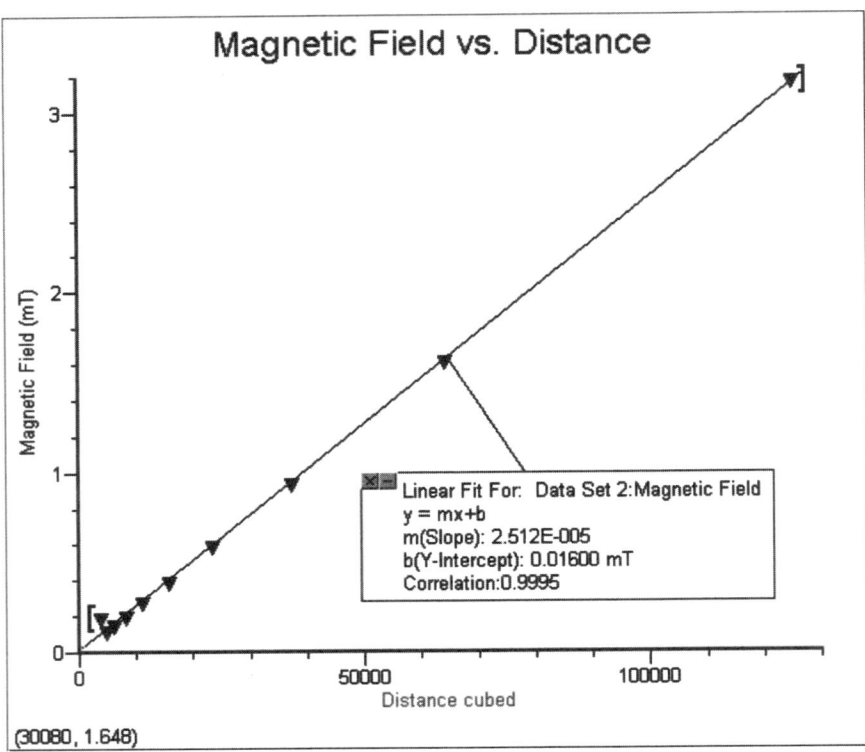

Computer Graphing Software
Using Graphical Analysis® 3

You will be asked to graph and analyze a vast amount of data in this course as well as future science courses. In the past, this was a daunting and time consuming task. Your school has purchased either Graphical Analysis® 3 (GA3) to greatly simplify analyzing and communicating the data you have collected during your laboratory exercises. The program also allows you to cut and paste your data tables and graphs into a word processing program such as Microsoft Word. This will be very helpful in generating lab reports. The site license your school has purchased allows you to load a copy of the software on any computer in your school or at home.

PURPOSE
In this activity you will learn the basic features of Graphical Analysis or Logger Pro computer graphing software so that you may use this valuable tool throughout this course and future science courses.

MATERIALS
Computer graphing software, Graphical Analysis® 3

PROCEDURE FOR GRAPHICAL ANALYSIS® 3 SOFTWARE

1. Start the program by double clicking either the Graphical Analysis icon.

2. The program may greet you with a Tip of the Day. Read the tip, close it and go to File and select Open or simply click on the second icon from the right to open files. Open the Tutorials folder and this dialogue box will appear:

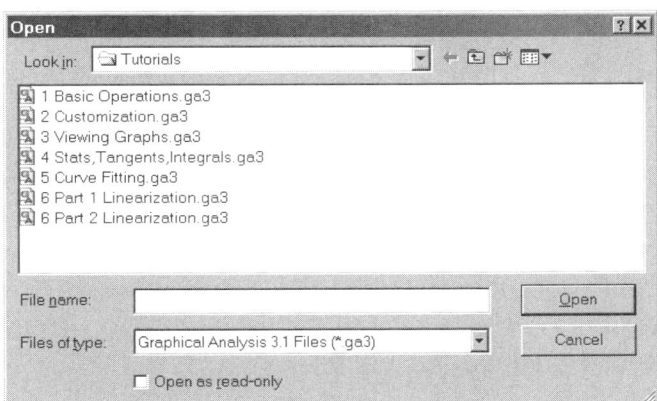

3. Double click on "1 Basic Operations.ga3" to open the file.

4. Without closing GA3, open a word processing program such as Microsoft Word®. Start a new file and title this document "Learning to Use Graphical Analysis". Your title should be centered with 18 point bold font. Type your name under your title making it centered with 16 point bold font. Type your course name and class period together under your name making it centered with 14 point bold font. Press the Enter key three times and type "Basic Operations Tutorial Results". Subtitles such as this one should be left justified, bold 14 point font. Press Enter twice, change the font to 12 point, and remove the bold. This will be the formatting for all of the body text in today's lab report. SAVE this document. Do not close the program; just minimize it since you will be going back and forth between GA3 and your document.

5. Go back to GA3 by clicking on it at the bottom of your screen. Read and follow all instructions contained within the tutorial. As you proceed through this tutorial, the pop up dialogue boxes may obscure the directions. You can move the dialogue boxes by clicking on the blue title bar of the box and dragging them to another part of the screen, out of your way. Don't forget that you will get tool tips if you let your mouse pointer hover over a button, these are very useful.
   a. When you reach page 5, you will be instructed to title your graph "Garbage vs. Year". This is not a title, but rather a restatement of the axes. Title your graph "Garbage Collected in Portland" instead.
   b. You will be asked questions throughout these tutorials about the data presented. In all cases, cut and paste the questions from GA3 by clicking on the box containing the question, highlighting the question, and pressing Ctrl+C to copy the question to the clipboard. Go to your word processing document and place your cursor under the appropriate subtitle heading and press Ctrl+V to paste the question into your document. Type your answer to the question. Pay special attention to the units of the answer. No naked numbers!

6. Continue through the tutorial until you reach the "Congratulations" page. Make sure you have cut and pasted (as well as answered) all questions throughout the tutorial. Once you are finished, go back 2 pages to your graph. Check to make sure it has a correct title, then click on the graph to select it and copy it onto the clipboard using Ctrl+C. Return to your document and press Enter twice to separate the graph from the questions asked. Use Ctrl+V to paste your completed graph from the first tutorial into this document. Resize the graph in your document so it takes up less space. Reduce it by a factor of *about* one half. Resize all of your graphs by a factor of *about* one half throughout the rest of your document.

7. Go back to GA3 and go to File, Open and select "2 Customization". **Do NOT save your changes to the first tutorial.** Proceed through this second tutorial.

8. Continue until you reach the "Congratulations" page. Once you are finished, go back 2 pages to your graph. Click on the graph to select it and copy it onto the clipboard using Ctrl+C.

9. Go back to your minimized word processing document. Press Enter twice and type the subtitle "Customization Tutorial Results". Press Enter once followed by Ctrl+V to paste your customized graph into this document. Resize the graph in your document. SAVE. Do not close the program, just minimize it.

10. Go back to GA3 and page back to page 3. Cut and paste the improved data table into your word processing document. Resize the data table in your document.

11. Go back to Graphical Analysis and go to File, Open and select "3 Viewing Graphs". **Do NOT save your changes to the second tutorial.** Proceed through this third tutorial.

12. Once you have rescaled the graph at the end of the first page, change its title to "EKG Rescaled Graph" and copy it to the clipboard.

13. Go back to your minimized word processing document. Press Enter once followed by Ctrl+V to paste your rescaled graph into this document. Resize the graph in your document. SAVE. Do not close the program, just minimize it.

14. Go back to Graphical Analysis and continue the tutorial. Once you have stretched the graph on page 2 of the tutorial, change its title to "EKG Stretched Graph" and copy it to the clipboard.

15. Go back to your minimized word processing document. Press Enter twice and type the subtitle "Viewing Graphs Tutorial Results". Press Enter once followed by Ctrl+V to paste your zoomed graph into this document. Resize the graph in your document. SAVE. Do not close the program, just minimize it.

16. Go back to Graphical Analysis and continue the tutorial. Once you have zoomed in on the graph on page 3 of the tutorial, change its title to "EKG Zoomed Graph" and copy it to the clipboard.

17. Go back to your minimized word processing document. Press Enter twice and type the subtitle "Viewing Graphs Tutorial Results". Press Enter once followed by Ctrl+V to paste your zoomed graph into this document. Resize the graph in your document. SAVE. Do not close the program, just minimize it.

18. Go back to Graphical Analysis and continue the tutorial. Continue until you reach the "Congratulations" page. Once you are finished, go back 1 page to your graph. Click on the graph to select it, change its title to "EKG Scrolled Graph" and copy it onto the clipboard using Ctrl+C.

19. Go back to your minimized word processing document. Press Enter once followed by Ctrl+V to paste your scrolled graph into this document. Resize the graph in your document. Resize the graph in your document. SAVE. Do not close the program, just minimize it.

20. Go back to Graphical Analysis and go to File, Open and select "6 Linearization Part 1" [Yes, we are skipping tutorials 4 & 5, they will be extra credit!]. **Do not save your changes to the third tutorial**. Proceed through this next tutorial. Be sure to cut and paste the two questions into your report under the subtitle "Linearization Part 1 Tutorial Results". There are no graphs to paste into your report, but you will need to use the skills presented in this tutorial to be successful in Part 2. You may want to print the last page of this tutorial for future reference.

21. Go to File, Open and select "6 Linearization Part 2". Follow all of the directions on page one and paste your data tables and your graphs into your report under the subtitle "Linearization Part 2 Tutorial Results". Be sure and run the linear regression on your data so that your graph shows the statistics box. Do this by highlighting the data on your graph and pressing the [button image] button.

22. SAVE. Continue to the end of the tutorial and paste your second graph into your report. [Hint: The shape of the second graph tells you it is an inverse function.] When you calculate your column you will need to use the reciprocal of one of the variables. Be sure and include the algebraic equation for your line of best fit once you have pasted the graphs into your document.

23. You may add the appropriate graphs from tutorials 4 & 5 to your report for extra credit. Be sure and include the proper subtitle headings and continue the proper format for the report.

Computer Graphing Software
Using Graphical Logger Pro® 3

You will be asked to graph and analyze a vast amount of data in this course as well as future science courses. In the past, this was a daunting and time consuming task. Your school has purchased either Logger Pro® 3 (LP3) to greatly simplify analyzing and communicating the data you have collected during your laboratory exercises. The program also allows you to cut and paste your data tables and graphs into a word processing program such as Microsoft Word. This will be very helpful in generating lab reports. The site license your school has purchased allows you to load a copy of the software on any computer in your school or at home.

PURPOSE
In this activity you will learn the basic features of Graphical Analysis or Logger Pro computer graphing software so that you may use this valuable tool throughout this course and future science courses.

MATERIALS
Computer graphing software, Logger Pro® 3

PROCEDURE FOR LOGGER PRO® 3 SOFTWARE

1. Start the program by double clicking the Logger Pro icon.

2. The program may greet you with a Tip of the Day. Read the tip then close it. You may also be asked about continuing without an interface attached. Choose to continue without an interface and click OK. Go to File and select Open or simply click on the second icon from the right to open files. Open the Tutorials folder and this dialogue box will appear:

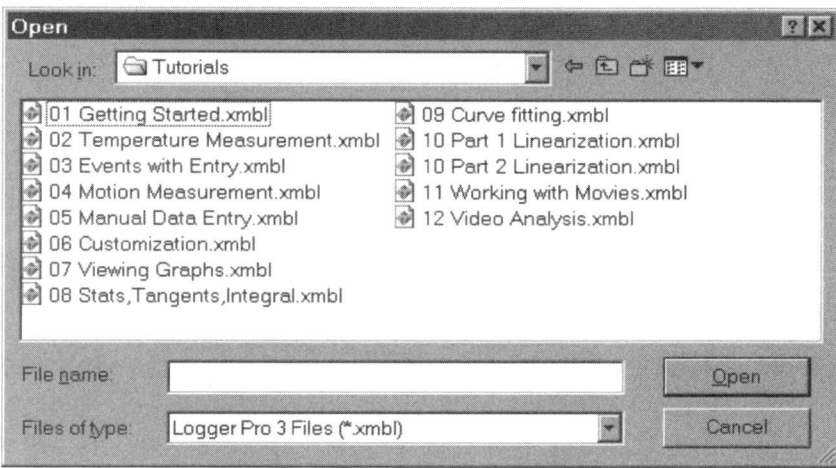

3. Double click "01 Getting Started.xmbl" to open the file.

4. Without closing LP3, open a word processing program such as Microsoft Word®. Start a new file and title this document "Learning to Use Logger Pro". Your title should be centered with 18 point bold font. Type your name under your title making it centered with 16 point bold font. Type your course name and class period together under your name making it centered with 14 point bold font. Press the Enter key three times and type "Manual Data Entry Tutorial Results". Subtitles such as this one should be left justified, bold 14 point font. Press Enter twice, change the font to 12 point, and remove the bold. This will be the formatting for all of the body text in today's lab report. SAVE this document. Do not close the program; just minimize it since you will be going back and forth between LP3 and your document.

5. Go back to LP3 by clicking on it at the bottom of your screen. Read and follow all instructions contained within the tutorial. Continue through this tutorial until you reach the "Congratulations" page.

6. Go to File, Open and select "05 Manual Data Entry". **Do NOT save your changes to the first tutorial.** As you proceed through this tutorial, the pop up dialogue boxes may obscure the directions. You can move the dialogue boxes by clicking on the blue title bar of the box and dragging them to another part of the screen, out of your way. Don't forget that you will get tool tips if you let your mouse pointer hover over a button, these are very useful.
   a. When you reach page 5, you will be instructed to title your graph "Garbage vs. Year". This is not a title, but rather a restatement of the axes. Title your graph "Garbage Collected in Portland" instead.
   b. You will be asked questions throughout these tutorials about the data presented. In all cases, cut and paste the questions from LP3 by clicking on the box containing the question, highlighting the question, and pressing Ctrl+C to copy the question to the clipboard. Go to your word processing document and place your cursor under the appropriate subtitle heading and press Ctrl+V to paste the question into your document. Type your answer to the question. Pay special attention to the units of the answer. NO naked numbers!

7. Continue until you reach the "Congratulations" page. Make sure you have cut and pasted (as well as answered) all questions throughout the tutorial. Once you are finished, go back 2 pages to your graph. Check to make sure it has a correct title, then click on the graph to select it and copy it onto the clipboard using Ctrl+C. Return to your document and press Enter twice to separate the graph from the questions asked. Press Ctrl+V to paste your completed graph from the first tutorial into this document. Resize the graph in your document so it takes up less space. Reduce it by a factor of *about* one half. Resize all of your graphs by a factor of *about* one half throughout the rest of the document.

8. Go back to LP3 and go to File, Open and select "6 Customization". **Do NOT save your changes to the first tutorial.** Proceed through the customization tutorial.

9. Continue until you reach the "Congratulations" page. Once you are finished, go back 2 pages to your graph. Click on the graph to select it and copy it onto the clipboard using Ctrl+C.

10. Go back to your minimized word processing document. Press Enter twice and type the subtitle "Customization Tutorial Results". Press Enter once followed by Ctrl+V to paste your customized graph into this document. Resize the graph in your document. SAVE. Do not close the program, just minimize it.

11. Go back to LP3 and page back to page 3. Cut and paste the improved data table into your word processing document. Resize the data table in you document.

12. Go back to Logger Pro and go to File, Open and select "07 Viewing Graphs". **Do NOT save your changes to the customization tutorial.** Proceed through the viewing graphs tutorial.

13. Once you have rescaled the graph at the end of the first page, change its title to "EKG Rescaled Graph" and copy it to the clipboard.

14. Go back to your minimized word processing document. Press Enter once followed by Ctrl +V to paste your rescaled graph into this document. Resize the graph in your document. SAVE. Do not close the program, just minimize it.

15. Go back to LP3 and continue the tutorial. Once you have stretched the graph on page 2 of the tutorial, change its title to "EKG Stretched Graph" and copy it to the clipboard.

16. Go back to your minimized word processing document. Press Enter twice and type the subtitle "Viewing Graphs Tutorial Results". Press Enter once followed by Ctrl+V to paste your zoomed graph into this document. Resize the graph in your document. SAVE. Do not close the program, just minimize it.

17. Go back to LP3 and continue the tutorial. Once you have zoomed in on the graph on page 3 of the tutorial, change its title to "EKG Zoomed Graph" and copy it to the clipboard.

18. Go back to your minimized word processing document. Press Enter twice and type the subtitle "Viewing Graphs Tutorial Results". Press Enter once followed by Ctrl+V to paste your zoomed graph into this document. Resize the graph in your document. SAVE. Do not close the program, just minimize it.

19. Go back to LP3 and continue the tutorial. Continue until you reach the "Congratulations" page. Once you are finished, go back 1 page to your graph. Click on the graph to select it, change its title to "EKG Scrolled Graph" and copy it onto the clipboard using Ctrl+C.

20. Go back to your minimized word processing document. Press Enter once followed by Ctrl+V to paste your scrolled graph into this document. Resize the graph in your document. Resize the graph in your document. SAVE. Do not close the program, just minimize it.

21. Go back to LP3 and go to File, Open and select "10 Linearization Part 1" [Yes, we are skipping tutorials 08 and 09, they will be extra credit!]. Do not save your changes to the tutorial. Proceed through this next tutorial. Be sure to cut and paste the two questions into your report under the subtitle "Linearization Part 1 Tutorial Results". There are no graphs to paste into your report, but you will need to use the skills presented in this tutorial to be successful in Part 2. You may want to print the last page of this tutorial for future reference.

22. Go to File, Open and select "10 Linearization Part 2". Follow all of the directions on page one and paste your data tables and your graphs into your report with the subtitle "Linearization Part 2 Tutorial Results". Be sure and run the linear regression on your data so that your graph shows the statistics box. Do this by highlighting the data on your graph and pressing the [button image] button.

23. SAVE. Continue to the end of the tutorial and paste your second graph into your report. [Hint: The shape of the second graph tells you it is an inverse function.] When you calculate your column you will need to use the reciprocal of one of the variables. Be sure and include the algebraic equation for your line of best fit once you have pasted the graphs into your document.

24. You may add the appropriate graphs from tutorials 4 & 5 to your report for extra credit. Be sure and include the proper subtitle headings and continue the proper format for the report.

Essay Writing Skills
Developing a Free Response

OBJECTIVE
This lesson is designed to introduce the students to the skill of planning appropriate free response essays.

LEVEL
All

NATIONAL STANDARDS
UCP.1, UCP.2, G.1, G.2

TEKS
6.2(C), 6.2(D)
7.2(D)
8.2(D)
IPC: 2(D)
Biology: 2(D)
Chemistry: 2(E)
Physics: 2(D)

CONNECTIONS TO AP
Writing appropriate free response answers is a fundamental skill needed in both AP Biology and AP Environmental Science.

TIME FRAME
45 minutes

MATERIALS
transparencies of Practice Essay #1-4
student copies of the practice pages

transparency of *Writing a Free Response*
strategies page

TEACHER NOTES
An appropriately written Free Response essay for an AP Biology or AP Environmental Science exam is markedly different from the type of essay that students are typically asked to write in English courses. For this reason, the skill of writing an essay in AP Biology and AP Environmental Science must be explicitly addressed. This activity is designed to help students understand how to set up a mechanical outline or plan to use when writing free response essays in AP science courses. It emphasizes the need for planning a response prior to writing. This planning and pre-thinking approach gives the students a tool to use and enables them to dissect complex prompts into manageable units. This practice activity focuses on the mechanics of dissecting the prompt rather than on writing the specific content. Once

students have mastered the skill of planning or outlining a free response prompt, they need to practice writing essays using specific content throughout the school year.

SUGGESTING TEACHING PROCEDURE:

1. Explain the need for knowing how to write appropriate essays in a science class and describe strategies for writing essays using the handout: Tips for Writing Free Response Essays as your guide.

2. Show the transparency of Practice Essay #1 and explain how to use the strategies covered in step 1 with the sample essay. Explain to students how a mechanical outline of a response to this prompt might look (see answer key to student pages). Note, at this point, students may not be able to provide the correct content for the response, so focus on the mechanics of what should be included.

3. Distribute copies of the Essay Writing Outline Practice pages and show the transparency of Practice Essay # 2 as you model how to outline an appropriate response for the students (see answer key for Practice Essay #2). Elaborate on tip 5 by showing the students a possible outline for the question.

4. Focus student attention on Practice Essay #3 and allow students time to work alone to outline the major items that should be included in an appropriate free response. After 2-3 minutes of individual planning, have students compare their outline to that of their partners. Call on two or three volunteer pairs to write their outlines on the board for all students to see.

5. Have students read the prompt for Practice Essay #4 and prepare an outline of the items that should be included in a well designed free response. Restate tip 5 from the strategies list.

6. Show the students the sample student answer to essay # 3. Ask them to look at the response and identify 4 things that this student could do to make this essay better. Use the annotated and revised student response to show the students how the incorrect answer could be written more appropriately.

7. Follow up this activity in future lessons by including the outlining and writing of free response type questions in your daily warm-ups, daily quizzes, homework assignments, and major tests. The following list of grading hints taken from Advanced Placement and TEKS: A Lighthouse Initiative for Texas Science Classrooms may assist you as you approach the inclusion of free response questions in your assessment strategies:
   a. Start early in the year writing a free response (over a simple topic) in class and going over the rubric for it. Sharing the rubric with the students helps them gain insight into what types of information should and could be included.
   b. Create a rubric with positive points when you write the question. Students are more willing to take a chance when writing an essay in which points are collected rather than lost.
   c. Highlight or check off correct parts of free response answers as you grade them to make it easier for you to add up the points.
   d. Grade all free response answers at one sitting to develop a flow/pattern and encourage consistency in your grading.

e. On math problems, a correct answer with correct unit and work shown clearly earns full credit. Give partial credit for
   i.   Set up of problem
   ii. Correct labels
   iii. Hint-look for final answer; if correct, just scan that work is present.

f. Encourage students to separate and label each section as this will help them organize their response and allow for easier grading.

g. Go over the rubric with the entire class to eliminate the need for making individual remarks. Alternately, you can use colored highlighter to mark the papers using a code such as blue — this statement scores a point, yellow — this statement included unnecessary or off topic information, and pink — this sentence contains incorrect information.

h. Use the College Board prompts and rubrics whenever possible

Tips for Writing Free Response Essays
AP Biology and AP Environmental Science*

1. Read the question twice.

2. Dissect the question to determine exactly what is being asked. (Highlight or underline)

3. Prepare a skeleton outline of the main components of your response.

4. Begin answering the question in the order it is written and DO NOT restate the question or write an introductory paragraph.

5. If the question says to 'discuss' or 'describe'
   a. Define the topic.
   b. Describe or elaborate on the topic.
   c. State an example of that topic.

6. If the question says to 'compare and contrast'
   a. Clearly state what the items have in common.
   b. Clearly state how items are different.

7. If the question asks for a graph to be made
   a. Label each axis with a name and units.
   b. Title the graph.
   c. Scale and number the axes correctly.
   d. Use the correct type of graphs (line or bar).

8. If the question asks a mathematical problem
   a. Show every single step of all work.
   b. Set up problems so that units cancel out (dimensional analysis).
   c. Write answers with units.
   d. If numbers are very large or very small, use scientific notation.

9. If the question asks for lab design
   a. State a hypothesis in the "if, then" format.
   b. Describe each step of a planned experiment in detail including what will be measured and how often the readings will be taken.
   c. Clearly identify the control(s).
   d. State that the experiment will have multiple trials for validity.
   e. Describe the expected results.

10. For ALL questions
   a. Answer in complete sentences, DO NOT use lists, charts, outlines in your final response.
   b. Label each section of your response as it is labeled in the question.
   c. Diagrams can support your statements but will not be scored.
   d. For every statement you write, ask yourself "why". If there is an answer to that why, keep on writing!
   e. Do not answer more than what is asked for. For example, if the question says to choose 3 out of 5 topics, ONLY answer 3 of the 5; If the question asked about RNA specifically, do not discuss DNA replication.
   f. Remember — this writing is timed. Use your time wisely.

*Edited from strategies list provided on page 85 of *A Lighthouse Initiative for Texas Science Classrooms.*

POSSIBLE ANSWERS TO THE CONCLUSION QUESTION AND SAMPLE DATA

1) Practice Essay#1-#4 could be dissected/outlined as follows:

Practice Essay #1

Main Topic: Carbon & Organic Compounds

    A) Characteristics of carbon atom
        * characteristic #1
        * characteristic #2
        * characteristic #3

    B) Structure and function of
        a. Lipids
            i. Structure
            ii. Function
        b. Proteins
            i. Structure
            ii. Function
        c. Nucleic Acids
            i. Structure
            ii. Function

Practice Essay #2

Main Topic: Usefulness of the Scientific Method

Components
    Name & describe

    Name & describe

    Name & describe

    Name & describe

1. How used in biological discovery # 1

2. How used in biological discovery # 2

Practice Essay #3

Main Topic: Changes in rate of Photosynthesis

A. Descriptions
How low levels of light affect the rate of photosynthesis.

How high temperature will affect the rate.

How low levels of water will affect the rate.

B. An adaptation to low levels of light
a. Describe adaptation
b. Give an example

Practice Essay #4

Main Topic: Growth curve fluctuations

A. Explain what is happening in phase A

B. Three factors that might cause changes in phase B
a. Factor one
b. Factor two
c. Factor three

C. Strategies
a. Explain (r) strategy
b. How (r) strategies effect population size
c. Explain (K) strategies
d. How (K) strategies effect population size

Annotated and Revised Student Response to Practice Essay #3

Sections labeled to match prompt

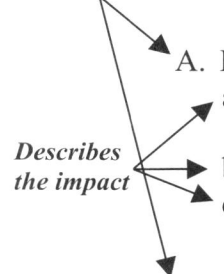

A. Descriptions

   a. Low light levels make it hard for plants to have enough solar energy to go through photosynthesis (so the rate of photosynthesis will go down in low light).

Describes the impact

   b. Low amounts of water will (also reduce the rate of photosynthesis).

   c. When temperatures get really high plants lose too much water when they open their stoma (causing photosynthesis rate to decline) so some plants are adapted to opening their stoma at night in order to take in carbon dioxide to use in photosynthesis.

B. Some plants have special adaptations that allow them to survive in extreme environments (such as low levels of light).

   a. Some plants have broad leaf surfaces to catch any available sun rays and can live in low light levels while some have thin leaves.

Gives an example

Revised Student Answer

A. Low light levels make it hard for plants to have enough solar energy to go through photosynthesis so the rate of photosynthesis will go down in low light. Low amounts of water will also reduce the rate of photosynthesis. hen temperatures get really high plants lose too much water when they open their stoma causing photosynthesis rate to decline so some plants are adapted to opening their stoma at night in order to take in carbon dioxide to use in photosynthesis.

B. Some plants have special adaptations that allow them to survive in extreme environments such as low levels of light. Some plants have broad leaf surfaces to catch any available sun rays and can live in low light levels while some have thin leaves.

REFERENCE

Jones, Kristen, Project Editor. *Advanced Placement and TEKS: A Lighthouse Initiative for Texas Science Classrooms*. Austin: Texas Education Agency, 2003. pg. 87

T E A C H E R P A G E S

Essay Writing Skills
Developing a Free Response

An appropriately written free response essay for an AP Biology or AP Environmental Science exam is markedly different from the type of essays that are typically written in English courses. For this reason, you can improve your free response writing skills through the practice of dissecting a prompt and preparing a brief outline of the components that should be included in a quality answer. The free response portion of an AP Exam is a timed exercise and as such requires efficient use of the allotted time. By making an outline prior to actually writing the essay, you will be much more likely to include the important parts of a good response. With practice you will become able to dissect even the most complex prompts into manageable pieces.

PURPOSE
In this activity you will practice the skill of dissecting a free response prompt and preparing a mechanical outline of an appropriate response.

MATERIALS
> copy of *Tips for Writing Free Response Essays*
> copy of Practice Essay Prompts #1-4

PROCEDURE
1. Read through Tips for Writing Free Response Essays as you teacher explains specific tips mentioned in the document.

2. Read the prompt for Practice Essay #1. Observe and record the sample mechanical outline shown by your teacher of an appropriate response for this prompt.

3. Read the prompt for Practice Essay #2, record the outline of an appropriate response as your teacher goes through this essay with the class.

4. Read the prompt for Practice Essay #3. For 2-3 minutes, work alone to use the tips discussed in Tips for Writing Free Response Essays to prepare a mechanical outline for this prompt. When the individual planning time expires, compare the outline you have designed with that of your partner's. You may be asked to share you outline with the class.

5. Read the prompt for Practice Essay #4 and prepare an outline of the items that should be included in a well designed free response. You will be working alone to prepare your outline. Refer to Tips for Writing Free Response Essays if needed.

6. Read through the sample student response to Practice Essay #4. This response is written incorrectly. In the space below the prompt identify 4 things that this student could have done to make this essay more appropriate.

Name _____

Period _____

Free Response:
Planning for Success

PRACTICE ESSAY # 1

Prompt:
Structure and function are closely related in living systems. For example, the structure of the carbon atom allows it to be the building block of a variety organic compounds.
A. Explain the characteristics of carbon that allow its atoms to provide molecular diversity.
B. Chose three of the following categories of organic compounds and describe each in terms of the compound's structure and function in living organisms.
  1. Carbohydrates
  2. Lipids
  3. Proteins
  4. Nucleic Acids

Main Topic: _____

A. _____
  * _____
  * _____
  * _____

B. _____

  _____

  _____

  _____

  _____

  _____

  _____

  _____

  _____

  _____

PRACTICE ESSAY # 2

Directions: Read the prompt and prepare an outline of the major components that should be included in a response using the lines and bullets as your guide.

> **Prompt:**
> The scientific method of problem solving is a useful tool for scientific investigation. Describe the components of the scientific method and cite two examples of how the scientific method has been used to make biological discoveries.

Main Topic: _____

_____

    _____

    _____

    _____

    _____

    _____

1. _____

2. _____

PRACTICE ESSAY #3

Directions: Read the prompt and prepare a mechanical outline of the major components that should be included in a response

---

Prompt:
The rate of photosynthetic activity may change in various environmental conditions.
a. Describe how each of the following environmental conditions could impact the rate of photosynthesis in a terrestrial plant.
  *low levels of light
  *high temperature
  *low availability of water
b. Select one of the conditions listed above and describe an adaptation that would allow a plant species to photosynthesize effectively in that specific environmental condition.

Main Topic: _____

  A_____

  _____

  _____

  _____

  B_____

  * _____

  * _____

Essay 4: Ecology question from 2003 exam

Directions: Read the prompt and prepare an outline of the major components that should be included in a response

Many populations exhibit the following growth curve:

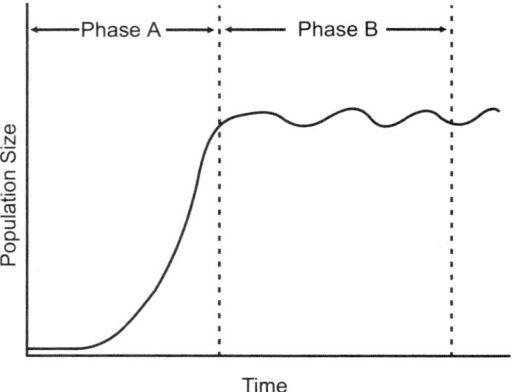

a. <u>Describe</u> what is occurring in the population during phase A.
b. Discuss THREE factors that might cause the fluctuations shown in phase B.
c. Organisms demonstrate exponential (r) or logistic (K) reproductive strategies. <u>Explain</u> these two strategies and <u>discuss</u> how they affect population size over time.

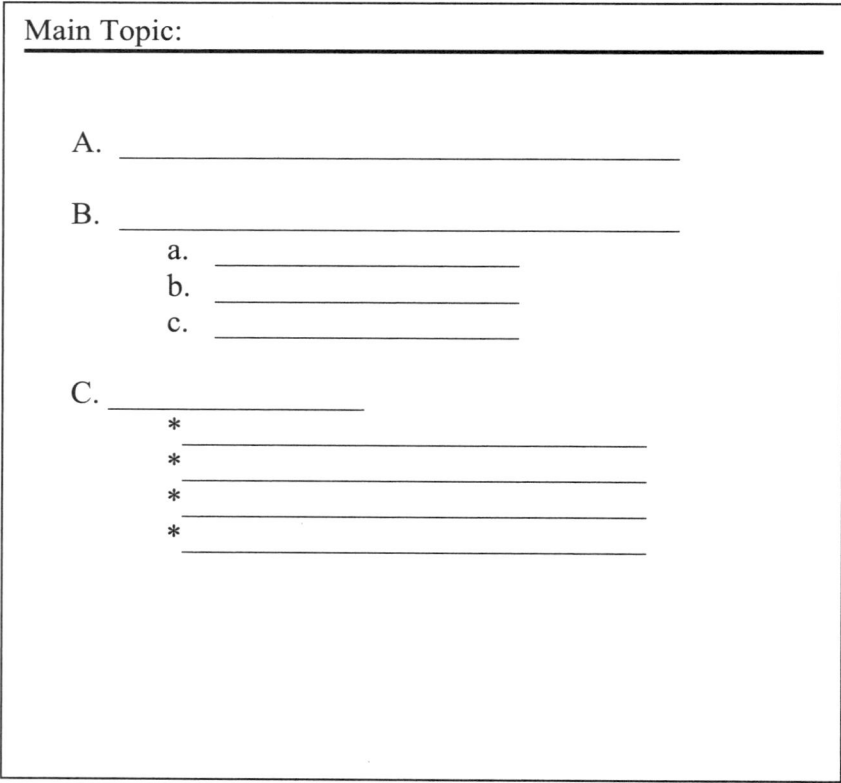

Read the following student's response to Essay #3. Identify 4 things that this student could do to make this essay better.

Prompt:
The rate of photosynthetic activity may change in various environmental conditions.
a. Describe how each of the following environmental conditions could impact the rate of photosynthesis in a terrestrial plant.
   *low levels of light
   *high temperature
   *low availability of water
b. Select one of the conditions listed above and describe an adaptation that would allow a plant species to photosynthesize effectively in that specific environmental condition.

Student's Answer
The rate of photosynthetic activity may change in various environmental conditions. Some plants have special adaptations that allow them to survive in extreme environments. Some plants have broad leaf surface to catch any available sun rays and can live in low light levels while some have thin leaves. Low light levels make it hard for plants to have enough solar energy to go through photosynthesis. Low amounts of water will harm plants. When temperatures get really high plants lose too much water when they open their stoma so some plants are adapted to opening their stoma at night in order to take in carbon dioxide to use in photosynthesis.

List 4 things this student could have done differently to improve the quality of this free response:

1.

2.

3.

4.

Content
Skills Chart

MIDDLE GRADES LIFE SCIENCE CONTENT SKILLS CHART

Scope	Activities	AP Connection

I. Introduction to the Science Classroom
  A. Classroom rules and laboratory safety
  B. Use of lab equipment
  C. Use of technology in the classroom
  D. Methods of scientific inquiry: scientific method
  E. Collecting Data/Data Analysis
  F. Biological diagrams
	Foundation Lesson I The Scientific Method: Exploring Experimental Design
	Can Mosquitoes Transmit HIV between Primates?: Simulating a Laboratory Experience Through a Role-playing Exercise
	Beetle Races!: Measuring Distance and Time and Calculating Rate in a Living, Moving Insect
	Picturing Life: Making a Biological Diagram
		AP Environmental Science
		I. Scientific Analysis A. Observing the natural world and developing hypotheses
		I. Scientific Analysis B. Collecting data 1. observation 2. controlled experiments
		I. Scientific Analysis C. Modeling
		I. Scientific Analysis D. Critical interpretation of data

II. Cells
  A. Eukaryotic-presence of nucleus
    1. Animal cells-structure and function
      a. lack of cell wall
      b. bound by cell membrane
    2. Plant cells-structure and function
      a. cell wall present
      b. presence of chloroplasts
  B. Prokaryotic-absence of nucleus
    1. Protists
    2. Monerans
      a. helpful species
      b. pathogenic species
	Basic introduction to microscope lab
	Plant and animal cell comparison microscope lab
	Protista observation microscope lab
	Monera observation microscope lab
		AP Biology
		I. Molecules and Cells B. Cells 1. Prokaryotic and eukaryotic cells
		I. Molecules and Cells B. Cells 2. Membranes
		I. Molecules and Cells B. Cells 3. Subcellular organization

III. Heredity
  A. DNA
    1. Structure
    2. Replication
    3. Protein translation
    4. Variability
  B. Genetics
    1. Simple monohybrid Punnett squares
    2. Complex monohybrid Punnett squares
      a. Incomplete dominance
      b. Codominance
    3. Dihybrid Punnett squares
    4. Chromosomes/ mitosis/meiosis
	Going Bananas for DNA: Isolating DNA, Deoxyribonucleic Acid, From Bananas
	Thumbs Up!: Measuring Phenotypic Variation in the Human Thumb
	Punnett Square Exercises: Solving Monohybrid Punnett Squares
	Gene Interactions: Solving Dihybrid and Complex Monohybrid Punnett Square Problems
	Onion root tip lab to view stages of mitosis
		AP Biology
		II. Heredity and Evolution B. Molecular Genetics 1. RNA and DNA structure and function
		II. Heredity and Evolution B. Molecular Genetics 2. Gene regulation
		II. Heredity and Evolution B. Molecular Genetics 3. Mutation
		AP Biology
		II. Heredity and Evolution A. Heredity 1. Meiosis and gametogenesis
		II. Heredity and Evolution A. Heredity 2. Eukaryotic chromosomes

Scope	Activities	AP Connection

C. Chromosomes
  1. Mitosis
  2. Meiosis
	Mitosis Square Dance
	  http://www.accessexcellence.org
	  /AE/newatg/Sullivan/
	"Reebops" meiosis modeling lab
		II. Heredity and Evolution A.
		  Heredity 3. Inheritance patterns

IV. Ecology
  A. Population Growth
    1. Arithmetic
    2. Exponential
    3. Logistic
    4. Carrying capacity
  B. Energy transfer
    1. Trophic pyramids
      a. Producers-Photosynthesis
      b. Consumers-Respiration
    2. Food webs
  C. Symbiotic relationships
    1. Mutualism
    2. Parasitism
    3. Commensalism
	Baby Dice Island: Modeling Exponential Growth
	Plotting Growth in Texas: Comparing Population Growth Rates in Texas: Dallas (Dallas County) and Houston (Harris County) and Your County for 150 Years
	Create a Species: Describing a New Imaginary Species
	Toothpick Birds: Modeling Predator Behavior in an Outdoor Lab
	Shed some Light on the Problem: Exploring the Relationship between Light and Photosynthesis
	Foundation Lesson VII
	Data Collection Devices
	Determining the Amount of Energy Found in Food
		AP Biology
		III. Organisms and Populations C. Ecology 1. Population dynamics
		III. Organisms and Populations C. Ecology 2. Communities and ecosystems
		AP Environmental Science
		I. Interdependence of Earth's Systems: Fundamental Principles and Concepts E. The Biosphere 1. organisms: adaptations to their environments
		I. Interdependence of Earth's Systems: Fundamental Principles and Concepts E. The Biosphere 2. populations and communities: exponential growth, carrying capacity
		I. Interdependence of Earth's Systems: Fundamental Principles and Concepts E. The Biosphere 3. ecosystems and change: biomass, energy transfer, succession
		II. Human Population Dynamics B. Carrying Capacity — Local, Regional, Global

V. Biomes
  A. Types of biomes
    1. Terrestrial
      a. Desert
      b. Tundra
      c. Tropical rain forest
      d. Temperate rain forest
      e. Coniferous forests
      f. Deciduous forest
      g. Savanna
    2. Aquatic
      a. Tidal pool
      b. Coral reef
      c. Abyss
      d. Estuary
      e. Streams and rivers
	Online Biome Exploration: Researching, Creating, and Presenting a Biome PowerPoint	AP Environmental Science
		I. Interdependence of Earth's Systems: Fundamental Principles and Concepts E. The Biosphere 1. organisms: adaptations to their environments
		I. Interdependence of Earth's Systems: Fundamental Principles and Concepts E. The Biosphere 3. ecosystems and change: biomass, energy transfer, succession

Scope	Activities	AP Connection

f. Ponds and lakes
B. Animal/Plant Adaptations to each biome
C. Abiotic effects on biome
  1. Temperature
  2. Rain
  3. Altitude
D. Human impact
		III. Renewable and Nonrenewable Resources: Distribution, Ownership, Use, Degradation A. Water 2. oceans: fisheries, industrial
		III. Renewable and Nonrenewable Resources: Distribution, Ownership, Use, Degradation D. Biological 1. natural areas
		IV. Environmental Quality A. Air/Water/Soil 2. effects of pollutants on: a. aquatic systems
		V. Global Changes and Their Consequences A. First-order Effects 3. biota: habitat destruction, loss of biodiversity, introduced exotics

VI. Evolution
A. Classification
  1. Use of shared characteristics
  2. Scientific nomenclature
B. Artificial and natural selection
C. Genetic variability
D. Mutation of DNA
	Caminalcules Discovery: Classifying Imaginary Animals by Analysis of Shared Characteristics
	Bean Bunny Evolution: Modeling Gene Frequency Change (Evolution) in a Population by Natural Selection
	Analysis of Peter and Rosemary Grant's study of Darwin's finches in the Galapagos
		AP Biology
		II. Heredity and Evolution C. Evolutionary Biology 2. Evidence for evolution
		II. Heredity and Evolution C. Evolutionary Biology 3. Mechanisms of evolution
		III. Organisms and Populations A. Diversity of Organisms 1. Evolutionary patterns
		III. Organisms and Populations A. Diversity of Organisms 2. Survey of the diversity of life
		III. Organisms and Populations A. Diversity of Organisms 3. Phylogenetic classification
		III. Organisms and Populations A. Diversity of Organisms 4. Evolutionary relationships

VII. Structure/Function of Plants and Animals
A. Animals
  1. Structure/function
  2. Animal adaptations and habitats
	Web Pages Made Easy: Creating Web Pages Using Netscape Composer
	Cyber Cephalopods on the Web: Researching and Creating Web Pages on Survival Strategies and Essential Functions of Class Cephalopoda
		AP Biology
		I. Molecules and Cells C. Cellular Energetics 2. Fermentation and cellular respiration
		I. Molecules and Cells C. Cellular Energetics 3. Photosynthesis

Scope	Activities	AP Connection

B. Plants
    1. Structure and function
      a. Leaf
      b. Seed
      c. stems
    2. Plant adaptations
	Planaria Chainsaw Massacre: Observing, Dissecting and Regenerating a Planarian
	Snails Versus Humans: Comparing Relative Pulling Strength Between Snails and Humans
	General survey of plants lab
		III. Organisms and Populations B. Structure and Function of Plants and Animals 2. Structural, physiological, and behavioral adaptations
		III. Organisms and Populations B. Structure and Function of Plants and Animals 3. Response to the environment

VII. The Human Body
  A. Levels of organization
    1. Cellular
    2. Tissue
    3. Organs
    4. System
  B. Body Systems
    1. Skeletal
    2. Digestive
    3. Muscular
    4. Endocrine
    5. Urinary
    6. Circulatory
    7. Integumentary
    8. Nervous
    9. Reproductive
	Antacid Analysis: Neutralizing Stomach Acids
	How Does Your Heart Rate?: Examining the Effects of Exercise on Heart Rate
	A Fishy (Tail) Tale: Observing the Circulatory System of a Goldfish with a Compound Light Microscope
	Dissection lab: earthworm
	Dissection lab: frog
	Dissection lab: squid
	Dissection lab: cow's eye
	Dissection lab: sheep heart
	Dissection lab: chicken leg
	Exploration of "My Gross and Cool Body" website:
	http://yucky.kids.discovery.com/flash/body/
		AP Biology
		III. Organisms and Populations B. Structure and Function of Plants and Animals 1. Reproduction, growth, and development
		III. Organisms and Populations B. Structure and Function of Plants and Animals 2. Structural, physiological, and behavioral adaptations

MIDDLE GRADES EARTH SCIENCE CONTENT SKILLS CHART

Scope	Activities	AP Connections

I. Weather and Climate
  A. Air movement
    1. air pressure
    2. winds
  B. Water cycle
    1. phases of matter
    2. relative humidity
  C. Seasons
    1. causes
  D. Climate zones
  E. Human Impact
	Are You Current On Convection?: Understanding Molecular Motion
	Evaporation and Condensation: Investigating Energy Transfer During a Phase Change
	The Watched Pot Never Boils….or Does it: Exploring Heat and Phase Changes
	Relative Humidity: Measuring the Amount of Water Vapor in the Air Versus the Total Amount the Air Can Hold
	Acid Rain Drops Keep Falling on My Head: Investigating Acid Rain
		AP Chemistry
		II. States of Matter A. Gases B. Liquids and solids
		III. Reactions A. Reaction types 1. Acid-base reactions
		
		AP Physics B
		II. Fluid Mechanics and Thermal Physics B. Temperature and Heat

II. Atmosphere
  A. Composition
  B. Structure
  C. Formation
  D. Human impact
	Emission Possible: Reporting for the President
	Greenhouse Effect: Investigating Global Warming
	Calculating Your Carbon Dioxide Footprint: Are You Meeting The Kyoto Protocol
		AP Environmental Science
		V. Global Changes and Their Consequences A. First-order Effects 1. atmosphere
		  B. Higher-order Interactions 1. atmosphere

III. Hydrology
  A. Unique properties of water
    1. polarity and uniqueness of as a solvent
    2. importance to life
  B. Distribution of water on earth
  C. Surface water resources
  D. Groundwater resources
  E. Human impact
	Aquifer Study: Investigating Porosity and Permeability
	That Sinking Feeling: Investigating Land Subsidence
	How Wet Is Our Planet?: Water, Water Everywhere, But Not a Drop to Drink
	Is Dilution the Solution to Pollution?: Washing Away Our Mess
	A Bug's Life: Testing Water Quality
		AP Environmental Science
		I. Interdependence of Earth's Systems: Fundamental Principles and Concepts B. The Cycling of Matter 1. water
		II. Renewable and Nonrenewable Resources: Distribution, Ownership, Use, Degradation A. Water 1. fresh

IV. Oceanography
  A. Waves
    1. Causes
    2. Diagraming and equations
  B. Currents
    1. surface currents
    2. density currents
      a. temperature
      b. salinity
      c. sediment
	Wave equations/diagramming activity
	Staying Out of Hot Water: Modeling A Thermocline
	Mapping surface versus deep currents
	Underwater Avalanche: Turbidity Currents: Investigating Density Currents
	Density lab: Salt vs. Fresh
		AP Environmental Science
		I. Interdependence of Earth's Systems: Fundamental Principles and Concepts A. Flow of Energy
		III. Renewable and Nonrenewable Resources: Distribution, Ownership, Use, Degradation A. Water 2. Oceans

Scope	Activities	AP Connections

C. Tides
    1. phases of the moon
D. Ocean floor topography
E. Human impact on ocean
    resources
	Lunar phases/tides worksheet
	Sonar Seas: Mapping the Ocean
	    Floor
	Slope of a Beach
	Ocean mining for minerals: project
		AP Physics
		IV. Waves and Optics A. Wave
		    motion (including sound) 1.
		    Properties of traveling waves
		    C. Geometric Optics 1.
		    Reflection and refraction

V. Geologic Cycles and Processes
A. Minerals
    1. formation
    2. identification
B. Rocks
    1. rock cycle
    2. weathering
C. Geologic Time
D. Plate Tectonics
    1. earthquakes and volcanoes
	Mineral identification lab
	Rock-n-Roll: Investigating the
	    Weathering of Rocks
	Rock identification lab
	Dynamic Planet: Discovering
	    Plate Tectonics
	Geologic time line lab
	Earthquake: Locating an Epicenter
	    lab
		AP Environmental Science
		I. Interdependence of Earth's
		    Systems: Fundamental
		    Principles and Concepts C. The
		    Solid Earth 2. Earth dynamics:
		    plate tectonics, volcanism, the
		    rock cycle

VI. Space Science
A. Earth-Moon System
B. Solar System
C. Stars
    1. characteristics of stars
    2. life cycle
D. the Universe
	Mapping the moons craters lab
	Reasons for the Seasons:
	    Exploring What Causes the
	    Seasons
	Not So Lost in Space: Building a
	    Model of the Solar System
	H-R Diagram: Interpreting Stars lab
	Black Holes and Beyond:
	    Modeling a Black Hole
		AP Physics
		I. Newtonian Mechanics F.
		    Oscillations and gravitation

National Standards

National Standards

The National Standard codes used in the lessons throughout the Laying the Foundation series are based on the following coding system. We encourage you to read the National Standards for your content area found at http://www.nap.edu/readingroom/books/nses/html.

Unifying Concepts and Processes

UCP.1 Systems, order, and organization
UCP.2 Evidence, models, and explanation
UCP.3 Change, consistency, and measurement
UCP.4 Evolution and equilibrium
UCP.5 Form and function

Science as Inquiry

A.1 Abilities necessary to do scientific inquiry
A.2 Understandings about scientific inquiry

Physical Science

B.1 Structure of atoms
B.2 Structure and properties of matter
B.3 Chemical reactions
B.4 Motions and forces
B.5 Conservation of energy and increase in disorder
B.6 Interactions of energy and matter

Life Science

C.1 The cell
C.2 Molecular basis of heredity
C.3 Biological evolution
C.4 Interdependence of organisms
C.5 Matter, energy, and organization in living systems
C.6 Behavior of organisms

Earth and Space Science

D.1 Energy in the earth system
D.2 Geochemical cycles
D.3 Origin and evolution of the earth system
D.4 Origin and evolution of the universe

Science and Technology

E.1 Abilities of technological design
E.2 Understanding about science and technology

Science in Personal and Social Perspectives

F.1 Personal and community health
F.2 Population growth
F.3 Natural resources
F.4 Environmental quality
F.5 Natural and human-induced hazards
F.6 Science and technology in local, national, and global challenges

History and Nature of Science

G.1 Science as a human endeavor
G.2 Nature of scientific knowledge
G.3 Historical perspective

Syllabi

Developing a Course Syllabus

There are many reasons to develop a course syllabus. The most important reason is to communicate to your students that you have carefully planned your course. It is often thought of as a contract between an instructor and a student. It conveys the message that each of you is committed to meeting the objectives of the course. Research has shown that students who are told what they are supposed to learn and how they are to be evaluated perform better than those who are not so instructed. The syllabus should specify the duties and responsibilities of the student while communicating a commitment from the instructor to respect a timeline. Once the syllabus has been published and distributed, it is important to do your very best not to change it. Students will feel that the rules have been changed in the middle of the game.

Other reasons to develop a course syllabus include:
- Clearly communicating course expectations to administrators and parents.
- Creating an organizational tool for yourself if you teach more than one kind of course.
- Organizing your laboratory time and resources for maximum efficiency.
- Enhancing cooperation and communication with your colleagues so that multiple teachers teaching the same course deliver the same quality of instruction to each student.
- Reducing the amount of time spent deciding what to do each class period; the plan has already been formulated making execution of the plan easier.
- Documenting what has been taught.
- Empowering students to take more responsibility for their education.
- Helping students better manage their time and set priorities while improving their study habits.

Constructing a syllabus is actually quite easy. A suggested procedure follows:
1. Realize that a school year usually consists of 36 weeks and that you are fortunate if you are able to instruct for 32 of those weeks. Factors such as semester exam weeks, TAKS testing, benchmark testing for TAKS, pep-rallies, assemblies and other unscheduled interruptions will consume up to 4 weeks of your instructional time.

2. Examine the course scope found in the front of this guide. Without regard for the fact that you only have 32 weeks, estimate how much time it would take for you to cover each major topic, including laboratory time and testing time.

3. Determine the mid-point of the course, which is the point you wish to reach by the end of the first semester. Tally the ideal number of weeks required for each semester.

4. Obtain an accurate district calendar and count the number of weeks you have for each semester. (The first semester is usually shorter if your district ends the semester before January.)

5. Compare. Do not be alarmed if your tally exceeds 32 weeks. Decide if you can shift the mid-point a bit. If not, it is time to adjust the ideal number of weeks so that they fit into your actual time frame. This may mean radically changing the structure of your course. You may have to change the number of activities that you do in a given unit, change the type of notes students are

expected to take, or other instructional methods to free up time to cover the entire scope of the content.

6. Now that you have determined the number of weeks, break each unit down by day and choose logical starting and stopping points for each class period. Allow a two day cushion at the end of each grading period for unforeseen obstacles such as personal illness, re-teaching, fire drills, etc…

7. If you find it daunting to publish the whole semester on a syllabus, consider publishing it in 3 week intervals. That way the plan stays more current as those unforeseen interruptions occur.

8. Now that you have a plan for student success, stick to the plan. Realize that you will have to teach bell-to-bell in order to successfully complete your plan.

The syllabi that follow were contributed by some of the best and most experienced Pre-AP teachers in the region, representing classes at both public and private schools. Each of these teachers has played an integral part in developing a successful Pre-AP and AP program at their school. Each of them would tell you there is no one right way to teach a Pre-AP science course. It is our hope that you will survey the collection of syllabi presented and formulate your own syllabus to share with your students and colleagues. Understand that your first attempt will require monitoring and adjusting. Keep careful notes about your timeline expectations and what worked and did not work on your initial syllabus. Do not be alarmed if it takes a couple of tries to create a syllabus that works consistently in your classroom year after year.

TEACHER

Lawanna Jenkins (6th)
Hodges Bend Middle School
Houston, Texas
Fort Bend ISD

SCHOOL PROFILE

School Location and Environment: Hodges Bend Middle School is located in southwest Houston in Fort Bend County. HBMS is a suburban/inner city environment with 1805 students enrolled, grades 6-8. The school is composed of neighborhood students and has a demographic representation from over 40 nations.

Grades the school contains: 6th-8th

Type of school: Public

Percentage of minorities: 85%

Do you teach the Pre-AP course at your school? No

If not, is it offered at your school? No

Number of students taking the PAP science course in your discipline: NA

Course sequence of the PAP courses in your school: NA

SYLLABUS

Safety (1 week, then ongoing for the rest of the year)

Metric Measurement (2 weeks, then ongoing for the rest of the year)

Transforming energy (3 weeks)
    Energy transformation
    Transforming energy into devices
    Magnetism
    Electromagnetism

Physical and Chemical Properties (6 weeks)
    Atoms and Elements
    Physical and Chemical Changes
    Classifying Substances
    States of Matter

Structure and Function of Earth Systems (6 weeks)
Rock Cycle
Plate Tectonics
Components of atmosphere
Water Cycle
Weather

Force and Motion (3 weeks)
Newton's Laws of Motion
Changes in Position
Graphing Motion

Solar System (3 weeks)
Characteristics of objects in space
Equipment and methods of space travel
Whole vs. part of a system

Structure and Function of Systems (3 weeks)
Differentiate between structure and function
Determine organism comprised of cells
Compare animal and plant cells

Traits of Species and Change (3 weeks)
Genetics
Changes in traits
Cells with genetic material
Role of genes in inheritance

Response to Stimuli (3 weeks)
Internal Stimuli
External Stimuli
Components of Ecosystem
Adaptation of Organisms
Survival/Extinction

Energy and the Environment (Matter and Energy in Ecosystems) (3 weeks)
Define matter and energy
Biomass
Food Chains
Energy Pyramid

TEACHER

Mr. Jesus "Chuy" Garcia (8th Earth Science)
Hyde Park Baptist Jr. / Sr. High School
3901 Speedway
Austin, Texas 78751

SCHOOL PROFILE

School Location and Environment: Hyde Park Jr. High School is a college preparatory school located in central Austin within close proximity to three universities, in Travis county. With an enrollment of 320 in grades 7-12, the school is composed of students from throughout central Texas and has a demographic representation from over 25 nations.

Grades the school contains: 7-12

Type of School: Private

Percentage of Minorities: 12%

Do You Teach the Pre-AP course at your school? Yes

If No, is it offered at your school?

Number of students taking the Pre-AP science course in your discipline: 11

Course sequence of the Pre-AP courses in your school: Pre-AP Earth Science, Pre-AP Biology, Pre-AP Chemistry, Pre-AP Physics, AP Biology, AP Chemistry or AP Physics.

COURSE SYLLABUS

1st Nine Weeks

    Metric Measurement
    Scientific Method/ Experimental Design
    Topographic Maps / Remote Sensing
    General Chemistry
    Rocks & Minerals
    Weathering & Water Movement

2nd Nine Weeks

    Fossils
    Geologic Time Scale
    Plate Tectonics
    Volcanoes
    Earthquakes
    Mountain Building
    Physical Oceanography

3rd Nine Weeks

    Telescopes
    Stars & Galaxies
    The Sun & Our Solar System
    Seasons
    Eclipses
    Comets, Asteroids & Meteoroids

4th Nine Weeks

    Structure of the Atmosphere
    The Water Cycle & Its Components
    Air Pressure & Wind
    Air Masses & Fronts
    Storms
    Weather Forecasting

TEACHERS

Jason Hook (7th), Lynn Kirby (8th), Tom Campbell (8th IPC)
Kealing Junior High
Austin, Texas

SCHOOL PROFILE

School Location and Environment: Kealing Junior High School is located in east central Austin, Texas. Kealing is an inner city environment with 963 students enrolled. The school is composed of students who are enrolled in the accelerated Magnet and Pre-AP programs, and also serves the neighborhood students.

Grades the school contains: 7th - 8th

Type of school: Public

Percentage of minorities: 71%

Do you teach the Pre-AP course at your school? Yes

If not, is it offered at your school? NA

Number of students taking the PAP science course in your discipline: 795

Course sequence of the PAP courses in your school:
7th Grade: Magnet Seventh Grade Science
8th Grade: Magnet Eighth Grade Science; some take 8th IPC concurrently for high school credit

MAGNET SEVENTH GRADE SCIENCE SYLLABUS: 2003-2004

I.   **The Nature of Science (~2 weeks)**
  A. Science and Discovery — Studying the Living World
    1. Observations and inferences
    2. Independent and dependent variables
    3. Validity of experiment design
    4. Methods of data analysis
  B. Ordering Life in the Biosphere
    1. Biological classification
    2. Kingdoms of organisms
    3. Classification schemes

II. **The World of Life: The Biosphere (~6 weeks)**
  A. Biomes of the World
    1. Climate and ecosystems
    2. Succession and evolution in ecosystems
    3. Adaptations and survival strategies
  B. Aquatic Ecosystems
    1. Standing fresh waters
    2. Flowing waters
    3. Ocean ecosystems
  C. Human Influences on Ecosystems
    1. Human culture and technology
      • All environmental issues are related to population growth.
    2. Ecosystem stability and human influences
      • Humans lower the stability of ecosystems, threaten diversity
      • Humans can help conserve ecosystems
    3. Change and the future
      • Sustainable agriculture
      • Decisions about land use
  D. The Web of Life
    1. Interactions among living things
      • Predator/prey relationships
      • Symbiosis: communalism, parasitism, mutualism
    2. Continuity through evolution
      • Diversity, variation and natural selection
  E. Communities and Ecosystems
    1. Niches
    2. Competition
    3. Ecosystem structure
  F. Populations
    1. Individuals, populations and environment
    2. Dispersal and barriers to dispersal
    3. Carrying capacity and limiting factors
    4. Uneven distribution of food and other resources
    5. Managing Earth requires global cooperation

G. Matter and Energy in the Web of Life
1. Matter and energy
2. Diffusion and osmosis

III. Diversity and Adaptation in the Biosphere (~20 weeks)
A. Cells
1. Cell structure and functions
2. Cell reproduction
3. Development and cell differentiation
B. Prokaryotes and Viruses
1. Bacteria
2. Vectors of disease
C. Eukaryotes: Protists and Fungi
1. Autotrophic and heterotrophic protists
2. Fungi
D. Eukaryotes: Plants
1. Evolution of land plants
2. Bryophytes and seedless vascular plants
3. Seed plants, form and function
4. Photosynthesis
E. Eukaryotes: Animals
1. Diversity and adaptation in animals
2. Life functions in animals including major body systems
3. Behavior, selection and survival
   - Hierarchical animal societies
   - Sexual selection

IV. The Human Animal (~8 weeks)
A. Food and Energy
1. Ingestion and digestion and excretion
2. Cellular respiration
3. Nutrition
B. Maintenance of Internal Environment
1. Circulation and Respiration
2. Immunity
3. Temperature regulation
C. Coordination
1. Human movement
2. Nervous system
3. Endocrine system
D. Heredity
1. Genes and DNA
2. Patterns of inheritance
3. Genetics and technology
E. Growth and Development
1. Human reproductive system
2. Responsible decision-making

MAGNET EIGHTH GRADE SCIENCE SYLLABUS: 2003-2004

It's Up In The Air (5 weeks)
- Weather
    - Heat Transfer
    - Convection
    - Density
    - Conduction
    - Evaporation
    - Condensation
    - Atmospheric Chemistry

It's Up In The Air (4 weeks)
- Climate
    - Global winds
    - Density
    - Effects of landmasses
    - Atmospheric Chemistry
    - Changes over time

Sweet, Fresh Water (4 weeks)
- Hydrology
    - Aquifers
    - Porosity
    - Permeability
    - Pollution
    - Usage
    - Chemistry of water
    - Freezing/boiling point
    - Density

The Deep Blue Sea (4 weeks)
- Oceanography Density
    - Convection
    - Thermo cline
    - Currents
    - Tides
    - Waves
    - Ocean floor features

A Molecular Moment (2 weeks)
- Chemistry
    - Periodic table
    - Elements, molecules and mixtures
    - Phases of matter
    - Intrinsic properties of matter
    - Metals & nonmetals
    - Ionic Bonding
    - Covalent Bonding
    - Density

Rock Steady (3 weeks)
> Geology
>> Minerals
>> Rocks
>> Density
>> Rock Cycle

The Dynamic Earth (4 weeks)
> Paleogeography and stratigraphy
>> Earthquakes
>> Volcanoes
>> Force and Motion
>> Geologic time
>> Plate Tectonics
>> Continental Drift
>> Changes over time

A Star Is Born (4 weeks)
> Astronomy
>> Our solar system
>> The universe
>> Star formation
>> Planet formation
>> Density
>> Time
>> Changes over time

Making Healthy Choices (2 weeks)
>> Drugs
>> Sex
>> Rock and roll

MAGNET IPC SCIENCE SYLLABUS: 2003-2004

Week 1 — 13-Aug
> Scientific Method, Measurements, Use of Computer

Week 2 — 20-Aug
> Matter; Volume and Density

Week 3 — 27-Aug
> Phases of Matter

Week 4 — 3-Sep
> Atomic models of matter

Week 5 — 10-Sep
> Periodic Table/Elements/Mneumonics

Week 6 — 17-Sep
> Atoms and bonding
> end 1st 6 weeks

Week 7 — 24-Sep
> Chemical reactions

Week 8 — 1-Oct
    Chemical reactions
Week 9 — 8-Oct
    Families of Chemicals
Week 10 — 15-Oct
    Families of Chemicals
Week 11 — 22-Oct
    Motion
Week 12 — 29-Oct
    Nature of forces
    end 2nd 6 weeks

Week 13 — 5-Nov
    Newton's Laws
Week 14 — 12-Nov
    Newton's Laws, Forces in fluids
Week 15 — 19-Nov
    Trial of the Roadrunner
Week 16 — 26-Nov
    Work
Week 17 — 3-Dec
    Energy: Forms and Changes
Week 18 — 10-Dec
    Heat and measurements
Week 19 — 17-Dec
    Fun Activity - Holiday theme-vacation
    end of 3rd 6 weeks

Week 20 — 24-Dec
    Vacation
Week 21 — 31-Dec
    Petrochemical technology
Week 22 — 7-Jan
    Radioactivity, nuclear reactions, stellar reactions
Week 23 — 14-Jan
    Stephen Hawkings book
Week 24 — 21-Jan
    Project Work
Week 25 — 28-Jan
    Work
Week 26 — 4-Feb
    Energy: Forms and Changes
Week 27 — 11-Feb
    Heat and measurements
    end of 4th 6 weeks

Week 28 — 18-Feb
     Radioactivity, nuclear reactions, stellar reactions
Week 29 — 25-Feb
     Stephen Hawkings book
Week 30 — 4-Mar
     Heat and measurements
Week 31 — 11-Mar
     Spring break!
Week 32 — 18-Mar
     Intro to Electricity - charges and currents
Week 33 — 25-Mar
     Intro to Electricity - charges and currents
Week 34 — 1-Apr
     Magnetism
     end of 5th 6 weeks

Week 35 — 8-Apr
     Electromagnetism
Week 36 — 15-Apr
     Electronics and Computers (diodes/transisters/photodiodes)
Week 37 — 22-Apr
     Electronics and Computers (diodes/transisters/photodiodes)
Week 38 — 29-Apr
     Sound and Light, characteristics of waves
Week 39 — 6-May
     Sound and Light, characteristics of waves
Week 40 — 13-May
     Light and its uses
Week 41 — 20-May
     End of year projects: Demonstrations
     end of 6th 6 weeks

TEACHER

Camie Fillpot (7th)
O. Henry Middle School
Austin, Texas

SCHOOL PROFILE

School Location and Environment: O'Henry Middle School is located in Austin, Texas, in an affluent neighborhood in west Austin.   849 students attend the O'Henry Middle School.

Grades the school contains: 6th-8th

Type of school: Public

Percentage of minorities: 39%

Do you teach the Pre-AP course at your school? Yes

If not, is it offered at your school? NA

Number of students taking the PAP science course in your discipline: 509

Course sequence of the PAP courses in your school:
6th Grade: Honors Science
7th Grade: Honors Science
8th Grade: Honors Science

O. Henry Middle School
Camie Fillpot
Room 504, 2610 W. 10th Street •
Austin, TX 78703
Phone 512.841.4520
FAX 512.477.7428
Web page: home.austin.rr.com/fillpot
E-mail: mrsfillpot@hotmail.com

In seventh grade, you will be doing field and lab experiments using scientific methods, critical thinking, and problem-solving to explain the world around you. You will understand a whole in terms of its parts and how these parts relate to each other and to the whole. In this class, you will use computers and technology to collect and analyze data. Technology will also be used to communicate experimental results.

Textbook: *Texas Science*, Glencoe/McGraw-Hill, New York, 2002.

Some things you will be learning…
forces and motion
phases of the Moon
reasons for seasons
biomes of the world
human body systems
chemistry

Class Supplies:
  pencil
  blue or black pen
  1-1" three-ring binder
  1-pkg. tabbed dividers
  notebook paper
  composition book
  markers or colored pencils

Optional supplies:
  scissors
  glue or tape
  highlighter
  pencil sharpener

Grading Policy:
  Daily/homework/warm-ups 30%
  Quizzes 20%
  Tests 50%

Late assignments will lose 10 points for each day late. Students will receive a grade of **zero** for assignments that are **4 or more** days late. No credit will be given on late assignments graded in class.

TENTATIVE SYLLABUS 2003-2004

1st Six Weeks
Methods of Science:
  Chapter 1 — Metrics, Safety, Scientific Method, and Scientific Method
    TEKS covered: (7.1) The student conducts field and laboratory investigations using safe, environmentally appropriate, and ethical practices.
      (7.2) The student uses scientific inquiry methods during field and laboratory investigations.
      (7.3) The student uses critical thinking and scientific problem solving.
  Chapters 3 and 4, sections 1 and 2 — Properties of Matter, Periodic Table
    TEKS covered: (7.7) Science concepts. The student knows that substances have physical and chemical properties.

2nd Six Weeks

    Chapter 5, sections 2 and 3 — Simple Machines, Force and Motion

        TEKS covered: (7.4) Scientific processes. The student knows how to use tools and methods to conduct science inquiry.

                (7.6) Science concepts. The student knows that there is a relationship between force and motion

                (7.8) The student knows that complex interactions occur between matter and energy

    Chapter 2, sections 1 and 2 — Interactions of the Sun, Earth, and Moon

        TEKS covered: (7.8) The student knows that complex interactions occur between matter and energy.

                (7.13) The student knows components of our solar system.

                (7.14) The student knows that natural events and human activity can alter Earth Systems.

3rd Six Weeks

    Chapters 24 and 25 — Catastrophic Events, Weathering and Erosion

        TEKS covered: (7.5) The student knows that an equilibrium of a system may change.

    Chapters 20 and 22 — Interactions of Life, Ecology, and Resource Conservation

        TEKS covered: (7.5) The student knows that equilibrium of a system may change.

                (7.8) The student knows that the responses of organisms are caused by internal or external stimuli.

                (7.12) The student knows that there is a relationship between organisms and the environment.

                (7.13) The student knows that natural events and human activity can alter Earth systems.

4th Six Weeks

    Chapter 12 — Cells, Mitosis and Meiosis

        TEKS covered: (7.9) The student knows the relationship between structure and function in living systems.

                (7.11) The student knows that the responses of organisms are caused by internal or external stimuli.

    Chapters 14 and 17 — Muscular, Skeletal, Integumentary Systems, Circulation, and Blood

        TEKS covered: (7.9) The student knows the relationship between structure and function in living systems.

                (7.11)The student knows that the responses of organisms are caused by internal or external stimuli.

5th Six Weeks

    Chapters 12, 14-18 — Circulation, Blood, Respiration, Excretion, Digestion, Nutrition, Brain and Senses,

        TEKS covered: (7.9) The student knows the relationship between structure and function in living systems.

                (7.11) The student knows that the responses of organisms are caused by internal or external stimuli.

6th Six Weeks
   Chapters 18 and 19 — Brain and Senses, Frog Dissection, Genetics and Human Reproduction
      TEKS covered: (7.3) The student uses critical thinking and scientific problem solving to make informed decisions.
      (7.10) The student knows that species can change through generations and that the instructions for traits are contained in the genetic material of the organisms.

Lessons

Picturing Life
Making a Biological Diagram

OBJECTIVE
Students will accurately and neatly draw a diagram of a biological specimen, according to basic rules.

LEVEL
Middle Grades: Life Science

NATIONAL STANDARDS
UCP.2, UPC.5, A.1, A.2

TEKS
6.1(B), 6.2(B), 6.2(E), 6.3(C)
7.1(B), 7.2(B), 7.2(E), 7.3(C)
8.1(B), 8.2(B), 8.2(E), 8.3(C)
IPC 1(A), 2(A), 2(B), 2(C), 2(D)

CONNECTIONS TO AP
AP Biology:

  III. Organisms and Populations: A. Diversity of Organisms 2. Survey of the diversity of life B. Structure and Function of Plants and Animals 2. Structural, physiological, and behavioral adaptations

TIME FRAME
50 minutes

MATERIALS
(For a class of 28 working in pairs)

28 pencils	14 biological specimens
28 sets of colored pencils	28 rulers

TEACHER NOTES
Picturing Life provides students with a standardized set of rules needed to make a consistent and uniform diagram of a biological specimen. *Picturing Life* can be a stand-alone activity or incorporated into a number of different labs and activities where the observation and creation of a biological specimen is required. Within this book, in the *Create a Species: Describing a New Imaginary Species* activity, students are specifically required to make a biological diagram of an organism as one part of the activity.

This activity is a great first day activity to "do some science" as opposed to a typical teacher directed activity. The straightforward and simple rules make it easy for this to be a very independent activity for students of all academic abilities. The activity can be as easy as handing each student a leaf, showing the students the parts of the leaf, and asking students to read the instructions and make a diagram.

One idea for a really motivating first day of science class is to give students a live *Tenebrio molitor* beetle (easily obtained and raised, see *Beetle Races!: Measuring Distance and Time and Calculating Rate in a Living, Moving Insect* in this book for more information) in a Petri dish, along with a magnifying lens. Students (and adults) are equally fascinated and repulsed by insects and this factor really adds to their inquiry level. The parts of the specimen that are labeled are not as important as the process of labeling and following the instructions, but, for *Tenebrio molitor*, the parts could include: head, cephalothorax, abdomen, antenna, leg, and elytra. An example of a student created *Tenebrio molitor* is included below.

SAMPLE STUDENT DIAGRAM OF TENEBRIO MOLITOR

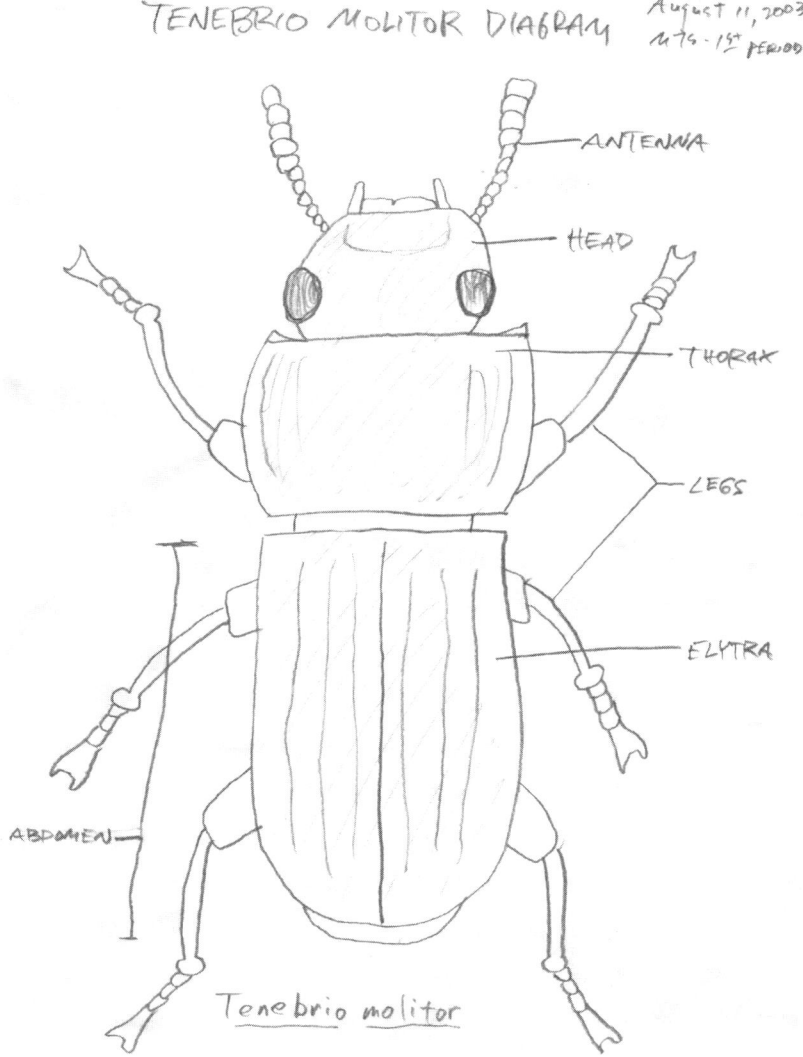

Picturing Life
Making a Biological Diagram

The ability to draw accurately and neatly is a useful skill, especially in science. Most sketch artists follow some basic rules to make their drawings attractive and, most importantly, easy to interpret. Below are some rules that will improve your drawing skills and make your sketches easier to understand. These rules should be followed whenever you make a biological diagram in your science class.

PURPOSE
In this activity you will accurately and neatly sketch a biological specimen according to basic rules.

MATERIALS

pencil   biological specimen
colored pencils   ruler

PROCEDURE

1. Read the rules for sketching a biological diagram below. As you read the rules, refer to the sample biological diagram that has been provided for you (Figure 1).

2. Obtain a biological specimen from your teacher.

3. Follow your teacher's instructions as to which parts to label on your biological diagram.

4. In the space provided on your student answer page draw a biological diagram with the required parts labeled accurately. Work neatly and carefully, following all the rules listed below. Your grade is determined by the neatness of your diagram and your ability to follow the instructions.

RULES FOR SKETCHING A BIOLOGICAL DIAGRAM

1. Whenever possible, use white unlined paper (provided on your student answer page for this activity).

2. Always draw in pencil. Begin drawing lightly, in case you have to erase. Try to erase as little as possible. Use colored pencils to make complex drawings easier to read.

3. Always print.

4. Leave at least a 2.5 cm (1 inch) margin on all four sides of the paper. [Only the heading — with your name, class and period, and date — should be written within the margin.]

5. Always title your drawing using all capital letters. Center the title above your sketch.

6. Center your drawing on the paper. [Don't forget to leave room for your labels.]

7. Always use a ruler to label the parts of your sketch.

8. Do not underline your labels.

9. Take care to keep the words horizontal; label lines, however, do not need to be kept horizontal.

10. Never allow two label lines to cross.

11. Place one end of the label line directly on the object you are labeling and the other end at the start of the word. Do not use an arrow.

12. Use the singular form of a word when pointing to a single object or part. If you need to point to two of the same parts (for clarity), use the plural form of the word.

13. Print your first and last name, class and period, in the upper right-hand corner of your paper in the space provided on your student answer page.

14. Capitalize the first letter of the genus (or first word) of a scientific name when using the scientific name of an organism in places other than the title. The species (or second word) of a scientific name should never be capitalized. For example, Canis familiaris is the scientific name for a dog. In a title, it would be CANIS FAMILIARIS. You should also remember that, unless you are typing and can italicize it, the scientific name should always be underlined.

15. Use subtitles ONLY when more clarification is needed. When using a subtitle, use upper and lower case letters as appropriate. The subtitle may be a sentence, information about the orientation of the sketch, the scientific name, etc.

SAMPLE BIOLOGICAL DIAGRAM

A TYPICAL PROTOZOAN

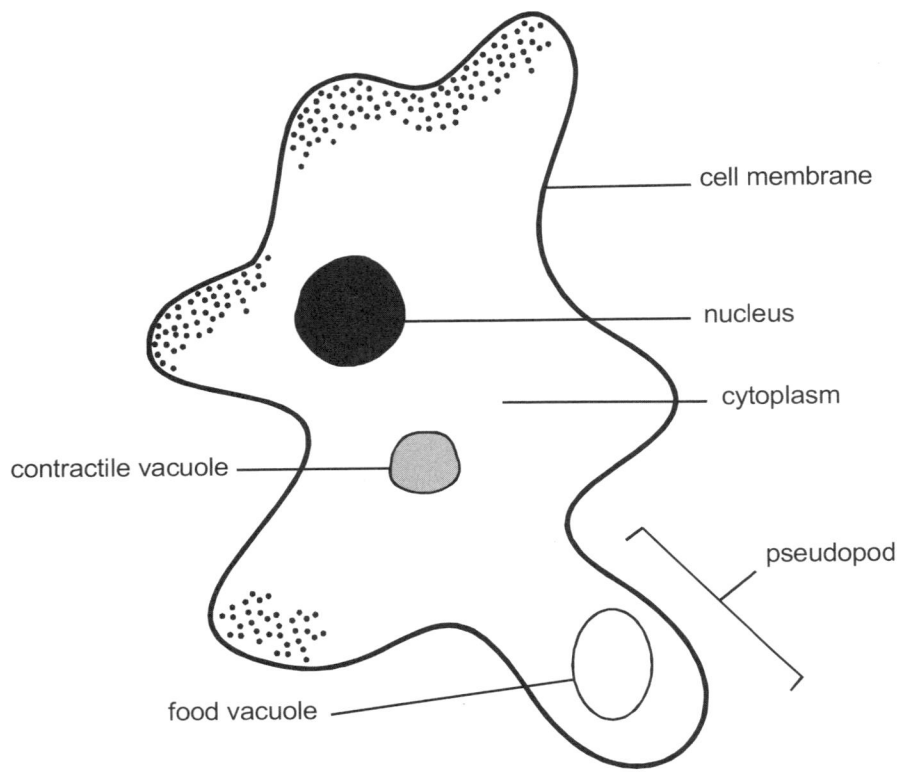

- cell membrane
- nucleus
- cytoplasm
- pseudopod
- contractile vacuole
- food vacuole

Some internal structures of <u>Amoebus proteus</u>

Name _____

Period _____

Picturing Life
Making a Biological Diagram

Using the rules, the sample diagram, and the instructions from your teacher, sketch and label a biological diagram in the space provided below.

OBSERVATIONS

Beetle Races!
Measuring Distance and Time and Calculating Rate in a Living, Moving Insect Objective

Students will measure distance and time as they race mealworm beetles, *Tenebrio molitor*, for 45 seconds. Students will graph their results, generating three lines of best fit and then calculating the actual beetle movement rates at different points in the trial.

LEVEL
Middle grades: Life Science

NATIONAL STANDARDS
UCP.3, A.1, A.2, C.6

TEKS
6.1(A), 6.2(B), 6.2(E), 6.4(A), 6.4(B)
7.1(A), 7.2(B), 7.2(E), 7.4(A), 7.4(B)
8.1(A), 8.2(B), 8.2(E), 8.4(A), 8.4(B)
IPC 1(A), 2(A), 2(B), 2(C), 2(D), 4(A)

CONNECTIONS TO AP
AP Biology:
    B. Structure and Function of Plants and Animals 2. Structural, physiological, and behavioral adaptations 3. Response to the environment
AP Physics:
    I. Newtonian Mechanics A. Kinematics 1. Motion in one dimension

TIME FRAME
90 minutes

MATERIALS
(For a class of 28 working in pairs)

14 metric rulers	14 stop watches
14 segments string ~30 cm in length	*Tenebrio molitor* beetle supply (darkling beetles)
14 overhead pens	28 pieces of metric graph paper (in appendix)
14 Petri dishes	28 cardboard paper strips

TEACHER NOTES
This activity serves as a fun introduction to the mathematical concept of rate. Students will measure the distance and movement of beetles at given time intervals. Students let beetles race across their desks, following and tracing the circuitous path of the beetle with an overhead pen, while timing the run with a stopwatch. An X drawn on the student's desk serves as the starting point. The beetles move in a roundabout path, so the students must measure the distance traveled with a string. This is accomplished

by overlaying the pen path with the string and subsequently measuring the distance traveled by measuring the length of the string with a meter stick or metric ruler. Students then graph their results on a distance versus time graph to calculate various rates. Emphasize that the slope of this graph IS the rate.

The students begin by selecting the fastest looking specimens from the container of bran. They should try to select the most active beetles. A small prize awarded to the pair of students having the fastest beetle will encourage students to be enthusiastic about getting the beetle to move quickly. Beetles that do not move at all on the table should be rejected and placed back into the container. Once a beetle is placed on the X, it should begin moving right away. If the beetle does not move immediately, the students can gently nudge the beetle to initiate its movement, but should not push it any way that will affect the rate.

Typically, the beetles move at approximately 1-3 cm/sec, but they could certainly move faster or slower. ANY lab that involves the use of live animals will have variable results. Animals simply will not do what we want them to, when we want them to, particularly insects. One way to make the lab more successful is to have a very large supply of *T. molitor* beetles on hand so students can have many specimens to choose from.

This activity can also serve as an introduction to the observation of living organisms in a laboratory setting. Students love working with the live beetles since they find them fascinating and repulsive all at the same time. You should initiate a discussion about ethical issues and animal cruelty before introducing students to the beetles, or any live creature in the lab.

Many teachers have a supply of these beetles in their rooms and you can easily start a culture of your own. This website explains the care and maintenance of your culture http://www.rygepetersen.dk/tenebrio_molitor.htm. These beetles are quite easy to maintain since the larval stage is the common mealworm. If you cannot find a teacher to help you supply a starting population, most of the scientific supply companies can send you a starting population of adults/pupae/larvae for about $20. If you have a few weeks before you want to start the investigation, you can purchase mealworms, the larval stage, at feed or pet stores.

Another interesting aspect of this lab is the humorous, and non-threatening, introduction of the idea of natural (or in this case, artificial) selection, and evolution. If you have a lizard or snake in the classroom, you can tell the students that the slow, rejected beetles are eaten, while the faster ones that got put back into the container of bran to continue breeding thus continuing the genetic line. If this experiment was repeated hundreds of times, then ideally, the population within the bran container could indeed pass along more "fast" genes, leading to a "faster" beetle population. This will inevitably lead to questions about natural and artificial selection and ultimately evolution. Somehow the idea of beetle evolution is less threatening to students (as well as parents and administrators) than human evolution but the same concepts of natural selection, artificial selection, genetic/phenotypic variation, and evolution can be taught to the students, in a non-threatening way.

To aid in the transferring of the beetle from place to place, two cardboard paper strips can be used, to gently lift and hold the beetle without injuring it. Each paper strip is just a thin strip cut from a 3 X 5 index card. The strip should probably be around 2 cm wide and is 3 inches long. Quite a few of the 3 inch long paper strips can be cut from one index card. With a little practice, students can become quite adept at moving the beetles around and flipping them over with the two cardboard paper strips.

POSSIBLE ANSWERS TO THE CONCLUSION QUESTIONS AND SAMPLE DATA

Beetle Races! Lab – Sample Data Table

	Time Trial #1		Time Trial #2		Time Trial #3	
	time (secs)	distance (cm)	time (secs)	distance (cm)	time (secs)	distance (cm)
1	0	0.00	0	0.00	0	0.00
2	5	4.22	5	12.32	5	8.58
3	10	9.20	10	20.86	10	10.32
4	15	13.75	15	30.53	15	20.06
5	20	20.00	20	40.10	20	27.11
6	25	26.64	25	45.70	25	35.87
7	30	34.27	30	53.23	30	45.46
8	35	36.23	35	61.31	35	52.30
9	40	37.51	40	75.80	40	60.13
10	45	44.03	45	80.02	45	69.94

Typical data table for this lab. This data table was generated using Graphical Analysis 3. Students are instructed to construct their own data table.

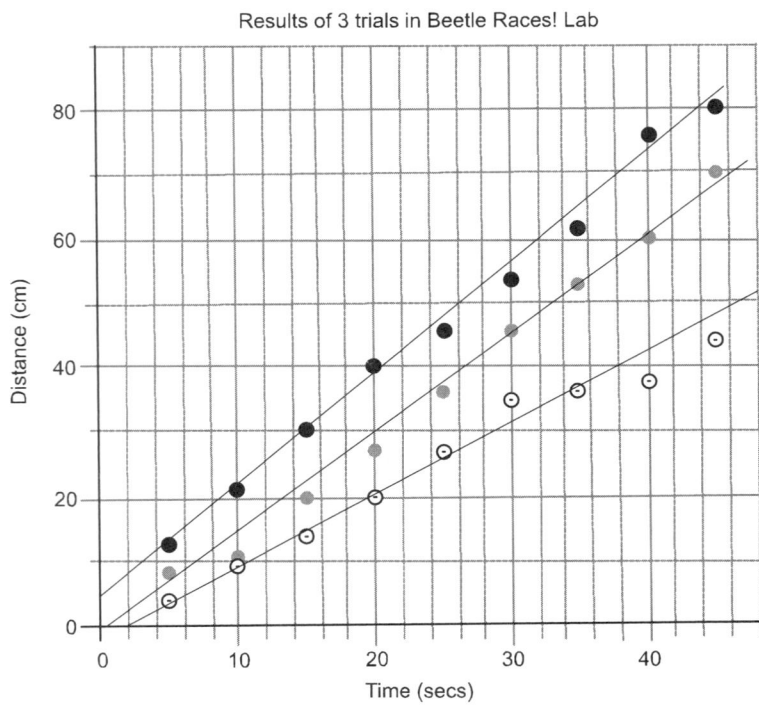

Typical graph of 3 data sets, representing the 3 trials.

Students should draw a line of best fit rather than "connect the dots". The slopes of these three lines represent the rates of the beetles for each of the three trials.

1. By looking at your graph, which trial (1, 2 or 3) was the fastest?
   - Answers will vary but the steepest curve will represent the fastest trial, representing the greatest distance traveled in the least amount of time. In other words, the line with the greatest, steepest slope.
   - In the sample graph above, the fastest trial would be the top line on the graph.

2. By looking at your graph, which trial (1, 2 or 3) was the slowest?
   - Answers will vary according to student data but students are looking for the trial where the beetle went the farthest distance in 45 seconds. (Time is always a constant in this experiment.)

3. What evidence from your graph allowed you to select the fastest trial?
   - Again the students will have to explain the idea that a particular line is steeper, versus one that is not as steep. The line with the steepest curve, or greatest slope represents the fastest rate.

4. Using your graph, calculate the average rate for each one of your trials. Show your work and label each calculation as "Trial 1", "Trial 2" and "Trial 3". The equation for rate is:

$$\text{rate} = \frac{\text{distance (cm)}}{\text{time (seconds)}}$$

Be sure to include units of cm/sec in your answers.
   - . . . Answers will vary but the average rate equals the total distance traveled divided by the total time, 45 seconds. If the beetle traveled 90 centimeters in 45 seconds, the rate would be 2 cm/sec.

5. Look at your graph and determine the fastest 5-second segment rate that your beetle traveled in all of the trials. (What is the farthest distance that a beetle traveled in 5 seconds?) Show your work in your answer.

$$\text{rate} = \frac{\text{distance (cm)}}{\text{time (seconds)}}$$

Be sure to include units of cm/sec in your answers.
   - Answers will vary, but students find on their data table the farthest distance traveled in any 5 second interval. Students divide the distance traveled in that segment by 5. A fast rate is probably 3 to 5 cm/sec but occasionally a student will record a rate of up to and possibly over 10 cm/sec.

6. Look at your data table and determine the slowest 5-second segment rate that your beetle traveled in all of the trials. (What is the shortest distance that a beetle traveled in 5 seconds?) Show your work in your answer.

$$\text{rate} = \frac{\text{distance (cm)}}{\text{time (seconds)}}$$

Be sure to include units of cm/sec in your answers
- . . . Answers will vary but students find on their data table the five second interval during which the beetle traveled the shortest distance. The student divides the distance traveled in that segment by 5. A slow rate is probably less than 1 cm/sec.

TEACHER PAGES

Beetle Races!
Measuring Distance and Time and Calculating Rate in a Living, Moving Insect

PURPOSE

In this activity you will measure the distance traveled by your selected mealworm beetle, *Tenebrio molitor*, during three 45 second races. You will collect data at five second intervals during each race. You will analyze your data to determine the five second intervals during which your beetle was the fastest and the slowest as well as determine the average rate of speed of your beetle for each race. You will also generate graphs to better communicate your results.

MATERIALS

metric ruler
piece of string
overhead pen
Petri dish

stop watch
Tenebrio molitor beetle
graph paper

PROCEDURE

1. Working with your partner, select what you believe to be the fastest adult *Tenebrio molitor* beetle from the mealworm container and put it into a Petri dish. Give your beetle a name. Do not torture or hurt your beetle or you will get no credit for today's activity.

2. Mark a small X with the overhead pen in the middle of your desk or table. This will serve as the starting point of the race.

3. Carefully transfer your beetle from your Petri dish onto the X. You can use two cardboard paper strips to transfer the beetle onto its legs to begin moving.

4. Student #1 will start the stopwatch **as soon as the beetle begins to move off the X**. At the same time, student #2 will trace a line following the movement of the beetle with the overhead pen. Student #1 will call time every 5 seconds while student #2 continues to trace the path of the beetle. At these 5 second intervals, student #2 will. also make a small mark on the line to indicate the position of the beetle when time was called. Your line may be squiggly or it might be straight.

5. You can also motivate your beetle by cheering its name, encouraging it go faster, **without touching it**! Also, you can very gently nudge the beetle with a small piece of paper to initiate its movement but do not push it in any way that it will affect its rate of movement. You must use the same beetle during all three trials.

6. When your beetle has traveled for 45 seconds, stop and make your final mark on the table with the pen. Gently place the beetle back into the Petri dish using a small piece of scratch paper.

7. Place your piece of string on top of the path you have drawn on the table. Measure in centimeters, to the closest 0.1 millimeter, the distance traveled by the beetle for each 5 second interval. Be sure to measure the ***total*** distance for each 5 second interval by measuring from the original starting point, the X you marked on your desk or table. The distance measurements should be increasing with each interval. It is critical that your distance measurements are as accurate as possible. Remember to estimate to the closest 0.1 mm, between the millimeter marks on your ruler for each measurement. For example, in the first segment, measure out to two decimal places: the measurement might read 10.35 cm (if this was how far the beetle traveled) and not 10.4 cm nor 10 cm.

8. Create a data table, using a ruler, showing how far the beetle traveled during each 5 second interval, up to 45 seconds. You will be collecting three sets of this data, one for each of the three races your beetle will run.

9. After you have recorded your data, use a damp paper towel to wipe off the overhead pen marks from your desk. Remove the marks completely from the desk. Once the desk is clean, re-mark another X for the next trial.

10. Be sure to record your data for this second race in your data table. Wipe down the overhead pen markings between each race.

11. Construct an appropriately-titled graph showing the progress of your *T. molitor* beetle during the three races.

12. Time in seconds is graphed on the *x*-axis.

13. Distance in centimeters is graphed on the *y*-axis.

14. Do not connect the points, but rather draw the best straight line through the points keeping an equal number of points above and below the best-fit line you are drawing. This line represents the average distance traveled for the 45 seconds of the race. The line will generally slope upward over time.

15. The 3 trials will be graphed as 3 separate lines on your graph paper. Make each line a different color so you can tell them apart. Also, label each line as Trial #1, Trial #2 and Trial #3.

16. The "faster" beetles will be put back into the gene pool and allowed to breed. The "slower" ones become a tasty meal for a lizard. This is known as artificial selection. This is done with race horses all the time. The "fast" genes will be selected for artificially rather than letting that animal choose its own mate from the total population! Maybe after a million generations, we will have the fastest beetles in the world!

Name _____

Period _____

Beetle Races!
Measuring Distance and Time and Calculating Rate in a Living, Moving Insect

CONCLUSION QUESTIONS

Each student will turn in an appropriately-titled graph with the axes correctly labeled and 3 labeled lines representing the rate of speed for the beetle during each of the 3 trials.
Each student will also answer the following questions on the back of their graph.

Answer the following questions on the back of your graph paper. Numbers 1 and 2 should be in complete sentences and numbers 3-5 should show calculations with the answers circled.

1. By looking at your graph, which trial (1, 2 or 3) was the fastest?

2. By looking at your data table, which trial (1, 2 or 3) was the slowest?

3. What evidence from your graph allowed you to observe and then recognize the fastest trial?

4. Using your graph, calculate the average rate of speed for each one of your trials. This rate is the slope of the line. It represents the total distance traveled divided by the total time measured. Show your work and label each calculation as "Trial 1", "Trial 2" and "Trial 3". The equation for rate is:

$$\text{rate} = \frac{\text{distance (cm)}}{\text{time (seconds)}}$$

Be sure to include units of cm/sec in your answers.

2 *Beetle Races!*

5. Look at your data table and determine the **fastest** 5-second segment rate that your beetle traveled in all of the trials. (What is the farthest distance that a beetle traveled in 5 seconds?) Show your work in your answer.

$$\text{rate} = \frac{\text{distance (cm)}}{\text{time (seconds)}}$$

Be sure to include units of cm/sec in your answers.

6. Look at your data table and determine the slowest 5-second segment rate that your beetle traveled in all of the trials. (What is the shortest distance that a beetle traveled in 5 seconds?) Show your work in your answer.

$$\text{rate} = \frac{\text{distance (cm)}}{\text{time (seconds)}}$$

Be sure to include units of cm/sec in your answers.

Write Your Notes and Ideas Here!

Can Mosquitoes Transmit HIV between Primates?
Simulating a Laboratory Experience
Through a Role-playing Exercise

OBJECTIVE

Students will design a valid experimental procedure to explore and suggest a solution for a complex, scientific, historical, and real-world question, by applying the steps of the scientific method.

LEVEL

Middle Grades: Life Science

NATIONAL STANDARDS

UCP.2, A.1, A.2, C.4, E.2, G.2

TEKS

6.2(A), 6.2(B), 6.2(C), 6.2(E)
7.2(A), 7.2(B), 7.2(C), 7.2(E)
8.2(A), 8.2(B), 8.2(C), 8.2(E)
IPC 1(A), 2(A), 2(B), 2(C), 2(D), 3(A)

CONNECTIONS TO AP

AP Biology:
   II. Heredity and Evolution: B. Molecular Genetics 4. Viral structure and replication
   III. Organisms and Populations: B. Structure and Function of Plants and Animals 1. Reproduction, growth, and development 2. Structural, physiological, and behavioral adaptations C. Ecology
   3. Global issues

TIME FRAME

6 or 7 days of 45 minute classes

MATERIALS

(For a class of 27 working in groups of 3)

9 overhead transparencies   9 overhead marking pens

TEACHER NOTES

Can Mosquitoes Transmit HIV between Primates? is a student inquiry lab simulation designed to be the culminating activity of a scientific method unit. Students role play as teams of research scientists suggesting possible solutions to the following complex question: Can mosquitoes transmit HIV between primates? Specifically, each group of "researchers" has to cooperatively plan and design a hypothetically valid experimental procedure that would test whether or not mosquitoes could transmit HIV between primates. On a more general level, this activity allows students to take a very complex scientific question and attempt to hypothetically answer it by reducing it down to a series of variables that are all accounted for and controlled by the experimenters in a scientific experiment.

Background Information:
The idea that mosquitoes could transmit HIV was first reported in 1985 in Bell Glades, Florida where evidence suggested that mosquitoes may have been responsible for the higher than average incidence rate of HIV infection. The National Center for Disease Control (CDC) demonstrated in scientific surveys that mosquito transmission of AIDS in that community appeared to be highly unlikely.

Today, scientists generally conclude that there is absolutely no evidence that mosquitoes can transmit HIV. There are three reasons generally accepted by scientists to explain why this is true:
(1) Mosquitoes digest HIV
(2) Mosquitoes do not ingest enough particles to transmit the virus
(3) The salivary glands of mosquitoes are separate from the blood-sucking pathway so they can inject saliva with its anti-coagulating agents into the bite without interrupting the flow of blood

The reasons above are cited in the official explanation from the CDC website that responds to the question, "Can I get infected with HIV from mosquitoes?"

http://www.cdc.gov/hiv/pubs/faq/faq32.htm

Another extremely detailed background article on the same subject by Wayne J. Crans, Associate Research Professor in Entomology, at Rutgers University, explaining "Why Mosquitoes Cannot Transmit AIDS" can be found at the following URL.

http://www.rci.rutgers.edu/~insects/aids.htm

More general information about the basics of HIV/AIDS can be found at the about.com or at the Centers for Disease Control websites referenced below.

http://aids.about.com/cs/aidsfactsheets/a/hivbasics.htm
http://www.cdc.gov/hiv/

Before beginning this activity with students, read through this entire document. This activity, although very rewarding and memorable for both teachers and students, is also very complex.

Suggested Teaching Procedure:
Day One:
Assign students the pre-lab assignment for homework before assigning the research groups and actually beginning the project. The pre-lab should be done individually. It encourages the students to engage their brains regarding the problem. Each student needs a copy of the pre-lab assignment as well as the student instruction page.

Day Two:
When the students return to the class, assign them into teams of 3, each group represents a team of researchers. Students, in their newly assigned groups, are to exchange each others' procedures and read through each one. The students are to read each procedure for about 5 minutes and then pass it on to the next student in the group, rotating each procedure until all students have read all procedures. Student should focus their attention on finding not only any hidden variables and potential problems but also on

good procedural ideas in each procedure. Students are then to discuss the best points of each proposal, deciding which steps from each procedure are the most scientifically sound and valid. Remind students that they are being graded not only on their scientific knowledge and capability but also on their cooperative skills. Instruct students to consult the *Cooperative Skills Assessment* rubric as well the *Experimental Design Assessment* rubric for expectations. Take the time to review the expectations of behavior assessed in the *Cooperative Skills Assessment* if you have not takeen the time to teach these skills previously in your class.

Day Three:
Give each student a copy of the student answer pages. Refer students to the job descriptions described in their student instructions: Procedural Author, Layout Designer and Materials Manager. Review with the entire class the responsibilities of each job. Students are to select one of the three jobs within the research group. Once the students have committed to a role, each member should record their names on their student answer page. By doing this, each student should know what is required from all participants and all group responsibilities are clearly outlined.

Initiate a discussion among the group members that will begin the procedure outline process. The Procedural Author will keep track of this discussion, outlining the group's ideas as the procedure develops. The Layout Designer should begin sketching a rough draft diagram of the laboratory room(s) and space needed to accomplish whatever procedure is decided upon. Simultaneously, the Materials Manager will begin listing the materials that will be needed to accomplish the research. Continue to monitor the groups as they work. Ask the students individually what their jobs are and what is required for that job. Monitor the groups' discussions and continually ask each group how the procedure that they have outlined is valid. Could anyone else follow their procedures and achieve the same results?

Days Four and Five:
Remind students that there is only a limited amount of time to get the procedure, diagram and overhead diagram completed and ready to present. Provide each group an overheard transparency and an overhead marker.

Refer students to the three grading rubrics (*Cooperative Skills Assessment*, *Experimental Design Assessment* and the *Oral Presentation Assessment*) that will be used to assign students grades during this project. Go through each set of descriptions on each of the rubrics so students will have a very clear picture of what is expected by you as the project unfolds.

Monitor the students as they prepare the final report, constantly asking questions about the procedures and troubleshooting with the students.

Also, an overhead transparency can be created of the sample student-created floor plan diagram below. Discussion can ensue about what a "good" diagram would look like and what a "not so good" diagram would look like.

An example of a student-created overhead diagram appears below.

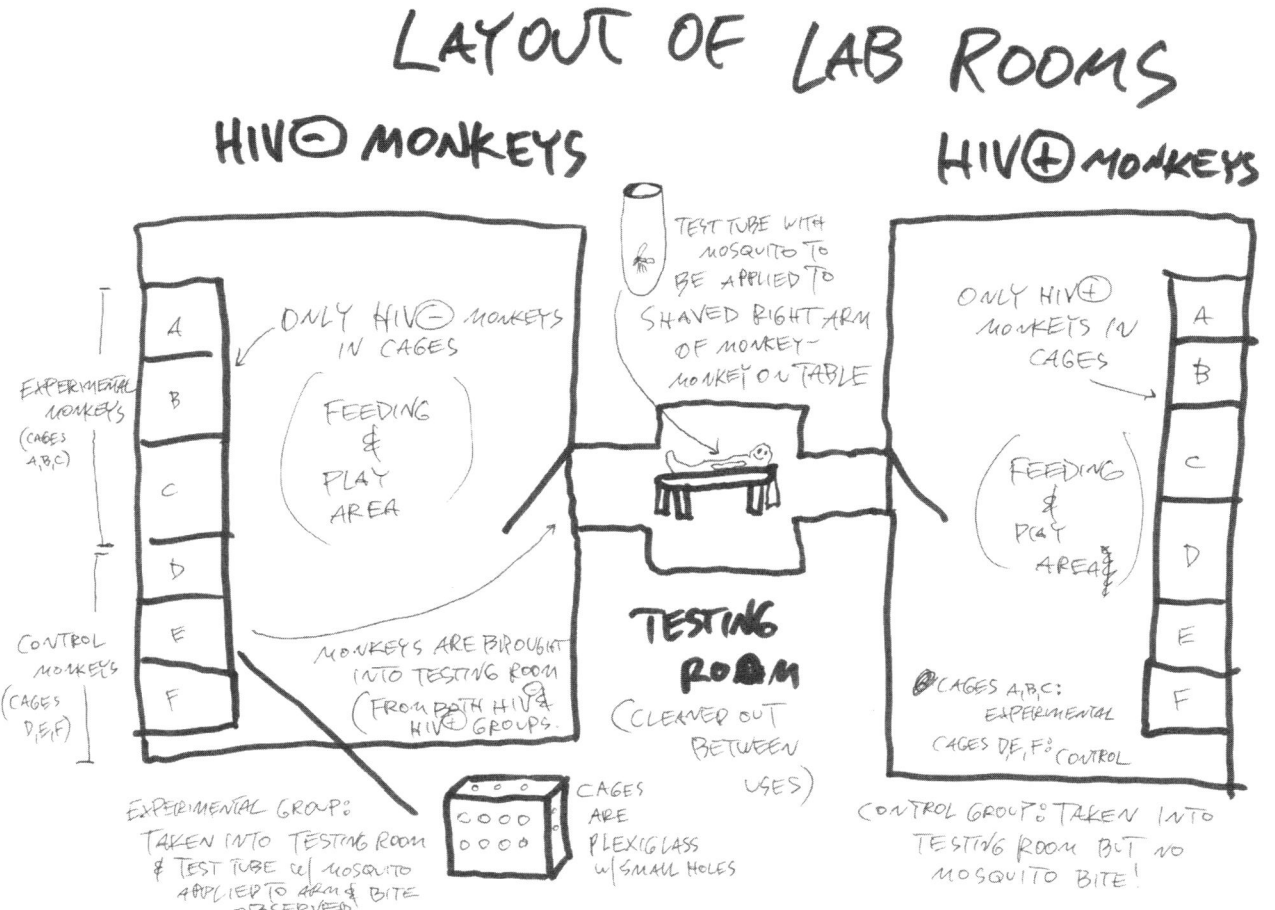

At the end of the 5th day, the written report should be completed, the overhead completely finished and the materials list finalized. Depending on how well the class is performing, more time can be given here.

Day Six:
Move from group to group and carefully look at each procedure and experimental layout diagram. Assign hypothetical results (new cases of HIV or no new cases of HIV) for each group, as you carefully try to find hidden variables in each procedure. Specifically, look for the groups that have not changed the default cages, where limbs could reach through and feces could be thrown through the spaces in the cages. This hidden variable would be especially problematic in a room that contains both HIV+ and HIV – monkeys. Also, if the students take monkeys to a testing room, play room, or some other central or communal area, make sure that the students have accounted for disinfecting the room between usages. When assigning the results, remind the students that today we know that mosquitoes do not transmit HIV so the students need to think about other ways that the HIV could be transferred, other than by mosquitoes. The chance of HIV being passed from one monkey to another by thrown feces would probably, in reality, be very minimal. The results that you assign should probably reflect this with only one or two primates being newly infected of the total group of monkeys tested (usually between 50 and 100), in the experimental group, and possibly and/or in the control group. Students are especially challenged when they get new HIV+ monkeys in the control group as they are not anticipating these results The idea here is to get students to look at their procedure in a critical way and try to account for flaws in their experimental design. After assigning results, each group should be able to explain their results and draw conclusions in the space provided on the student answer page and also in the oral presentation.

Day Seven:
Each group should begin practicing their oral presentation, with each student presenting the section that they wrote. Again, refer the students to the *Oral Presentation Assessment* rubric.

After about 20 minutes of practicing and clarifying the oral presentation, you should COLLECT ALL WRITTEN LAB REPORTS (So the students cannot change the procedure once another group has presented and hidden variables are discovered.). Students will present their experiment to the rest of the class, dividing the presentation between the 3 assigned jobs. After presenting, there should be time for other students to ask questions and get clarification on each procedure. You should consider how well the students in each group answered the questions when using the Oral Presentation Assessment.

POSSIBLE ANSWERS TO THE CONCLUSION QUESTION AND SAMPLE DATA

Students will generally create one of two procedural scenarios to answer the question. Students will "move the monkeys" or students will "move the mosquitoes". Typically in the moving-the-monkeys scenario, an HIV+ monkey (or a group of monkeys) is transferred into a room with mosquitoes, exposed to the mosquitoes for a certain amount of time and then removed. A HIV – monkey (or a group of monkeys) is then brought in and exposed to those same (presumably blood-filled) mosquitoes. The other scenario involves students moving the mosquitoes from one room, filled with one or more HIV+ monkeys to another room with HIV– monkeys. For each of those scenarios, more detailed analysis is provided below.

When students opt to transfer the monkeys from one room to another, you can question the groups on the logistics of this move.
- "How did you move the monkeys?"
- "Did you move the monkeys in a cart?"
- "Were the monkeys' cages on wheels?"
- "Were the monkeys anesthetized? If so, how did you accomplish this in a humane way?"

- "Were the monkeys restrained? Again, how was this accomplished in a humane way?"
- "How did you get the monkeys to cooperate with the transfer?"
- "Did the monkeys have any contact with each other during the transfer?"
- "Could bodily fluids have been left behind by a group of monkeys in a "play area" or a "testing area"?"
- "How did you make sure that the mosquito actually bit the monkey?"

Typically the students will move the monkeys in their cages, and the cages will be on wheels. For the move-the-mosquitoes scenario, when transferring mosquitoes from room to room, students may use the following strategies:

- use of a fan to transfer the mosquitoes by blowing or sucking them
- use of UV light to attract the mosquitoes from one room to another
- use of lactic acid to attract mosquitoes from room to room
- use of a net to transfer the mosquitoes from room to room
- use of a test tube to transfer a single or a few mosquitoes from one monkey to another
- use of a "bloody sponge" on a conveyor belt to lead the mosquitoes from room to room

The logistics of moving these mosquitoes can be quite humorous for the teacher, from a realistic logistics perspective. It is up to your judgment whether more creative but probably "unrealistic" scenarios are allowed.

Questions that can be asked of students about the move-the-mosquito type of scenario include:

- "How did you move the mosquito(s) from room to room without harming or injuring them?"
- "How did you know that the mosquito(s) that you transferred was one that had already bit a monkey(s)?"
- "How did you observe that the mosquito(s) had already bit the monkey(s)?"
- "How did you make sure that the mosquitoes actually went into the cage of the monkey(s)?"
- "How did you capture the mosquito that had just bit an HIV+ monkey and quickly transfer it to the HIV– monkey?"
- "How did you make sure that the mosquito actually bit the monkey?"

One of the most critical student misconceptions is that students will put the monkey(s) in the presence of the mosquitoes and assume that the mosquitoes have bitten the monkeys. Typically, when explaining their procedure, students will say, "We put the monkey in the room with the mosquitoes and since we put 400 mosquitoes in the room, we were sure that the mosquitoes had bitten the monkey." Therefore, there is an assumption that the mosquito's bite, i.e., the independent variable, has been applied. But this is not necessarily so. Another common misconception is that a when a group of mosquitoes has been released around the monkeys, there is an assumption that the same exact mosquito has bitten both a HIV+ monkey then that same mosquito has bitten an HIV– monkey.

Hidden variables:
A major hidden variable students will need to identify is that the default cage that has been provided to the student that has the bars spaced 10 cm apart. Bored primates will probably engage each other through the cages. One reason to have students completely design the cage is to limit any variables about monkeys exchanging bodily fluids, including semen, blood, and fecal material. When the cage is composed of bars that are spaced 10 cm apart, monkeys could get arms or other body parts through the

spaces in the cages. Students who recognize this variable can ask for another type of cage. Also, bodily fluids can easily be transferred, or "flung" across the room; monkeys have been observed to throw feces on numerous occasions, especially when in captivity. It is up to the teacher's discretion whether or not to make the students aware of this hidden variable before they turn in the final written report and present to the class.

A common misconception is that students believe that in the control group, nothing is done to the control, that is, the monkeys just stay in their cages and do nothing else. The control group must have the same steps of the procedure applied to it, leaving out the independent variable, i.e., the mosquito bite. So if the experimental group of monkeys were taken to a testing room to be tested for HIV after being exposed to the mosquitoes, the control group would need to be taken to the testing room and tested as well. This is to eliminate the variable that something in the testing room transmitted the HIV, like feces or blood left behind by an infected monkey. The ideal control group would have all other procedural steps applied, including shaving or any other step that is applied to the experimental group.

When assigning results, you will have to trouble shoot each procedure and ascertain very quickly whether there are hidden variables. The major hidden variables can include:

- Cages sitting close enough that monkeys could interact with each other
- Use of the default cages, easily allowing monkeys to interact or transmit bodily fluids through the cages
- Allowing monkeys to interact, outside of the cages
- Use of one area where monkeys could interact (like a social area) where they have not disinfected between each use by HIV+ or HIV− monkeys
- Use of a testing area where both HIV+ and HIV− monkeys are tested in some way, but not disinfected in between.
- Use of any testing materials like a razor that was used between both HIV+ and HIV−

There could be many other hidden variables so use your discretion here!

To assign a grade for this complex activity three rubrics have been provided: *Experimental Design Assessment, Cooperative Skills Assessment* and *Oral Presentation Assessment.* These rubrics are located in the student answer pages. Familiarize yourself with them before attempting this activity to have a clear idea in your mind as to how to assess the students. As you monitor the groups while they are planning, use the rubrics to ask questions to students. For example, a major part of the students' grade in experimental design involves: "Design shows students have analyzed the problem, have designed and conducted a thoughtful experiment, have carefully applied independent variable and controlled obvious and hidden variables." Continue to ask students, did you carefully apply the independent variable? How did you know the independent variable has been applied?

The best procedures, those that will be awarded the most points, are the ones that involve some sort of direct observation of the independent variable being applied, that is, the mosquitoes biting the monkey.

Some ways how students have been observed to appropriately apply the independent variable:

- shaving a part of the monkey, typically the arm, and applying a test tube or some other clear tube with one mosquito inside it, directly observing the mosquito biting the monkey
- shaving the monkey so the mosquito bites can be observed as red bumps

- use of cameras to directly observe the mosquito biting the monkey and keeping track of each mosquito as it moves to another monkey

Of the above procedures, the test tube procedure is usually awarded the most points. Ideally, the students take one mosquito, place it in clear test tube-like device, place the open end of a test tube on the shaven smooth skin of the arm of the HIV+ monkey and allow the mosquito to bite the monkey's arm. The bite can be directly observed through the glass. The test tube can easily be transferred to a HIV− monkey in another room and allowed to bite its shaven arm.

The best part about this activity is that it is open-ended. The scenarios presented are by no means complete. Your own students will develop creative ways to answer this question and you will be amazed each time you do it.

POSSIBLE ANSWERS TO PRE-LAB QUESTIONS

1. What is the problem question presented in the lab?
   - Can mosquitoes transmit HIV between primates?

2. What is your hypothesis?
   - Answers will vary.

3. What is the independent variable in this experiment?
   - Mosquitoes biting the monkeys or not.

4. What is the dependent variable in this experiment?
   - HIV being transmitted between primates or not.

5. What are three controlled variables that you held constant in this experiment?
   - Answers will vary as there are many (same amount of food, water, same type of monkeys, etc…)

6. Explain how you would design a control group in this experiment?
   - Answers will vary but essentially the control group should have all the same procedural steps applied to it as the experimental group, except for the independent variable being applied (the mosquito bite).

7. Write a procedure in the space provided below that would test your hypothesis, using the equipment provided in the materials section of the lab procedure and any other equipment that you think that you would need. Your procedure should have as many numbered steps as necessary that clearly explains your procedure.
   - Answers will vary widely! A real student procedure has been included elsewhere in the teacher's notes.

8. Include a labeled diagram of your laboratory floor plan in the space provided below. Be sure to include storage rooms, testing rooms, cages, monkeys and equipment.
   - Answers will vary widely! A real student laboratory floor plan diagram has been included elsewhere in the teacher's notes.

Can Mosquitoes Transmit HIV between Primates?
Simulating a Laboratory Experience
Through a Role-playing Exercise

Imagine the scenario:

Your teacher is president of the prestigious Albert Einstein University and the year is 1984. You are a Ph.D. researcher are part of a team that has been hand selected by the university president to suggest a solution to a real world complex problem. Media reports from Bell Glades, Florida are beginning to surface and propose that mosquitoes can transmit HIV. The reports suggest that mosquitoes may have been responsible for the higher than average incidence rate of HIV infection among the people that live in that community.

Albert Einstein University has received a federal grant of $3,000,000 dollars from Centers for Disease Control (CDC) in Atlanta, Georgia to study this potential epidemic. The CDC wants the university research team to conduct controlled scientific experiments that will answer the question: Can mosquitoes carry the human immunodeficiency virus (HIV) from an infected primate to an uninfected primate? As a member of a 3 person research team, your task is to design a valid experiment that will answer the CDC's question.

The university president must decide which research team will earn the right to do the research and receive the $3,000,000 grant. To help make this decision, the president has requested that each team submit its proposed research plan.

Good luck, your competition is tough. Only the best plans will be used and will receive highest grades for the lab.

PURPOSE
In this activity you will design a valid scientific experiment by applying the steps of the scientific method to suggest a solution to a complex, scientific, historical, and real-world problem.

PROCEDURE
Day One:
1. Carefully read this entire procedure and complete the pre-lab questions on the student answer page. Each student is to complete the pre-lab individually and return it the following day.

Day Two:
2. Gather the members of your research group and present your completed pre-lab assignment. You will have 5 minutes to read each person's pre-lab plan. As you read each one, decide which parts of each procedure are best suited to answer the problem question. Be objective in determining the best parts from each procedure, regardless of who wrote it.

3. After brainstorming and discussing which ideas you thought were the best, discuss with your team members how your team should proceed with the problem.

4. After all group members have expressed their ideas about the design of the experiment, the group should cooperatively decide on the best plan for the group procedure. An overriding question that you should be asking at all times is, "Is this procedure valid?" Could anyone repeat this experiment and obtain the same results?

NOTE: You will be graded on how well you work together as a group according to the Cooperative Skills Assessment on your student answer page.

NOTE: You will be graded on how well you design your experimental procedure according to the Experimental Design Assessment on your student answer page.

Day Three:
5. Decide who will perform each job in your group. Read each one of the descriptions below and decide who will do which job. Write each student's name on your student answer page in the blank provided next to the job title.

A detailed list of job functions follows below.

Layout Designer: Design an easy-to-read diagram on a teacher-provided overhead transparency that shows the layout of the testing, observation, and storage facilities in your lab. Be sure to design a legend to show how each item is represented in your diagram. A clear, accurate diagram, neatly and correctly labeled, should make your experimental setup easier to understand. Each person in the room should be able to understand your experiment clearly by looking at your diagram. Also, include a close-up diagram of a typical cage with a monkey in it. (If your group wants another type of cage instead of the default cage, then you must request it from the president.) Take great care to show how and where the monkeys will be housed and be sure to include both the experimental group as well as the control group.

Procedural Author: Write your group's hypothesis to answer the question: *Can mosquitoes transmit HIV between primates?* in the space provided on your student answer page. After deciding on a plan of action, write your group's procedural steps down in a logical and legible fashion on your student answer page, labeling each sequential step in order with a number. Include in your procedure a detailed explanation of how your team will use a control in your experiment. The control must be carefully thought out and planned. The only variable that should be different from the experimental group is the independent variable. *NOTE:* Be sure to write your experimental procedure with enough detail so that it could be repeated. Also state your independent variable, dependent variable, controlled variables, and control in your procedure. If you have access to a computer, type your procedure and print it.

Materials Manager: Determine the complete list of equipment that is going to be needed to test your hypothesis. Include in your list a labeled diagram of the cage that is used to house your monkeys. (This should match the cage design drawn by the Layout Designer on the overhead transparency.) The following list is a starting point for resources, supplies, and organisms that you will be provided. You may not add anything to the list without the permission of the Albert Einstein University President, however, the president is usually generous as long as the items are justified as reasonable and necessary. If you have access to a computer, type your procedure and print it.
*   2 identical, empty rooms (30m x 30m) with air-tight doors and windows, supplied with fresh air and air conditioning

- 100 primates (50 are HIV negative, 50 are HIV positive)
- 100 identical cages, with metal bars, spaced 10 cm apart, on all sides of the cage
- water and feed bowls
- 400 female mosquitoes in water, capable of laying viable eggs
- Grade A primate food, "Monkey Chow", served daily at your cafeteria
- Water
- 100 HIV tests (capable of detecting HIV instantaneously)

6. After you have decided on each person's job, you will have two class days to prepare your final report which includes the procedure, diagram, and list of materials. Don't forget that you are being graded on your ability to work together cooperatively as a team. Each individual person in the group should begin doing the tasks that are required for that job.

Day Four and Five:
7. Continue to work on your final product (procedure, floor-plan, and list of materials), carefully refining your final product to be a meticulously thought out and valid scientific procedure.

When your group has finished creating your final product (procedure, diagram, and list of materials), your teacher will study your diagrams and procedure and he/she will assign you some hypothetical results (the number of new HIV or no new HIV cases), as if you had really done your experiment and tested all the monkeys for HIV. Write your teacher-generated results in the results section of your student answer page. Be ready to state these results in your oral presentation.

Your next task, as a group, is to explain your results in a conclusion. Remember, results answer "what" questions and conclusions answer "why" questions. If your teacher states that you have new HIV cases, then explain your results. What caused these new cases? Since we now know that mosquitoes CANNOT transmit HIV, how do you explain your results? Was it a hidden variable? Think about potential flaws in your experiment to address theses results. Also, if your teacher states that you have no new cases, you will need to justify these results in your conclusion. Write your conclusion in the space provided on your student answer page.

Day Six:
After you have finished the written product, including your results and conclusion, as a team you will present your procedural design to the class. Work cooperatively to decide how your presentation will occur and be sure to reference your diagram and list of materials.

NOTE: You will be graded on your oral presentation according to the Oral Presentation Assessment on your student answer page.

There will be a brief period of time after your presentation for other research teams to ask questions about your procedure. Everyone in your group should be able to answer *any* question about *any* aspect of your procedural design. Part of your oral presentation grade will be based on how well you answer the questions. Prior to the first group's presentation you will be required to turn in your entire laboratory report. You may not make changes during the presentations.

8. As you listen to each group present their experimental design, think critically about the validity and logic of their procedure. Keep your attention, intellectual power, and questions focused on how each group has applied the steps of the scientific method. Is the procedure valid? Could the experiment be repeated with the same results? Do not dwell on questions such as, "What did you feed your monkeys?", "How did you keep your mosquitoes alive?" and so on. These are questions that do not really address the scientific method or influence the experimental design. Good questions that you can ask a group include: "How did you treat the experimental group differently from the control group?", "How did you verify that you applied the independent variable?", "Where is your control group?", "Why didn't you disinfect the testing cage between each use?", and so on.

Name _____

Period _____

Can Mosquitoes Transmit HIV between Primates?
Simulating a Laboratory Experience
Through a Role-playing Exercise

GROUP MEMBERS

Layout Designer _____

Procedural Author _____

Materials Manager _____

HYPOTHESIS

PROCEDURAL DESIGN STEPS

MATERIALS LIST

RESULTS

CONCLUSION

Can Mosquitoes Transmit HIV between Primates?
Simulating a Laboratory Experience

A Role-playing Exercise

Carefully read the lab paper, "Can Mosquitoes Carry HIV between Primates?". After reading the lab procedure, answer all questions on this pre-lab answer sheet.

PRE-LAB QUESTIONS

1. What is the problem question presented in the lab?

2. What is your hypothesis?

3. What is the independent variable in this experiment?

4. What is the dependent variable in this experiment?

5. What are three controlled variables that you held constant in this experiment?

6. Explain how you would design a control group in this experiment?

7. Write a procedure in the space provided below that would test your hypothesis, using the equipment provided in the materials section of the lab procedure and any other equipment that you think that you would need. The procedure will have as many numbered steps in sequential order steps as necessary but that clearly explains your procedure.

8. Include a labeled diagram of your laboratory floor plan in the space provided below with storage rooms, testing rooms, cages, monkeys and equipment.

PRE-LAB PROCEDURE DESIGN BY STUDENT

PRE-LAB LAYOUT DIAGRAM BY STUDENT

RUBRICS

Experimental Design Assessment	Super	Good	Fair	Poor	Omitted
Experimental Design: Design shows students have analyzed the problem, have designed and conducted a thoughtful experiment, have carefully applied independent variable and controlled obvious and hidden variables.	5	4	3	2	0
Scientific Concepts: Report/presentation illustrates an accurate and thorough understanding of scientific concepts underlying the lab.	5	4	3	2	0
Diagram/Drawing: Clear, accurate diagram is included and make the experiment easier to understand. Diagrams are labeled neatly and accurately.	5	4	3	2	0
Use of Control: Control was carefully thought out and planned. The only variable that was that was different from the experimental group was the I.V.	5	4	3	2	0
Total Points					

Cooperative Skills Assessment	Super	Good	Fair	Poor	Omitted
Helping: The teacher observed the students offering assistance to each other.	5	4	3	2	0
Listening: The teacher observed students working from each other's ideas	5	4	3	2	0
Persuading: The teacher observed the students exchanging, defending, and rethinking ideas.	5	4	3	2	0
Questioning: The teacher observed the students interacting, discussing, and posing questions to all members of the team.	5	4	3	2	0
Respecting: The teacher observed the students encouraging and supporting the ideas and efforts of others.	5	4	3	2	0
Sharing: The teacher observed the students offering ideas and reporting their findings to each other.	5	4	3	2	0
Total Points					

Oral Presentation Assessment	Super	Good	Fair	Poor	Omitted
Preparedness: ability to speak without reading, knowledge and understanding of information, fielding of questions, organization	5	4	3	2	0
Content: accuracy, clarity	5	4	3	2	0
Thoroughness: all required information included, amount of research, length of speech	5	4	3	2	0
Overall Presentation: volume, use of language, stance, eye contact	5	4	3	2	0
Total Points					

Going Bananas for DNA
Isolating DNA, Deoxyribonucleic Acid, from Bananas

OBJECTIVE
Students will extract DNA from bananas using kitchen reagents.

LEVEL
Middle grades: Life Science

NATIONAL STANDARDS
A.1, A.2, B.3, C.1, C.2

TEKS
6.1(A), 6.1(B), 6.2(A), 6.4(A), 6.7(A), 6.11(B)
7.1(A), 7.1(B), 7.2(A), 7.4(A), 7.7(A)
8.1(A), 8.1(B), 8.2(A), 8.4(A), 8.9(D)
IPC 1(A), 2(A), 2(B), 2(C), 2(D)

CONNECTIONS TO AP
AP Biology
    I. Molecules and Cells: A. Chemistry of Life 2. Organic molecules in organisms B. Cells
    Prokaryotic and eukaryotic cells 2. Membranes
    II. Heredity and Evolution: A. Heredity 2. Eukaryotic chromosomes B. Molecular Genetics
    1. RNA and DNA structure and function 5. Nucleic acid technology and applications

TIME FRAME
50 minutes

MATERIALS
(For a class of 28 working in teams of 4)

7 thumb-size pieces of fresh banana	7 10 mL graduated cylinder
7 plastic 15 mL reaction tubes (labeled: reaction tube)	14 1 mL disposable transfer pipets
ICE COLD 95% Ethanol in freezer	7 plastic cups of liquid detergent
7 plastic 15 mL reaction tube (labeled: salt)	7 plastic 15 mL reaction tubes (labeled:
7 paper weighing paper boats	meat tenderizer)
7 beakers of distilled water	7 mortar and pestles
28 microcentrifuge tubes (to store DNA)	7 paper clips, bent

TEACHER NOTES
The banana DNA extraction lab serves as a simple and almost fail-proof way to extract DNA (and RNA) from a living cell, in this case a banana. (Other fruits will work as well: reportedly, strawberries work quite well as they are polynucleated. Citrus fruits such as oranges, lemons and grapefruits do not work well for this lab.) The results are stunning even if students make errors on the ingredients. The volumes and masses do not have to be exact and the procedure still works. This lab not only serves as a good

introduction to basic lab equipment, but it can be extended to an in-depth study of the chemistry of the DNA molecule to include details about the chemistry of the DNA molecule. Students are typically very excited about this lab and find isolating this important molecule highly rewarding.

Before doing this lab, review the basic structure of the DNA molecule with students. There are many paper or toothpick models of DNA available that can be created by middle or junior high school students to provide a visual image for this molecule. The website, http://www.accessexcellence.com has an assortment of these diagrams.

Two critical ingredients for this lab are Adolph's Unseasoned Meat Tenderizer and 95% Ethanol. All of the ingredients that are needed for this lab, with the exception of the 95% Ethanol, can be purchased at your local grocery store. The meat tenderizer should be purchased within a week of the lab and used immediately. The older the meat tenderizer, the less active its protease. The 95% ethanol can be ordered from a science supply company. The cost, as of this writing, is approximately $25 for 4 liters, which should last you a couple of years. Do not try to substitute isopropyl alcohol (rubbing alcohol) as it never seems to precipitate the DNA as well as ethanol. As well, any type of alcohol that is less than 95% pure will get less than anticipated results. The ethanol must be put in the freezer before class and kept there at all times when not being used. It is a good idea to place all glassware in the freezer as well. To keep the ethanol near freezing during class you can store the ethanol in a squeeze bottle and keep it on ice in a Styrofoam cooler.

Time needs to be taken in class to explain the basic chemistry behind this reaction, either before or after the lab. The explanations are reiterated in the student procedure but should include:

- Macerating the banana with a mortar and pestle breaks down the tissue of the banana, exposing the individual cells.

- The liquid soap binds to the lipid portion of the phospholipid bi-layer comprising the cell membrane and keeps it from reassociating. The cell membrane is dissolved as a result.

- Adding non-iodized salt helps to further disrupt the cell membrane. It also helps to isolate and precipitate the DNA by shielding the phosphate groups along the sugar/phosphate backbone of the DNA molecule. Specifically, the Na^+ ions interacts with the negatively charged phosphates on the side chains of the DNA molecule. This forms a "shield", insulating the molecule and offsetting the attractions between the negative phosphates and the polar water molecules that would occur in the absence of salt.

- Chromatin is made up primarily of histone proteins and DNA. When chromatin condenses into a chromosome during the earliest stage of mitosis, it does so by wrapping and coiling itself around these histone proteins. Adding meat tenderizer releases Papain, a protease enzyme that breaks down the histone protein which frees the DNA. Papain also denatures or breaks down the DNase enzymes present. DNase is an enzyme that digests or breaks down and destroys the DNA. DNase is contained in all cells in nature and will be released into your solution when maceration occurs.

- Adding ice-cold ethanol will precipitate the DNA. DNA is water soluble as explained earlier. In the presence of salt and alcohol, the DNA forms a precipitate and makes a viscous, white, stringy solid that can be spooled onto a pipet or a bent paper clip. Remember, the colder the ethanol, the less soluble the DNA, the greater you product yield.

The salt and the meat tenderizer can be kept at each lab table. It is important to keep them in re-sealable containers. One option is to use a 150 mL conical tube with a screw-on lid. The conical tubes should be labeled "salt" and "meat tenderizer" as they are both white granular powders.

A very small amount, 5 mL or so, of liquid dishwashing soap can be put in a disposable small cup at each table. The original formulation of Dawn dishwashing soap, the plain blue variety, appears to work well, as does Palmolive, the green variety.

Always use distilled water, available at any grocery store. Tap water is filled with impurities, especially DNase, and will cause your DNA not to spool.

Use the disposable transfer pipet to pull the DNA up from the alcohol layer. Alternately, straighten out a paper clip and bend one end (about 3 mm from the end) to make a hook that can be used to fish out the DNA.

One way to enhance student motivation is to provide inexpensive microcentrifuge test tubes to each student to store the DNA after it is extracted. Each student can leave the lab with a microtube of DNA (topped off with ethanol) to show his/her friends and parents. Doing science can indeed be "cool".

Once students have completed the lab, explain to them the type of assignment that is required as a conclusion. They are to create a comic strip depicting themselves executing the procedure for this lab. Each step of the procedure and the equipment used should be illustrated, including an explanation of the following steps: maceration, adding salt, adding soap, adding meat tenderizer, and adding ice-cold ethanol. The comic strip should be neat, colored and labeled in some way. Some students will make their comic as a "one panel" comic with all the steps labeled. Other creative and motivated students will make multi-panel comics. It is up to you to decide if stick figures are acceptable in the comic, but it is always interesting to see how the students view themselves in an artistic rendition. Another idea is to have students include you in the comic. This also can provide some humor.

Some student examples of comics have been included to highlight the incredible variation among the products that each student can create. Since the comic is a creative product, any reasonable comic should be accepted, as long as it labels the equipment, and the chemistry behind the reaction is explained. A grading rubric has been provided for you below to aid in grading these tremendously variable products!

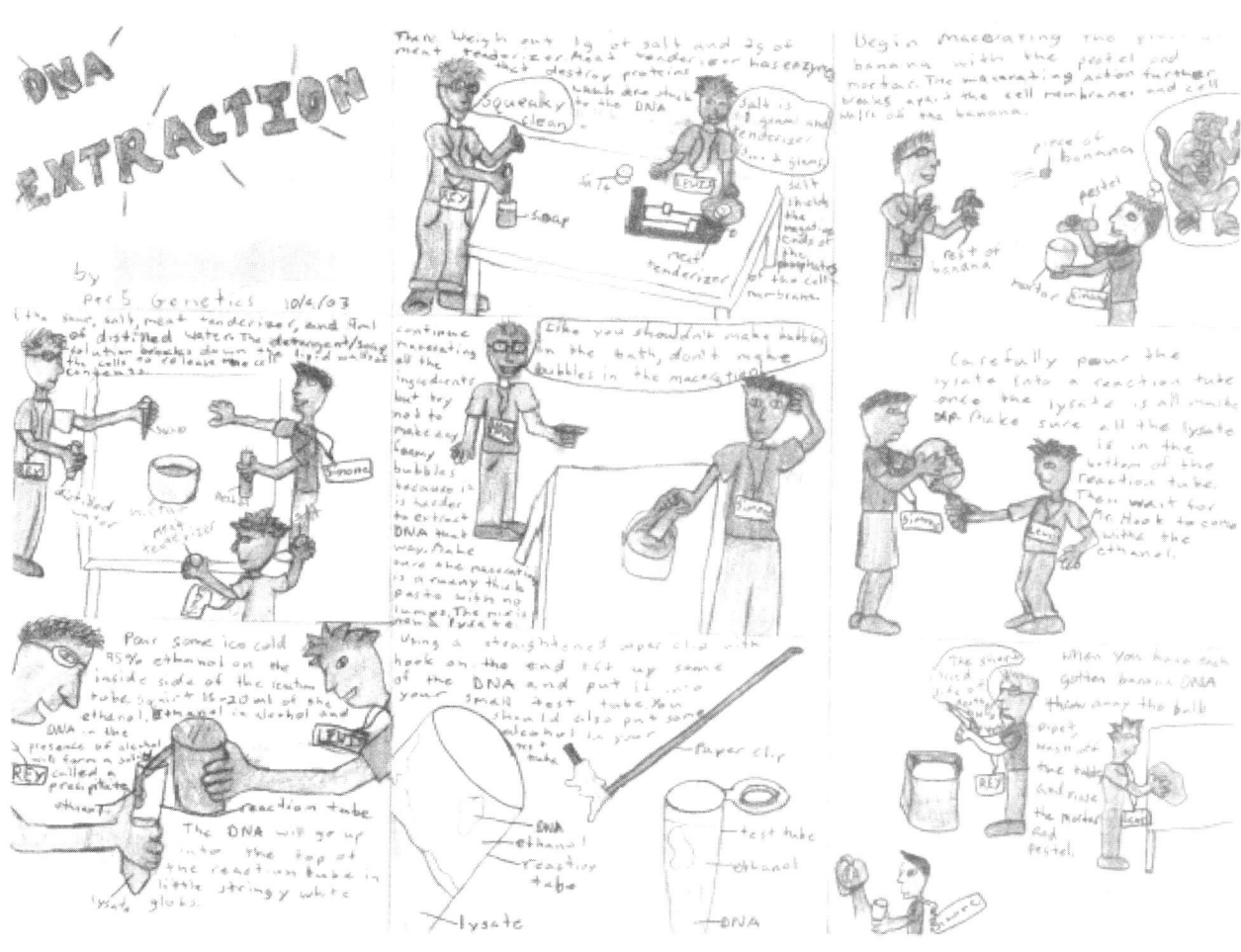

1st came Marcos, macerating the BANANA with his trusty mortar and pestle!

Then, there was Pranay, with his carefully weighed 1 gram of salt and 2 grams of meat tenderizer

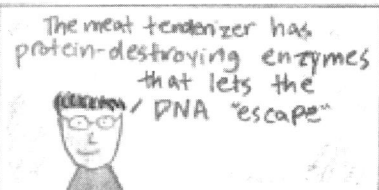

The meat tenderizer has protein-destroying enzymes that lets the DNA "escape"

Alicia was next, adding her bulb Pipet of liquid detergent

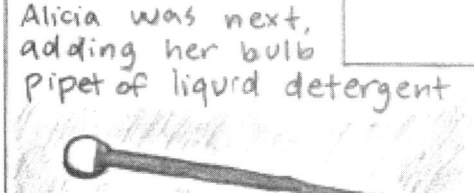

The soap & salt breaks down the cell walls to release it's contents.
The salt shields the cell membrane.

After Marcos mixed the ingredients well with his pestle... Sharonte arrived!!...

She poured the lysate into a clean reaction tube...

And added the final ingredient — 95% ICE-COLD ETHANOL!

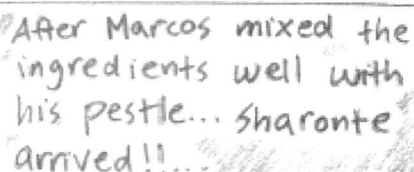

The group crowded around expectantly, cheering on the DNA

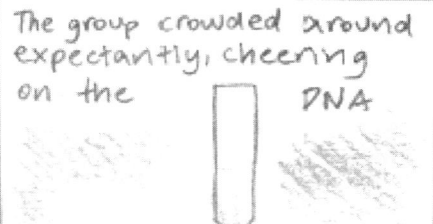

And it worked!! The Challenged have succesfully extracted BANANA DNA !!

Genetics 5th
The End

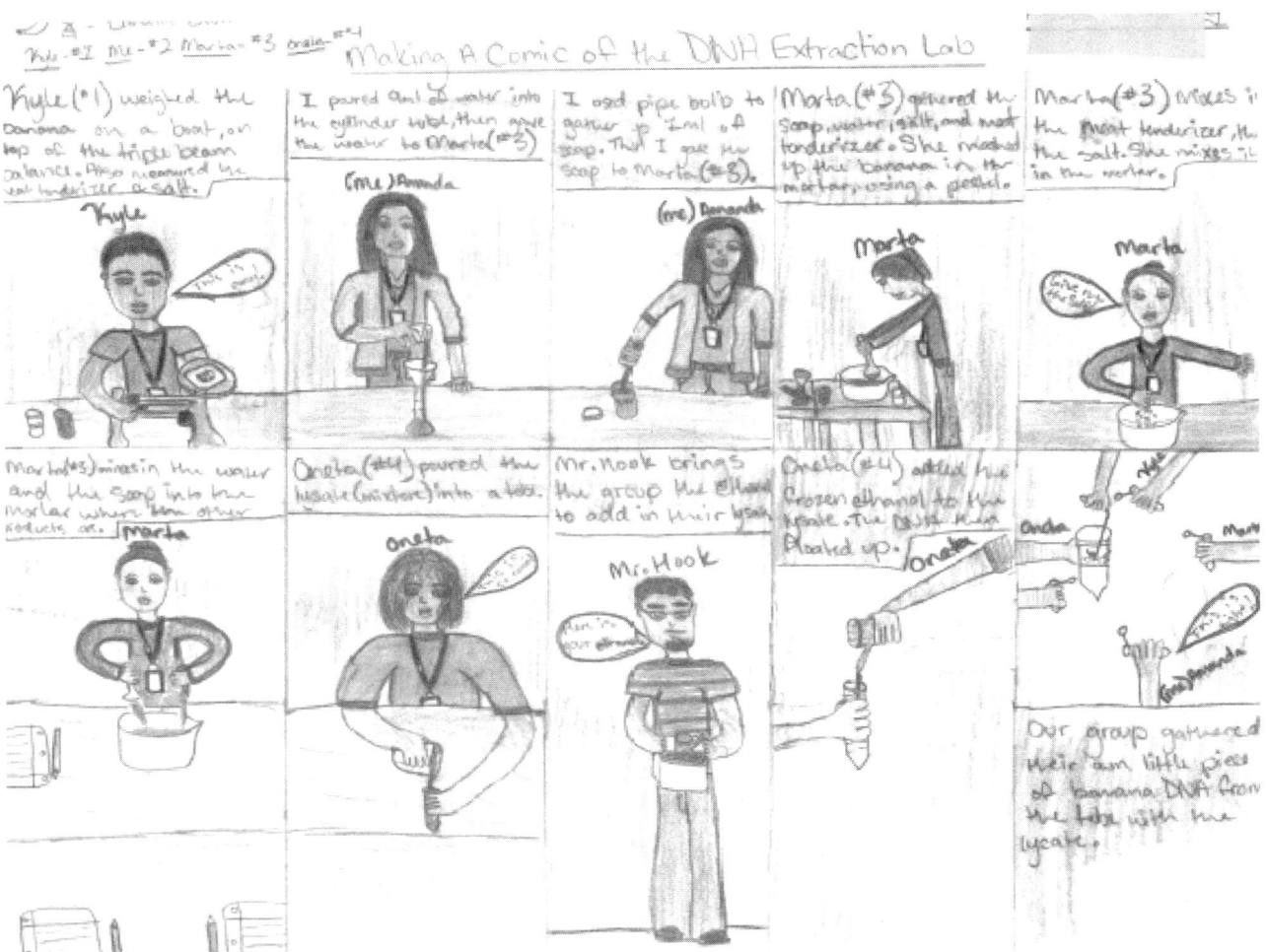

Rubric for DNA extraction lab comic				
	Excellent	Well done	Poor	None
Cartoon has enough panels to detail all steps	32	28	18	0
Cartoon is neatly colored	15	12	7	0
Cartoon is neatly drawn	15	12	7	0
Cartoon has labeled equipment	11	8	6	0
All group members drawn and labeled	12	10	7	0
Each step of procedure and reagents (why we added each one) explained				
• Macerating with the mortar and pestel	3	2	1	0
• Adding the meat tenderizer	3	2	1	0
• Adding the salt	3	2	1	0
• Adding the soap	3	2	1	0
• Adding the 95% ethanol to the lysate	3	2	1	0

As a possible extension to the lab, students can be asked to determine a quantitative lab protocol that produces the greatest yield of DNA. Each group can be given the same exact amount of banana (its mass measured on a balance) and are then allowed to experiment and design a procedure that maximizes the yield of DNA. The precipitated and isolated DNA, can be dried on filter paper (or a small precut pieces of paper towel) and weighed on an analytical balance after it has dried. Be sure and tell students to mass the filter paper before placing the DNA on it and to write the mass of the paper near one edge in pencil. Some inks will be water soluble and bleed making it difficult to read the measurement.

POSSIBLE ANSWERS TO THE CONCLUSION QUESTIONS AND SAMPLE DATA

HYPOTHESIS:

If banana cells are treated with *salt* , *soap*, and *meat tenderizer*, then when exposed to ethanol, the lysate will precipitate DNA

DATA AND ANALYSIS

Description of lysate's appearance:

- Descriptions will vary but will probably include runny, yellowish paste, and so on…

Labeled sketch of lysate, DNA precipitate and alcohol layers in the reaction tube:

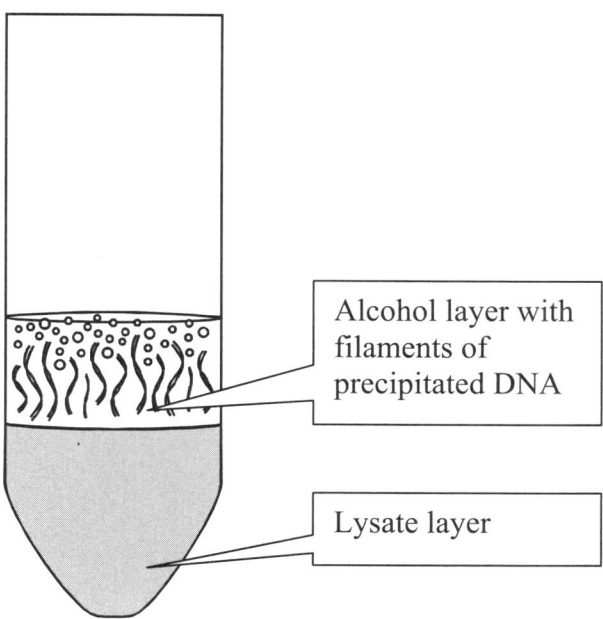

Alcohol layer with filaments of precipitated DNA

Lysate layer

Chemical added to banana	Role of chemical in the reaction
Salt	Adding non-iodized salt helps to further disrupt the cell membrane. It also helps to isolate and precipitate the DNA by shielding the phosphate groups along the sugar/phosphate backbone of the DNA molecule. Specifically, the Na+ ions interacts with the negatively charged phosphates on the side chains of the DNA molecule. This forms a "shield", insulating the molecule and offsetting the attractions between the negative phosphates and the polar water molecules that would occur in the absence of salt.
Meat tenderizer	Adding meat tenderizer releases Papain, a protease enzyme that breaks down the histone protein which frees the DNA. Papain also denatures or breaks down the DNase enzymes present. DNase is an enzyme that digests or breaks down and destroys the DNA. DNase is contained in all cells in nature and will be released into your solution when maceration occurs.
Detergent	The liquid soap binds to the lipid portion of the phospholipid bi-layer comprising the cell membrane and keeps it from reassociating. The cell membrane is dissolved as a result.
Alcohol	Adding ice-cold ethanol will precipitate the DNA. DNA is water soluble as explained earlier. In the presence of salt and alcohol, the DNA forms a precipitate and makes a viscous, white, stringy solid that can be spooled onto a pipet or a bent paper clip. Remember, the colder the ethanol, the less soluble the DNA, the greater you product yield.

Going Bananas for DNA
Isolating DNA, Deoxyribonucleic Acid, from Bananas

PURPOSE
In this laboratory exercise you will extract DNA from bananas using kitchen reagents.

MATERIALS

thumb-size pieces of fresh banana	10 mL graduated cylinder
plastic 15 mL reaction tubes (labeled: reaction tube)	2 1 mL disposable transfer pipets
ICE COLD 95% Ethanol in freezer	1 plastic cup of liquid detergent
plastic 15 mL reaction tube (labeled: salt)	plastic 15 mL reaction tubes (labeled: meat tenderizer)
paper weighing boat or paper	
beaker of distilled water	mortar and pestle
microcentrifuge tubes (to store DNA)	paper clip, bent

Safety Alert
EtOH is flammable, avoid contact with flames

PROCEDURE
Each person needs to read and carefully do their part. If one person does not do his/her part, then the experiment will not work.

Person #1
1. Weigh the empty weigh boat or a small square of paper on a triple beam balance.

2. Write down the mass of boat or paper in grams on the student answer sheet.(to the nearest 1/10th, or .1 of a gram)

3. Remember to consider the mass of the paper or boat when measuring out your reagents. If the paper or boat weighs 1.1 grams, then to obtain 1.0 gram of salt you would need to add enough salt to the paper or boat until the balance reads 2.1 grams.

4. Weigh out 1.0 gram of salt onto the paper or boat on the balance. Give the salt boat to Person #3 and wait for that student to return the empty boat. Be sure and share your numbers with all group members so they can record them on their student answer sheet.

5. Again place the paper or boat on the balance and weigh out 2.0 grams of meat tenderizer. Give the meat tenderizer to Person # 3 and wait for that student to return the empty boat.
   - Meat tenderizer contains enzymes that will destroy the proteins, called histones. Histones stick to and bind up the DNA and protein enzymes called DNases. DNases are enzymes that destroy DNA.

Person #2

6. Measure 9.0 mL of distilled water in the 10mL graduated cylinder and give it to person #3.

7. Use one of your disposable transfer pipets to add approximately 1 mL (fill to the bottom of the bulb) of liquid detergent soap into the mortar. Be careful the soap is very messy.

8. Dispose of disposable transfer pipets into the lined trash can, without dripping!

9. When water and soap have been added to the mortar, let person #3 continue with his/her job.
   - The detergent/salt solution breaks down the lipid membrane of the cells to release the cytoplasm and the cell contents. The salt shields the negative ends of the phosphate molecules from interacting with water molecules on the side chains of the DNA molecule. This will make the DNA less water soluble and thus easier to precipitate or capture.

Person #3

10. Get a thumb-sized piece of banana from your teacher.

11. Place banana in the mortar and pestle. Macerate (mash up) the banana in the mortar.
    - The macerating action further breaks apart the cell membranes of the banana cells.

12. Try not to macerate too hard as you will get foam and bubbles, making it harder to extract the DNA.

13. As team members #1 and #2 hand you reagents, carefully pour all the reagents into the mortar, return the container, and continue to macerate while mixing all the reagents.

14. Once all of your reagents are added, the resulting mix should be a runny thick paste, called a lysate. If it is not a runny paste, add a couple more milliliters of water, measuring the water in the graduated cylinder. Show the lysate to all members and ask them to describe it in the space provided on their student answer sheet.

Person #4

15. Carefully pour the lysate from the mortar into a clean plastic reaction tube. Wipe off lysate from the sides of the reaction tube so all the lysate is in the bottom of the tube.

16. If you have any foam or bubbles on top of your lysate or in the bottom of the reaction tube, try to wipe some of them out with a paper towel. Dispose of paper towels in a lined trash can.

17. Examine your reaction tube. Notice it has volume markings. Carefully, using a squeeze bottle, squirt 15-20 mL of ice cold ethanol down the side of the reaction tube to form a visible alcohol layer on top of the lysate. Do not touch the squirt bottle tip to the reaction tube and do not squirt directly into the lysate. The ethanol must enter the lysate gently.
    - The 95% ethanol must be ICE COLD. It should be left in a plastic squeeze bottle in the freezer or on ice. DNA, in the presence of alcohol, will become an insoluble mucous-appearing stringy solid, called a precipitate.

18. You should immediately begin to see bubbles forming in the alcohol layer (showing that the DNA is becoming a solid.). DNA will form a mucous looking solid with small bubbles, causing it to rise from the lysate layer to the alcohol layer. Gently twirl your test tube to allow more DNA to come up into the alcohol layer. Allow all group members to observe the tube and sketch its appearance on their student answer page.

19. Taking turns, use a clean pipet to suck the DNA out of the alcohol layer and put it into a small DNA microcentrifuge test tube that your teacher provided.

20. Person #4 should make sure that each member gets a small amount of DNA in his/her microcentrifuge tube. After each student gets a small sample of the DNA, Person #4 can add a little ethanol from the alcohol layer to each person's microcentrifuge tube.

21. If the DNA is too thick for the bulb pipet, use a paperclip with a bent tip to pull the DNA out of the alcohol layer.
    • Congratulations! You have extracted DNA!!

22. Person #4: After you have extracted DNA, dispose of the contents of the reaction tube into a lined trash can. Do not throw away the reaction tube.

23. Answer the analysis chart and conclusion activity on the student answer page.

CLEANING DUTIES:

Person #1 → Wipe down balance pan with a damp paper towel, help Person #2.

Person #2 → With Person #1, wipe down the lab table with wet paper towels.

Person #3 → Wipe off all the items in the container that contains the lab equipment, including the container itself with damp paper towels.

Person #4 → Rinse the mortar and pestel in sink. Rinse and clean out the reaction tube.

Going Bananas for DNA
Isolating DNA, Deoxyribonucleic Acid, from Bananas

HYPOTHESIS:

If banana cells are treated with _____, _____-_, and _____ then when exposed to ethanol, the lysate will precipitate DNA

DATA AND ANALYSIS

Weigh boat mass from step : _____

Description of lysate's appearance:

Labeled sketch of lysate, DNA precipitate and alcohol layers in the reaction tube:

Chemical added to banana	Role of chemical in the reaction
Salt	
Meat tenderizer	
Detergent	
Alcohol	

CONCLUSION

1. Draw a comic strip illustrating each step of the procedure, with labeled equipment and explanations of each reagent's purpose. Be sure to include both you and your teacher in the comic strip. The comic strip must be neat and use colors. It can be one big scene or multiple smaller scenes.

Thumb Variation Lab
Measuring Human Phenotypic Variation

OBJECTIVE
Students measure the phenotypic variation of thumbs in a classroom population. Students then use Microsoft Excel to calculate the average measured thumb length and standard deviation within their sample. Students will then write and analyze reasons for variation in a human population.

LEVEL
Middle grades: Life Science

NATIONAL STANDARDS
UCP.3, UCP.4, UCP.5, A.1, C.2, C.3

TEKS
6.1(C), 6.2(A), 6.2(B), 6.2(C), 6.2(D), 6.2(E), 6.4(A), 6.4(B), 6.11(A), 6.11(C)
7.1(A), 7.2(A), 7.2(B), 7.2(C), 7.2(D), 7.2(E), 7.4(A), 7.4(B), 7.10(C)
8.1(A), 8.2(A), 8.2(B), 8.2(C), 8.2(D), 8.2(E), 8.4(A), 8.11(B), 8.11(C)
IPC 1(A), 2(A), 2(B), 2(C), 2(D)

CONNECTIONS TO AP
AP Biology
   II. Heredity and Evolution: A. Heredity 3. Inheritance patterns
   III. Organisms and Populations: A. Diversity of Organisms 1. Evolutionary patterns

TIME FRAME
Two 50 minute periods

MATERIALS
(For a class set of 28)

thumbs	28 metric ruler
paper	28 computers with MS Excel

TEACHER NOTES
Exploring genetic diversity in a middle school classroom can be quite challenging. No child wants to be different than his/her peers, especially in a physical way. The *Thumb Variation Lab* is an ideal way to measure this variation in a non-threatening manner.

The thumb bending trait is determined by a gene represented by a pair of alleles. The two thumb alleles include the more common dominant allele for straight thumbs (H) and a recessive allele for hitchhiker thumb (h). Those people with two recessive alleles for hitchhiker's thumb are able to bend the distal joint of the thumb at an angle of 45 degrees or more. This is also called hyperextension of the thumb.

Fig. 1: Diagram showing angle of thumb

People who possess at least one dominant, (H) allele for a thumb that is straight, or close to being straight, express the dominant phenotype. (See the diagram below.)

Figure 2 represents an ideal hitchhiker thumb. The hitchhiker thumb phenotype would have a homozygous recessive genotype (hh).

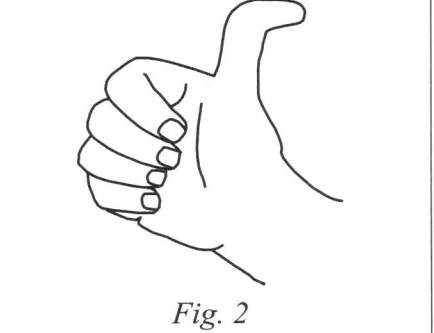

Fig. 2

Because the h allele is recessive, it takes a homozygous recessive individual to express the hitchhiker phenotype. A simple Punnet square analysis shows that about 25% of the population should express this phenotype.

	H	h
H	HH	Hh
h	Hh	hh

However, because our classrooms represent such a small population (as far as population genetics goes), the results will vary widely. Generally the number of straight thumb individuals in a population is greater than the number of hitchhiker thumb individuals due to the dominant gene that expresses itself in the straight phenotype. If you feel it is appropriate, demonstrate a simple monohybrid Punnet square on an overhead to illustrate how 25% of the population is expected to show the recessive phenotype.

The major challenge of this lab is the identification of straight thumb versus hitchhiker thumb in a single individual. By definition, if the distal joint can be bent 45 degrees or more, then that person is said to be hitchhiker. But what if a person's thumb can only be bent to about 35 degrees? Since there is so much phenotypic variation within a population of humans, including thumbs, a true hitchhiker thumb is hard to observe. A certain amount of teacher and student judgement will have to be entertained to determine if a particular thumb is hitchhiker or not. One way to address this difficulty is to have all the students raise their hands, or specifically their thumbs, to see if they have the phenotype early on in the class. You can begin by having the possible hitchhiker students put their thumbs up and have all the students look around and see what sort of variation exists in the hitchhiker thumb phenotype. Once you have observed a student with an obvious hitchhiker thumb, that student can be pointed out as having an "ideal phenotypic expression" Be sensitive to singling out the student in a way that may be interpreted as negative.

After the hitchhiker and straight thumb issues have been clarified, students will make measurements of each student's thumb, determine the phenotype (hitchhiker thumb or non-hitchhiker thumb) and genotype (H_ or hh), analyze the results with Microsoft Excel and explain their findings. Be sure to remind students to keep track of each person's thumb measurement by adding the person's initials next to the measurement. Some teachers have reported that it is actually easier for students to take an accurate measurement if they draw a small line, in pen, on the midpoint of joint 2 of their thumbs, as illustrated in Figure 3.

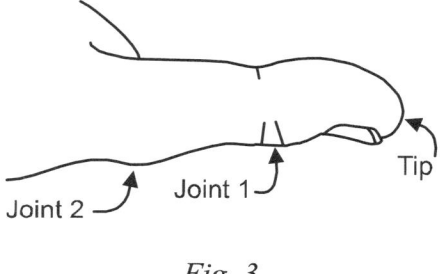

Fig. 3

Also, it is recommended that students practice taking the measurements of each others' thumbs, under the watchful eye of the teacher, in a guided practice, to ensure that accurate measurements are taken.

One way to calculate variation within a population is to calculate the standard deviation of the population. When using Excel, this is a simple formula that is entered into the cell, =STDEV(B2:B28), where B2 through B28 are the spreadsheet cells that display the range of numbers that represent the measurements of the thumbs. The real challenge comes in explaining what standard deviation means! A link to a web site has been included here (http://www.robertniles.com/stats/stdev.shtml) that is for journalists (non-scientists) that does a very adequate job of explaining standard deviation in very simple terms. This website is essentially for you to be able to explain to the students what standard deviation is; however, the website could also be assigned as a pre-lab reading assignment. Once the students calculate the standard deviation, it is important to discuss the significance of the number and its relevance to population diversity. A larger standard deviation represents more diversity in the population and a smaller calculation of standard deviation represents a smaller amount of deviation in the population.

There are many extensions that can be done within the Excel spreadsheet. The data could be sorted by sex, age, or race. Then those subpopulations could be compared. There are an almost unlimited number of extensions to this activity resulting from subdividing into many different categories and analyzing the data in many different ways.

It is highly recommended for students to measure the thumbs of students in another classroom. Increasing the sample size results in a smaller standard deviation and thus the data collected better fits the results predicted by the monohybrid Punnet square. As another option, require the students measure the thumbs of all the teachers in the school. You could send a group of 3 or 4 students to get each teacher's thumb length and a determination of hitchhiker phenotype. The students return to the classroom and then compile the results on a computer or overhead so everyone can get all the teacher data. The results of the teachers can then be put in spreadsheets and shared with the faculty. Students and faculty alike share in the joy of learning!

Included here is an image of a spread sheet that contains data from real teachers. This data was collected by students in a junior high school.

```
Microsoft Excel - teacher_thumb_data.xls
  File   Edit   View   Insert   Format   Tools   Data   Window   Help

  D34            fx
```

	A	B	C	D	E
1	**thumb variation lab**				
2	**teacher data**				
3	**measurements in mm**				
4					
5					
6	hitchhiker thumb (hh)			non-hitchhiker thumb (HH or Hh)	
7	initials of each teacher	length (mm)		initials of each teacher	length (mm)
8	C.V	69.2		M.W.	70.4
9	A.C.	70.6		D.S.	65.8
10	G.C.	65.7		T.W.	50.9
11	F.D.	70.2		J.N.	65.6
12	J.B.	71.3		C.H.	80.1
13	B.D.	58.4		S.W.	64.5
14	M.G.	60.5		K.W.	61.7
15	P.H	60.2		J.M.	60.4
16	K.G.	70.6		G.O.	70.1
17				S.R.	59.3
18	average of HHT-teachers	66.30		D.V.	56.1
19	std dev of HHT-teachers	5.23		A.H.	62.0
20				M.G.	59.7
21				J.G.	63.1
22				B.B	55.2
23				S.G.	67.1
24				S.R.	65.2
25				R.D.	65.7
26				J.E.	68.6
27				S.F.	64.3
28					
29				average non-HHT-teachers	63.79
30				std dev of non-HHT-teachers	6.29
31					

This lab has been modified significantly from the original "Thumb Variation Lab" that is posted at http://www.accessexcellence.com.

POSSIBLE ANSWERS TO THE CONCLUSION QUESTION

Using the entire front side of a piece of notebook paper write an essay explaining why there is variation (measurable differences) in our experiment/measurements. What sort of results did you observe? Were there any patterns that you could detect? Address issues such as: Why are students in junior high so different? Are there reasons that students at your particular school would be even more or less different than other schools? What are some sources of error? If you measured the teachers' thumbs, were they different from the students' measurements? What would explain variation in different aged populations?

Each person should turn in the following:
   a. Your student answer page(s) containing data tables (from one or two classes depending on how much time you had to spend in class on the lab) with each class member's initials and appropriate thumb length in millimeters from one class (or two classes if you had time to get data from another class). (30 pts, (15 pts each if two tables))
   b. 1 printed copy of your spreadsheet table(s). (print it twice, one for each student) (30 pts)
   c. 1 essay on a piece of notebook paper answering the CONCLUSION QUESTION above. (40 pts)
   d. Staple each of these papers together in the order above. Make sure that your name and period are written on the top page.

   • Consider discussing with the students the idea of variation before they attempt to answer the question. This can include (but is certainly not limited to): puberty (and how puberty is so variable at the junior high/middle school level), racial diversity, age and also gender. How can each of these affect thumb length?
   • Certainly the students in the typical junior high school classroom exhibit a great deal of variation in their physical traits. There are students that look as if they are 10 years old and those that look as if they are 18 years old. Because of this amazing amount of variation in maturity (and consequently in thumb length), a large standard deviation calculation is expected and usually calculated.
   • Also, some other things to consider include differences between the genders. Do males have larger than average thumbs than females? Probably in nature, with adults, that is true. But in junior high, when females typically mature at an earlier age, this might not be completely correlated. Also, racial diversity and other factors can be considered, depending on your comfort level. Of course, these are all sensitive topics that you must treat with the utmost sensitivity and respect.

5

Thumb Variation Lab
Measuring Human Phenotypic Variation

This lab will investigate some of the variation of genetic traits in a population. Thumb length and bending ability will be examined and then statistically analyzed. The thumb bending trait is determined by a pair of alleles. The two thumb alleles include the more common dominant allele for straight thumbs (H) and a recessive allele for hitchhiker thumb (h). Those people with two recessive alleles for hitchhiker's thumb are able to bend the distal joint of the thumb at an angle of 45 degrees or more. This is also called hyperextension of the thumb. People possessing at least one dominant H allele for a thumb that is straight, or close to being straight, express the dominant phenotype.

PURPOSE
You will examine the thumb phenotype of each member of your class. Once you have determined everyone's thumb type, you will measure and record each person's thumb length. The data will then be analyzed using the spreadsheet program, Excel.

MATERIALS
thumb

paper

metric ruler

computer with MS Excel

PROCEDURE
1. In the section labeled hypothesis on your student answer sheet record the percentage of people in your class that you expect to display the hitchhiker thumb phenotype.

2. In the space provided on your student answer page, draw a data table with three columns and as many rows as there are students in your class. The columns should be titled: Initials, straight thumb length (mm) and hitchhiker thumb length (mm).

3. Determine whether or not you have a hitchhiker's thumb by comparing to the diagram below.

4. By using the picture below, measure your thumb in millimeters from joint 2 to the tip of your finger. Do not include the measurement of your fingernail. Be as accurate as you can in your measurement, to the nearest 0.1 millimeter. Your answer might be 80.1 millimeters but not 80 millimeters.

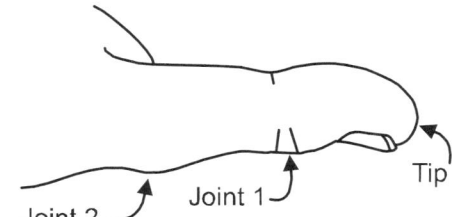

(Use this picture when determining what to measure.)

5. Record the measurement of your own thumb in millimeters in the appropriate thumb type column. Record your initials in the first column.

6. Record the measurements of your classmates' thumbs in the appropriate columns. Be sure you have the same number of rows as there are students in your class. If you have difficulty determining the thumb type, ask your teacher for help.

7. If your teacher instructs, repeat step 6 for another group of people. Use a separate piece of paper for this second data table.

8. Working with a partner, use a computer to start the program, MS Excel (Start>Programs>Microsoft Excel), to create a spreadsheet of your data from the two classes.

9. Save the empty spreadsheet onto your hard drive or into a network folder if you have that capability. Record the file name on your student answer page.

10. Transfer the data from your data table into columns A, B, and C on the Excel worksheet. If you measured a second group of people use a different sheet in the same workbook.

11. Once your values are entered, skip a row and title a new row AVERAGE and the following row STANDARD. DEVIATION. Then, use the spreadsheet to calculate the following: (Your teacher will demonstrate how to start this.)
   a. The mean (average) thumb lengths of H_ individuals in your class.
      • The command for averaging a column is =AVERAGE(B2:Bx), where B or C is the column of numbers you wish to average and where x is the number of your last row on the spreadsheet.
   b. The mean (average) thumb lengths of hh individuals in your class.
   c. The mean (average) thumb lengths of H_ individuals in the second test group.
   d. The mean (average) thumb lengths of hh individuals in the second test group.
   e. The standard deviation in thumb lengths of H_ individuals in your class
      • The command for calculating the standard deviation is =STDEV(B2:Bx) where x is the number of your last row on the spreadsheet.
      • Standard deviation is a statistic showing how much variation from the mean exists in your sample.
   f. The standard deviation in thumb lengths of hh individuals in your class
   g. The standard deviation in thumb lengths of H_ individuals in the second test group.
   h. The standard deviation in thumb lengths of hh individuals in the second test group.

i. If your teacher allows, determine how you would calculate the standard deviation for the H_ and hh trait for BOTH classes on your spreadsheet. Extra credit may be awarded for successful completion.

A sample worksheet with formulas is shown below:

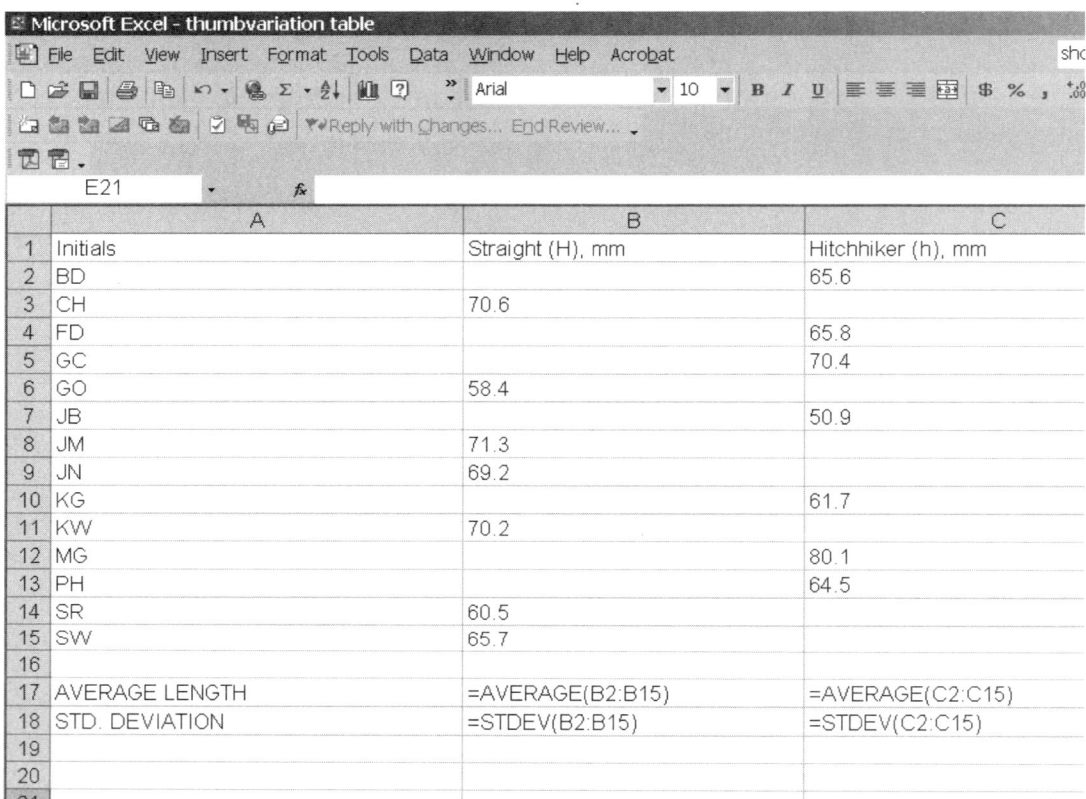

	A	B	C
1	Initials	Straight (H), mm	Hitchhiker (h), mm
2	BD		65.6
3	CH	70.6	
4	FD		65.8
5	GC		70.4
6	GO	58.4	
7	JB		50.9
8	JM	71.3	
9	JN	69.2	
10	KG		61.7
11	KW	70.2	
12	MG		80.1
13	PH		64.5
14	SR	60.5	
15	SW	65.7	
16			
17	AVERAGE LENGTH	=AVERAGE(B2:B15)	=AVERAGE(C2:C15)
18	STD. DEVIATION	=STDEV(B2:B15)	=STDEV(C2:C15)
19			
20			
21			

Your single class spreadsheet will look something like the sample below when it is finished. To round calculations to 2 decimal places highlight the calculated cells, and right click. Select **format cells** from the pop up menu. In this dialog box the **number** tab will allow you to change the number of decimal places used for these cells.

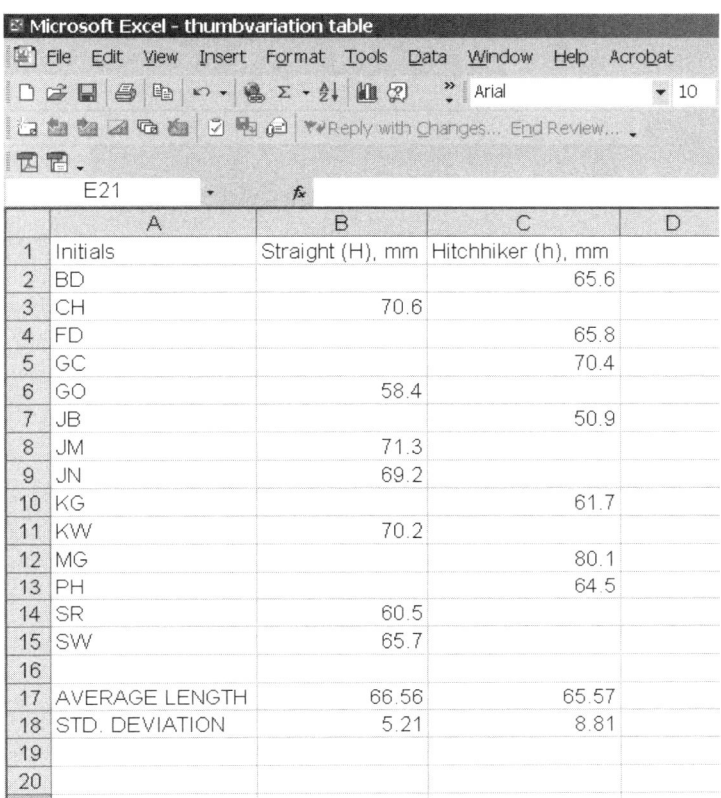

	A	B	C	D
1	Initials	Straight (H), mm	Hitchhiker (h), mm	
2	BD		65.6	
3	CH	70.6		
4	FD		65.8	
5	GC		70.4	
6	GO	58.4		
7	JB		50.9	
8	JM	71.3		
9	JN	69.2		
10	KG		61.7	
11	KW	70.2		
12	MG		80.1	
13	PH		64.5	
14	SR	60.5		
15	SW	65.7		
16				
17	AVERAGE LENGTH	66.56	65.57	
18	STD. DEVIATION	5.21	8.81	
19				
20				

5

Name _____

Period _____

Thumb Variation Lab
Measuring Human Phenotypic Variation

HYPOTHESIS _____

DATA AND OBSERVATIONS _____

Create your data table here. Use a piece of notebook paper if necessary.

Excel file name:_____

CONCLUSION QUESTION

Using the entire front side of a piece of notebook paper write an essay explaining why there is variation (or measurable differences) in our experiment/measurements. What sort of results did you observe? Were there any patterns that you could detect? Address issues such as: Why are students in junior high so different? Why are students at your school so different? What are some sources of error? If you measured the teachers' thumbs, were they different from the students' measurements? What would explain variation in different aged populations?

Each person should turn in the following:

  a. Your student answer page with 2 data tables, drawn neatly with a ruler, one for each class set of data (your class and another class) with each class member's initials and appropriate thumb length in millimeters. (30 pts, 15 pts each table for the two tables representing two classes)

  b. 1 printed copy of your spreadsheet table(s). (print it twice, one for each student) with calculations for average and standard deviation. (30 pts)

  c. 1 essay on a piece of notebook paper answering the CONCLUSION QUESTION above (40 pts)

  d. Staple each of these papers together in the order above. Make sure that your name and period are written on the top page.

Punnett Square Exercises
Solving Monohybrid Punnett Squares

OBJECTIVE
Students will use Punnett squares to determine possible gene combinations in the offspring for different sets of parents.

LEVEL
Middle grades

NATIONAL STANDARDS
C.2

TEKS
6.11(A), 6.11(B), 6.11(C)
7.10(C)
8.11(C)
IPC 2(C), 2(D)

CONNECTIONS TO AP
AP Biology:
    II. Heredity and Evolution: A. Heredity 1. Meiosis and gametogenesis 2. Eukaryotic chromosomes
    3. Inheritance patterns

TIME FRAME
90 minutes

MATERIALS
(For a class of 28 working individually)

    28 *Genetics: Helpful Terms and Key Concepts* sheets

TEACHER NOTES
Punnett Square Exercises is an introductory activity in solving Punnett squares and Mendelian analysis. This exercise may serve as a beginning activity for a middle grades genetics unit. It is designed to be a stand-alone lesson. The directions are written very specifically to teach the concepts with minimal teacher guidance. Students will become quite proficient at solving Punnett squares using the steps outlined in the student procedure pages. Students begin the lesson by reading, *Genetics: Helpful Terms and Key Concepts* pages. Students should pay close attention as the teacher works through one example on the board or overhead. They should then work independently to solve the problems given on the student answer page. You should monitor the students as they work, clarifying misconceptions or confusion.

POSSIBLE ANSWERS TO THE CONCLUSION QUESTIONS

1. Explain why there are at least two alleles that make up the gene coding for a particular trait.
   - There are at least two alleles for every gene because each parent contributes equally to the inheritance of the offspring.

2. Explain the term "heterozygous". How can an organism have two different alleles for one single trait but only show one observable trait?
   - Heterozygous refers to two different alleles representing a single trait. For each gene, there are two alleles, one from each parent. Sometimes these alleles can be the same (homozygous) and sometimes they can be different (heterozygous). If the alleles are different, then one must be dominant, and the other recessive. The recessive allele will be masked by the expression of the dominant allele when only one trait is observable.

PRACTICE PUNNETT SQUARES: Draw all Punnett Squares with a ruler. <u>Each side of the square should be at least 4 cm long</u>.

3. A common recessive trait in dogs is deafness. A homozygous line of normal-hearing dogs was crossed with a homozygous line of deaf dogs. F_1 and F_2 generations were produced.
   a. What percentage of the F_1 generation is expected to have normal hearing?
   b. What is the phenotypic ratio of the F_2 generation?

F_1 Generation:

	d	d
D	D d	D d
D	D d	D d

F_2 Generation:

	D	d
D	D D	D d
d	D d	d d

a. 100%

b. 3:1 ratio

4. Black color in horses is dominant over chestnut color. If a homozygous black horse is mated to a chestnut horse, what percent of the offspring will be chestnut colored? Use a Punnett square to show how you derived your answer.

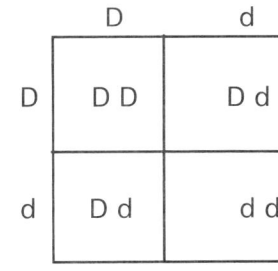

0% will be chestnut colored.

(All are heterozygous; i.e., black.)

5. Construct a Punnett Square to show all possible combinations of gametes that could result from the cross of a black guinea pig that is heterozygous for black fur and a guinea pig that is homozygous recessive for black fur (which results in white fur coloration).
   a. What is the probability that the offspring will have black fur? [Answer should be a ratio.]
   b. What percent of the offspring will be homozygous recessive?

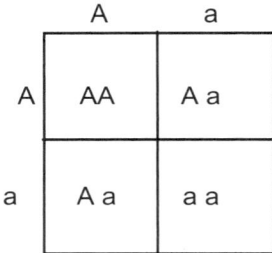

    a. 1:1 ratio of black to white fur.

    b. 50% will be homozygous recessive.

6. George is the youngest child in a family that includes 4 children. All of the children have the recessive phenotype — attached earlobes. However, both parents have unattached earlobes, the dominant trait. What are the possible genotypes for the parents? Use a Punnett square to justify your answer.

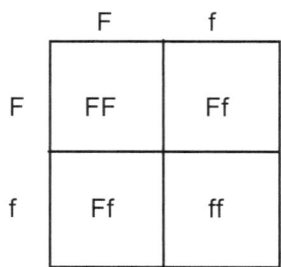

Both parents must be heterozygous for this trait since their offspring express the recessive trait.

7. Explain why it is possible for two parents with a dominant trait, like freckles, to produce 10 children, all of whom do not have freckles. Use a Punnett square to illustrate your answer.

	F	f
F	FF	Ff
f	Ff	ff

Both parents must be heterozygous for this trait since their offspring express the recessive trait.
If even one parent was homozygous dominant, then all the children would also have freckles. Because both parents are heterozygous, however, then there is a one-in-four chance (or 25%) that any child born will not have freckles. For those heterozygous freckled parents to have 10 children in a row without freckles is highly unlikely but IS possible.

8. A woman is homozygous dominant for short fingers. Will any of her children have long fingers? Use a Punnett square to justify your answer.

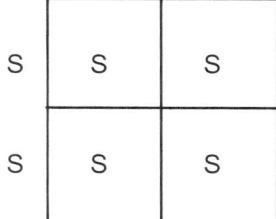

None of her children will have long fingers. The Punnett square shows that, regardless of the father's genotype, all of her offspring will receive at least one dominant allele.

9. In guinea pigs, black fur color is dominant over white fur. Describe the parents of a family of 7 black and 2 white guinea pigs. Use a Punnett square to illustrate your answer. [Hint: The F_1 ratio is 7:2 or approximately 3:1.]

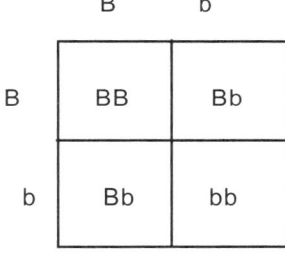

The F1 ratio of 7:2 is close to the ratio 3:1. The students should realize by now that a 3:1 ratio of phenotypes is produced by crossing heterozygous parents. The genotype of both parents, therefore, is Bb.

Punnett Square Exercises
Solving Monohybrid Punnett Squares

PURPOSE
In this activity you will use Punnett squares to determine possible gene combinations for the offspring from several sets of parents.

PROCEDURE

1. Read and study Genetics: Helpful Terms and Key Concepts pages.

2. To start, pay close attention to the VOCABULARY section. Genetics is in many ways, like a foreign language with many new terms. Your teacher will review the concepts with you.

3. Read and study the next section, KEY CONCEPTS, paying very close attention to the Gene-Chromosome Theory section.

4. Finally, read through USING PUNNETT SQUARES TO PREDICT OFFSPRING section, studying the diagram and instructions. Your teacher will assist you with an example..

5. Work through the remainder of the Punnett square problems on the student answer page. Be sure to provide a Punnett square for each answer as well as all written work to show how you solved each problem.

Name _____

Period _____

Punnett Square Exercises
Solving Monohybrid Punnett Squares

CONCLUSION QUESTIONS

1. Explain why there are at least two alleles that make up the gene coding for a particular trait.

2. Explain the term "heterozygous". How can an organism have two different alleles for one single trait but only show one observable trait?

PRACTICE PUNNETT SQUARES: Draw all Punnett Squares with a ruler. Each side of the square should be at least 4 cm long.

3. A common recessive trait in dogs is deafness. A homozygous line of normal-hearing dogs was crossed with a homozygous line of deaf dogs. F_1 and F_2 generations were produced.
   a. What percentage of the F_1 generation is expected to have normal hearing?
   b. What is the phenotypic ratio of the F_2 generation?

4. Black color in horses is dominant over chestnut color. If a homozygous black horse is mated to a chestnut horse, what percent of the offspring will be chestnut colored? Use a Punnett square to show how you derived your answer.

5. A gene that codes for black fur color in guinea pigs has two alleles. Two recessive copies of this allele will result in a guinea pig that cannot produce black fur; it will therefore have white fur. Construct a Punnett Square to show all possible combinations of gametes that could result from the cross of a heterozygous black guinea pig with a white guinea pig.

Construct a Punnett Square to show all possible combinations of gametes that could result from the cross of a black guinea pig that is heterozygous for black fur and a guinea pig that is homozygous recessive for black fur (which results in white fur coloration).
   a. What is the probability that the offspring will have black fur? [Answer should be a ratio.]
   b. What percent of the offspring will be homozygous recessive?

6. George is the youngest child in a family that includes 4 children. All of the children have the recessive phenotype — attached earlobes. Yet both parents have unattached earlobes, the dominant trait. What are the possible genotypes for the parents? Use Punnett squares to prove your answer.

7. Explain why it is possible for two parents with a dominant trait like freckles to produce 10 children, all of whom do not have freckles. Use a Punnett square to illustrate your answer.

8. A woman is homozygous dominant for short fingers. Will any of her children have long fingers? Use a Punnett square to prove your answer.

9. In guinea pigs, black fur color is dominant over white fur. Describe the parents of a family of 7 black and 2 white guinea pigs. Use a Punnett square to illustrate your answer. [Hint: The F_1 ratio is 7:2 or approximately 3:1.]

Punnett Square Exercises
Solving Monohybrid Punnet Squares

Genetics: Helpful Terms and Key Concepts

VOCABULARY

- **heredity**: passing traits from parent to offspring.

- **genetics**: the study of heredity and variation in organisms.

- **DNA or deoxyribonucleic acid**: a double-stranded helical nucleic acid molecule that codes and stores genetic information; it determines the inherited structure of a cell's proteins

- **chromosomes**: threadlike strands of DNA and protein in a cell nucleus that carry the code for the inherited characteristics of an organism.

- **gene**: a distinct unit of hereditary material found in chromosomes; the region of DNA that directs the making of a specific protein, thus controlling traits that are passed to offspring.

- **allele**: different forms of a gene controlling a particular trait; e.g., one allele for height may be short, while another allele may be tall.

- **dominant allele**: the form of a gene that appears to dominate or mask another form of the same trait. The form of the allele that is expressed in the phenotype.

- **recessive allele**: the form of a gene that is often masked by the dominant allele and is therefore often not expressed in the phenotype.

- **homologous chromosomes**: a pair of chromosomes having the same size and shape and carrying alleles for the same traits.

- **incomplete dominance**: a type of inheritance in which two contrasting alleles contribute to the individual a trait not exactly like either parent; blending inheritance; when both alleles are expressed in offspring and neither is dominant.

- **codominance**: a type of inheritance in which two dominant alleles are expressed at the same time without blending of traits.

- **heterozygous**: organism that has two different alleles for a trait.

- **homozygous**: organism that has two identical alleles for a trait.

- **genotype**: the genetic makeup of an individual.

- **phenotype**: the physical traits that appear in an individual as a result of its genetic makeup.

- **gamete**: a reproductive cell that joins with another in fertilization; most often, the joining of an egg and a sperm cell; a haploid cell.

- **mitosis**: the process by which the nucleus of a cell divides, while maintaining its correct chromosome number; in mitosis, the cell makes an exact (though smaller) copy of itself.

- **meiosis**: cell division in diploid cells that results in haploid cells; reduction division; in meiosis, the sex cells are produced; sex cells are nonidentical.

- **haploid**: having only one chromosome from each pair of homologous chromosomes; sex cells are haploid and are designated as *n*.

- **diploid**: having two sets of chromosomes, designated as $2n$; the cells of the body are diploid with the exception of the sex cells.

- **cross**: gamete union resulting in offspring.

- **P generation**: parent generation; the starting generation in a breeding experiment.

- **F$_1$ generation**: first filial generation; the first generation produced in a breeding experiment.

- **F$_2$ generation**: second filial generation; the second generation produced in a breeding experiment; offspring of the F$_1$ generation.

- **Punnett square**: diagram in which all possible types of gametes from one parent are lined up vertically and all possible types of gametes from the other parent are lined up horizontally and every possible fertilization combination is considered; a tool that shows how genes can combine; used to predict results in genetics.

KEY CONCEPTS

- **The Law of Dominance**: When an organism is heterozygous for a pair of contrasting traits, only the dominant trait can be seen in the organism.

- **The Law of Segregation**: Genes that occur in pairs are separated from each other during gamete formation and recombined at fertilization.

- **The Law of Probability**: If there are several possible events that could occur, and one is no more likely to happen than any other, then the events will all occur equally when a large number of trials are studied. This law allows you to predict the results of breeding experiments. However, remember that these predictions apply only when large numbers of individuals are involved.

- **Gene-Chromosome Theory**:

Gregor Mendel conducted his experiments with pea plants from 1857 to 1865. He stated that each trait in an individual was controlled by a pair of "factors." The idea that Mendel's "factors" might be genes carried by homologous chromosomes was suggested first in 1903 by an American graduate student,

W. S. Sutton. He observed the pairs of homologous chromosomes in diploid grasshopper cells separate during sperm formation (spermatogenesis). After reviewing Mendel's work, Sutton concluded that the factors of Mendel's theory were carried on the chromosomes since the chromosomes were separating and recombining, as Mendel's factors were theorized to do. Figure 1 shows how Sutton's chromosome theory would apply to a Mendelian cross of two tall heterozygous plants.

As you can see in figure 1, the separation of homologous chromosome pairs during meiosis and their recombination during fertilization would account for the separation and recombination of the Mendelian factors.

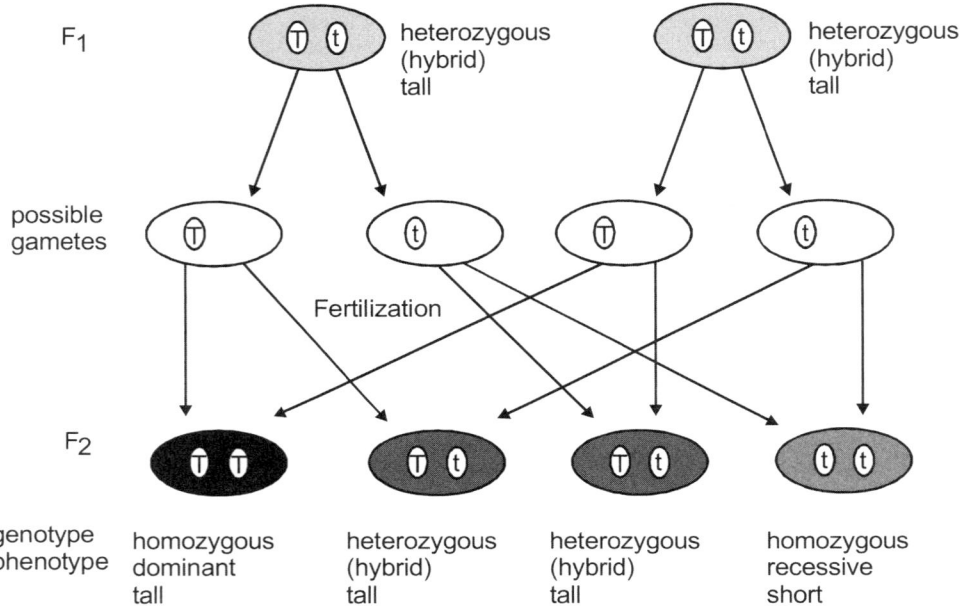

Figure 1. The Gene-Chromosome Theory. T represents the allele (factor) for tallness and t the allele for shortness. Each allele is on a homologous chromosome. When the pairs of homologous chromosomes separate during gamete formation, they form two kinds of gametes: one with T and the other with t. Upon fertilization chromosomes from each parent reunite to form a new homologous pair.

Following the publication of Sutton's paper, titled "The Chromosomes in Heredity," in 1903, many experiments showed that this hypothesis was correct. At that time, the term *gene* was used in place of Mendel's "factor." Research showed not only that chromosomes carry genes but also that the genes are in a definite order along each chromosome. This work led to the modern gene-chromosome theory of heredity.

USING PUNNETT SQUARES TO PREDICT OFFSPRING

A Punnett square is used to determine all possible gene combinations in the offspring of a set of parents. Remember that each gene is made up of at least two alleles — one allele from the mother and one from the father. Biologists represent a particular allele in a Punnett square by using a symbol. The dominant allele is represented by a capital letter. The recessive allele is represented by the corresponding lowercase letter. In Mendel's pea plant experiments, for example, T represents the dominant allele for tallness and t represents the recessive allele for shortness.

The Punnett squares in Figure 2 below show the types of reproductive cells, or *gametes,* produced by the F_1 generation parents along the top and left-hand side of the square. It includes directions on how to complete the square. Finally, it shows each possible gene combination for the F_2 offspring in the four boxes that make up the square.

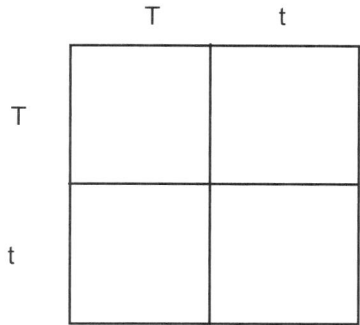

Step 1. Place the alleles of the parents on the outside of the square.

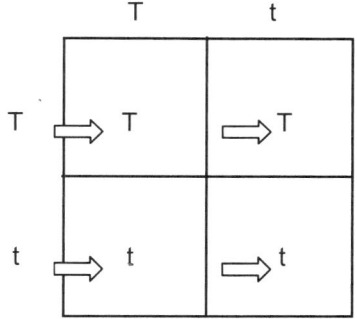

Step 2. Fill in each square with the allele that is to the left on the outside of the box.

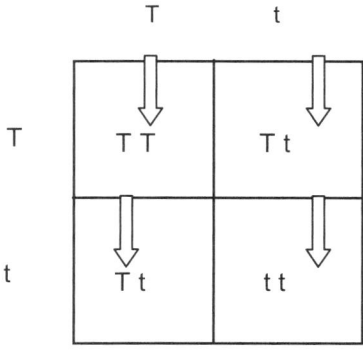

Step 3. Fill in each square with the allele that is above on the outside of the box.

Figure 2. Steps used in constructing a Punnett square. Note that the final square shows all possible gene combinations for the offspring. In addition, the Punnett square allows you to predict the probability of obtaining any particular outcome from a given cross. In this example, 2 out of 4, or 50%, of the offspring will have the Tt genotype.

Figure 2

You can see the probable results of the cross of two F_1 generation plants from the Punnett square: 1/4 of the F_2 generation plants have two dominant alleles (TT); 2/4 or 1/2 of the F_2 plants have one dominant allele and one recessive allele (Tt); and 1/4 of the F_2 plants have two recessive alleles (tt). Because tall is dominant over short, 3/4 of the F_2 plants would be tall and 1/4 of the F_2 plants would be short. These numbers are often expressed as ratios. For example, there are 3 tall plants for every 1 short plant in the F_2 generation. Thus the phenotypic ratio of tall plants to short plants is 3:1.

Look again at figure 2 You will see that three of the possible combinations result in tall plants. Because all these plants appear tall, we can say that they have the same *phenotype,* or physical characteristic. They do not, however, have the same *genotype,* or genetic makeup. The genotype of 1/3 of the tall plants (or 1/4 of all offspring) is pure dominant (homozygous dominant), or TT, whereas the genotype of 2/3 of the tall plants is mixed dominant and recessive (heterozygous dominant), or Tt. The short plants, tt, are known as homozygous recessive.

Organisms that have two identical alleles for a particular trait (TT or tt in our example) are said to be *homozygous.* Organisms that have two different alleles for the same trait are *heterozygous.* In other words, homozygous organisms are pure-bred for a particular trait and heterozygous organisms are *hybrid* for a particular trait. See Table 1 for a comparison of genotypes and phenotypes.

Table 1. Comparison of genotype to phenotype.		
Genotype Abbreviation	Genotype	Phenotype
TT	Homozygous dominant	Tall
Tt	Heterozygous	Tall
tt	Homozygous recessive	Short

Gene Interactions
Solving Dihybrid and Complex Monohybrid Punnett Square Problems

OBJECTIVE
Students will solve dihybrid and complex monohybrid Punnett squares.

LEVEL
Middle grades: Life Science

NATIONAL STANDARDS
UCP.1, UCP.2, UCP.5, C.2

TEKS
6.2(C), 6.11(A), 6.11(B), 6.11(C)
7.3(C), 7.10(B), 7.10(C)
8.2(C), 8.3(C), 8.11(B), 8.11(C)
IPC 2(C), 2(D)

CONNECTIONS TO AP
AP Biology:
    I. Molecules and Cells: B. Cells 3. Subcellular organization
    II. Heredity and Evolution: A. Heredity 1. Meiosis and gametogenesis 3. Inheritance patterns
    II. Heredity and Evolution: B. Molecular Genetics 1. RNA and DNA structure and function

TIME FRAME
60 minutes

TEACHER NOTES
Gene Interactions: Solving Dihybrid and Complex Monohybrid Punnet Square Problems is an intermediate to advanced activity in solving Punnett squares. This exercise should follow in sequence after the *Punnett Square Exercises* lesson. This lesson is designed to be a stand-alone lesson, once simple Punnett squares are mastered. The directions are written very specifically to teach the concepts with minimal teacher guidance. Students will become quite proficient at solving Punnett squares using the steps outlined in the student procedure.

The lesson also assumes that you are proficient in solving dihybrid and other complex Punnett squares. You should work through all these problems before attempting to teach them to the students as they are quite complex. To brush up on your background knowledge, there are also a number of good Internet websites devoted to Mendelian inheritance, genetics and Punnett squares. A search on Google, http://www.google.com for "punnett square tutorial" will provide hundreds of results.

One exceptional site worth noting is *The Biology Project-Mendelian Genetics* site.

http://www.biology.arizona.edu/mendelian_genetics/mendelian_genetics.html

Here you, and possibly your students, will find exceptionally helpful tutorials, showing step-by-step solutions with clear phenotype diagrams of simple monohybrid, complex monohybrid crosses (including incomplete dominance and codominance), and a heterozygous dihybrid cross. It is highly recommended you and/or your students visit this site.

Students begin the lesson by reading, *Genetics: Key Concepts* pages, including the vocabulary. Genetics in one respect is like studying a foreign language, with the amount of new and difficult-to-pronounce words/terminology. After the students have read the key concepts pages, you should work a dihybrid Punnett square problem on the overhead or chalkboard. The major stumbling block for students when performing a dihybrid cross is generating the gametes from the parental genotype. Consequently, you should spend quite some time reinforcing this concept. Students should pay close attention and take notes as you work through one example on the board or overhead. They should then work independently to solve the problems given on the student answer page. You should monitor the students as they work, clarifying misconceptions or confusion.

POSSIBLE ANSWERS TO THE CONCLUSION QUESTION AND SAMPLE DATA

1. Using the Punnett square in figure 3, list the genotype, genotype abbreviation, phenotype and predicted fraction of offspring for each different gene combination.

Abbrev	Genotype	Phenotype	Predicted Fraction
RRYY	homozygous dominant for both	round and yellow	1/16
RRYy	homozygous dominant-heterozygous	round and yellow	2/16 or 1/8
RrYY	heterozygous-homozygous dominant	round and yellow	2/16 or 1/8
RrYy	heterozygous-heterozygous	round and yellow	4/16 or 1/4
RRyy	homozygous dominant-homozygous recessive	round and green	1/16
Rryy	heterozygous-homozygous recessive	round and green	2/16 or 1/8
rrYY	homozygous recessive-homozygous dominant	wrinkled and yellow	1/16
rrYy	homozygous recessive-heterozygous	wrinkled and yellow	2/16 or 1/8
~~Rryy~~ rryy	homozygous recessive for both	wrinkled and green	1/16

2. What is the phenotypic ratio of the offspring that results from crossing two individuals that are heterozygous for two nonlinked traits?
   - The ratio is 9:3:3:1. Using your answer to #1 above, you can see that 9 of 16 will be round and yellow, 3 of 16 will be round and green, 3 of 16 will be wrinkled and yellow, and 1 of 16 will be wrinkled and green.

3. Construct a Punnett square showing all possible gamete combinations and all possible offspring for a cross between a woman who is heterozygous for widow's peak and short fingers (WwSs) and man who has a continuous hairline and short fingers (wwSs).

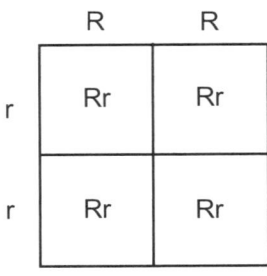

	WS	Ws	wS	ws
wS	WwSS	WwSs	wwSS	wwSs
ws	WsSs	Wsss	wwSs	wwss
wS	WsSS	WsSs	wwSS	wwSs
ws	WsSs	Wsss	wwSs	wwss

WwSs — WS, Ws, wS, ws
wwSs — wS, ws, wS, ws

a. 1/8th will be homozygous recessive for both traits.
b. 1/8th will have a widow's peak with long fingers.
c. 1/4th will have the identical genotype as the mother.

4. How did Kölreuter know that the genes of the parent flowers had not blended together but that the alleles for red color and white color still existed separately?
   - Because a cross between pink offspring (members of the F_1 generation) produced offspring that were red, white or pink.

5. Construct a Punnett squares to show the F_1 and F_2 generations from a P generation cross of red with white carnations.

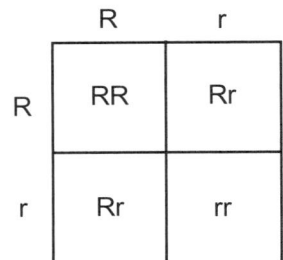

	R	R
r	Rr	Rr
r	Rr	Rr

F_1 generation

	R	r
R	RR	Rr
r	Rr	rr

F_2 generation

TEACHER PAGES

a. For the F$_2$ generation:

Abbrev.	Genotype	Phenotype	Predicted Fraction
RR	homozygous (incomplete) dom.	red	1/4
Rr	heterozygous	pink	2/4 or 1/2
rr	homozygous (incomplete) rec.	white	1/4

6. Construct a Punnett square to show the cross between a pink and a white carnation.

	R	r
r	Rr	rr
r	Rr	rr

a. 2/4 or 1/2 will be pink.
b. 0/4 or none will be red.

7. Construct a Punnett square to show the possible offspring of erminette chickens.

	F^B	F^W
F^B	F^{B}F^B	F^{B}F^W
F^W	F^{B}F^W	F^{W}F^W

a. 1/2 will be pink erminette chickens.

8. Dimples and freckles are dominant traits in people. A person who has dimples and freckles marries someone who does not have freckles or dimples. This couple produces a child that does not have dimples or freckles. What is the genotype of both parents? (Use a Punnett square to prove your answer.)

- The parent with dimples and freckles must have been heterozygous for both traits. The child, like the other parent, is homozygous recessive for both traits.

DdFf

	DF	Df	dF	df
df	DdFf	Ddff	ddFf	ddff
d f	DdFf	Ddff	ddFf	ddff
d f	DdFf	Ddff	ddFf	ddff
d f	DdFf	Ddff	ddFf	ddff

9. A woman with dark coloration (AABb) marries a man with light coloration (Aabb). Construct a Punnett square to show all possible genetic combinations of their gametes.

Very Dark Coloration   =   AABB
Dark Coloration   =   AABb or AaBB
Medium Coloration   =   AaBb or AAbb or aaBB
Light Coloration   =   Aabb or aaBb
Very Light Coloration   =   aabb

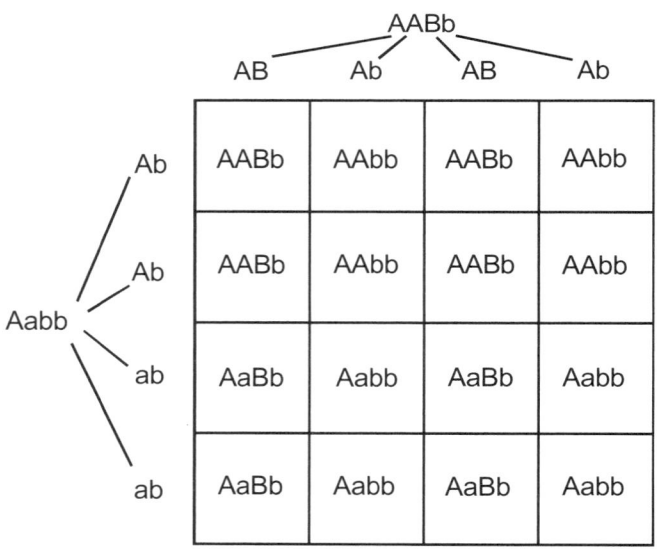

AABb

	AB	Ab	AB	Ab
Ab	AABb	AAbb	AABb	AAbb
Ab	AABb	AAbb	AABb	AAbb
ab	AaBb	Aabb	AaBb	Aabb
ab	AaBb	Aabb	AaBb	Aabb

Aabb

10. What are the phenotypic ratios of the offspring between two people of medium coloration where the man is AaBb and the woman is AAbb?

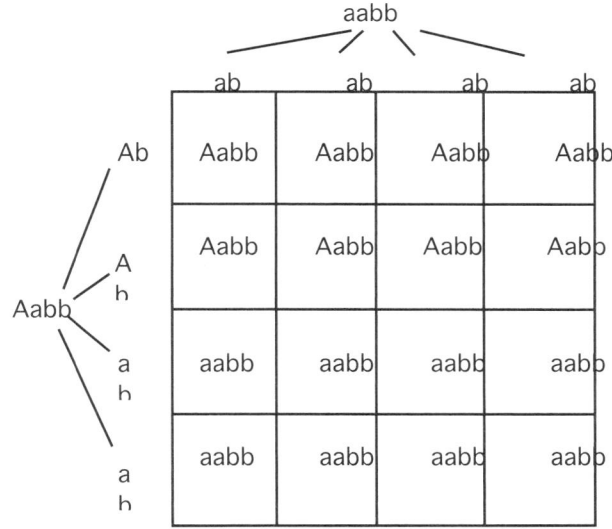

AaBb

	AB	Ab	aB	
Ab	AABb	AAbb	AaBb	Aabb
A h	AABb	AAbb	AaBb	Aabb
Ab h	AABb	AAbb	AaBb	Aabb
Ab h	AABb	AAbb	AaBb	Aabb

AAbb

4/16 or 25% offspring of dark coloration

8/16 or 50% offspring of medium coloration

11. What is the darkest coloration of a child that would result from a mating between a light-colored individual and a very light colored individual.
   - Darkest coloration would be light-colored: either Aabb or aaBb. Punnett square is shown only for the light colored parent that is Aabb.

aabb

	ab	ab	ab	ab
Ab	Aabb	Aabb	Aabb	Aabb
A h	Aabb	Aabb	Aabb	Aabb
a h	aabb	aabb	aabb	aabb
a h	aabb	aabb	aabb	aabb

Aabb

Gene Interactions
Solving Dihybrid and Complex Monohybrid Punnett Square Problems

PURPOSE

In this activity you will solve dihybrid and complex monohybrid Punnett squares.

PROCEDURE

1. Read the *Genetics: Key Concepts* page that your teacher has given you. It explains the background and major concepts needed to solve complex Punnett square problems.

2. Once you have read the *Key Concepts* pages, your teacher will demonstrate how to construct a dihybrid, or 16 square, Punnett square problem on the overhead or chalkboard. Take very careful notes so you can refer to them as you are completing the rest of the problems.

3. Work the rest of the problems on the student answer page. Continue to work these in the classroom, as your teacher monitors your progress.

Gene Interactions
Solving Dihybrid and Complex Monohybrid Punnet Square Problems

CONCLUSION QUESTIONS

1. Using the Punnett square in figure 3, list the genotype, genotype abbreviation, phenotype and predicted fraction of offspring for each different gene combination.

2. What is the phenotypic ratio of the offspring that results from crossing two individuals that are heterozygous for two nonlinked traits?

3. Construct a Punnett square showing all possible gamete combinations and all possible offspring for a cross between a woman who is heterozygous for widow's peak and short fingers (WwSs) and man who has a continuous hairline and short fingers (wwSs).

4. How did Kölreuter know that the genes of the parent flowers had not blended together but that the alleles for red color and white color still existed separately?

5. Construct a Punnett squares to show the F_1 and F_2 generations from a P generation cross of red with white carnations.

6. Construct a Punnett square to show the cross between a pink and a white carnation.

7. Construct a Punnett square to show the possible offspring of erminette chickens.

8. Dimples and freckles are dominant traits in people. A person who has dimples and freckles marries someone who does not have freckles or dimples. This couple produces a child that does not have dimples or freckles. What is the genotype of both parents? (Use a Punnett square to prove your answer.)

9. A woman with dark coloration (AABb) marries a man with light coloration (Aabb). Construct a Punnett square to show all possible genetic combinations of their gametes.

Very Dark Coloration	=	AABB
Dark Coloration	=	AABb or AaBB
Medium Coloration	=	AaBb or AAbb or aaBB
Light Coloration	=	Aabb or aaBb
Very Light Coloration	=	aabb

10. What are the phenotypic ratios of the offspring between two people of medium coloration where the man is AaBb and the woman is AAbb?

11. What is the darkest coloration of a child that would result from a mating between a light-colored individual and a very light colored individual.

Gene Interactions
Solving Dihybrid and Complex Monohybrid Punnett Square Problems

GENETICS: KEY CONCEPTS

DIHYBRID CROSSES AND INDEPENDENT ASSORTMENT

After establishing that genes segregate during the formation of gametes (reproductive cells), Mendel began to explore the question of whether they do so independently. In other words, does the gene that controls one trait have anything to do with the gene that controls a different trait? For example, does the gene that determines whether a seed is round or wrinkled in shape have anything to do with the gene for seed color? Must a round seed also be yellow? To answer these questions, Mendel first crossed purebred plants that produced round yellow seeds with purebred plants that produced wrinkled green seeds.

The Dihybrid Cross: F_1. In this cross, the two kinds of plants would be symbolized like this:

> Round yellow seeds   RRYY
> Wrinkled green seeds   rryy

Because two traits are involved in this experiment, it is called a *dihybrid cross*. As you examine the cross, keep in mind that you are looking at the kind of seeds the plant produces. These seeds are not necessarily the same as the seeds from which the plants grew.

The plant that bears round yellow seeds produces gametes that contain the alleles R and Y, or RY gametes. The plant that bears wrinkled green seeds produces ry gametes. A RY gamete and a ry gamete combine to form a fertilized egg with the genotype RrYy. Thus only one kind of plant will show up in the F_1 generation — plants that are heterozygous for both traits (see Figure 1). What is the phenotype of the F_1 plants? That is, what will the seeds produced by the F_1 plants look like? Because we know that round and yellow are dominant traits, we can conclude that the F_1 plants will produce seeds that are round and yellow. Remember that the concept of dominance tells us that the dominant traits will show up in a heterozygous, whereas the recessive traits will seem to disappear.

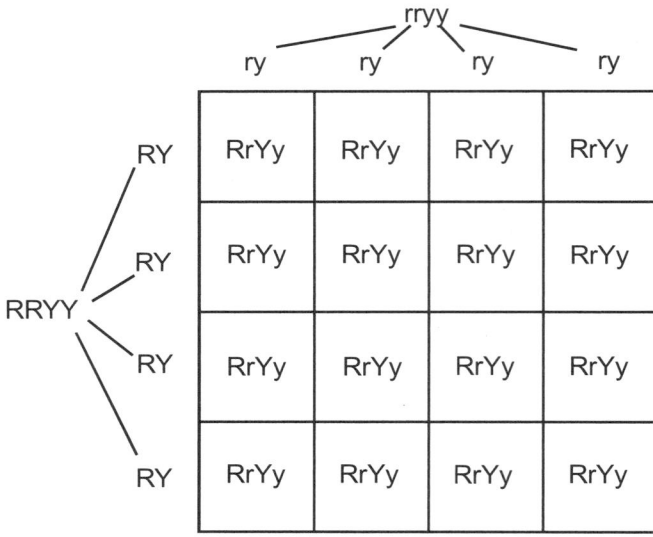

Figure 1. When an individual that is homozygous dominant for two traits is crossed with an individual that is recessive for the same two traits, all of the offspring are heterozygous for those two traits. (Note that the dominant allele is always written first.)

This cross does not indicate whether the genes assort, or segregate, independently. However, it provides the heterozygous plants needed for the next cross — the cross of F$_1$ plants to produce the F$_2$ generation. The seeds from the F$_2$ plants will show whether the genes for seed shape and seed color have anything to do with one another.

<u>The Dihybrid Cross</u>: F$_2$. What will happen when F$_1$ plants are crossed with each other? If the genes for seed shape and color are connected in some way, then the dominant R and Y alleles (which came from one parent) and the recessive r and y alleles (which came from the other parent) will be segregated as matched sets into the gametes. Thus, the gametes could only contain one of two possible gene combinations: RY or ry, as seen in Figure 2.

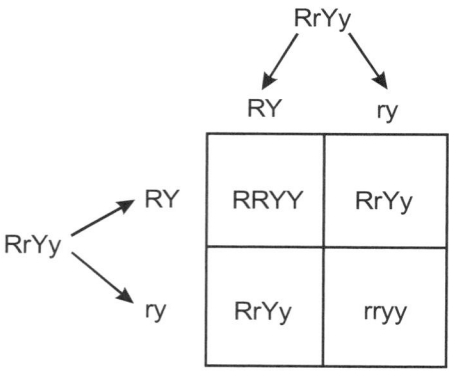

Figure 2. If the genes for two traits are connected in some way, only two combinations of the traits are possible in the offspring.

If the genes are not connected, then they should segregate independently, or undergo *independent assortment*. This produces four possible types of gametes: RY, Ry, rY and ry. In addition, if the genes assort independently, some of the seeds produced by the F$_2$ plants will have new combinations of traits — they may be wrinkled and yellow or round and green.

This dihybrid cross is examined in Figure 3. Now that we have four possible gamete types (and sixteen possible offspring types) the square is especially useful. If the genes for seed shape and seed color are inherited independently, then the seeds produced by the offspring should be in the same proportions as that predicted by the Punnett square.

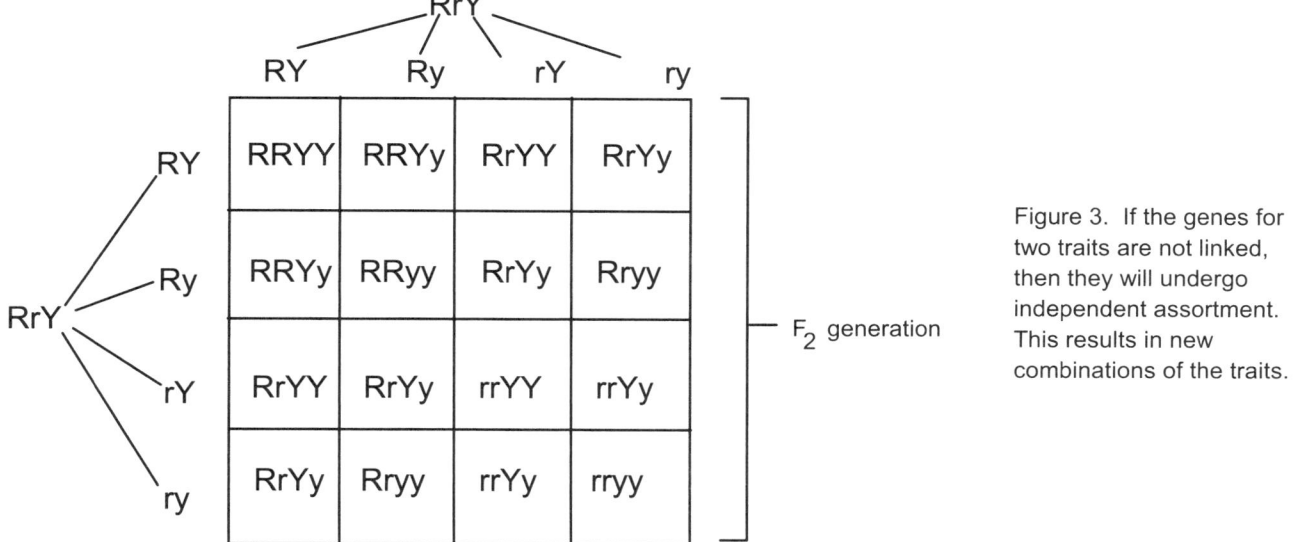

Figure 3. If the genes for two traits are not linked, then they will undergo independent assortment. This results in new combinations of the traits.

Mendel actually carried out this exact experiment, and his results were very close to the ratio predicted by the Punnett square in Figure 3. From these results Mendel concluded that genes could segregate independently during the formation of gametes. (As we shall see later, there is an important exception to independent assortment. Genes located on the same chromosome are linked and may not undergo independent assortment.)

WHAT CAUSES DOMINANCE?

Dominance is the simplest example of how genes interact with each other. We have already learned that the effects of the dominant allele are seen even when the recessive allele is present. But what causes dominance?

A gene is a section of DNA on a chromosome, and the DNA codes for a *polypeptide,* or string of amino acids. In many cases, the dominant allele codes for a polypeptide that works, whereas the recessive allele codes for a polypeptide that does not work. For example, suppose that the allele *B* codes for an enzyme that makes a black pigment in a mouse's fur and allele *b* codes for a defective enzyme that cannot make the pigment. A mouse that has the genotype *bb* will have white fur because it lacks any working enzyme that makes the black pigment. But a mouse that has the genotype *BB* or *Bb* will have black fur because it possesses the enzyme that makes the black pigment. Although each cell in the *Bb* animal has just one copy of the functioning allele, that single copy can code for thousands of molecules, each of which can code for thousands of enzymes. This is the reason the *B* allele is dominant over the *b* allele.

INCOMPLETE DOMINANCE

In 1760 the German scientist Josef Kölreuter reported on experiments in which he crossed white carnations (rr) with red carnations (RR). Kölreuter found that all of the offspring from his crosses had pink flowers (Rr). In other words, the heterozygous had a phenotype that was intermediate between those of the parents. At first glance, it might appear as if the parents' genes had blended together. But when Kölreuter crossed his pink F_1 heterozygous with each other to form an F_2 generation, the parents' phenotypes reappeared. In the F_2 generation, 1/4 of the plants had red flowers, 1/2 had pink flowers, and 1/4 had white flowers, a 1:2:1 ratio.

In carnations, the R allele, which codes for an enzyme that makes red pigment, is incompletely dominant over the r allele, which codes for a defective enzyme that cannot make pigment. In *incomplete dominance* the active allele does not compensate for the inactive allele, and the heterozygous phenotype is somewhere in between the homozygous phenotypes.

CODOMINANCE

Many genes display *codominance,* a condition in which both alleles of a gene are expressed. In other words, both alleles are active. (Remember that in incomplete dominance only one of the alleles is active.) Codominant alleles are written as capital letters with subscripts (for example, B_1 and B_2) or superscripts (for example, R and R').

Codominance is seen in many organisms. For example, red hair (H^R is codominant with white hair (H^W) in cattle and horses. Cattle that have the genotype $H^R H^W$ are roan, or reddish brown, colored because their coats are a mixture of red and white hairs. This is different from incomplete dominance because neither allele is inactive. The hair is not pink; instead, each individual hair is either red or white. Much the same thing happens in certain varieties of chickens. Black feathers (F^B) are codominant with white feathers (F^W). Erminette chickens ($F^B F^W$) are speckled black and white.

POLYGENETIC INHERITANCE

Two or more genes may affect the same trait in an additive fashion. Let's consider skin color. When a very dark-skinned person has children by a very light-skinned person, the skin color of the children is some variation of is medium color. Two individuals with medium coloration can produce children who range in skin color from very dark to very light. This can be explained if we assume that there are two genes that control skin color and that only alleles indicated with a capital letter contribute equally to skin color. (In actuality, skin color may be controlled by more than two genes — the actual number of genes involved has not been finally determined.)

Very Dark Coloration	=	AABB
Dark Coloration	=	AABb or AaBB
Medum Coloration	=	AaBb or AAbb or aaBB
Light Coloration	=	Aabb or aaBb
Very Light Coloration	=	aabb

Polygenetic inheritance can cause the distribution of human traits according to a bell-shaped curve, with most individuals exhibiting the average phenotype. The more genes that control the trait, the more continuous the distribution will be. Just how many genes actually control skin color and height (another example of polygenetic inheritance) is not known.

Baby Dice Island
Modeling Exponential Growth

OBJECTIVE
Students will use dice rolling to model unrestricted exponential growth in an imaginary population of baby dice.

LEVEL
Middle Grades: Life Science

NATIONAL STANDARDS
UCP.2, C.3, F.2

TEKS
6.2(B), 6.2(C), 6.2(D), 6.2(E), 6.4(B)
7.2(B), 7.2(C), 7.2(D), 7.2(E), 7.4(B)
8.2(B), 8.2(C), 8.2(D), 8.2(E), 8.4(B)
IPC 2(B), 2(C), 2(D), 4(A)

CONNECTIONS TO AP
AP Biology:
  III. Organisms and Populations: C. Ecology 1. Population dynamics 3. Global issues
AP Environmental Science:
  I. Interdependence of Earth's Systems: Fundamental Principles and Concepts E. The Biosphere
  2. populations and communities: exponential growth, carrying capacity
  II. Human Population Dynamics A. History and Global Distribution 1. numbers 2. demographics, such as birth rates and death rates B. Carrying Capacity — Local, Regional, Global

TIME FRAME
150 minutes or 3 class periods

MATERIALS
(For a class of 28 working in groups of 4)

  140 dice (20 dice per group of 4 students)

TEACHER NOTES
Baby Dice Island: Modeling Exponential Growth is an excellent introduction to the concept of unrestricted exponential population growth. This lab can be used to introduce discussions about population growth, ecological issues, or biomes and fits easily into an ecology unit.

Students model the growth of an imaginary population of individuals on a make-believe island by using dice to model births and deaths through a series of years, starting with an initial population, equal to the number of dice that the students have been assigned. Let us assume that the initial population is 20, as each group ideally has 20 dice. If the students roll a "three" or a "six", this represents a birth within their initial population. If the students roll a "one", this represents a death within their initial population.

After the 20 dice have been rolled, the newly modeled births (3's and 6's that were rolled) are added to the initial population and the newly added deaths (1's that were rolled) are subtracted from the initial population, and the final population for that "year" (A year occurs when all the individual dice within a initial population have "mated", i.e., have been rolled.) is determined. The students continuing to roll the dice, modeling and recording the number of "births" and "deaths" of their population for each year until the final population reaches 500 individuals.

Depending on the background of your students you may need to review proper graphing procedures outlined in *Foundation Lesson IV—Graphing Skills*. In a graph representing growth, time is displayed on the *x*-axis and population numbers are displayed on the *y*-axis. The *x*-axis has units of years, hours, or seconds. The *y*-axis represents the population numbers from zero to some maximum population number and typically has units of "number of individuals". At each point in time on the graph, there is a corresponding value of population. A line of best fit may be drawn in order to use the graph for extrapolation of population data. The most common trend seen in a growth graph typically starts low and generally increases over time.

Growth can occur according to different mathematical patterns, including arithmetic growth and exponential (also called geometric) growth. Arithmetic growth, represented by a straight line, increases at a constant rate represented by the equation $y = mx + b$. Exponential growth is represented by a much more complex mathematical equation and is beyond the scope of most middle school/junior high school students. This type of growth does not increase at a constant rate. In nature, almost any organism provided with ideal conditions for growth and reproduction will experience a rapid increase in its population, much more so than is represented by the arithmetic growth pattern. More importantly, the larger the population gets, the faster it grows. If nothing stops or limits the population from growing, it will continue to increase at a faster and faster rate. This kind of growth pattern creates a graph with an exponential growth curve. Below is diagram showing the differences between arithmetic and exponential growth curves. You might want to make an overhead transparency of this diagram to show to your students.

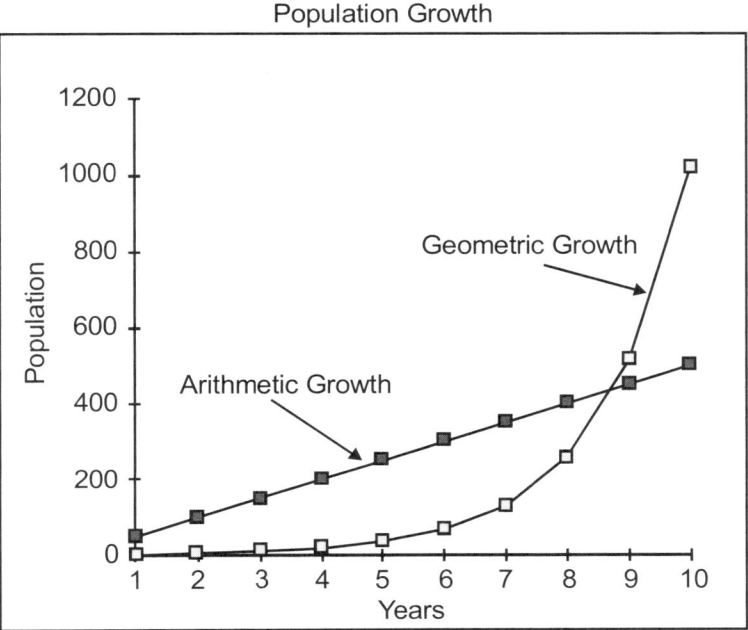

The lab is for groups of four students and the student jobs of each group are as follows:
- *dice roller*: rolls dice
- *recorder*: records all final data per year in the data table
- *death tracker*: keeps track of number of deaths each year on tally sheet
- *birth tracker*: keeps track of number of births each year on tally sheet

Some assumptions built into lab that should be explained to the students:
- the initial population of year one is 20 (or the number of dice that you are using) individuals
- one-year-old dice on the imaginary island can have babies
- each individual (or individual die) can have an offspring (sex of the die does not matter)

The major logistical concern for you to perform this lab with a class of students is finding enough dice. The lab is designed for students working in groups of four. If you have a class of 28 students, then 7 lab groups of materials will be needed. Each lab group ideally needs 20 dice but 15 or even 10 will produce adequate results. Keep in mind that using smaller numbers of dice will increase the time needed to reach a final population of 500.

One suggestion for managing classroom chaos during this activity is to instruct the students that for every die that hits the floor, 1 point will be deducted from the 100 point final grade.

Students will be faced with the logistical challenge of tallying all the births and deaths per roll until the final population reaches 500. You should suggest that the death tracker and the birth tracker use a temporary tally sheet to track their "hits" and then transfer their data to the final data table once the year's worth of rolls are completed.

To illustrate how the tally sheets may be used, the following example is presented. If the students have reached a final population of 60 at the end of year 8, the initial population for year 9 would also be 60. With only 20 dice the students will have to roll the dice three different times to model the population change for those 60 individuals during year 9. The group would start by rolling the 20 dice one time. The birth tracker would count how many "3's" and "6's" were rolled and record this on their tally sheet. The death tracker would simultaneously keep track of the number of "1's" that were rolled and record this on their tally sheet. This procedure would need to be repeated 2 more times to account for the total of 60 individuals. At the end of the 3 rolls representing year 9 the total number of births and deaths can be transferred to the official data table.

Examples of typical tally sheets are presented below, indicating how many deaths and births occurred, in total, for year 9, with an initial population of 60.

Tally sheet for birth tracker			
Year	New births—1st roll (of 20 dice)	New births—2nd roll (of 20 dice)	New births—3rd roll (of 20 dice)	Total births for year
9	6	7	5	18

Tally sheet for death tracker				
Year	New deaths— 1st roll (of 20 dice)	New deaths— 2nd roll (of 20 dice)	New deaths— 3rd roll (of 20 dice)	Total deaths for year
9	2	4	3	9

As the birth tracker and the death tracker keep tally of the births and deaths for each year, the recorder only transfers the number of the total births and deaths in a year to the final data table.

Once transferred to the final data table, the year 9 data will appear as follows:

Data Table for Modeling Exponential Growth Lab					
Year	Initial Population	Births	Deaths	Change	Final Population
9	60	18	9	+9	69

Students are instructed to continue rolling dice and adding years until a final population of at least 500 is reached. It typically takes anywhere from 16-23 years before a population of 500 is reached by the students.

The data table is to be written in ink to discourage students from changing/falsifying data. Students are to turn this into you each day after doing the lab. By doing this the recorder will not have a chance to lose the data table. You can then return the data table each day.

Calculating doubling time:
To find the doubling time of a population growing at a constant annual growth rate, divide the growth rate into 70. This represents number of years required for the population to double. For example, with a 10% annual growth rate, the doubling time of a population would be 70/10=7, or 7 years.

The math behind the rule of the "70" used in the doubling time equation is very complex and beyond the scope of this book. However, there is a detailed explanation at the "Exponential Growth and The Rule of 70" website found here:

http://www.ecofuture.org/pop/facts/exponential70.html

REFERENCES:
http://www.ecofuture.org/pop/facts/exponential70.html
http://www.balance.org/population.html
http://www.census.gov

SAMPLE DATA

Data Table for Modeling Exponential Growth					
Year	Initial Population	Births	Deaths	Change	Final Population
1	15	4	1	+3	18
2	18	4	4	+0	18
3	18	5	1	+4	22
4	22	4	7	-3	19
5	19	8	2	+6	25
6	25	6	7	-1	24
7	24	10	4	+6	30
8	30	9	3	+6	36
9	36	12	5	+7	43
10	43	25	4	+21	64
11	64	31	8	+23	87
12	87	26	15	+11	98
13	98	23	8	+15	113
14	113	38	24	+14	127
15	127	47	25	+22	149
16	149	45	24	+21	170
17	170	47	21	+26	196
18	196	76	26	+50	246
19	246	64	36	+28	274
20	274	110	27	+83	357
21	357	142	36	+106	463
22	463	156	55	+101	564

ANALYSIS: SAMPLE GRAPHS

Part I: Unrestricted Exponential Growth of the Baby Dice Island Population

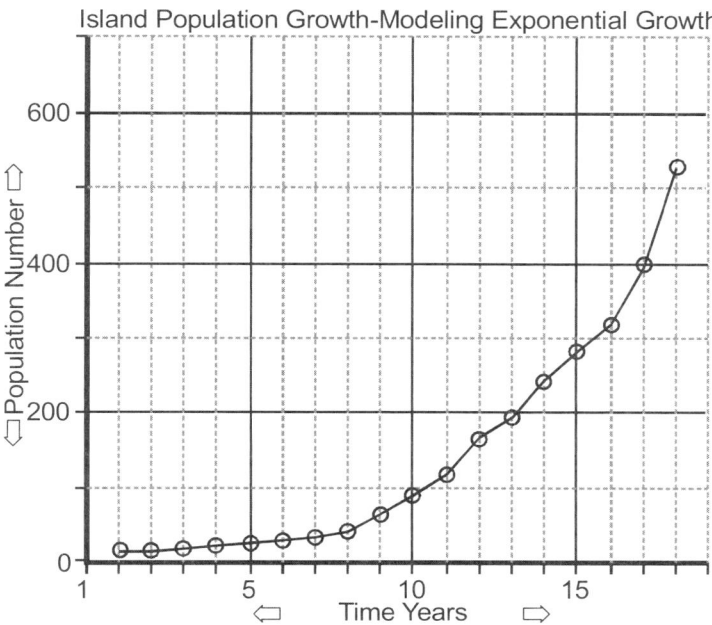

Part II: World Population Trends

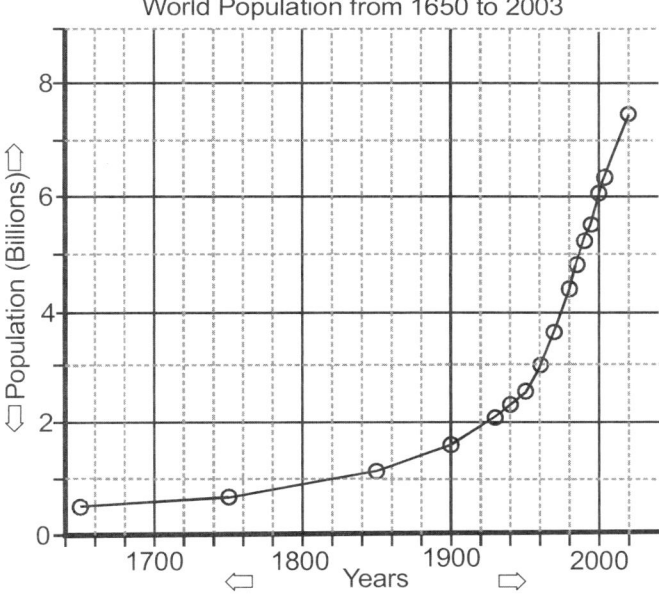

TEACHER PAGES

Use your graph to predict world population for the year 2020 (Hint: Use dotted lines to extend your graph into the future.)

- Answers will vary here according to each student's hand drawn graph. The current estimate, according to the United States Census Bureau (http://www.census.gov) website is that the population will be approximately 7.5 billion.

POSSIBLE ANSWERS TO THE CONCLUSION QUESTIONS
Part I: Unrestricted Exponential Growth of the Baby Dice Island Population
1. What is the ratio of births to deaths in this model population?
- 2 to 1. When rolling the dice, there are two different ways to model a birth, rolling a 3 or a 6. To model a death, there is only one way, by rolling a 1.

2. How many "years" did it take you to reach a population of 100?
- Answers will vary but typically it takes between 9-13 years for students to reach a population of 100.

3. After you reached a population of 100, how many more "years" did it take to reach a population of 200? How many **more** years to reach 300? 400? 500?
- Again, answers will vary but because of the exponential growth, the amount of time it takes to increase by each increment of 100 individuals decreases. So its takes less years to get to 200 and even less time to get to 300 and so on.

4. Using this experiment, define exponential (or geometric) growth.
- Answers will vary but exponential population growth occurs with slow growth at the start turning into faster and faster growth reaching an almost vertical spike.

5. In what way do exponential (or geometric growth rates) differ from arithmetic growth rates?
- Arithmetic growth is represented by a straight line increasing to the right and exponential growth increases at a much faster rate and is represented by a curved line showing rapid increase.

Part III: Doubling Time of a Population
6. Calculate the growth rate for the U.S. (Show your work.)
As of 2002, the population of the US was approximately 292,000,000. The increase in the US population is approximately 3,300,000 a year, from immigration and new births.
- Growth rate=amount of change in a population/total population
- Growth rate(USA)=3,300,000/292,000,000
- Growth rate (USA)= 1.13%

7. Calculate the doubling rate for the population of the U.S. (Show your work.)
- To determine doubling time, divide 70 by the populations' growth rate.
- Doubling time (USA)=70/growth rate
- Doubling time (USA)=70/1.13%
- Doubling time (USA)=61.95 years

Baby Dice Island
Modeling Exponential Growth

In this experiment you will roll dice to model population growth of the individuals on *Baby Dice Island*. Each die represents a living organism, capable of reproducing. You will start out with an initial population equal to the number of dice that you have at your table. Every time you roll a "three" or a "six", this represents the birth of an offspring, adding an individual to your initial population. Each time a "one" is rolled, a death has occurred, decreasing your initial population by one. After all the dice for the initial population have been rolled (representing one year) you will determine your final population for that year, by adding the numbers of births and subtracting the number of deaths from your initial population. You will be rolling dice over a series of "years", adding births, and subtracting deaths from your initial population, until you finally reach a final population of 500 individuals.

Each member of the four person lab group should perform one of the following roles. The roles must be changed each day that this lab is performed.
- *dice roller*: rolls dice
- *recorder*: records all final data per year in the data table
- *death tracker*: keeps track of number of deaths each year on tally sheet
- *birth tracker*: keeps track of number of births each year on tally sheet

PURPOSE
In this activity you will use dice to model exponential growth of an imaginary population.

MATERIALS
10 to 20 dice

PROCEDURE
1. *Dice roller*: Put all dice into the cup, shake the cup and carefully pour the dice onto the table.

   Death tracker: Remove and count all the "ones" that appear. A "one" represents a death and will be subtracted from your initial population. Record the number of deaths on a tally sheet, according to the instructions provided by your teacher.

   Birth tracker: Determine the number of "threes" and "sixes" that appear. This number corresponds to births and will be added to your initial population. Record the number of births on a tally sheet, according to the instructions provided by your teacher.

   Recorder: Use a ruler to carefully draw, in ink, a data table similar to the one below that can be used to keep track of all the accumulated deaths and births that occur. Draw the table in the space provided on your student answer page. Your initial population will vary depending on the number of dice you were assigned.

Data Table for Modeling Exponential Growth Lab					
Year	Initial Population	Births	Deaths	Change	Final Population
1	20				

2. Continue rolling and tallying the births and deaths until total population exceeds 500 individuals.

3. Complete the analysis section and the conclusion questions on the student answer page.

Baby Dice Island
Modeling Exponential Growth

DATA

Using a PEN and RULER draw a data table as described in step 5 of the procedure. The table should contain space for at least 25 years of data.

ANALYSIS

Part I: Unrestricted Exponential Growth of the Baby Dice Island Population

Use your data table to plot a graph demonstrating the unrestricted exponential growth of the dice population. Use "Years" for the x-axis and "Total Population" for the y-axis. Be sure to give your graph an appropriate title and to label the axes of the graph.

Part II: World Population Trends

1650 = 0.5 billion	1940 = 2.3 billion	1985 = 4.8 billion
1750 = 0.7 billion	1950 = 2.5 billion	1990 = 5.3 billion
1850 = 1.1 billion	1960 = 3.0 billion	1995 = 5.5 billion
1900 = 1.6 billion	1970 = 3.6 billion	2000 = 6.1 billion
1930 = 2.1 billion	1980 = 4.4 billion	2003 = 6.3 billion

Using the above information, plot a graph of world population versus time from 1650 to 2003 in the space below. Use your graph to predict world population for the year 2020. (Hint: Use dotted lines to extend your graph into the future.)

8 *Baby Dice Island*

CONCLUSION QUESTIONS

Part I: Unrestricted Exponential Growth of the Baby Dice Island Population

1. What is the ratio of births to deaths in this model population?

2. How many "years" did it take you to reach a population of 100?

3. After you reached a population of 100, how many more "years" did it take to reach a population of 200? How many **more** years to reach 300? 400? 500?

4. Using this experiment, define exponential (or geometric) growth.

5. In what way do exponential (or geometric growth rates) differ from arithmetic growth rates?

Part III: Doubling Time of a Population

6. Calculate the growth rate for the U.S. (Show your work.)

    As of 2002, the population of the U.S. was approximately 292,000,000. The increase in the U.S. population is approximately 3,300,000 a year, from immigration and new births.

7. Calculate the doubling rate for the population of the U.S. (Show your work.)

Write Your Notes and Ideas Here!

Plotting Growth in Texas
Comparing the Population Growth Rates in Texas: Dallas (Dallas County) and Houston (Harris County) and Your County for 150 Years

OBJECTIVE
Students will compare the rate of population growth between 3 urban Texas areas by accessing data from several websites. Students will then present the population data in a concise, easy-to-comprehend data table. Students will also graph data of population growth in an accurate and easy-to-read graph and calculate rate of growth for the different populations.

LEVEL
Middle Grades: Life Science

NATIONAL STANDARDS
UCP.2, A.1, A.2, E.1, F.2

TEKS
6.2(C), 6.2(E), 6.4(A), 6.4(B)
7.2(C), 7.2(E), 7.4(A), 7.4(B)
8.2(C), 8.2(E), 8.4(A), 8.4(B)
IPC 2(C), 2(D), 4(A)

CONNECTIONS TO AP
AP Biology:
   III. Organisms and Populations C. Ecology 1. Population dynamics
AP Environmental Science:
   II. Human Population Dynamics A. History and Global Distribution 1. numbers 2. demographics, such as birth rates and death rates B. Carrying Capacity - Local

TIME FRAME
50+ minutes

MATERIALS
(For a class of 28 working in pairs)

14 computers with web access   colored pencils or markers

TEACHER NOTES
Plotting Growth in Texas has two functions for students: First and foremost, to compare the growth rates of three urban Texas counties: Dallas County (representing the Dallas metro area), Harris County (representing Houston) and their own county. (If you and your students reside in Dallas or Harris County, students are instructed to pick any county in Texas to compare.) The activity also allows students to search for specific pieces of data within some large data base websites.

TEACHER PAGES

Since finding the data in these websites is quite easy, this activity can be an introduction to using websites for gathering specific information. The websites are provided for students within their student instruction page, however, if you would like the activity to have an advanced internet searching aspect students could use a search engine to find their own websites containing the relevant population data.

Graph paper has been provided for students within their student answer page, however, if you prefer students can use computer graphing software such as Excel or Graphical Analysis to generate their graphs. Refer to *Foundation Lesson V: Graphing with Excel* and *Foundation Lesson VIII: Computer Graphing Software* for more information about these graphing programs.

Years	Dallas Co.	Harris Co.	Travis Co.
1850	2,743	4668	3138
1860	8665	9070	8080
1870	13314	17375	13153
1880	33488	27985	27028
1890	67042	37249	36322
1900	82726	63786	47386
1910	135748	115693	55620
1920	210551	186667	57616
1930	325691	359328	77777
1940	398564	528961	111053
1950	614799	806701	160980
1960	951527	1243158	212136
1970	1327321	1741912	295516
1980	1556390	2409547	419573
1990	1852810	2818199	576407
2000	2218899	3400578	812280

Figure 1. Here is a data table with the correct population numbers that were obtained from the indicated websites for Dallas, Harris and Travis County.

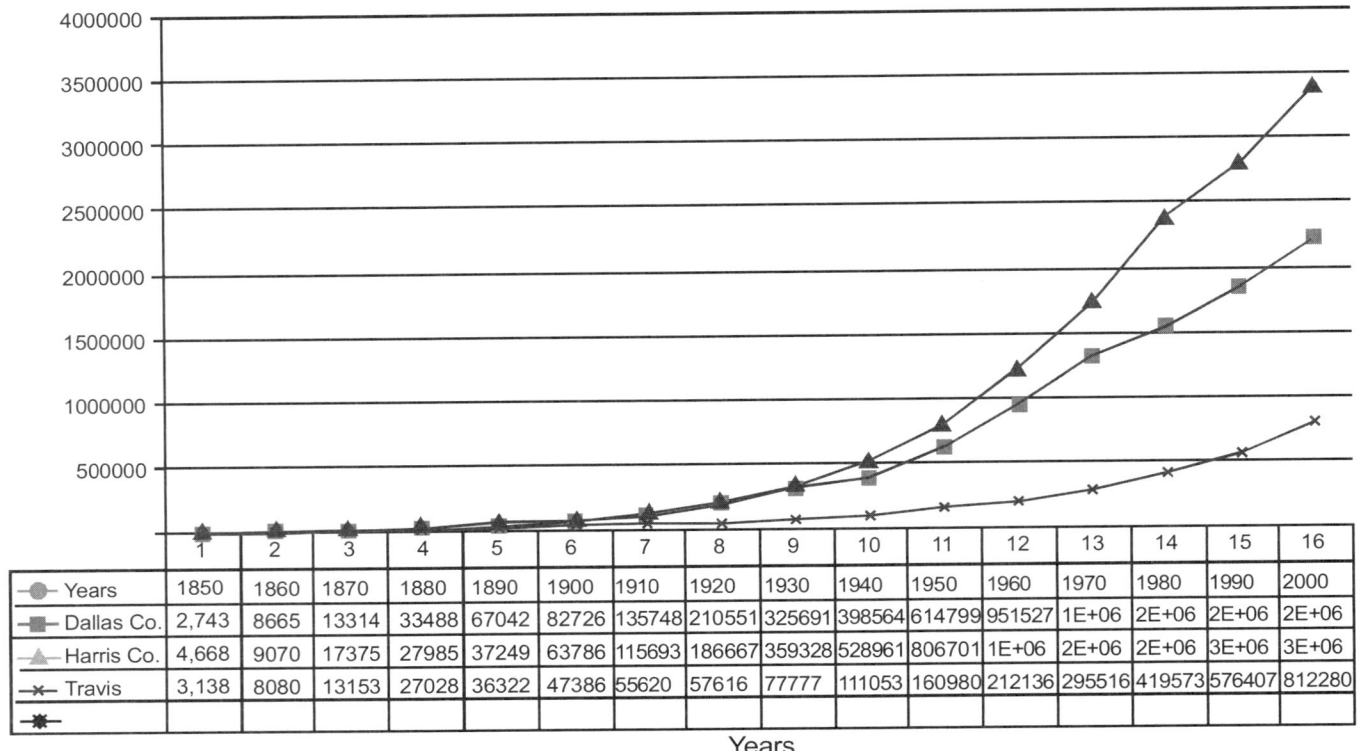

Comparison of Texas Counties-Population Growth

	1	2	3	4	5	6	7	8	9	10	11	12	13	14	15	16
Years	1850	1860	1870	1880	1890	1900	1910	1920	1930	1940	1950	1960	1970	1980	1990	2000
Dallas Co.	2,743	8665	13314	33488	67042	82726	135748	210551	325691	398564	614799	951527	1E+06	2E+06	2E+06	2E+06
Harris Co.	4,668	9070	17375	27985	37249	63786	115693	186667	359328	528961	806701	1E+06	2E+06	2E+06	3E+06	3E+06
Travis	3,138	8080	13153	27028	36322	47386	55620	57616	77777	111053	160980	212136	295516	419573	576407	812280

Years

Figure 2. This sample graph, created in MS Excel, shows the growth rate for Dallas, Harris and Travis Counties.

POSSIBLE ANSWERS TO THE CONCLUSION QUESTION AND SAMPLE DATA

1. Which county had the most rapid growth rate?
   - Harris County.

2. Why did the comparison begin with 1850?
   - Texas only became a state in 1845. The census records for Texas begin in 1850. Also, the populations of all counties in the state of Texas were relatively low, especially compared to today's populations.

3. Using only your graph, how can you tell that the county that you answered in #1 has the greatest rate of population growth?
   - All lines (populations) start at basically the same point, almost zero. The line representing the population of Harris County has the steepest rise or slope.

4. Which county, as of 2000, has the greatest population?
   - 3,400,578, Harris County

5. By looking at your graph, how did your county compare in its rate of growth versus Harris and Dallas counties?
   - Answers will vary but probably your county grew at a slower rate than either Harris or Dallas counties.

   Use your data table and graph to answer questions 6-8: **Show your work**.

   Average rate of growth = $\dfrac{\text{(Final population} - \text{initial population)}}{150 \text{ years}}$ The units will be people/year.

6. Calculate the average rate of growth from 1850 until 2000 in Dallas County. Round your answer to the nearest whole number.
   - Dallas: $\dfrac{(2218899 - 2743)}{150}$ =14,774 people/year

7. Calculate the average rate of growth from 1850 until 2000 in Harris County. Round your answer to the nearest whole number.
   - Harris: $\dfrac{(3400578 - 4668)}{150}$ =22, 639 people/year

8. Calculate the average rate of growth of your county from 1850 until 2000. Round your answer to the nearest whole number.
   - Answers will vary according to your county.

9. What are some potential problems that the county that you answered in question #1 can look forward to by having such rapid rate of growth?
   - Overpopulation, resource depletion, habitat destruction, pollution of air and water, etc…

10. Using the your graph, extrapolate your growth rate line to 2010 and predict the number of people that will be living in your county and in the county with the highest growth rate. (Hint: Use dotted lines to extend your graph into the future.) Write an estimate of both populations on the graph paper at that extrapolated point.
    - Because the students are creating the graphs by hand, their predictions will be varied. You should look for a reasonable hypothesized increase in population, shown by the student as a dotted line, continuing the increasing slope of the rate line progression.

11. Thought question: Why do you think the population rose so much faster after 1940? (You can use the internet to research this question. Also, talk to your parents and your history teacher!)
    - Although there are many reasons, a major factor in the population boom after World War II involves "baby boomers", etc… Also, with advances in fertility, medicine and technology, the population has increased.

TEACHER PAGES

Plotting Growth in Texas
Comparing the Population Growth Rates in Texas: Dallas (Dallas County) and Houston (Harris County) and Your County for 150 Years

PURPOSE
In this activity you will compare the population growth between 3 urban Texas areas by accessing data from several websites. You will then present the population data in a concise, easy-to-comprehend data table. You will also graph data of population growth in an accurate and easy-to-read graph and calculate the rate of growth for the different populations.

MATERIALS
computer with Internet access   graph paper

PROCEDURE
Note: Historically population data has been recorded by county. We will be using county data to represent the major city population and calculate its growth rate in the state of Texas since 1850. In 1845, Texas became a state in the United States. All state census records for Texas began in 1850.

1. Use a ruler to create a data table with 4 columns on your student answer page. The first column will have the heading "Year", the second column "Dallas Co.", the third "Harris Co." and the fourth with your county name. Your teacher can help you determine what county you live in if you do not already know. Also, if you live in Dallas or Harris counties, add a county of your choice.

   Underneath the Year column, write in the following dates: 1850, 1860, 1870, 1880, 1890, 1900, 1910, 1920, 1930, 1940, 1950, 1960, 1970, 1980, 1990, and 2000. You will be finding the total population for each of the selected counties in those years on the websites below.

2. Go to the following website: http://fisher.lib.virginia.edu/census, the United States Historical Census Data Browser site.

3. Once there, click on the link for "1850".

4. Scroll down to the total population section and select "total aggregate population" (aggregate means mixed) by clicking on it, making sure that it stays highlighted.

5. Scroll down to the bottom of the page and click the Browse 1850 Data button.

6. Scroll down and select Texas by clicking on the box next to it. Then click the "View Counties" button.

Here is an image of what you should see after you select Texas.

7. The page that loads should have all the data for every Texas county. Find the population data for Dallas, Harris and your county and record the values in your data table on the student answer page.

8. Continue finding the rest of the data for the years 1860 to 1960 by repeating the steps above for each year. (The data on this website only goes up to1960.) For the years from 1860-1900, the value that you are looking for is the "total population".

9. To find the 1970 and 1980 data for each Texas county, go to this website: http://www.census.gov/population/cencounts/tx190090.txt.

10. Continue to fill in the data for the population values for1970 and 1980 for all 3 counties.

11. For the 1990 and 2000 data, go to http://www.tdh.state.tx.us/dpa/popdata/menup.htm. Click on "HTML" to view the total population data for each Texas county for each year.

12. Continue to fill in the data for the population values from 1990 until 2000 for all 3 counties.

13. Once you have filled in your data table, you will construct a line graph to display your results. On your student answer page, in the space provided, label the *x*-axis with the years and the *y*-axis with population numbers and appropriately title your graph. Make sure to use units for population that are logical and make your graph use as much of the graph paper as possible. Use a different color pencil or marker to construct the line for each county. Connect your dots to make a line graph representing the growth rates for each of the 3 counties. Using different colors will allow you to quickly compare growth rate lines. Create a key to correlate the color to a county. Each student will turn in a graph at the end of the period.

Name _____

Period _____

Plotting Growth in Texas
Comparing the Population Growth Rates in Texas: Dallas (Dallas County) and Houston (Harris County) and Your County for 150 Years

DATA TABLE

GRAPH

CONCLUSION QUESTIONS

Using your graph and data table, answer the following questions in complete sentences. Each student will turn in their own data table, graph and conclusion questions.

1. Which county had the most rapid growth rate?

2. Why did the comparison begin with 1850?

3. Using only your graph, how can you tell that the county that you answered in #1 has the greatest rate of population growth?

4. Which county, as of 2000, has the greatest population?

5. By looking at your graph, how did your county compare in its rate of growth versus Harris and Dallas counties?

Use your data table and graph to answer questions 6-8: ***Show your work***.

$$\text{Average rate of growth} = \frac{\text{(Final population - initial population)}}{150 \text{ years}}$$ The units will be people/year.

6. Calculate the average rate of growth from 1850 until 2000 in Dallas County. Round your answer to the nearest whole number.

7. Calculate the average rate of growth from 1850 until 2000 in Harris County. Round your answer to the nearest whole number.

8. Calculate the average rate of growth of your county from 1850 until 2000. Round your answer to the nearest whole number.

9. What are some potential problems that the county that you answered in question #1 can look forward to by having such rapid rate of growth?

10. Using the your graph, extrapolate your growth rate line to 2010 and predict the number of people that will be living in your county and in the county with the highest growth rate. (Hint: Use dotted lines to extend your graph into the future.) Write an estimate of both populations on the graph paper at that extrapolated point.

11. Thinking question: Why do you think the population rose so much faster after 1940? (You can use the internet to research this question. Also, talk to your parents and your history teacher!)

Create a Species
Describing a New Imaginary Species

OBJECTIVE
Students will apply ideas and concepts of ecology, classification, and biological diagrams in a creative manner to describe a new imaginary species in a pseudo-scientific report.

LEVEL
Middle Grades: Life Science

NATIONAL STANDARDS
UCP.1, UCP.4, UCP.5, C.3, C.4, C.5, C.6

TEKS
6.3(C), 6.8(C), 6.10(A), 6.10(B), 6.10(C), 6.11(A), 6.12(A), 6.12(B), 6.12(C)
7.3(C), 7.9(B), 7.10(B), 7.10(C), 7.11(B), 7.12(A), 7.12(B), 7.12(C)
8.3(C), 8.6(B), 8.6(C), 8.11(A), 8.11(B)

CONNECTIONS TO AP

TIME FRAME
150 minutes, or 3 class periods

MATERIALS
(For a class of 28 working individually)

28 sets of colored pencils

TEACHER NOTES
Create a Species can be the culminating activity to units in the following areas: ecology, characteristics of living things, and classification. The *Animicules Discovery* (classification exercise), *Picturing Life* (biological diagram exercise) and *Online Biome Exploration* (creating a biome Powerpoint presentation exercise) lessons, all found elsewhere in this book, should be completed prior to this activity. In *Create a Species*, students write a pseudo-scientific report about a newly discovered species created entirely within their imaginations, bound only by the biological concepts of life that have been previously studied. The report includes a written description, a labeled biological diagram of the species and a diagram of the food web in which the new species lives. The activity will require approximately 3 class periods to complete. The grading rubric is included at the end of the student answer page.

Below you will find a few comments related to each section of the student report.

A. NAME OF SCIENTISTS
No comment is really needed in this section.

B. TITLE OF SCIENTISTS

Your academic credentials and area(s) of research. Make it up! Do you have a Ph.D. in molecular genetics? Do you have a master's degree in alien life forms?

Students often have difficulty with this section, especially if they are not familiar with specialized degrees. You can give several examples of other specialized degrees until they understand this concept. The main focus is for the students to envision themselves as very specialized scientists.

C. PROPOSED SCIENTIFIC NAME OF ORGANISM

Binomial nomenclature is the standard convention for naming species. Every species can be unambiguously identified with just two words, the combination of the genus and the species. The genus name is always capitalized while the species part is all lower case. In print format the scientific name is always in italics but when hand-written the scientific name is underlined. Using this format, the identifying name for the human species is *Homo sapiens*, or <u>Homo sapiens</u>. This is also described for students in the background information.

D. CLASSIFICATION OF YOUR ORGANISM

The students are to create names for each level of our modern classification scheme. The kingdom name should be chosen from one of the five kingdoms listed. Have students write Latin sounding names by adding a -a, -um, or -us to the end of the word. For example, a phylum could be named Predatora if the student's creature was predatory in nature. Information about the five kingdom classification system is included in the student background information.

E. PHYSICAL DESCRIPTION/CHARACTERISTICS AND BEHAVIORS OF ORGANISM

Although students do not really have a difficult time with physical descriptions, they do sometimes have a difficult time describing behaviors. If you give some examples of behaviors, it should clear up confusion. For example, "The organism circles its prey, prior to eating it", might be a behavioral characteristic.

F. ADAPTATIONS

This section describes how the physical characteristics help the organism to survive in its environment. The idea is that the students take an inherited physical (or behavioral) characteristic and then describe how that characteristic helps the organism survive in its environment. There is a common misconception among students that somehow an individual creature can, within its life span, change its characteristics and then "get used to" or "adapt" to its environment. Adaptations are genetic and inherited from parents. The creature does not "get used to" its environment but possesses inherited adaptations already inscribed within its genetic code. You should address this concept and potential misconception with the students. In the student sample, there is an excellent example of an inherited adaptation:

"Has four tentacle-like ears colored a light green to camouflage with the grass. This adaptation helps Acuta rodentia survive, because not only does it rely on its sense of hearing, the ears help it camouflage into the grass and avoid predators."

G. MEANS OF ENERGY INTAKE/FEEDING STRATEGIES

Typically students will choose some sort of animal as their organism and many students will choose to make the creature somehow predatory. However, others will come up with very creative ways of obtaining energy. It is up to your discretion as to how much freedom you will allow your students in this area. The student background information contains considerable discussion about the autotrophic/heterotrophic nature of organisms.

H. REPRODUCTIVE STRATEGIES

Addressing this section depends on the maturity of your students. For many middle school students, as soon as you mention this section there will probably be giggling in the classroom. One possible way to subvert some of this behavior is to comment on how reproduction is an essential, yet mature, idea and that only mature students can handle discussing it. You may also choose to delete this section if you feel necessary. If you choose to include it, the major challenge is that the students are to describe this section in a non-obscene way. Although many students will use the standard "sperm and egg, internal fertilization" description, some students will generate some very interesting responses.

I. DESCRIPTION/CHARACTERISTICS OF ORGANISM'S HABITAT AND ENVIRONMENT

List the characteristics of the environment/habitat where the organism lives. What is the average temperature? What is the average amount of rainfall? What type of biome is it: tropical rain forest, tidal pool, desert, etc...?

The students should describe in what biome the organism will live. Specifically the temperature and amount of precipitation if a terrestrial biome is chosen and specifically the amount of light and temperature if an aquatic biome is chosen. The student will have extensive background knowledge in this area from the Online Biome Exploration activity.

J. DESCRIPTION(S) OF INTERRELATIONSHIPS WITH OTHER ORGANISMS

Some background knowledge will be needed here. Included in the student background information of this activity are descriptions of the major types of symbiotic relationships: mutualism, commensalism and parasitism.

BIOLOGICAL DIAGRAM OF ORGANISM: On the student answer page, draw, color and label, according to the rules of a biological diagram, a diagram of your organism, labeled with the 5 adaptations you described earlier.

The student should have completed the Picturing Life lesson (elsewhere in this book) to be able to draw an appropriate biological diagram. A student sample has been included in the following section. An overhead could be made of this diagram to show students an example of a correctly drawn biological diagram.

DIAGRAM OF ORGANISM INTERACTING IN ITS FOOD WEB: On the student answer page, neatly draw, color and label a food web for your organism. Be sure to include arrows in your food web indicating the flow of energy. Your food web diagram must include pictures/diagrams of at least five (5) other organisms.

In the student background section, students will find an explanation and discussion about how food webs are drawn and how arrows are used to indicate the energy flow within the web. A student sample has been included in the following section. An overhead could be made of this diagram to show students an example of a correctly drawn food web diagram.

SAMPLE STUDENT WORK

VERY NICE 10

A. NAME OF SCIENTIST: Alicia
B. TITLE OF SCIENTIST: PhD in Ecology from Kealing University, works in magnet science and genetic fields
C. PROPOSED SCIENTIFIC NAME OF ORGANISM: *Acuta rodentia*
D. CLASSIFICATION OF YOUR ORGANISM: Kingdom Animalia, Phylum Mammalus, Class Leadista, Order Herbitera, Family Trecentae, Genus Acuta, Species Rodentia
E. PHYSICAL DESCRIPTION/CHARACTERISTICS AND BEHAVIORS OF ORGANISM:
1) Small body about half a foot tall
2) Small eyes and nose, has relatively good eyesight but relies on ears
3) Has four tentacle-like ears colored a light green to camouflage with the grass
4) Has sharp spines on its back used to scrape bark off of trees
5) Body colored a light tan to camouflage with the ground
6) Has short feet that are very efficient at digging, lives in burrows
7) Curls up into a ball and fold ears over its body so that its spines protect its body when spotted by a predator
8) Hibernates for a long period of time over the winter
9) Lives in groups of six or seven, not very defensive of its territory
10) Has one to two babies at a time, about two inches tall when born
11) Takes care of young until age two when they are fully mature
12) Has an average life span of about ten years
13) Uses small pebbles outside of its burrow to collect dew in the morning
14) Is an herbivore – eats mainly soft tree bark, will eat wild grass when tree bark is scarce
15) Is a marsupial – females have a small pouch on its stomach it uses to carry babies
F. ADAPTATIONS:
a) Has four tentacle-like ears colored a light green to camouflage with the grass
    This adaptation helps *Acuta rodentia* survive, because not only does it rely on its sense of hearing, the ears help it camouflage into the grass and avoid predators
b) Has sharp spines on its back to scrape bark off of trees
    The spines on its back are very important, because without the bark to eat, *Acuta rodentia* would lose an important food source and would not have enough nutrients it needs to survive.
c) Is a marsupial – females have a small pouch on its stomach it uses to carry babies
    This pouch is important to the female, because without it, the female would not be able to go anywhere with her young.
d) Curls up into a ball and fold ears over its body so that its spines protect its body when spotted by a predator
    This is an important defensive tactic, because without it, this animal doesn't have another weapon against predators.

e) Has short feet that are very efficient at digging, lives in burrows
   Acuta rodentia lives in burrows and doesn't have another way to make a home.

G.   MEANS OF ENERGY INTAKE/FEEDING STRATEGIES: *Acuta rodentia* is a heterotrophic herbivore that eats mainly soft tree bark and sometimes wild grasses. It cannot use photosynthesis and is not chemosynthetic.

H.   REPRODUCTIVE STRATEGIES: *Acuta rodentia* reproduces with sperm and egg. It uses internal fertilization, like other mammals, and gives birth to live young.

I.   DESCRIPTIONS/CHARACTERISTICS OF ORGANISM'S HABITAT AND ENVIRONMENT: The *Acuta rodentia* lives in grasslands and forests, where there is a moderate climate of about 30 to 40 Celsius. The amount of rainfall is average – there is more rainfall in their environment than there would be in the desert, but less then there would be in a tropical rainforest. The grasslands are mostly flat plains covered in thick grass and scattered trees. My organism uses the ground as its home by burrowing into it, and uses pebbles to collect dew to drink. They always make their home near an area with trees, and stay by their burrow for most of their lives.

J.   DESCRIPTION OF INTERRELATIONSHIPS WITH OTHER ORGANISMS: *Acuta rodentia* doesn't seem to have too much competition with other animals, but it does have a lot of predators. Because it is a primary consumer, it has a lot of secondary and tertiary consumers to watch out for. It isn't in any type of symbiotic relationship, unless parasites happen to choose it as their host.

Alicia
Science ~2
10/29/2003

ACUTA RODENTIA

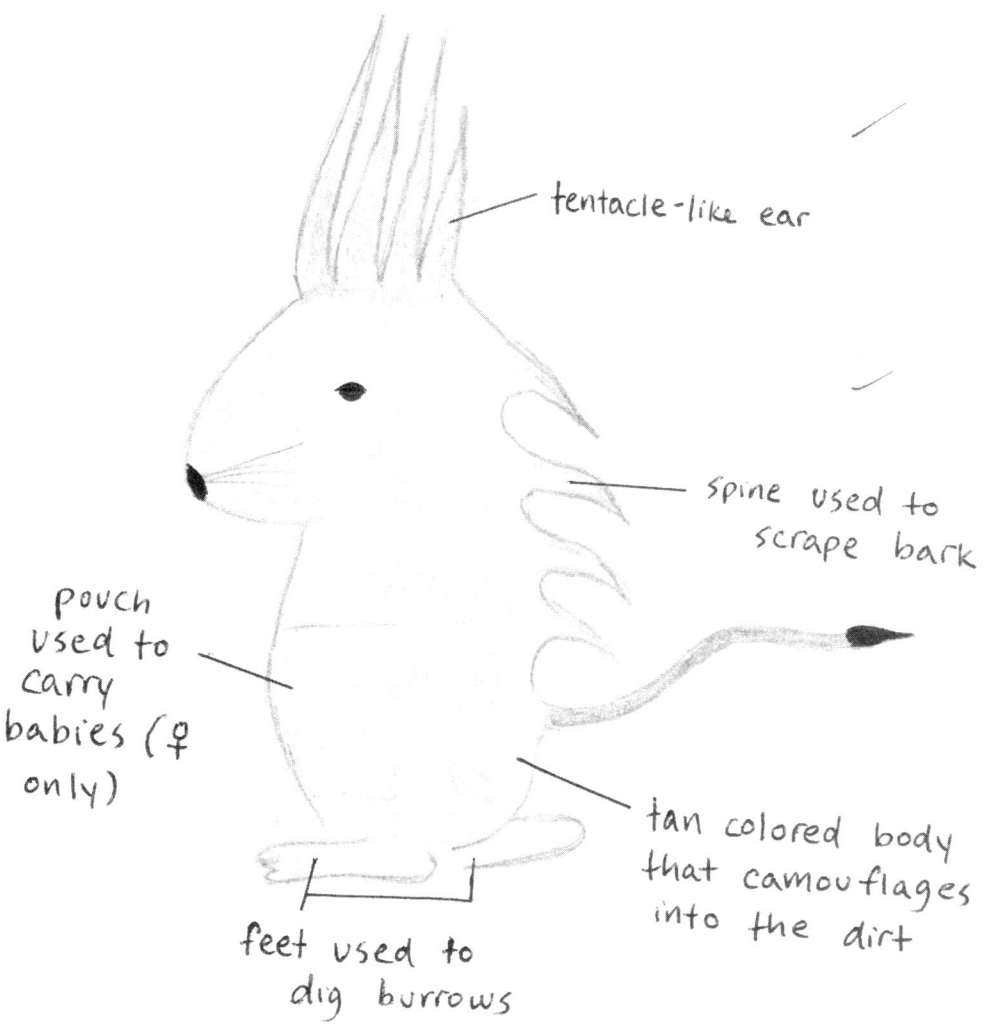

tentacle-like ear

spine used to
scrape bark

pouch
used to
carry
babies (♀
only)

tan colored body
that camouflages
into the dirt

feet used to
dig burrows

cool!

FOOD WEB DIAGRAM

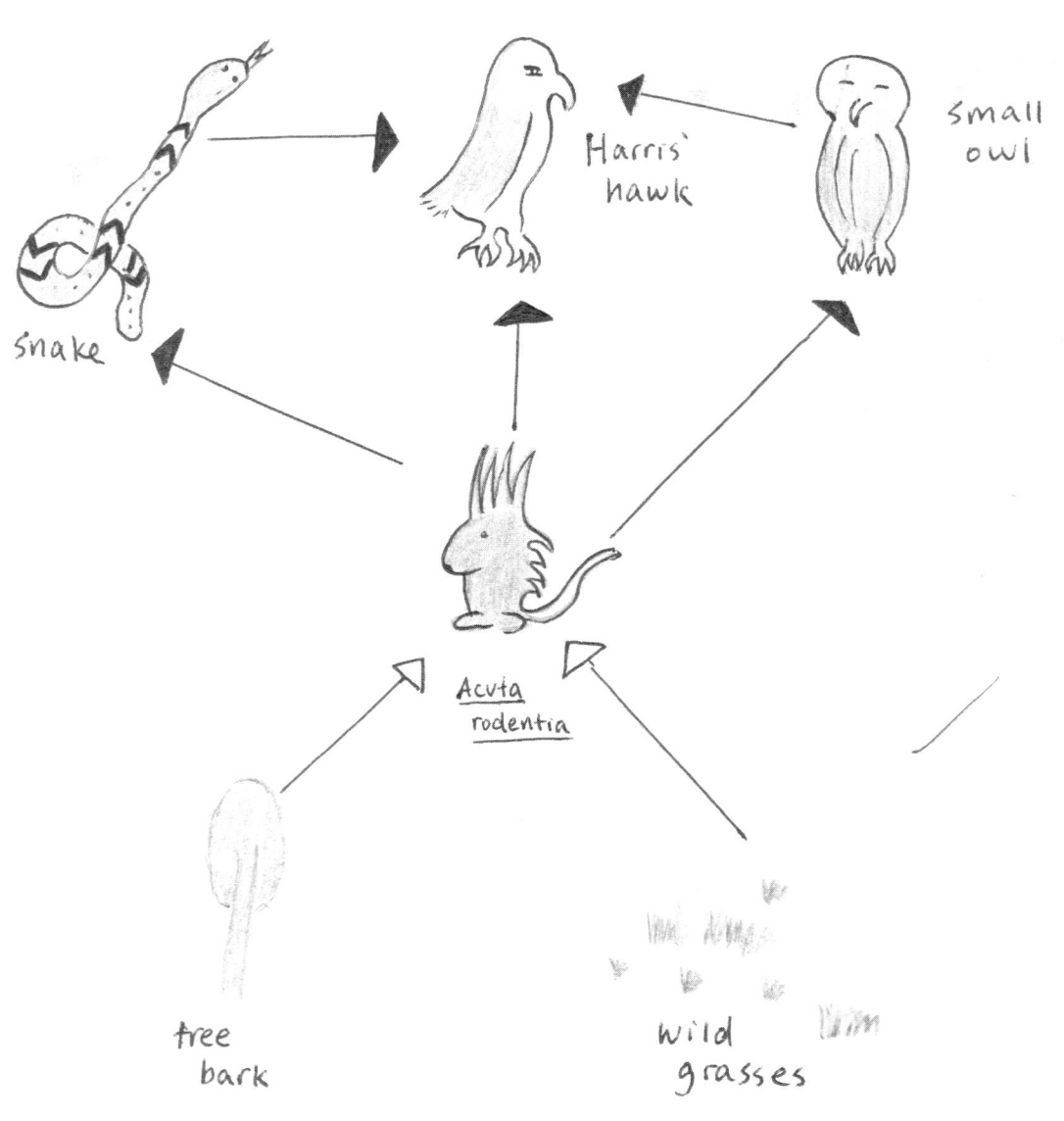

snake

Harris' hawk

small owl

Acuta rodentia

tree bark

wild grasses

Create a Species
Describing a New Imaginary Species

You are a famous ecologist returning from some far and remote location. During your travels, you discovered a new species of life. Upon your return, you prepare a scientific report on the newly discovered life form to present to your research peers. Below is some background information regarding living things as well as evolutionary relationships.

PURPOSE

In this activity you will apply many ideas and concepts of ecology, classification, and biological diagrams in a creative manner to describe a new imaginary species in a pseudo-scientific report.

PROCEDURE

1. Read through the background information on the paper that your teacher has provided. Carefully read the *CHARACTERISTICS OF LIVING THINGS* section. Although this is an activity designed to inspire your creativity, your organism must have these characteristics.

2. Begin to imagine some creature that you would like to create. It can be a member any one of the five kingdoms described in the background information. Be sure to carefully read *THE FIVE KINGDOMS OF LIVING THINGS* section.

3. It might be easier to begin with a sketch of your creature than by trying to describe it in a report. Let your imagination go wild! But try keeping it within the confines of what we know about living organisms.

4. Be sure to read the 3 other sections of the background information, *BINOMIAL NOMENCLATURE, SYMBIOTIC RELATIONSHIPS*, and *FOOD CHAINS AND FOOD WEBS* before beginning your report as you will have to apply concepts from each of these sections in your report/diagrams.

5. Include the required written information in each section on your student answer page.

6. Draw your biological diagram and your food web diagram, according to the instructions given on your student answer page, in the space provided. Use colored pencils to make your diagrams neat, easy to read, and in color. You will have three class periods to finish this activity. Let your imagination go wild!

7. Review the grading rubric at the end of the student answer page to understand how your grade will be determined.

Create a Species
Describing a New Imaginary Species

BACKGROUND INFORMATION

CHARACTERISTICS OF LIVING THINGS

What are the characteristics of living things? (1) All living organisms are made of cells containing cytoplasm. The cytoplasm is made up of the same basic components in all forms of life — proteins, carbohydrates, lipids (fatty materials), water and nucleic acids. Some simple forms of life have only one cell, but most have many cells; the human body has trillions of them. (2) All living organisms exhibit growth. Living things utilize food and convert some of it into their own cytoplasm. As a result, they grow larger, at least during some periods of their lives. (3) All living organisms have a restricted size and shape. All members of a species stay within certain limits of size and body shape, though individual differences do occur. (4) All living organisms take in and release energy. The very act of living requires energy so all forms of life must have some means of obtaining it. Plants as a rule absorb their energy from sunlight, carbon dioxide and water. They store this energy in the form of food they manufacture and can draw upon this food for their energy needs. Animals, as a rule, must eat food from plants or other animals that have eaten plants to provide for their energy needs. Plants generally produce much more food than they need, otherwise, animal life could not exist. (5) Living things adapt to their environment. Living things respond to their environment in ways that are favorable to them. (6) Living things develop inherited adaptations; they can change over many generations so they can better survive. (7) Living organisms can reproduce and have a limited life span. (8) Living things contain nucleic acids. One nucleic acid is DNA (deoxyribonucleic acid) which is the substance of genes of heredity that make possible the transmission of traits from one generation to the next.

THE FIVE KINGDOMS OF LIVING THINGS

Today, the most generally accepted classification system of living organisms contains five kingdoms: Kingdom Monera, Kingdom Protista, Kingdom Fungi, Kingdom Plantae and Kingdom Animalia. (Some classification schemes further subdivide Monera into two groups: Eubacteria and Archaebacteria.)

Kingdom Monera. Bacteria are placed in the Kingdom Monera, and in fact, the terms "bacteria" and "moneran" are interchangeable. Monerans are unicellular organisms (organisms that consist of only one cell). A moneran's cell does not have its hereditary material enclosed in a nucleus, a trait unique to monerans. They also lack many other membrane-bound structures found in other cells. (Cells without membrane-bound organelles are called prokaryotic.) Because of their unique characteristics, monerans are considered to be very distantly related to the other kingdoms. Like other organisms, monerans can be placed into two categories based on how they obtain energy. Organisms that obtain energy by making their own food are called autotrophs. This name makes sense because the prefix auto- means self and the root word -troph means food. Organisms that cannot make their own food are called heterotrophs. The prefix hetero- means other. Heterotrophs may eat autotrophs in order to obtain food or they may eat other heterotrophs. But all heterotrophs ultimately rely on autotrophs for food. Scientists have evidence that monerans were the earliest life forms on Earth. They first appeared about 3.8 billion years ago (bya).

Kingdom Protista. Like all other kingdoms except Kingdom Monera, the protists are eukaryotic, which means that their cells contain a nucleus. The nucleus controls the functions of the cell and also contains the cell's hereditary material. In addition, the cell of a protist has special structures that perform specific functions for the cell. Protists are unicellular but include some species that live together in large colonies that give the appearance of being multicellular. A number of protists are capable of animal-like movement but also have some distinctly plantlike characteristics. Specifically, some protists are green in color from chlorophyll, and can use the energy of light to make their own food from simple substances. However, they are neither plants nor animals. Protists were the first kind of cells that contained a true nucleus. Ancient types of protists that lived millions and millions of years ago are probably the ancestors of fungi, plants, animals and the modern protists. The protozoa — a group of organisms commonly studied in life science that includes amoeba, paramecium and euglena — are classified in this kingdom.

Kingdom Fungi. These are multicellular organisms that lack photosynthetic pigments and absorb nutrients directly from their surroundings. Mushrooms and toadstools are fungi. Molds that sometimes grow on leftover foods also belong to this kingdom. The mildews that may appear as small black spots in damp basements and bathrooms are also fungi. For many years, fungi were classified as plants (some out-of-date textbooks call them "nongreen plants"). However, they are quite different from plants in some basic ways. Their cell wall, a tough protective layer that surrounds the cell, is made of chitin rather than cellulose, the material that composes the cell wall of plants. And most importantly, unlike plants, fungi are not able to make their own food (heterotrophic versus autotrophic).

Kingdom Plantae. Plants make up this kingdom. These organisms are multicellular autotrophs. You are probably quite familiar with members of this kingdom, which includes flowering plants, mosses, ferns and trees.

Kingdom Animalia. Animals are multicellular heterotrophic organisms with specialized tissues that reproduce sexually and must take in oxygen to respire. Animals are heterotrophs, obtaining the nutrients and energy they need by feeding on organic compounds that have been made by other organisms. Animals are multicellular, which means that their bodies are composed of more than one cell. Animal cells are eukaryotic, containing a nucleus and membrane-enclosed organelles. Unlike plant, fungus and bacterial cells, animal cells do not have cell walls. Additionally, animals reproduce sexually (by exchanging sperm and eggs); some animals can also reproduce asexually.

BINOMIAL NOMENCLATURE
Binomial nomenclature is standard notation for naming species. Every species can be unambiguously identified with just two words, the combination of the genus and the species. The genus name is always capitalized while the species name is all lower case. In print format the scientific name is always in italics but when hand written the scientific name is underlined. Using this format, the identifying name for the human species is *Homo sapiens*, or Homo sapiens.

SYMBIOTIC RELATIONSHIPS
Symbiosis is a term used to describe a long-term relationship between two different species that live in close association with each other and in which at least one of the organisms benefits. There are three types of symbiotic relationships: mutualism, commensalism and parasitism.

In mutualism, both organisms benefit from their association. For example, termites have cellulose-digesting bacteria living in their digestive tracts. Without these microbes, termites could not obtain nutrients from the wood they eat. In turn, the termites provide the bacteria with food and a place to live. Cows have a similar association with the microbes that live in their digestive tracts.

Lichens consist of algal and fungal cells. Both types of cells benefit from this association. It allows them to live in environments in which neither could survive alone. Through photosynthesis, the algae produce food for themselves and for the fungi. The fungi provide moisture and the structural framework and attachment sites in which the algae grow.

Peas, clover and alfalfa are legumes. Legumes have nodules on their roots in which nitrogen-fixing bacteria grow. These bacteria convert nitrogen gas from the atmosphere into usable nitrogen compounds for plants. In this relationship, the plants are supplied with the nitrogen compounds they need, while the bacteria are given an environment in which they can grow and reproduce.

In commensalism, one organism benefits from the symbiotic relationship and the other is not affected. For example, pilotfish are small fish that live with sharks. They eat the scraps left over from the shark's feeding. Thus, the shark provides the pilotfish with food. As far as is known, the pilotfish neither helps nor hurts the shark. Barnacles may attach themselves to the large body surface of a whale. Barnacles are sessile (unable to move on their own) and rely on water currents to bring them food. The movements of the whale provide them with a constantly changing environment and food supply. The whale is not affected by the presence of the barnacles.

In parasitism, one organism benefits from the symbiosis and the other is harmed. The organism that benefits is called the parasite, while the organism that is harmed is called the host. Some parasites cause only slight damage to their hosts, while others kill the host. Tapeworms, for example, are parasites that live in the digestive tracts of various animals. There, they are provided with nutrients and an environment in which to grow and reproduce. However, the host is harmed by the presence of the tapeworms. The loss of nutrients and tissue damage caused by the worm can cause serious illness. There are also parasitic plants that grow on other plants. Two examples of plant parasites are mistletoe and Indian pipe.

FOOD CHAINS AND FOOD WEBS
Within an ecosystem, there is a pathway of energy flow that always begins with the producers. Energy stored in organic nutrients synthesized by the producers is transferred to consumers when the plants are eaten. Herbivores are the primary consumers, or first-level consumers. The carnivores that feed on the herbivores are secondary consumers. For example mice feed on plants and are primary consumers. The snake that eats the mouse is a secondary consumer, while the hawk that eats the snake is a tertiary consumer, or third-level consumer. Since many consumers have a varied diet, they may be second-, third-, or higher-level consumers, depending on their prey. Each of these feeding relationships forms a food chain, a series of organisms through which food energy is passed. A simple food chain, with arrows showing the direction of energy flow, can be shown as:

grass ———————— field mouse ———————— great horned owl

where the grass is a producer, the field mouse is a primary consumer, and the owl is a secondary consumer.

Feeding relationships in an ecosystem are never just simple food chains, however. There are many types of organisms at each feeding level, and there are always many food chains in an ecosystem. Usually, each organism is part of several different food chains. These food chains are interconnected to form a food web. A simple food web could be shown as:

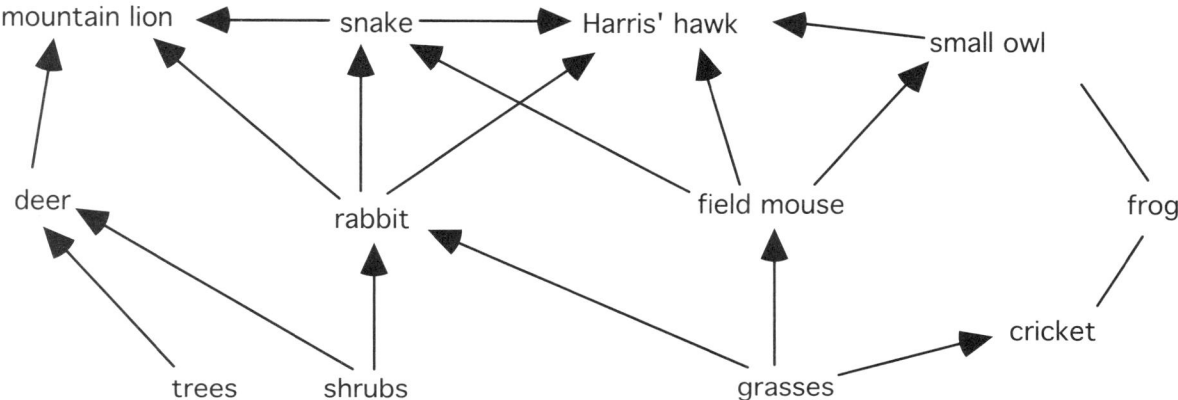

At every level in an ecosystem, there are organisms that act as decomposers. The decomposers make use of the wastes and remains of all organisms in the system. They use the energy they find in these materials for their own metabolism (life processes). At the same time, they break down organic compounds into inorganic compounds and make substances available for reuse. The decomposers are the final consumers in every food chain and food web.

Create a Species
Describing a New Imaginary Species

NEW SPECIES DESCRIPTION REPORT

A. NAME OF SCIENTIST:

B. TITLE OF SCIENTIST:

C. PROPOSED SCIENTIFIC NAME OF ORGANISM:

D. CLASSIFICATION OF YOUR ORGANISM:

Kingdom:

Phylum:

Class:

Order:

Family:

Genus:

Species:

E. PHYSICAL DESCRIPTION/CHARACTERISTICS AND BEHAVIORS OF ORGANISM:

PHYSICAL CHARACTERISTICS
1.
2.
3.
4.
5.
6.
7.
8.
9.
10.

BEHAVIORAL CHARACTERISTICS

1.

2.

3.

4.

5.

F. ADAPTATIONS:

1.

2.

3.

4.

5.

G. MEANS OF ENERGY INTAKE/FEEDING STRATEGIES:

H. REPRODUCTIVE STRATEGIES:

I. DESCRIPTION/CHARACTERISTICS OF ORGANISM'S HABITAT AND ENVIRONMENT:

J. DESCRIPTION(S) OF INTERRELATIONSHIPS WITH OTHER ORGANISMS:

BIOLOGICAL DIAGRAM OF ORGANISM: In the space below, draw, color and label, according to the rules of a biological diagram, a diagram of your organism, labeled with the 5 adaptations you described earlier.

DIAGRAM OF ORGANISM INTERACTING IN ITS FOOD WEB: In the space below neatly draw, color and label a food web for your organism. Be sure to show the arrows in the food web, indicating flow of energy. Your food web diagram must include pictures/diagrams of at least five (5) other organisms in your biome/ecosystem.

Create a Species Rubric	Super	Good	Fair	Poor	Omitted
Written Report Describing Species: sections A-J completed, describing all aspects of the imaginary species.	60	52	42	35	0
Biological Diagram of Species: clear, accurate, colored and neatly drawn diagram is included, with 5 species adaptations labeled and identified.	20	16	14	10	0
Food Web Drawing: clear, accurate, colored and neatly drawn diagram is included, with the new species and its ecological interaction with 5 other species within its ecosystem/biome.	20	16	14	10	0
Total Points	100	84	70	55	0

Write Your Notes and Ideas Here!

Toothpick Birds
Modeling Predator Behavior in an Outdoor Lab

OBJECTIVE
Students will observe how protective coloration helps some animals to survive in nature by modeling predatory behavior while feeding on toothpick "insect" prey in an outdoor area.

LEVEL
Middle Grades: Life Science

NATIONAL STANDARDS
UCP.2, UPC4, UPC5, A1, A2, C4, C6

TEKS
6.2(B), 6.2(E), 6.3(C), 6.4(B), 6.11(A), 6.12(C)
7.2(B), 7.2(E), 7.3(C)
8.2(B), 8.2(E), 8.3(C), 8.11(A)
IPC 1(A), 2(A), 2(B), 2(C), 2(D)

CONNECTIONS TO AP
AP Biology:
    II. Heredity and Evolution: C. Evolutionary Biology 3. Mechanisms of evolution
    III. Organisms and Populations: B. Structure and Function of Plants and Animals 2. Structural, physiological, and behavioral adaptations 3. Response to the environment
    C. Ecology 1. Population dynamics 2. Communities and ecosystems
AP Environmental Science:
    II. Interdependence of Earth's Systems: Fundamental Principles and Concepts D. The Biosphere

TIME FRAME
50 minutes

MATERIALS
(For a class of 28 working individually)

50 uncolored wooden toothpicks	50 red toothpicks
50 green toothpicks	50 blue toothpicks
50 yellow toothpicks	watch or other timing device (for timing one minute)
50 yellow with red stripe toothpicks	plastic cups labeled for each round of play
whistle for teacher	

TEACHER NOTES

In *Toothpick Birds*, students model the behavior of a predatory bird, selecting "prey insects" (colored toothpicks) and demonstrating that protective coloration helps an organism to survive in a specified outdoor area.

The student "predators" visually attempt to feed by locating "prey" (toothpicks) in their "natural" environment. Results will differ slightly depending on the location and environment of your school. Fall, winter, or spring seasons will also affect the outcome as the toothpicks are more or less camouflaged by the changing ground colors. Ideally, this activity should be done in the spring, and the background should be mixed, some areas of dirt, some areas of green, and so on.

You will need to prepare the toothpicks in advance. A total of 275 toothpicks will need to be counted, organized and divided in the following fashion:

- 50 "wood" (natural uncolored toothpicks)
- 50 yellow
- 50 green
- 50 blue
- 50 red
- 50 yellow with a red stripe (colored by the teacher, described below)

The yellow with red stripe needs to be colored in the following fashion:

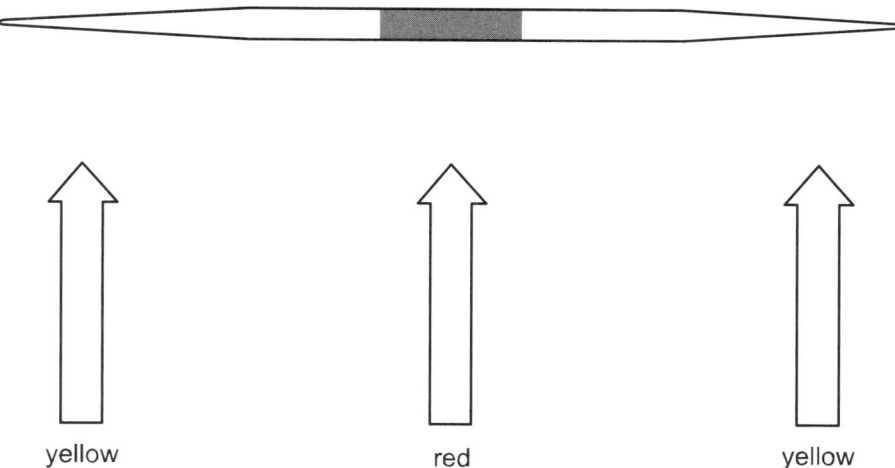

yellow   red   yellow

The red should be colored with a permanent red marker so the color does not bleed, especially if it is damp or wet outside. Twenty-five red with yellow stripe toothpicks are recommended because if the full total of 50 is used, all the students will notice the ones with red stripes and ask questions about them. It is up to you to decide whether or not to reveal to the students what the yellow striped toothpicks represent during the middle of the activity. Some teachers may decide to reveal what they are and some may decide to ignore the questions during the activity and address that issue once the data analysis begins back in the classroom. The yellow and red striped toothpicks have been intentionally left off the list of materials for the students so as not to spoil the discussion of the concept of mimicry.

TEACHER PAGES

The "wood" or uncolored toothpicks are easily found in any grocery store and one box of 500 is enough for a few years. The colored toothpicks might be a bit more difficult to locate but the colors in the box are typically the colors that are needed for this lab: yellow, blue, red, and green. It is a good idea to get two boxes of 500 to make sure that you have enough of each color for all your classes. It will take about 30-45 minutes to count, divide, and bundle the toothpicks with a rubber band into colored groups of 50 for all of your classes. All the toothpicks for each class can be put into a labeled cup or a beaker.

The "area of predation", where the modeling activity is to be performed, is generally between 100 square meters and 200 square meters. The model ecosystem can be varied according to the level of challenge for the students. Ideally, the "area of predation" will have a varying amount of background material, including grass, straw, dirt, and so on.

Before going outside to start the activity, have the students read the entire procedure so that they understand the rules. Students should then hypothesize which color will be selected the most by the "birds" and which color will be selected the least. Typically, they will choose green to be the least selected because of the camouflage effect but many times this is not the color that is selected least due to the "unnatural" shade of green dye. Usually, most or all students choose red or blue to be selected the most. Most students have a very clear idea of the concept of camouflage from previous knowledge.

To begin the activity, outline the boundaries for your students by using natural landmarks at each edge such as a tree or a sidewalk. You could also use spray paint or chalk to outline the edges of your area. Have students line up along one side of this area, facing outward. Spread all the toothpicks randomly around the "area of predation". Once toothpicks are spread, blow a whistle to signal the students to begin selecting the "prey". Allow students to "feed" for 1 minute. Blow the whistle again to signal the students to stop feeding. Each student needs 4 toothpicks to survive each round. Instruct students to line up in their original starting place, along the outside of the "area of predation", again facing out. Walk by the line of students with a cup labeled "round 1", stopping at each student to collect toothpicks. "Survivors", those with 4 toothpicks, remain in position, standing on the boundary. The "non-survivors", those with less than 4 toothpicks, are directed to sit down in another area, near the "area of predation" but separated from it. The survivors compete in "round 2", competing for the now decreased food source. After one minute of feeding, the survivors and non-survivors are determined and sorted as before and the toothpicks are collected in a cup labeled "round 2". Again, the same process, as described for rounds 1 and 2 is followed for rounds 3, 4 and so on until one "super predator" remains. You may choose to assign the super predator extra credit or a homework exemption or some other reward.

Once the super predator has been determined, you can assign a few of the students to sort and count the toothpicks within the cups. Record the toothpick numbers in the data table. If time permits the entire process can be repeated, going through the rounds to select another super predator. You may or may not choose to collect data for the second scenario.

Behaviors will arise during the activity that model the aggressive and competitive predatory behaviors seen in nature when predators compete for a limited food source. Students will typically playfully push each other out of the way to get a toothpick. Also, students will sometimes try to take toothpicks from each other. This is also a natural behavior in a competitive predation. Some students will also purposefully move away from the other predators, "foraging" in areas where there is not as much competition. Although you should always discourage these behaviors, they will arise in this activity. More often than not, they are playful in nature as students get very excited while doing this activity. You can choose how to address these inevitable behaviors.

The data analysis is most efficiently done back in the classroom, away from the outside distractions. Discussion should focus on which colors were selected and which were not and why. Also, discussion of the competitive behaviors that were witnessed can lead to very lively discussions.

Major ecological concepts:

Mimicry: when a species mimics the traits of a poisonous species. One classic example in nature of mimicry includes the nonpoisonous viceroy butterfly mimicking the poisonous monarch butterfly. In toothpick birds, the yellow with red stripe mimics the yellow one.

Camouflage: when a species blends in to the background environment. Camouflage is probably the most well-known of animal adaptations. There are hundreds of examples in nature. Some of the most dramatic can be found in the Arthropod phylum, particularly in the order Homoptera, commonly called the leaf hopper insects, where many insects have characteristics of leaves, even down to leaf venation pattern on their wings.

Other ecological concepts that arise include natural selection and competition. You can choose how much detail to cover with the students on these topics as well.

POSSIBLE ANSWERS TO THE CONCLUSION QUESTIONS AND SAMPLE DATA

Provide a data table similar to the one below on a chalkboard or overhead for students to copy.

Color of prey	R1	R2	R3	R4	R5	R6	Total selected	% Selected	Reason for Selection or non-selection
red									
blue									
green									
wood									
yellow									

Generally the toothpicks will be selected and not selected according to the following patterns.

Color of toothpick	Typical frequency of selection	Reason for selection or not
Red	High frequency of selection, typically 80-90%	"Stands out" on typically any background found outside.
Blue	High frequency of selection, typically 80-90%	"Stands out" on typically any background found outside.
Yellow	Very low degree of selection, typically 0-5%.	Poisonous.
Green	Medium frequency of selection, typically 50-60%	The green found in toothpicks is "unnaturally" green and many times will stand out. If the background is extremely green grass, then of course the degree of selection is decreased as they are more camouflaged against the green background.
Wood	Typically lower frequency of selection in a typical mixed dirt and grass environment. Typically lower selection than expected.	Camouflage.
Yellow with red stripes	Very low frequency of selection.	Mimics poisonous.

1. Which color of prey was selected more than any other color and therefore survived the least?
   - Usually red or blue is selected with the highest frequency.

2. Why was the color of prey in question #1 selected the most?
   - Red or blue is not camouflaged in any natural environment.

3. Excluding the yellow poisonous prey, which color of prey was selected least and therefore survived the most?
   - Depending on the environmental conditions, either green or wood color will usually be the colors selected the least.

4. Why was the color of prey in question #3 selected the least?
   - These colors are camouflaged in most natural environments.

5. Why didn't the predators select the yellow "insects"?
   - They are poisonous.

6. Why didn't the "birds" eat yellow-and-red-striped "insects"?
   - They mimic the poisonous prey items.

7. Which color(s) of insect is an example of camouflage?
   - Wood or green.

8. Name one example of camouflage in the natural world.
   - Accept any reasonable response. There are many well-documented examples of camouflage in nature.

9. Which color of insect is an example of mimicry?
   - The yellow with red stripe is an example of mimicry.

10. Give an example of mimicry in the natural world.
    - Accept any reasonable example of mimicry. The most common ones are probably the milk snake (non-poisonous)/coral snake (poisonous) and the viceroy butterfly (non-poisonous) and the monarch butterfly (poisonous).

11. If this simulation activity is similar to what occurs in nature, then what survival strategies are most effective in avoiding predation?
    - Camouflage and poisonous advertising.

12. Explain how, in nature, the color of an insect may determine whether it will be preyed upon or not?
    - If a prey item is able to be seen by the predator, then it has a high probability of being eaten. If it is camouflaged, there is a good chance it will not be seen and, therefore, eaten by the predator. If a prey item advertises its poisonous nature, and predators learn that those prey are poisonous, then they will not be eaten. The non-poisonous ones that mimic the poisonous ones are typically not eaten as well.

13. Explain how predators help to "select" which animals will survive in nature.
    - After many generations of selection, the colors that are selected and eaten with the highest frequency are eventually eliminated from the population. The color of organisms that are not selected are able to transmit their "color genes" onto the subsequent generations, causing the populations of those colors to increase.

Toothpick Birds
Modeling Predator Behavior in an Outdoor Lab

PURPOSE
In this activity you will observe how protective coloration helps some animals to survive in nature. You will model predatory behavior by feeding on toothpick prey items in an outdoor area.

MATERIALS
wood colored toothpicks
green toothpicks
yellow toothpicks
plastic cups labeled for each round
writing utensil

red toothpicks
blue toothpicks
student instruction page
student answer page

Safety Alert
1. Be careful when picking up the toothpicks as they are sharp and can scratch and/or poke your skin.
2. Do not push, trip, tackle or engage in other horse play in the lab.

PROCEDURE
1. Read through the entire procedure before beginning.

2. In the space marked HYPOTHESIS on the student answer page, formulate a hypothesis as to which "insect" will be selected the least and which one will be selected the most.

3. In the space marked PURPOSE on the student answer page, write the purpose of the lab.

4. In the space marked DATA TABLE on the student answer page copy the data table as your teacher instructs. You must draw the lines with a ruler or straight-edge. (R1 is an abbreviation for round one, R2 for round two, and so on…)

5. You are going to be a predatory "bird" feeding on "insects" to survive. Colored toothpicks will represent your insect prey. When instructed to begin, you will have one minute to "feed" on at least four "insects" in the feeding area outlined by your teacher. If you do not capture at least four prey toothpicks, you will die from starvation and have to sit out for the remaining rounds. After several rounds, only one "Super Predator" will survive.

Here are the rules:
- You must use your "beak" (forefinger and thumb) of one hand only.
- You may "eat" only one "insect" at a time.
- You must place the "insect" in your "crop" (your other hand) before you can reach for another insect.
- You must stop feeding when time is called. If you are reaching for an insect and your teacher has called time, do NOT finish reaching for it and stand up straight.
- You must line up along the boundary that your teacher has designated, with your back turned to the area of predation.
- Yellow "insects" are poisonous. Do not eat a yellow insect. If you accidentally pick up a yellow "insect" into your beak, put it back onto the ground. If you put the yellow insect into your "crop", then you will be poisoned and you will have to stop feeding and sit out the duration of this round.

6. After time is called for each round, put all captured insects into the plastic cup labeled, "ROUND 1". The "dead" predators will then sit in a designated area. All of the surviving birds will feed again in "ROUND 2", with each bird again eating at least 4 "insects". The eaten "insects" for Round 2 will go in the cup labeled, "ROUND 2". The class will do as many rounds as are necessary to decrease or "select" the population down to one surviving bird, the "Super Predator".

7. When there is only one "Super Predator" left, return to the room and complete your data table. Use the cups collected during each round to determine how many toothpicks of each color were consumed.

8. Use the following formula to determine the % selected for each color toothpick.

$$\% \text{ selected} = \frac{\text{total number of a particular color collected}}{\text{number of that color of toothpick your teacher distributed}} \times 100$$

9. In the space marked EXPLANATION on your data table, provide a specific reason as to why you and the class collected the particular insects that you did. Do not state reasons that are not factual; for example, do not say that the insects tasted good since you did not really taste them.

Name _____

Period _____

Toothpick Birds
Modeling Predator Behavior in an Outdoor Lab

PURPOSE

HYPOTHESIS

DATA TABLE

CONCLUSION QUESTIONS

1. Which color of prey was selected more than any other color and therefore survived the least?

2. Why was the color of prey in question #1 selected the most?

3. Excluding the yellow poisonous prey, which color of prey was selected least and therefore survived the most?

4. Why was the color of prey in question #3 selected the least?

5. Why didn't the predators select the yellow "insects"?

6. Why didn't the "birds" eat yellow-and-red-striped "insects"?

7. Which color(s) of insect is an example of camouflage?

8. Name one example of camouflage in the natural world.

9. Which color of insect is an example of mimicry?

10. Give an example of mimicry in the natural world.

11. If this simulation activity is similar to what occurs in nature, then what survival strategies are most effective in avoiding predation?

12. Explain how, in nature, the color of an insect may determine whether it will be preyed upon or not?

13. Explain how predators help to "select" which animals will survive in nature.

Online Biome Exploration
Researching, Creating, and Presenting a Biome PowerPoint

OBJECTIVE
Students will research, create and present a PowerPoint presentation on a particular biome, focusing on climate, floral and faunal adaptations, ecological interactions, interesting facts, and human impact.

LEVEL
Middle Grades: Life Science

NATIONAL STANDARDS
UCP.2, C.4, C.5, E.1, F.3, F.4, F.5, F.6

TEKS
6.8(C), 6.12(C)
7.2(E), 7.12(C), 7.14(C)
8.2(E), 8.6(A), 8.14(B)
IPC 2(C), 2(D)

CONNECTIONS TO AP
AP Biology:
    AP Environmental Science: I Scientific Analysis, II; II Interdependence of Earth's Systems: Fundamental Principles and Concepts, D. The Biosphere

TIME FRAME
5 class periods of 50 minutes each

MATERIALS
(For a class of 27 working in groups of 3)

   9 computers with Internet access   Microsoft PowerPoint software on computers

TEACHER NOTES
Online Biome Exploration is designed to be a group research project on the components of a biome. It will incorporate technology through internet usage and PowerPoint construction. Students will also present their PowerPoint presentation to their peers.

A night or two before the project is to be started in class, divide the students into groups of 3. Have groups randomly select a biome (as described below). As a home work assignment that night, instruct the students to check out a book or find a magazine article on their assigned biome. It is important for students to find some non-internet sources of information so that students who are not using the computer can have research material. You may also wish to pull the relevant books from your school library to have available in your classroom.

To randomize the assignment of biomes, print the included list of biomes and cut the paper into strips. Fold the strips into tiny squares and put the squares into a container of some type and have each group select a strip of paper. The group will then be responsible for that biome.

Here is a list of biomes that have information readily available.
- tundra
- coniferous forest (a.k.a. boreal forest, a.k.a. taiga)
- deciduous forest
- grassland
- Antarctica
- desert
- tropical rainforest
- temperate rainforest
- freshwater wetland: (student must choose bog, marsh or swamp)
- estuaries (a.k.a. coastal wetland, a.k.a. saltwater wetland, a.k.a. tidal pools)
- abyssal zone
- coral reef (a.k.a. neritic zone)
- stream/river
- pond/lake

There is a wealth of information online, accessible through the search engine Google found at http://www.google.com. Students search from Google and sift through this wealth of information to find specific information for their biome.

A note on Internet searching:
Many teachers and students may have felt frustration when presented with finding specific information on the Internet. Google has a very useful set of help files found at http://www.google.com/help. As a teacher it is worth visiting and reviewing. You may wish to present this information to your students or have students review this page on their own. Especially worth noting is the "Basics of Search" link. All the basics of searching are presented here. The links for "Interpreting Results" and the "Advanced Search Tips" links are also worth reading. Taking 15 minutes to read through these three pages will save hours of frustration for both you and your students.

Once you are ready to begin creating the PowerPoint presentation, it is assumed that all students have a basic understanding of creating a PowerPoint presentation. This lesson is designed to serve as an application of PowerPoint rather than as a starting point for creating a PowerPoint.

PowerPoint presentations are a tool for presenting basic information and act as a visual tool to enhance a verbal presentation. Each slide in a PowerPoint should not be a paragraph of text but rather a bulleted list of phrases that summarize the information that is being orally presented. You will need to emphasize this point over and over to students as they are creating their Power Points. Students will tend to be extremely verbose and overly detailed on each slide. As well, students will want to spend all their time finding pictures as opposed to concentrating on text and content. It is plausible to require that students find the entire content first and then insert the required pictures at a later time.

The following PowerPoint tips can be presented to students either orally or through a PowerPoint of your own.

PowerPoint Tips

Organize Your Information
- Keep information short and to the point.
- Don't put too much information on one slide: no more than 6 bullets per slide and no more than 6 words per bullet is a good rule of thumb
- Remember to use metric measurements.

Keep Things Visually Pleasing
- Make sure text is very visible and contrasts against the background
- Do not over use animated effects as this can be annoying

General Reminders
- Use your time wisely.
- Concentrate on research first, beautification (and pictures) later.
- Make sure all your power point slides are saved together in one file.

Once students begin researching, it is a good idea to get an inexpensive egg timer or some other device with an alarm to keep track of 15 minute time intervals. Each student should have 15 minutes to conduct research on the internet. The other two students in the group should be conducting their research with the library books and magazines available. Keeping the computer time limited to 15 minutes helps the student on the Internet stay focused.

It is recommended that you take 3 class periods for research and PowerPoint construction and another 1-2 periods for student presentations. You may find it necessary to adjust this time frame given your individual computer accessibility and/or student load.

To ensure student attentiveness during the presentations you may wish to have students take notes in preparation for a quiz or other graded event. You might find it helpful to standardize these notes by having each student create a table like the one below for recording notes as each group presents.

Biome	General Climate	General Info	Faunal Adaptations	Floral Adaptations	Human Impact	Interesting Facts
Desert						
Grassland						

It is also suggested that a question and answer session follow each presentation. The way the groups answer and respond to the other students' questions should be included in the oral presentation grade.

There are two rubrics provided for assessing student performance: a rubric for biome content and a rubric for the oral presentation. Both of these rubrics are included on the student answer page. Familiarize yourself with both of these before starting the project. It is suggested that you score the groups' rubrics as you watch the presentation.

Left margin: TEACHER PAGES

Online Biome Exploration
Researching, Creating, and Presenting a Biome PowerPoint

PURPOSE

In this activity you will research, create and present a PowerPoint presentation on a particular biome, focusing on climate, floral and faunal adaptations, ecological interactions, interesting facts, and human impact.

MATERIALS

computer with Internet access   Microsoft PowerPoint software on computers

PROCEDURE

1. Read through the information in the BIOME PROJECT: CONTENT section to determine what is required for your oral presentation.

2. Read through the information in the BIOME PROJECT: DIVISION OF LABOR to determine who is responsible for each part of the presentation.

3. On the BIOME PROJECT: DIVISION OF LABOR page, write the name of your selected biome and also the name of each person assigned to each section. Turn this page in to your teacher.

4. Read carefully through the content and presentation rubrics on your student answer page.

5. Once it is time to research, each person should rotate being on the computer every 15 minutes. Your teacher will remind you when 15 minutes have passed. The other two members should be using the other provided resources for collecting biome information.

6. As each person completes their part of the research, write that person's initials on the content rubric in the space provided.

7. You will have three class periods to research. Work quickly and do not waste time.

8. You will be orally presenting your PowerPoint to the rest of the class. You are responsible for presenting the part of the project that you researched. Your teacher will review with you some tips for making effective PowerPoint presentations.

BIOME PROJECT: CONTENT

Each number below corresponds to an individual slide and must be included in your PowerPoint presentation (except for #4 and #5, which include 4 slides each). Also, next to each slide description is the number of points that particular slide is worth. When a slide requires a picture, your group can find one on the Internet (usually takes longer) or scan one from a book/magazine.

1. **TITLE PAGE**: (2 pts.)
   Slide must include the title of your PowerPoint presentation and the names of all your group members.

2. **GENERAL DESCRIPTION OF BIOME**: (6 pts. for description, 3 pts. for map)
   Slide must include a general description of biome type including locations worldwide. (Must include the use of a global map in some way.)

3. **CLIMATE DATA FOR YOUR BIOME**: (6 pts. for graph, 3 pts. for data)
   Slide must contain climate data in a graphical format. If you do not include this data in a an easy-to-interpret graphical format, you will not receive any credit in this section. Make sure that your data points are easy to read/understand.
   a. If terrestrial, your biome must include the air temperature ranges during the year in degrees Celsius. Ask your teacher how to convert from Fahrenheit to Celsius if needed. Monthly rain totals must also be included.
   b. If aquatic, your biome must include water temperature ranges during the year in degrees Celsius. The depth range (in meters) must also be included.

4. **FLORAL/FUNGAL ADAPTATIONS**: (20 pts., 5 pts. for each slide)
   a. First slide must contain a list of the typical flora found in your biome.
   b. The next 3 slides must include a description of 3 of the flora (one per slide) in your particular biome with special emphasis on how these plants are adapted for survival in your biome. [Go for diversity — that is, try to find plants that have unique methods of survival.]
   c. Each slide must include a picture of the plant and the adaptation on the picture. You can substitute one fungal adaptation for a floral if you prefer.

5. **FAUNAL ADAPTATIONS**: (20 pts., 5 pts. for each slide)
   a. First slide must contain a list of the typical fauna found in your biome.
   b. The next 3 slides must include a description of 3 of the fauna (one per slide) in your particular biome with special emphasis on how these animals are adapted for survival in your biome. Describe at least three animals and their particular adaptations. Each slide must include a picture of the animal and the adaptation on the picture. [Go for diversity — that is, try to find animals that have different methods of survival. Don't choose all mammals or all reptiles.]
   c. For aquatic biomes, you may substitute a protist or moneran for one of the animals.

6. **FOOD WEB FOR YOUR BIOME**: (15 pts.)
   Slide must include a food web with at least 8 animals and 4 plants, using arrows to show the direction of energy flow within the food web in your biome. You can create this diagram with a simple graphics program such as MS Paint.

a. If you are researching the abyss biome, there are no plants, but you must include the microorganisms that are the producers in your biome.
b. If you are researching the coral reef biome, you need to include the zooxanthellae algae that have the symbiotic relationship with the coral polyps.

7. **HUMAN IMPACT**: (15 pts.)
Slide(s) must describe the impact of humans on your biome. For example: How do we use the land? Have we affected the biome detrimentally in any way? What has been done or may be done to correct problems caused by humans? Has the size of the biome changed over time? Do we receive any benefits from this biome?
 a. Include at least one photo documenting the impact of humans on your biome.
 b. This can be presented on one or two slides.

8. **INTERESTING FACTS**: (8 pts)
Slide(s) must describe a minimum of at least four other facts that you find interesting about your biome. You must include 2 photos illustrating these interesting facts.

9. **CONCLUSION/REFERENCES**: (2 pts.)
Slide must end your presentation and thank your audience. See the section about references on the *Division of Labor* page for notes about referencing.

BIOME PROJECT: DIVISION OF LABOR

Team members should write their name beside the section of the report that they agree to research. This paper should be turned in to your teacher before beginning research.

BIOME: _____

TEAM MEMBER   REPORT SECTIONS

_____   1.

- description of the biome, including an introductory paragraph

- interesting facts

- reference list

- concluding paragraph

_____   2.

- description of faunal survival strategies

- human impact

- reference list

_____   3.

- description of conditions to which organisms must be adapted

- description of floral survival strategies

- reference list

NOTE: The reference list must be combined from all team members and listed in alphabetical order by author. For references to websites, use the rules found at the *Cyberbee, Electronic Referencing* site: http://www.cyberbee.com/citing.html.

Name _____

Period _____

Online Biome Exploration
Researching, Creating, and Presenting a Biome PowerPoint

RUBRICS

BIOME PROJECT POWERPOINT CONTENT RUBRIC

Slide	Point distribution	Number of points	Write the person's initials that completed this part of the slide show in the blank.
1. TITLE SLIDE	a. includes the title b. names of all your members	a. 1 point b. 1 point	a. _____ b. _____
2. GENERAL DESCRIPTION OF BIOME	a. general description including locations worldwide b. use of a global map	a. 6 points b. 3 points	a. _____ b. _____
3. CLIMATE DATA FOR YOUR BIOME	a. data present b. table or graph	a. 6 points b. 3 points	a. _____ b. _____

4. FLORAL/FUNGAL ADAPTATIONS	a. list of the typical flora
	
	b. floral adaptation #1
	   (1) description
	   (2) picture
	
	c. floral adaptation # 2
	   (1) description
	   (2) picture
	
	d. floral adaptation # 3
	   (1) description
	   (2) picture
		a. 4 points
		
		b. 4 points
		  (1) 2 points
		  (2) 2 points
		
		c. 4 points
		  (1) 2 points
		  (2) 2 points
		
		d. 4 points
		  (1) 2 points
		  (2) 2 points
			a. _____
			
			b.
			  (1) _____
			  (2) _____
			
			c.
			  (1) _____
			  (2) _____
			
			d.
			  (1) _____
			  (2) _____

5. FAUNAL ADAPTATIONS	a. list of the typical fauna
	
	b. faunal adaptation #1
	  (1) description
	  (2) picture
	
	c. faunal adaptation # 2
	  (1) description
	  (2) picture
	
	d. faunal adaptation # 3
	  (1) description
	  (2) picture
		a. 4 points
		
		b. 4 points
		  (1) 2 points
		  (2) 2 points
		
		c. 4 points
		  (1) 2 points
		  (2) 2 points
		
		d. 4 points
		  (1) 2 points
		  (2) 2 points
			a. _____
			
			b.
			(1) _____
			(2) _____
			
			c.
			(1) _____
			(2) _____
			
			d.
			(1) _____
			(2) _____

6. FOOD WEB FOR YOUR BIOME	a. Include at least 8 animals and 4 plants b. arrows showing energy flow	a. 12 points b. 3 points	a. _____ b. _____
7. HUMAN IMPACT	a. description of human impact on biome b. picture of human impact	a. 10 points b. 5 points	a. _____ b. _____
8. INTERESTING FACTS	a. interesting fact #1 b. interesting fact #2 c. interesting fact #3 d. interesting fact #4	a. 2 points b. 2 points c. 2 points d. 2 points	a. _____ b. _____ c. _____ d. _____
9. CONCLUSION	a. slide is present	a. 2 points	a. _____
Total Points			

BIOME ORAL PRESENTATION RUBRIC

	Great	Good	Average	Poor	Omitted
Preparedness: ability to speak without reading, knowledge and understanding of information, fielding of questions, organization	40	36	34	28	0
Content: accuracy, clarity	20	18	16	12	0
Thoroughness: all required info included, amount of research, length of speech	20	18	16	12	0
Overall Presentation: volume, use of language, stance, eye contact	10	9	7	4	0
Use of PPT in Presentation: effective use of power point	10	9	7	4	0
Total Points					

Caminalcules Discovery
Classifying Imaginary Animals by Analysis of Shared Characteristics

OBJECTIVE

Students reinforce the concept of classification by grouping imaginary organisms with similar characteristics. Students will also assign scientific names to the organisms to indicate the degree of relatedness based on similarities and differences in their physical traits.

LEVEL

Middle Grades: Life Science

NATIONAL STANDARDS

UCP.1, UCP.2, UCP.4, UCP.5, A.1, A.2, C.3, C.5

TEKS

6.1(A), 6.2(B), 6.2(C), 6.2(D), 6.2(E), 6.3(C), 6.10(A), 6.10(C), 6.11(A)
7.1(A), 7.2(B), 7.2(C), 7.2(D), 7.2(E), 7.3(C), 7.10(B)
8.1(A), 8.2(B), 8.2(C), 8.2(D), 8.2(E), 8.3(C), 8.11(B)
IPC 2(C), 2(D)

CONNECTIONS TO AP

AP Biology:

III. Organisms and Populations: A. Diversity of Organisms 1. Evolutionary patterns 2. Survey of the diversity of life 3. Phylogenetic classification 4. Evolutionary relationships
III. Organisms and Populations: B. Structure and Function of Plants and Animals 2. Structural, physiological, and behavioral adaptations

TIME FRAME

90 minutes

MATERIALS

(For a class of 28 working individually)

28 photocopies of 29 imaginary organisms (Caminalcules)   glue sticks
scissors   colored pencils

TEACHER NOTES

The imaginary organisms presented here are called Caminalcules, named after evolutionary taxonomist, Joseph Camin, who created them. *Caminalcules Discovery* is a complex inquiry activity where students devise a system of classification for a group of imaginary organisms. The students divide the Caminalcules into groups of families, genera, and species and create scientific Latin-sounding names for each group. Because there are 29 different individual Caminalcules, there are many potential ways to organize them into different classification schemes. *Caminalcules Discovery* requires a good deal of higher order thinking and should follow an introductory classification activity. Introductory classification activities may be found in most textbook ancillaries or on the Internet.

As an introduction to this activity you may wish to walk the students through a simple classification scheme using their shoes as objects. Have all students remove their left shoes and place them on the floor in the middle of the room. Have the students look at the shoes in terms of characteristics. "What are some characteristics we see in all those shoes?" The characteristics observed could be: leather, shoelaces, Nikes, athletic, white, red, black, etc… Ask the question, "What is one way that we could divide the shoes into two large groups?" Students could suggest something like athletic shoes and non-athletic shoes. Have two or three students come up to the front and put all the athletic shoes in one pile and all the non-athletic into another pile. Suggest a Latin-sounding name for the athletic pile, perhaps "Athletica" or something similar and "Dressupica" for the non-athletic shoes. Then within the "Athletica" have students suggest another characteristic to further sub-divide the shoes, such as: has laces or not; ("Laceia" and "Velcroa"), leather or non-leather ("Leatherum" and "Nonleatherum", by brand name, or some other way that your students create. Within one of the subdivisions, such as "Laceia", have students suggest further subdivisions based on characteristics: "Whiteus" versus "Blackeus" and so on, again assigning Latin-sounding names to these sub-groupings. Then continue with the non-athletic and further subdivide into the smaller groups, to where you eventually have the smallest group of division.

Once the groups of shoes have been subdivided, ask students to describe characteristics of the groups of shoes and how those characteristics are different from other subgroups. If time permits, pile all the shoes back together in a pile, pick 2 or 3 different students, and have them come up with another classification scheme, this time using different characteristics to divide the groups of shoes. The idea can then be presented that all the students in the Caminalcules activity can come up with logical, yet different schemes of classification.

The method of subdividing the shoes by shared characteristics can be then expanded to include how all living species are classified and subdivided in biology, referred to as cladistics, which hypothesizes relationships among organisms.

The pictures of the Caminalcules are copyrighted by the journal *Systematic Biology* and Robert R. Sokal. They are used here with permission.

SAMPLE DATA

A sample of a student's work appears in Figure 1. The student work represents just one way of classifying the 29 Caminalcules. Perhaps you could show students this example on an overhead to give them a concrete idea of how to begin classifying the Caminalcules into different groups. This student has divided all her Caminalcules into 4 families (from top to bottom): Dotsidae, Blobidae, Potatodae, and Wingidae. Within the Dotsidae family, she has subdivided her Caminalcules into 2 genera, which contain 1 species each: *Wavidae feettogetherum* and *Bentidae feetapartum*. These names reflect the characteristics of the Caminalcules in those groups as well. Blobidae family is represented by 5 different species (also 5 genera): *Smallheadidae roundtailum*, *Longneckidae pointum*, and *Neckfoldidae antennum*, *Forktailidae armsuppum*, and *Mermaidtailidae bulging eyesum*. Again the names reflect the physical characteristics. (An alternative way of classifying these would be to put the *Smallheadidae roundtailum*, *Longneckidae pointum*, and *Neckfoldidae antennum* into one genus, all three being different species. The names could perhaps be *Roundplainbodium roundtailum, Roundplainbodium pointium, Roundplainbodium antennum* as these 3 groups have similar enough characteristics that they could be possibly be grouped together in one genus.) Her family division, Potatodae, has been divided into *Smalltailidae* and *Bigtailidae* genera, again, with each genera being a separate species. She has finally divided the last family, Wingidae, into 3 genera: *Normalidae, Longidae* and *Shortidae*. This is not the perfect student assignment but represents a very well thought-out classification scheme. All her genus and species names are underlined, her family names end in -idae. One criticism for this particular student's work is the use of -idae in her genera naming scheme, as opposed to -a, -us, or -um, as -idae is the Latin ending for family names. Included on the sample student example is the list of characteristics that describe each one of her species groups. Figure 2 summarizes the previous written explanation into an organizational flow chart.

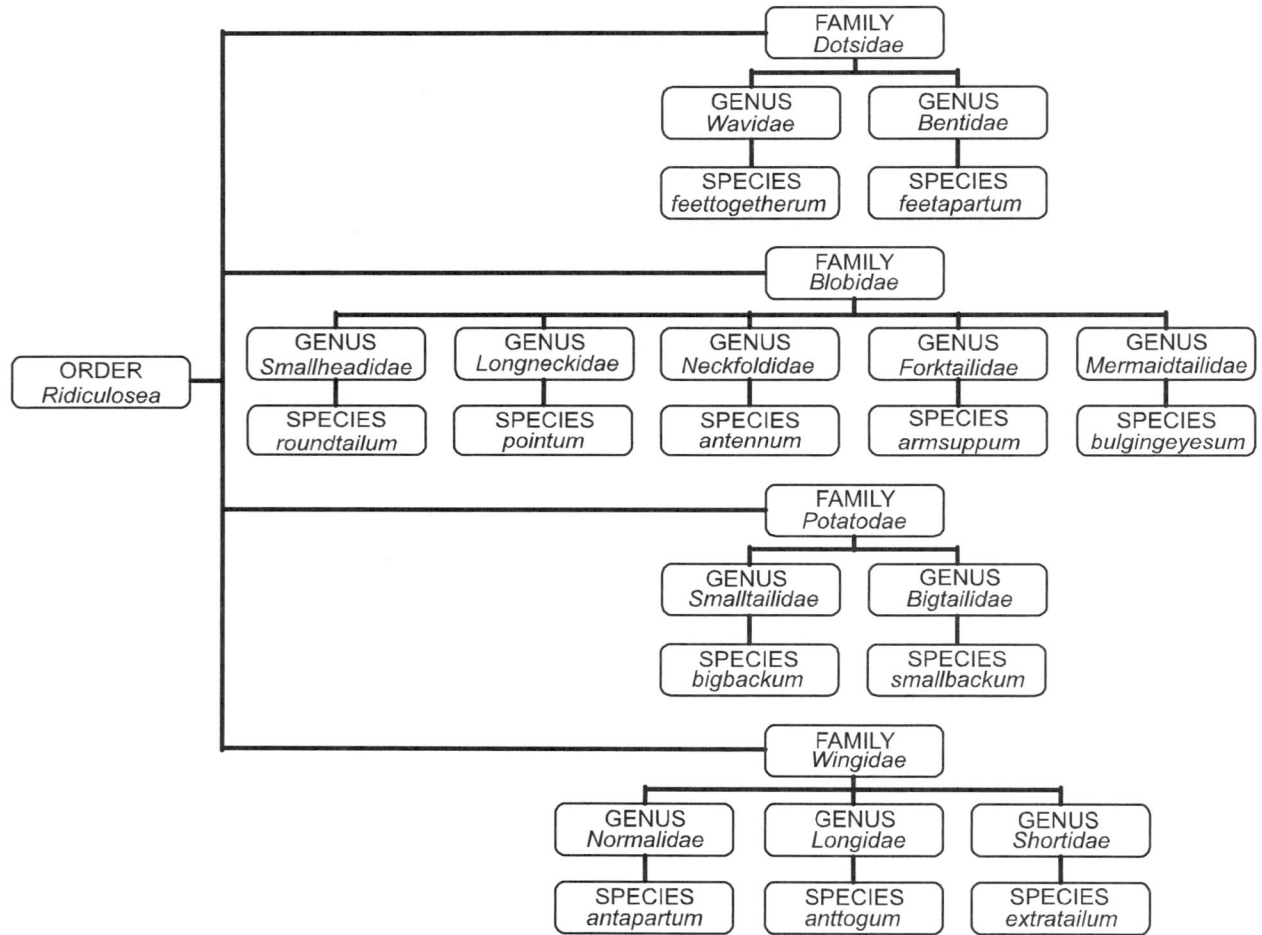

Also included is her taxonomic tree, in the student sample, which was done correctly and awarded extra credit.

The obvious "gray area" in this activity is deciding where the variation ends and begins between different species and genera. As long as the student can explain and justify their scheme, then full credit should be awarded.

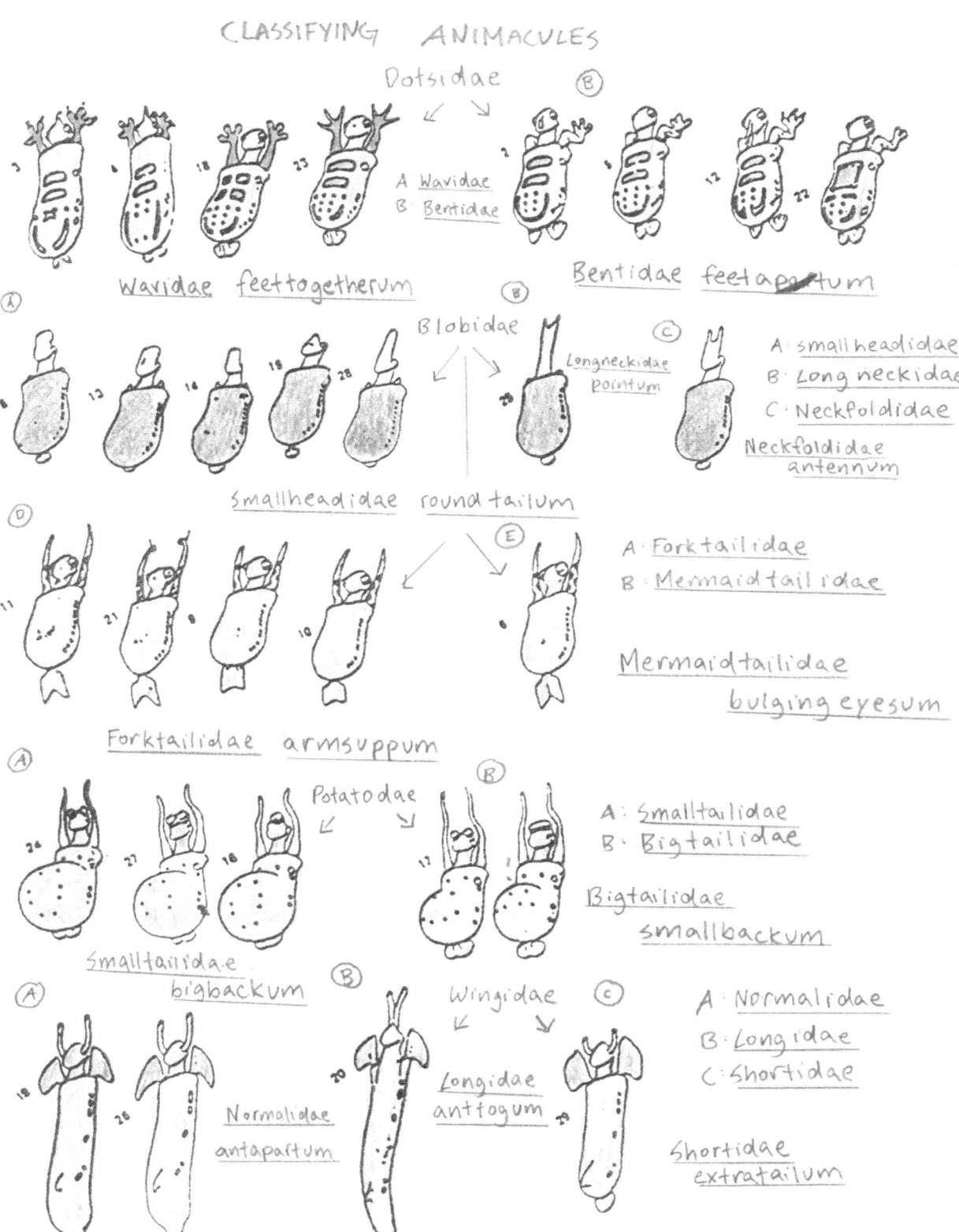

Fig. 1

EXTRA CREDIT

POSSIBLE ANSWERS TO THE CONCLUSION QUESTIONS

1. What difficulties did you experience in developing your classification system? Describe at least two.
   - Probably the greatest difficulty is deciding how to differentiate the variation between a genus and a species. The variation among a group of individuals could represent several different species within the same genus, it could represent the variation that is naturally present within one species, or it could possibly be interpreted as more than one genera.
   - Another difficulty that students find is naming the groups with Latin-sounding names.

2. Explain why all Caminalcules are not placed in the same family.
   - The number of differences in the characteristics between all the Caminalcules is too great to represent just one biological family.

3. Explain how this activity illustrates the concept: "Classification is the grouping of objects based on similarities and differences."
   - The students organize the groups of Caminalcules on the basis of shared characteristics (and differences). The characteristics that are present in an individual Caminalcule determine how closely related it is to another individual.

4. What is one inherent problem with a classification scheme based only on different physical characteristics?
   - Different scientists could see different ways to organize and classify the same group of organisms.

5. The technique of DNA sequencing has enabled scientists to create large data bases of genomic sequences for different organisms. How do you think the ability to sequence an organism's entire genome could help to more accurately classify groups of organisms?
   - Genome sequencing of biological groups of organisms has revolutionized the field of classification. Now entire DNA sequences can be analyzed by computers to determine, mathematically, which organisms are more closely related, by comparing the number of shared nucleotide bases. Using DNA sequence comparisons between different species can tell a much more accurate story of relatedness and evolutionary history than just observing external characteristics.

Caminalcules Discovery
Classifying Imaginary Animals by Analysis of Shared Characteristics

PURPOSE
In this activity you will reinforce the concept of classification by grouping imaginary organisms with similar characteristics.

MATERIALS

28 photocopies of 29 imaginary organisms   glue sticks
scissors   colored pencils

Safety Alert
1. Always use care when handling scissors.

The imaginary organisms presented here are called Caminalcules, named after evolutionary taxonomist, Joseph Camin, who created them. *Caminalcules Discovery* is a complex inquiry activity where you devise a system of classification for a group of imaginary organisms. You will divide the Caminalcules into groups of families, genera, and species and create scientific Latin-sounding names for each group.

PROCEDURE

1. Study the set of imaginary organisms on the "Caminicules" sheet and carefully cut out each individual organism with your scissors.

2. In this activity, you are the scientist that discovered a new group of animals, which you have named "Caminalcules." You have researched their physiology and behavior very thoroughly and have identified them as members of the following groups: Kingdom Animalia, Phylum Mollusca, Class Imaginata, Order Ridiculosea. It will be your responsibility to classify and create names for the family, genus and species levels for these organisms.

3. Organize the Caminalcules into families as you see fit. Within each family further subdivide the group into genus and species on the basis of similar characteristics. Even though no two Caminacules are identical, remember there is wide variation within the human species as well.

4. Once you have sorted the organisms into their respective species, neatly glue each group of organisms onto your student answer page. Leave plenty of room to write in family and scientific names above each group and three shared traits below each group. To best organize your groups place different species of the same genus next to each other on your page.

5. You will need to invent family names for each family you created. (Note: Family names always end in: "idae.") Write the family name above each family.

6. For each of the species groups create a "scientific name" in the form of *Homo sapiens* and write it below the family name on your paper. You should create six to ten scientific names, depending upon how many groups you have in your classification system. All individuals in the same species should have the same scientific name. In other words, even though there are some minor individual differences, all organisms in the same group should be enough alike to be placed in the same "species" When naming the species, try to make them "sound" Latin; e.g., *Burritos longus*. Remember, the genus name is capitalized, the species name is not, and the scientific name (genus + species) is underlined or italicized. Also, some Caminacule species may be closely related (very similar) and may be placed in the same genus.

7. Below each species grouping, list at least three characteristics that all organisms in the species have in common.

8. For extra credit, create an evolutionary tree with lines drawn between your species to show how your different species are related to each other. Your teacher will show you an example of an evolutionary tree that a student has drawn previously.

Caminalcules Discovery
Classifying Imaginary Animals by Analysis of Shared Characteristics

ANIMICULE CLASSIFICATION

Once you have organized and classified your Caminalcules into groups, use the glue stick to adhere the individuals to this page and the next. Be sure to glue the Caminalcules that share the most characteristics more closely together.

CONCLUSION QUESTIONS

1. What difficulties did you experience in developing your classification system? Describe at least two.

2. Explain why all Caminalcules are not placed in the same family.

3. Explain how this activity illustrates the concept: "Classification is the grouping of objects based on similarities and differences."

4. What is one inherent problem with a classification scheme based only on different physical characteristics?

5. The technique of DNA sequencing has enabled scientists to create large data bases of genome sequences for different organisms. How do you think the ability to sequence an organism's entire genome could help to more accurately classify groups of organisms?

CAMINALCULES

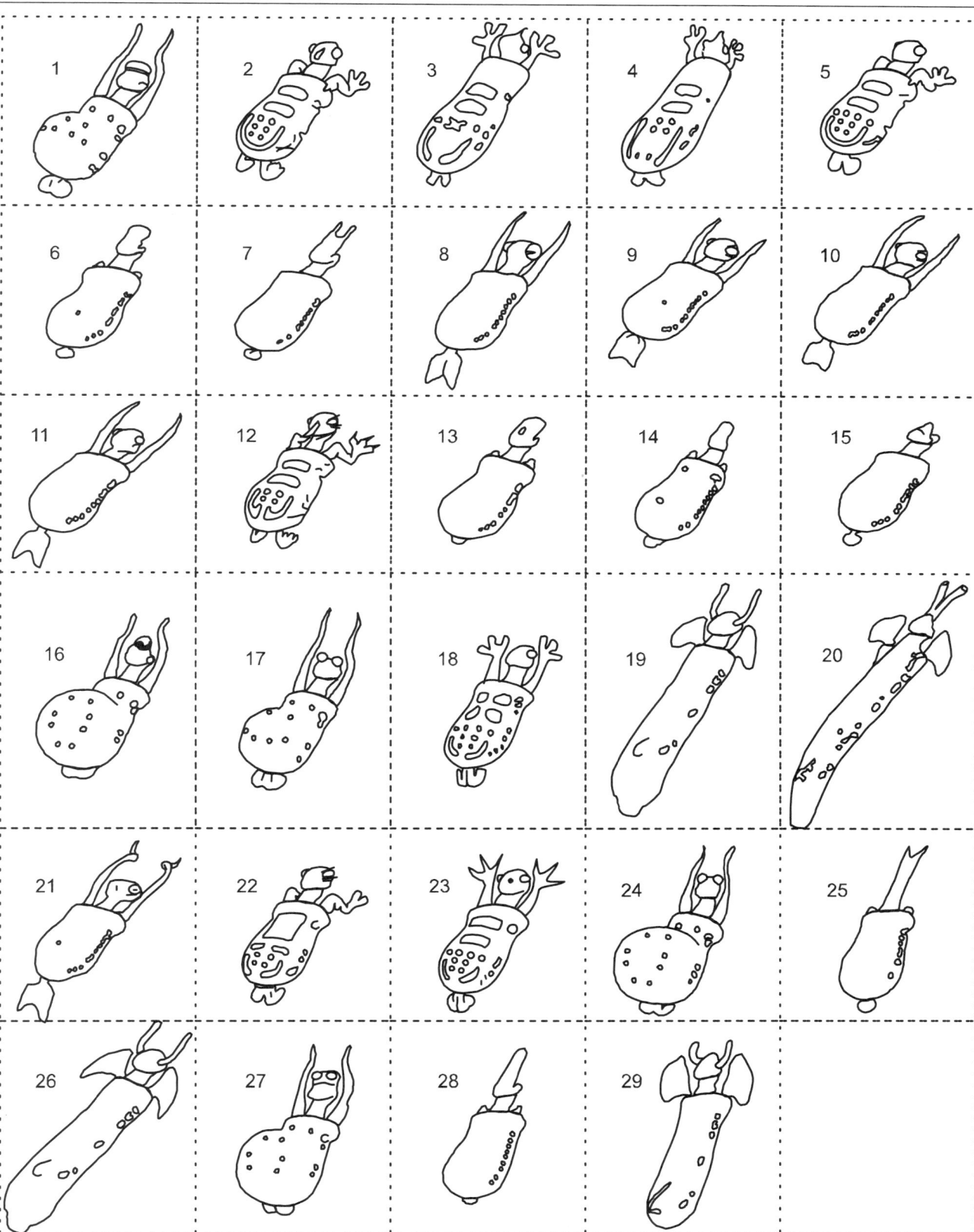

Bean Bunny Evolution
Modeling Gene Frequency Change (Evolution) in a Population by Natural Selection

OBJECTIVE
Students will model evolution by natural selection of the gene frequency of two alleles in a population of organisms. Students will also calculate the gene frequency of the alleles for each generation and then graph the frequency of the two alleles over 10 generations.

LEVEL
Middle Grades: Life Science

NATIONAL STANDARDS
UCP.1, UCP.2, UCP.4, A.1, A.2, C.2, C.3

TEKS
6.1(A), 6.2(A), 6.2(B), 6.2(C), 6.2(D), 6.2(E), 6.3(A), 6.3(C), 6.4(A), 6.4(B), 6.10(C), 6.11(A)
7.1(A), 7.2(A), 7.2(B), 7.2(C), 7.2(D), 7.2(E), 7.3(A), 7.3(C), 7.4(A), 7.4(B), 7.10(B), 7.10(C), 7.12(C)
8.1(A), 8.2(A), 8.2(B), 8.2(C), 8.2(D), 8.2(E), 8.3(A), 8.3(C), 8.4(A), 8.11(A), 8.11(B), 8.11(C)
IPC 1(A), 2(A), 2(B), 2(C), 2(D)

CONNECTIONS TO AP
AP Biology:
II. Heredity and Evolution: A. Heredity 1. Meiosis and gametogenesis 3. Inheritance patterns
II. Heredity and Evolution: C. Evolutionary Biology 3. Mechanisms of evolution
    III. Organisms and Populations: A. Diversity of Organisms 1. Evolutionary patterns
    III. Organisms and Populations: B. Structure and Function of Plants and Animals 2. Structural, physiological, and behavioral adaptations 3. Response to the environment
III. Organisms and Populations: C. Ecology 1. Population dynamics

TIME FRAME
90 minutes

MATERIALS
(For a class of 28 working in pairs)

1 bag of small pinto beans, or "red beans"	1 bag of small lima beans, or "white beans"
14 paper bags	42 Petri dishes
14 overhead marking pens	

TEACHER NOTES

Bean Bunny Evolution provides a straight forward and simple exploration of a natural selection evolutionary model using only different colored beans, a bag and some Petri dishes!

This activity should be done only after students have had a good introduction to genetics, Punnett squares and alleles. Also, the students should have a thorough understanding of natural selection and evolution.

In *Bean Bunny Evolution,* there are several simplifications. We have assumed that rabbits mate once and do not reenter the breeding pool. (Of course, this is not true with real rabbits.) For this reason, the population declines at an artificially rapid rate. The activity helps show the change in frequency of a lethal recessive allele over time, but is not intended to represent exactly what would happen with a population of real rabbits.

Dried small pinto beans and dried small lima beans can be found at any grocery store at minimal cost. The beans should be as close to the same size as possible so the students cannot feel or detect a difference between the beans when they are in the paper bag. Students may try, after a few generations, to "kill off" the recessive alleles once they begin to figure out the pattern.

To help students begin, it is a recommended that you model for students how to select the beans from the bag and how to mark the data table. Also, an explanation of how to calculate the gene frequency on the overhead or the chalkboard is recommended. The text of the student lab procedure explains how to calculate the frequency:

"To find the gene frequency of F, *divide the number of* F *by the total, and to find the gene frequency of* f, *divide the number of* f *by the total. Express results in decimal form."*

The sample data below came from two real student lab groups. It demonstrates that students can possibly eliminate all f alleles in a few generations or they might continue on through 10 generations with a few recessive alleles remaining in the population.

When the students start the lab, carefully and vigilantly monitor the students as they perform the 1st generation to make sure that they are following the procedure correctly. Once they finish the 1st generation, they can (and probably will) work independently.

This activity, used with permission, was only slightly modified from the original activity, *Breeding Bunnies*, written by Joseph Lapiana and displayed at the PBS (http://www.pbs.org) website.

SAMPLE DATA

Here are two examples of student generated results.

Student Data Table #1

Generation	Number of *FF* Bunnies	Number of *Ff* Bunnies	Number of *ff* Bunnies	Number of *F* Alleles	Number of *f* Alleles	Total Number of Alleles	Gene Frequency of *F*	Gene Frequency of *f*
1	12	26	12	50	26	76	.66	.34
2	19	12	7	50	12	62	.81	.19
3	22	6	3	50	6	56	.89	.11
4	23	4	1	50	4	54	.93	.07
5	25	0	2	50	0	50	1	0
6	25	0	0	50	0	50	1	0
7	25	0	0	50	0	50	1	0
8	25	0	0	50	0	50	1	0
9	25	0	0	50	0	50	1	0
10	25	0	0	50	0	50	1	0

Student Data Table #2

Generation	Number of *FF* Bunnies	Number of *Ff* Bunnies	Number of *ff* bunnies	Number of *F* alleles	Number of *f* alleles	Total Number of Alleles	Gene Frequency of *F*	Gene Frequency of *f*
1	10	30	10	50	30	80	.63	.37
2	15	20	10	50	20	70	.71	.29
3	18	14	3	50	14	64	.78	.22
4	18	14	0	50	14	64	.78	.22
5	19	12	1	50	12	62	.81	.19
6	19	12	0	50	12	62	.81	.19
7	21	8	1	50	10	60	.83	.17
8	21	8	1	50	8	58	.86	.14
9	22	6	2	50	6	56	.89	.11
10	22	6	0	50	6	56	.89	.11

TEACHER PAGES

Example of Typical Graph, Using Data from Student Data Table #2

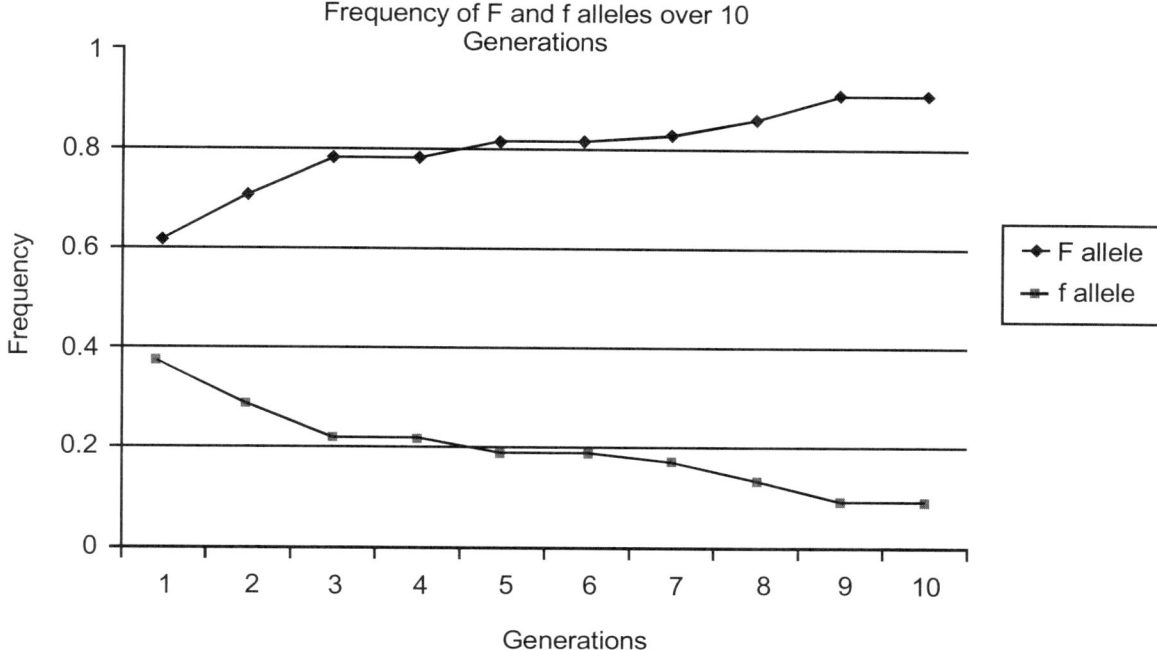

Frequency of F and f alleles over 10 Generations

POSSIBLE ANSWERS TO THE CONCLUSION QUESTIONS

1. Based on your lab data, do you need to modify your hypothesis? Explain.
   - Answers will vary. Most students will correctly hypothesize, however, that the gene frequency of the alleles will change over time.

2. What was the original gene frequency of the F and the f alleles before selection occurred?
   - The original gene frequency of *F* and *f* was .5 each. There were 50 red beans, representing F alleles, and 50 white beans, representing f alleles, for a total of 100 beans, or alleles.

3. How did the gene frequency of the *F* and *f* alleles change by the 10th generation?
   - By the 10th generation, most groups of students will have selected out all of the white beans, or f alleles. If they have not selected them all out, by the 10th generation, the frequency of *F* will probably be approximately 0.9 or more while the frequency of *f* will be approximately 0.1.

4. How do you explain that both alleles, *F* and *f*, changed in frequency over time in the lab?
   - The *f* allele, coding for the furless phenotype, is a lethal allele. When two *f* alleles are combined in a rabbit as a homozygous recessive genotype, and expressed as the hairless phenotype, the rabbit will die, as the lethal condition is naturally selected against. This rabbit can no longer pass along its genes to the next generation. Over time and after multiple generations, the number of rabbits passing along *f* alleles will decrease. Concurrently, the *F* alleles will increase in the population. This causes the f alleles to decrease in frequency and the *F* alleles to increase in frequency.

5. In a real rabbit habitat new animals often come into the habitat (immigrate), and others leave the area (emigrate). How might emigration and immigration affect the gene frequency of F and f in this population of rabbits? How might you simulate this effect if you were to repeat this activity?
   - Immigration and emigration will change the gene frequencies of both F and f, depending on what types of alleles migrate in or out of the population.
   - To simulate this effect in the modeling lab, students could add or take away beans from the bag, representing new alleles coming in or out of the population.

6. How do your group's results compare with the class data? If significantly different, why are they different?
   - Each group's F alleles will increase in frequency and each group's f alleles will decrease in frequency. The number of generations it takes to decrease is what varies from group to group. Some groups might eliminate the f alleles within 4 generations or less, but other groups will still have a few f alleles even after 10 generations.

7. How is this simulation an example of evolution by natural selection?
   - Evolution can be defined as biological change over time. As the ff (furless) individuals die off because of natural selection, the gene frequency of both the F and f alleles changes, as there are more F alleles in the population. As the genotypic frequency within the population changes over time, the phenotypic variation also changes. Genotypic and phenotypic change is "biological change". As biological characteristics of a population change over the generations, the population evolves.

8. In nature, how is it that lethal alleles, like furlessness, are still passed along through the generations and not completely selected out of the population?
   - The fact that all genotypes are represented by two alleles (one from each parent) means that when a recessive lethal allele in combination with a dominant normal or wild type allele, the recessive condition will not be expressed. This allows the "silent killer" recessive genes to be passed down through the generations.

Bean Bunny Evolution
Modeling Gene Frequency Change (Evolution) in a Population by Natural Selection

In this activity, you will examine natural selection in a small population of wild rabbits. Evolution, on a genetic level, is a change in the frequency of alleles in a population over a period of time. Breeders of rabbits have long been familiar with a variety of genetic traits that affect the survivability of rabbits in the wild, as well as in breeding populations. One such trait is the trait for furless rabbits (naked bunnies). This trait was first discovered in England by W.E. Castle in 1933. The furless rabbit is rarely found in the wild because the cold English winters are a definite selective force against it.

In this lab, the dominant allele for normal fur is represented by *F* and the recessive allele for no fur is represented by *f*. Bunnies that inherit two F alleles or one *F* and one *f* allele have fur (*FF* or *Ff*), while bunnies that inherit two *f* alleles have no fur (*ff*).

PURPOSE
In this activity you will model evolution by natural selection of the gene frequency of two alleles in a population of organisms. You will also calculate the gene frequency of the alleles for each generation and then graph the frequency of the two alleles over 10 generations.

MATERIALS
(For each pair of students)

    50 red beans   50 white beans
    1 paper bag   3 Petri dishes
    overhead marking pen

PROCEDURE
1. Your teacher will divide all the students into pairs and pass out the materials for today's lab.

2. Working with a partner, read the following problem question for today's lab:

    How does natural selection affect gene frequency over several generations?

3. From your previous knowledge about natural selection, evolution and genetics, answer the problem question on your student answer page in the space marked HYPOTHESIS. State your hypothesis in an "If…then…" format. Include your predictions regarding how you think natural selection will affect the gene frequency in a population. State what you would predict about the frequency of *F* alleles and *f* alleles in the population of rabbits after 10 generations, where *ff* bunnies are selected against and do not survive.

4. The red beans represent the F allele for fur, and the white beans represent the f allele for no fur. The paper bag represents the English countryside, where the rabbits randomly mate.

5. Using the overhead marking pen, label one Petri dish FF for the homozygous dominant genotype. Label a second Petri dish Ff for the heterozygous condition. Label the third Petri ff for those rabbits with the homozygous recessive genotype.

6. Place the 50 red and 50 white beans (alleles) in the paper bag and shake up (mate) the rabbits. (Please note that these frequencies have been chosen arbitrarily for this activity.)

7. Without looking at the beans, select two at a time, and record the results in the data table next to "Generation 1." or instance, if you draw one red and one white bean, place a mark in the chart under "Number of Ff individuals." Continue drawing pairs of beans and recording the results in your chart until all beans have been selected and sorted. As you draw out the bean pairs place the "rabbits" into the appropriate dish: FF, Ff, or ff. To determine the number of individual rabbits produced for each phenotype count the total number of beans in the appropriate Petri dish and divide this number by two since an individual is represented by a pair of beans. Record you results in the data table.

8. The ff bunnies are born furless. The cold weather kills them before they reach reproductive age, so they can't pass on their genes. Place the beans from the ff container aside before beginning the next round. You will not count these beans.

9. Count the F and f alleles (beans) that were placed in each of the "furred rabbit" dishes in the first round and record the number in the data table under the columns labeled "Number of F Alleles" and "Number of f Alleles." Total the number of F alleles and f alleles for the first generation and record this number in the column labeled "Total Number of Alleles."

10. Place the alleles of the surviving rabbits (FF and Ff dishes) back into the container and mate them again to get the next generation.

11. Repeat steps five through nine to obtain generations two through ten. Make sure that each partner has a chance to either select the beans or record the results.

12. Determine the gene frequency of F and f for each generation and record them in the chart in the columns labeled "Gene Frequency F" and "Gene Frequency f." To find the gene frequency of F, divide the number of F by the total, and to find the gene frequency of f, divide the number of f by the total. Express results in decimal form. The sum of the frequency of F and f should equal one for each generation.

13. If there is time, your teacher may have you record your group's frequencies on the board so your classmates can see them.

14. Complete the conclusion questions on your student answer page.

15. Graph your frequencies of the F allele and the f allele on the graph paper provided. Prepare a line graph with the horizontal x-axis as the generation number and the vertical y-axis as the frequency in decimals. Plot all frequencies on one graph. Use a solid line for F and a dashed line for f.

16. If your teacher requests it, plot the class totals on the same graph. Use a different color for line lines representing the class data and make a legend illustrating which colors represent your individual data and which ones represent the class data.

Bean Bunny Evolution
Modeling Gene Frequency Change (Evolution) in a Population by Natural Selection

HYPOTHESIS

DATA TABLE

Generation	Number of *FF* Bunnies	Number of *Ff* Bunnies	Number of *ff* Bunnies	Number of *F* Alleles	Number of *f* Alleles	Total Number of Alleles	Gene Frequency of *F*	Gene Frequency of *f*
1								
2								
3								
4								
5								
6								
7								
8								
9								
10								

DATA ANALYSIS

CONCLUSION QUESTIONS

1. Based on your lab data, do you need to modify your hypothesis? Explain.

2. What was the original gene frequency of the *F* and the *f* alleles before selection occurred?

3. How did the gene frequency of the *F* and *f* alleles change by the 10th generation?

4. How do you explain that both alleles, *F* and *f*, changed in frequency over time in the lab?

5. In a real rabbit habitat new animals often come into the habitat (immigrate), and others leave the area (emigrate). How might emigration and immigration affect the gene frequency of *F* and *f* in this population of rabbits? How might you simulate this effect if you were to repeat this activity?

6. How do your results compare with the class data? If significantly different, why are they different?

7. How is this simulation an example of evolution by natural selection?

8. In nature, how is it that lethal alleles, like furlessness, are still passed along through the generations and not completely selected out of the population?

Web Pages Made Easy
Using Netscape Composer to Create Web Pages as a Research Product

OBJECTIVE

Students will learn the basics of creating a web page using Netscape Composer. Subsequent activities within this book will apply this lesson to make more complex and content specific web pages.

LEVEL

All

NATIONAL STANDARDS

E.1, E.2

TEKS

6.4(A)
7.4(A)
8.4(A)
IPC 2(A)

TIME FRAME

50 minutes

MATERIALS

(for a class of 28 working in teams of four)

    7 computers with Internet access   Netscape Communicator software 7.0 or higher

TEACHER NOTES

Students can easily create web pages on a computer with the free software, Netscape Communicator, downloaded at http://www.downloads.netscape.com.

The lesson that follows is a tutorial designed to serve as an introduction to creating a web page. The student-created web page serves as the product of this exercise, as opposed to a traditional report, poster, or Power Point presentation. The web page(s) is intended to be uploaded to the school or district web server but it is likely you will need to converse with your building or district network administrator as to how this can be accomplished.

The students will complete the student answer page(s) as they explore the different options within the software, and hand it in by the end of the period.

POSSIBLE ANSWERS TO THE CONCLUSION QUESTIONS

1. Outline the steps for beginning a web page starting with the launch of Netscape Communicator.
   - From the tool bar menu, choose File, choose New, choose Blank Page (or Composer page in later versions of Netscape).

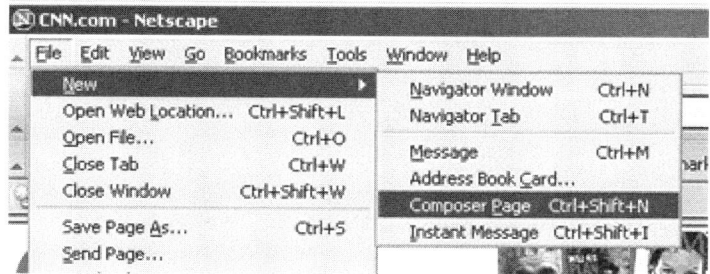

2. How do you alter the formatting of the text in Netscape Composer? Be sure and address how to make text bold, change the font style, and so on.
   - To format text on the page use the font options from the tool bar found at the bottom of the browser. These options are very similar to MS Word and offer choices for font size, color, style and alignment.

3. Where is the title of the web page found when viewing a screen in Netscape Composer?
   - The title bar on a web page is the bar that runs along on the top of the browser window. See the example below:

The title of this web page is "MSNBC Cover"

4. What would be a good file name for a web page about the snakes of Texas?
   - Keep your file names simple. Use all lower case letters, without dashes, underscores, or spaces. Sometimes, when uploading a page to a server, there will be problems with spaces, dashes, etc., in the file names. An example of a good file name would be:

     texassnakes.html
     or
     snakestexas.html

     Examples of a poor file name would be:
     Snakestexas.html
     or
     Snakes of texas.html
     or
     TEXASNAKES.html
     and so on…

5. How do you save an image from a web site?
   - You can save images locally from the WWW by right clicking on the image you want and then choosing Save Image As and placing the image (picture) in the folder you created and named for your web page on your computer.

6. Explain two ways to insert an image into a web page?
   - To insert an image, you can either click the Image button on the tool bar or choose Insert from the menus and then select Image from the drop down list. A dialogue box will appear and ask you to choose file, which is where the picture file is located.

7. How do you alter the default format of the background color, link color and font style on your web page?
   - To format the colors for your page, click Format and choose Page Colors and Properties from the drop down menu. The dialogue box that appears allows you to decide the color of your text, links and background.
     - Link text color is the color of the link before someone clicks on it.
     - Active link text is the color that the link turns as it is clicked.
     - Visited link is the color of the link after it has been clicked.

8. How do you make an image the background of your web page?

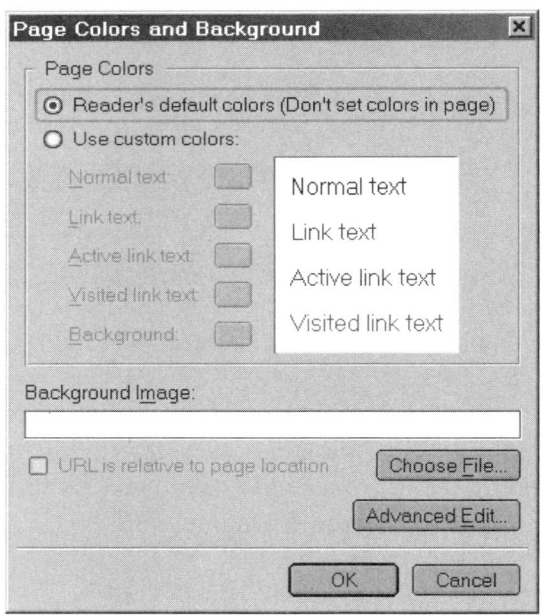

   - Select Format, choose Page Colors and Background, and the dialogue box above appears. You can apply an image to your background from this dialogue box by clicking Choose File and selecting the image from its stored folder.

9. How do you create a link from your web site to another web page or another web site?
   - To create a text link on your page, type the text that is to serve as the link. Next, highlight the link text and click the Link icon on the toolbar. The dialogue box that appears will ask you to enter a web page or file you wish to link to. You must list the correct URL (web address) for

your link, and it may be best to copy the address from the web site and then paste it into the dialogue box to avoid typing errors. Then, click OK which will apply the changes and close the dialogue box.

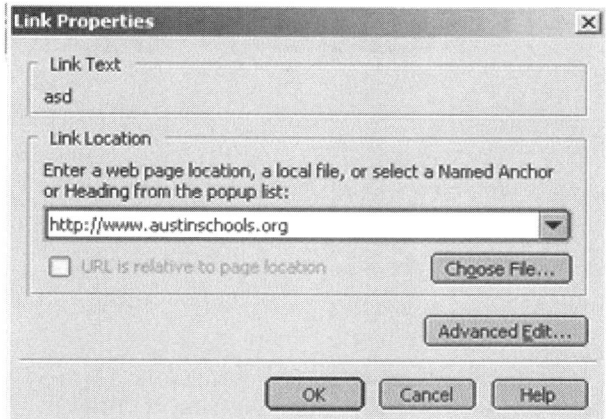

10. How do you make an image a link to another web page or web site?
  - To create an image link on your page, first insert the image and position it on the page. Next, highlight the image that is to be the link and click the Link icon on the toolbar. The dialogue box that appears will ask you where the link should connect to. You must list the correct URL (web address) for your link and again, it is best to copy the address from the web site and then paste it into the dialogue box to avoid typing errors. Finally, in the dialogue box, click OK.

11. How do you view your web page in the Netscape browser and test your links?
  - To check to see if your link works, save your web page in the folder you created on your computer. Next, open Netscape Navigator and choose File▼, Open Page and then in the dialogue box choose the new page you just saved.
  - Alternatively, click on the Browse button at the top of the page in Composer. This will open the page in the Browser allowing you to view your page and test your links.

YOU CANNOT EDIT OR CHANGE YOUR PAGE IN NETSCAPE NAVIGATOR. YOU MUST RE-OPEN YOUR PAGE IN NETSCAPE COMPOSER in order to edit it.

12. If you have additional questions about how to do a specific task when making a web page, how would you find further information?
  - The help files associated with Netscape Composer (and Navigator) are quite helpful. You can press F1 or click on Help and select Help and Support Center to answer further questions.

Web Pages Made Easy
Using Netscape Composer to Create Web Pages as a Research Product

PURPOSE
To learn the basics of creating a web page using Netscape Composer.

MATERIALS
computer with Internet access   Netscape Communicator software 7.0 or higher

PROCEDURE
1. Open Netscape. (This tutorial is based on using Netscape version 7.0. The newest version, as of October 2003, is 7.1 and works just the same as far as this tutorial is concerned.)

2. This activity is designed to help you acquire the skills necessary to create a web page. You will create a web page based on an educationally appropriate subject. This can be anything that you could expect to learn in your science classroom. For this exercise, the content is not as important as your ability to acquire the skills needed to create a web page. Soon, you will apply these newly acquired skills to design a web page with very specific content.

3. To open a new blank webpage, locate the File menu from to top menu bar. Select New ▶, then Blank Page (or Composer page in later versions of Netscape).

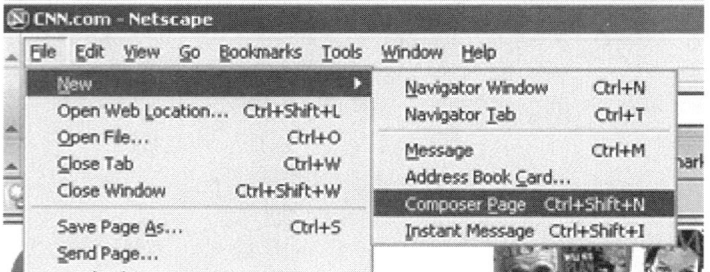

Your blank page will then appear on the screen. To create text on the page use the font options from the tool bar. These options are very similar to MS Word and offer choices for font size, color, style and alignment.

Before continuing, it is best to save your new page with an appropriate title. To do this choose Save As… from the File drop down menu. Enter an appropriate and descriptive title for your page. It can be multiple words.

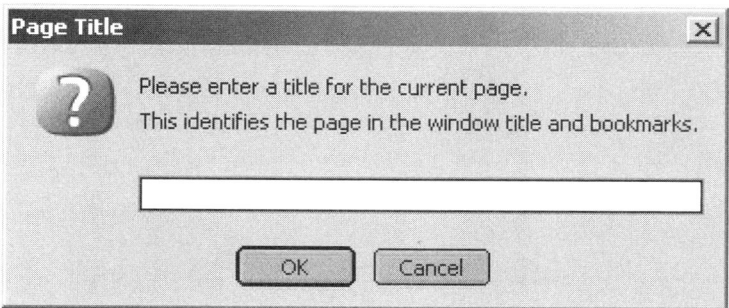

The title on a web page is found on the bar that runs along on the top of the browser window. See below for an example:

The title of this web page is "MSNBC Cover"

Save your .html file into a folder as directed by your teacher. If you teacher allows, it is easier to find your folder if it resides on the Desktop. It might also be easier to create a folder for each separate web page that you generate. This folder can then house the page as well as all images and other additions.

Note: It is a good idea to keep your file names simple. Use all lower case letters, without dashes, underscores, or spaces. Sometimes, when uploading a page to a server, there will be problems with spaces, dashes, etc., in the file names. An example of a good file name would be:

sciencelinks.html

and not:
Sciencelinks.html
or
Science links.html
or
science.links.html

The same rules for naming files will also apply to naming the images that you collect from the WWW or any other source.

4. To gather images for your webpage you may save them from the WWW by right clicking on the image and selecting Save Image As… from the menu. You may also obtain pictures from other sources, but remember to save the images with simple names in the folder you created to house your webpage.

5. To insert an image into your webpage, you can either click the Image icon located on the tool bar or choose Insert ▶ Image from the drop down menu bar. A dialogue box will appear and ask you for the location of the image file. In the dialogue box you click Choose File and then find the image file you saved earlier. When you click on the image's file name it should appear in the dialogue box. You must also type a description of the image in the Alternate text: box for text only browsers.

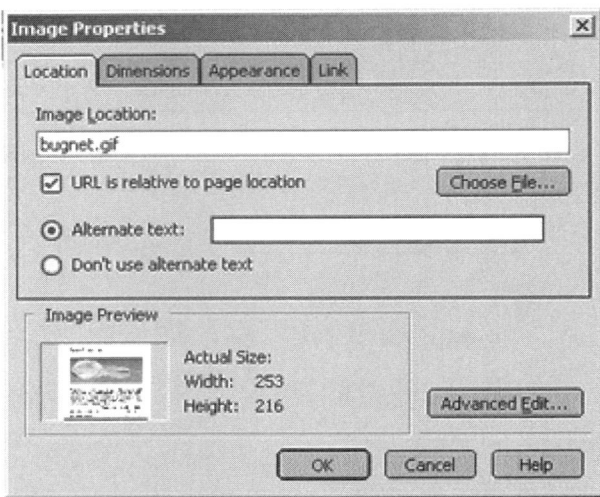

You then click OK which applies the changes. Your image should now appear on your page.

6. To format the colors on your page choose Format from the menu bar. Next, choose Page Colors and Properties from the drop down menu. The dialogue box that appears allows you to decide the color of your text, links, and background. There are three types of link colors to select:
   - *Link text color* is the color of the link before someone clicks on it
   - *Active link text* is the color that the link turns as it is being clicked on
   - *Visited link* is the color of the link after it has been clicked on

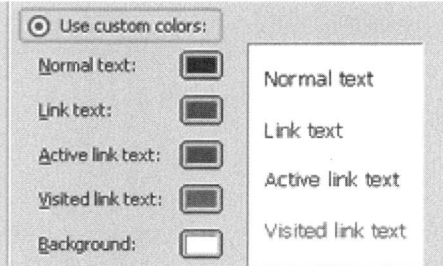

For the background you have two choices: solid color or image. Select a solid color as you did for the links or select an image by clicking the Use Image option at the bottom of the dialogue box. When you have finished choosing your colors, click OK and the changes will be applied to your web page.

7. To create a text link on your page, begin by typing the text that will serve as the link. Next, highlight this link text and click the Link icon on the toolbar. The dialogue box that appears will ask you to enter a web page or file for linking. You must list the correct URL (web address) for your link. It is best to copy the address from the web site and then paste it into the dialogue box to avoid typing errors. When you are finished, click OK and close the box.

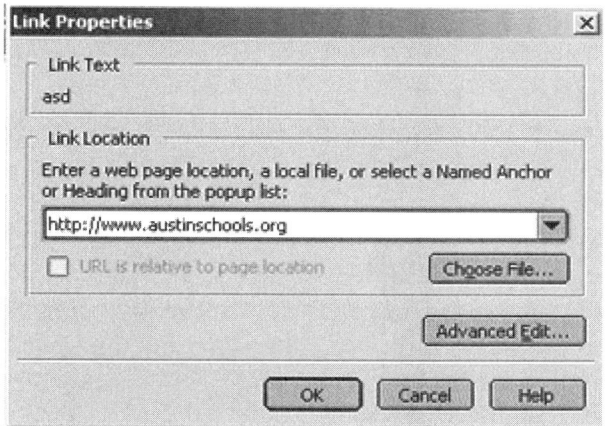

8. To create an image link on your page, first insert the image and position it on the page. Next, highlight the image that is to be the link and click the Link icon on the toolbar. The dialogue box that appears will ask you where the link should connect to. You must list the correct URL (web address) for your link and again, it is best to copy the address from the web site and then paste it into the dialogue box to avoid typing errors. When you are finished click OK.

9. To verify that your link works, save your web page in its appropriate folder. Next, open Netscape Navigator and choose File▼, Open Page. In the dialogue box choose the new page you just saved. Alternatively, click on the Browse button at the top of the page in Composer. This will open the page in the Browser allowing you to view your page and test your links.

YOU CANNOT EDIT OR CHANGE YOUR PAGE IN NETSCAPE NAVIGATOR. YOU MUST RE-OPEN YOUR PAGE IN NETSCAPE COMPOSER in order to edit it.

Go have fun! The help files associated with Netscape Composer (and Navigator) are indeed quite helpful. You can press F1 or select Help ▼, Help and Support Center to answer further questions.

The best way to learn is to explore!

Additionally, here are some web page addresses that contain cool, free, webpage resources:

http://www.davesite.com
This is the best interactive HTML tutorial site available. Click on the link for the HTML Interactive tutorial. You can send your creation on a form to see your results right away. Instant gratification! If you want to actually learn HTML, this is where to do it.

http://www.coolarchive.com
TONS of resources here! The best part of this site is the logo generator which generates banners and logos with your text online and for free. Also contained here are background images, fonts, additional images for your website.

http://www.flamingtext.com
This is a great site for making animated banners with moving flames on the text! They look really cool! Many other resources can also be found here.

10. Answer the conclusion questions on the student answer sheet. Each student will turn in an answer sheet at the end of class.

Name _____

Period _____

Web Pages Made Easy
Using Netscape Composer to Create Web Pages as a Research Product

CONCLUSION QUESTIONS

1. Outline the steps for beginning a web page starting with the launch of Netscape Communicator.

2. How do you alter the formatting of the text in Netscape Composer? Be sure and address how to make text bold, change the font style, and so on.

3. Where is the title of the web page found when viewing a screen in Netscape Composer?

4. What would be a good file name for a web page about the snakes of Texas?

5. How do you save an image from a web site?

6. Explain two ways to insert an image into a web page?

7. How do you alter the default format of the background color, link color and font style on your web page?

8. How do you make an image the background of your web page?

9. How do you create a link from your web site to another web page or another web site?

10. How do you make an image a link to another web page or web site?

11. How do you view your web page in the Netscape browser and test your links?

12. If you have additional questions about how to do a specific task when making a web page, how would you find further information?

Cyber Cephalopods on the Web
Creating Web Pages as a Research Product

OBJECTIVE

Students will create a professional looking web project consisting of one or multiple webpage that celebrates the life, survival strategies and essential functions of cephalopods (Class: Cephalopoda): squid, octopi and cuttlefish.

LEVEL

Middle Grades: Life Science

NATIONAL STANDARDS

UCP.1, UCP.5, C.3, C.5, C.6

TEKS

6.4(A), 6.10(A)
7.4(A), 7.9(A), 7.10(B), 7.12(C)
8.2(E), 8.4(A), 8.11(A)
IPC 2(C), 2(D)

CONNECTIONS TO AP

AP Biology:
    II. Heredity and Evolution A. Heredity 3. Inheritance patterns
    III. Organisms and Populations A. Diversity of Organisms 2. Survey of the diversity of life
    3. Phylogenetic classification B. Structure and Function of Plants and Animals 1. Reproduction, growth, and development 2. Structural, physiological, and behavioral adaptations 3. Response to the environment
AP Environmental Science:
    I. Interdependence of Earth's Systems: Fundamental Principles and Concepts E. The Biosphere
    1. organisms: adaptations to their environments

TIME FRAME

2 50 minute classes

MATERIALS

(For a class of 27 working in teams of 3)

9 computers with WWW access	9 Netscape Communicator/Composer software
27 Rubrics for Cyber Cephalopods on the Web	27 Phylum Mollusca, Class Cephalopoda ("Head-Foot" Mollusks) readings

TEACHER NOTES

Cyber Cephalopods on the Web should be assigned as a follow up to the *Creating Web Pages Using Netscape Composer* lesson as there are concepts and technical terms though out this lesson that are explained in the *Composer* lesson. Students are to work in groups of 3 (or 2 if you have access to enough computers), with each group researching and creating a web project for a particular cephalopod. (This lesson could also be done in a "low tech" manner with students creating posters or other displays to demonstrate the research.) Cephalopods were chosen for this lesson because they are widely misunderstood and maligned organisms. They are unique among animals typically studied in a science classroom, and for invertebrates, they are quite intelligent. This lesson can be adapted if you want to choose another group of animals. You could rewrite this lesson quite easily as all animals have the seven life functions (feeding/digestion, respiration, internal transport, excretion, response to stimuli, movement and reproduction) as well as survival strategies and other interesting facts.

Having students share one computer is a great challenge, especially in the middle grades. One way to make this task a little less complicated is to give the students specific assigned tasks to complete and a limited amount of time in which to complete those tasks. On the student answer page, there is a job description for each of the three team members with a checklist of responsibilities. As one student is typing and searching for images, the others can be researching in a book, magazine, or pages printed from the Internet. The best way to monitor students using the computer is to activate a timer that sounds an alarm every 10 or 15 minutes. Students can be trained to know that when the alarm sounds, it is time for one of the other team members to rotate to the computer and work on the project.

To gather content for the web page, students should be required to bring in material from other sources such as library books, magazines, or web print outs prior to class. This should be a homework grade. Much of the general content that is needed for the 7 functions is printed in the *Phylum Mollusca, Class Cephalopoda ("Head-Foot" Mollusks)* reading. Give this to each student when the project is first assigned. The website, http://is.dal.ca/~ceph/TCP/index.html, *The Cephalopod Page*, is a wonderful starting place for the students. This site may help them choose which cephalopod to research as it has a great deal of both content and images. Google image search, http://images.google.com/, is a valuable tool when searching for images. Students should be encouraged to search for content and images only after thoroughly examining The Cephalopod Page since valuable time can be wasted randomly searching the web for content.

Explain the expectations of the project and the point distributions for the project as outlined on the *Rubric for Cyber Cephalopods on the Web*. The rubric was designed for ease of use and primarily to check for the presence of the required content. If the students have the required content on the page, then they get the credit. An effective way to have students aid in the editing process is to have students critique each others' web pages while they look for missing content and checking for spelling errors. If this is done before assigning a final grade your job will be easier!

A major problem for middle school and junior high school students is spelling, or more specifically, misspelling. Having to check multiple classes for misspelled words is too time consuming for any teacher. Netscape Composer has a built in spell check! Simply click on the spell check icon on the tool bar within Composer:

Words are easily checked in each document.

POSSIBLE SAMPLE WEBPAGE

An actual student product is displayed below.

The online example of this student website can be found at:
http://www.webmutations.com/ceph1/)

These students chose to do 5 pages that covered all the content requirements in the rubric.

A general introduction page:

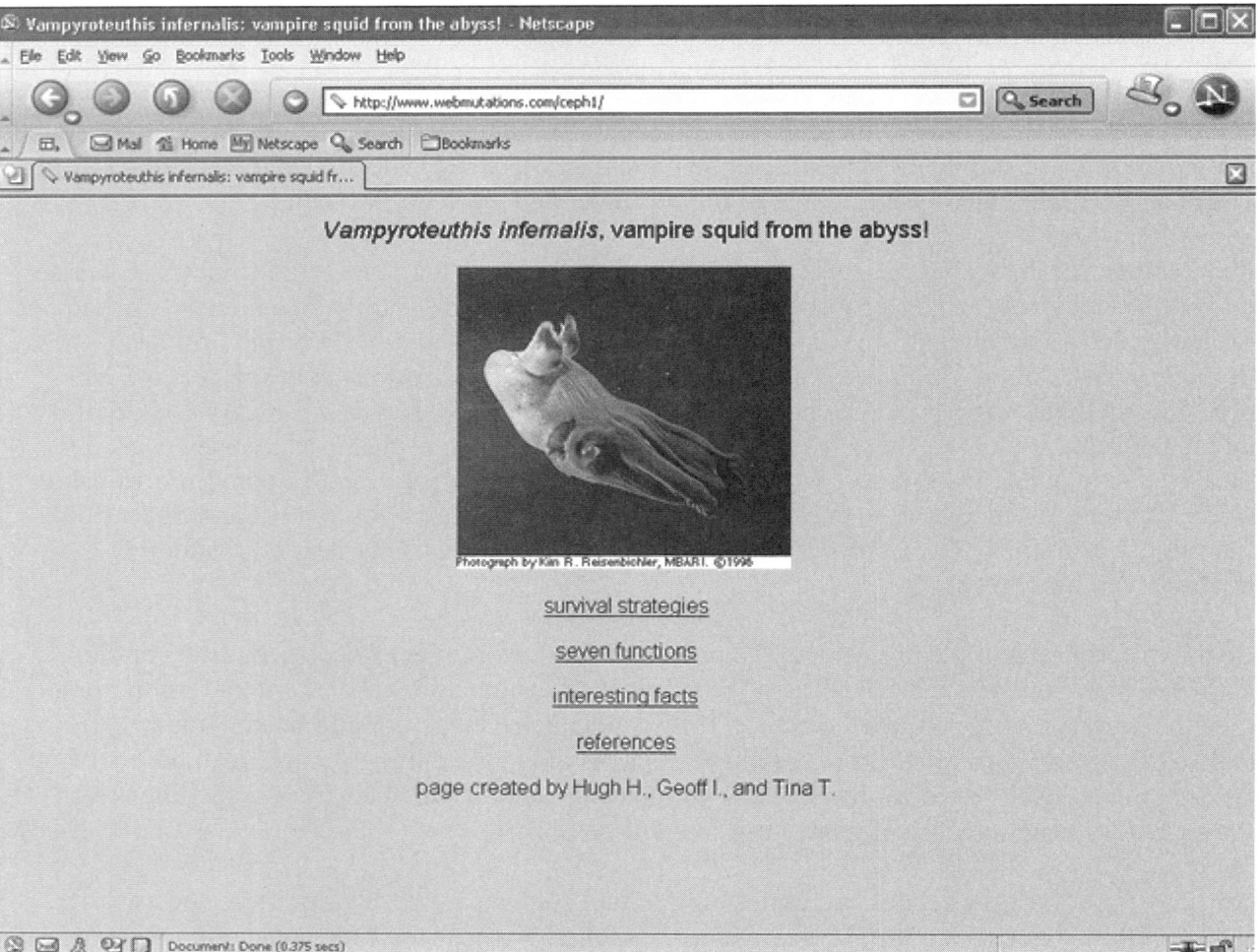

(left margin) TEACHER PAGES

A page that describes the survival strategies:

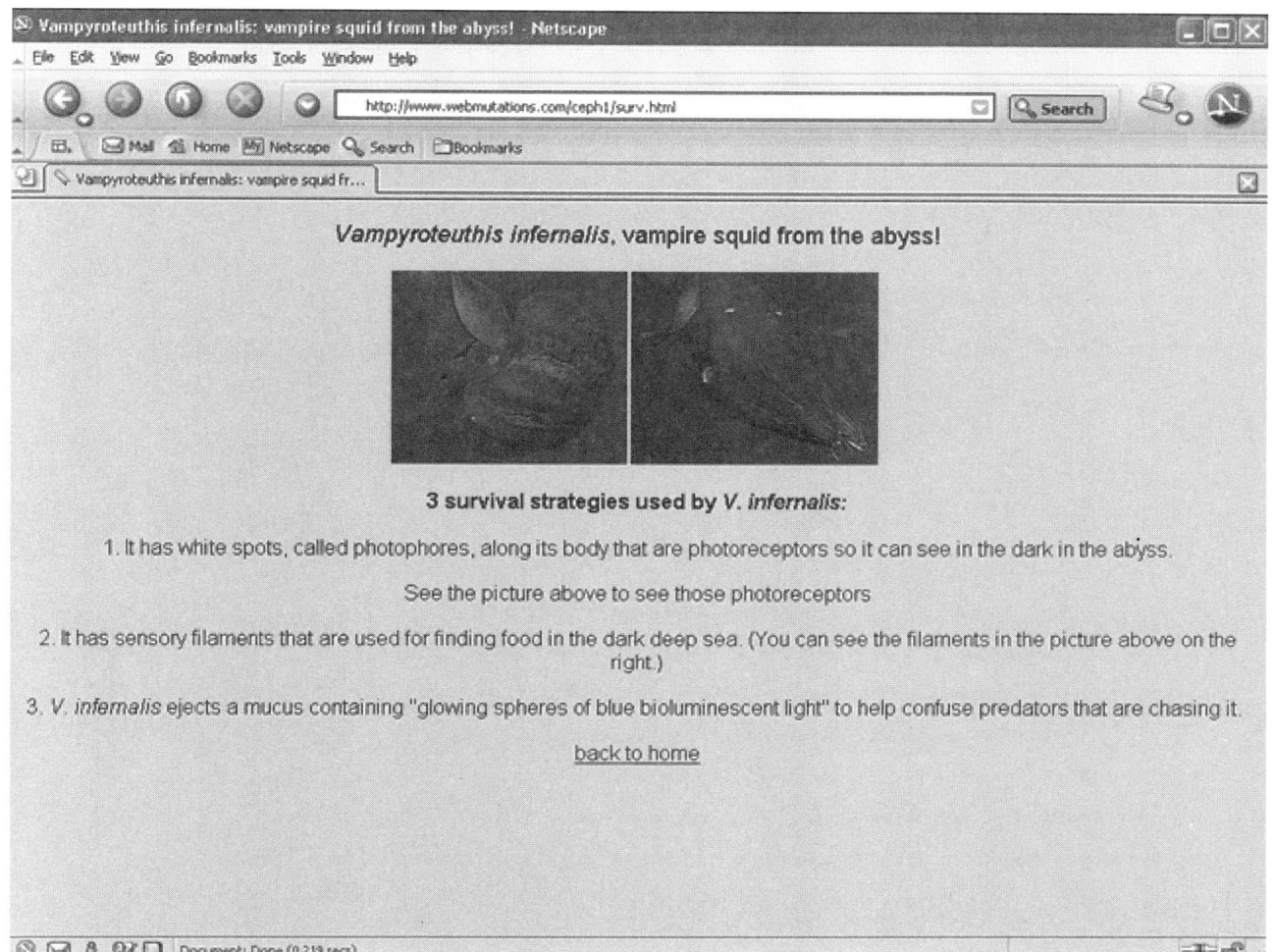

A page that describes how the cephalopod accomplishes the seven functions:

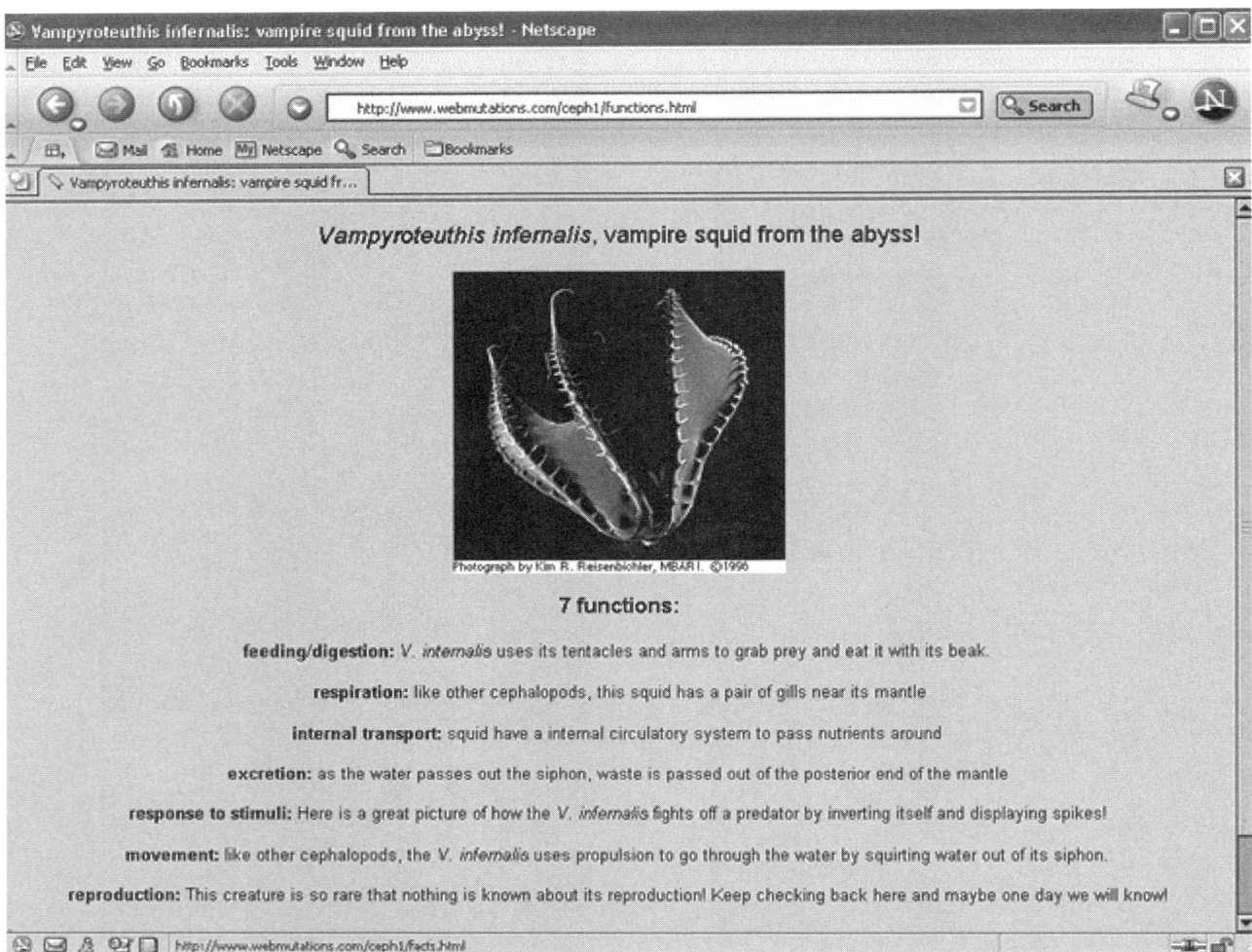

A page for the interesting facts:

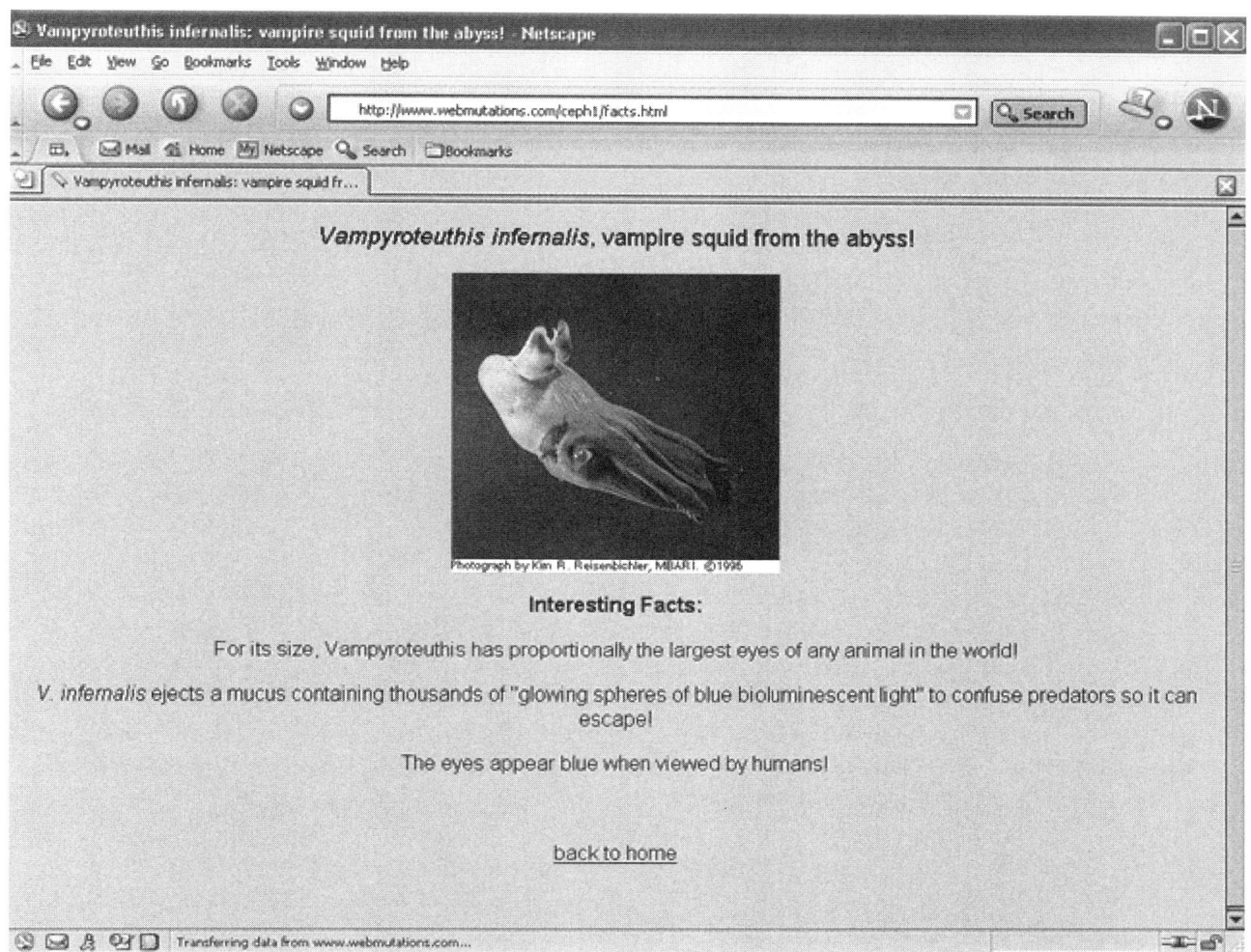

And finally here is the group's reference page:

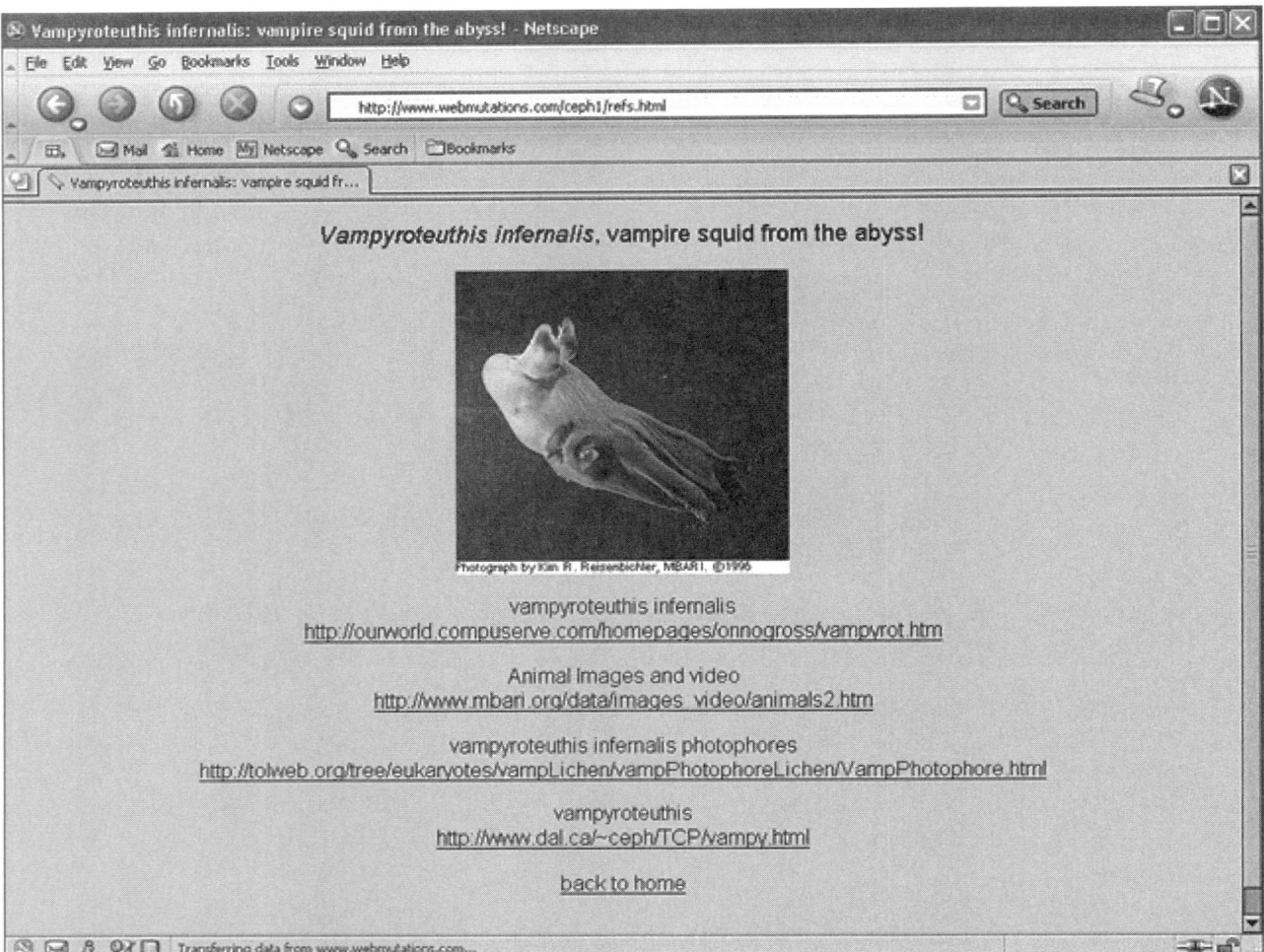

These web pages can be uploaded to your district's server if your network administrator allows. Students are really motivated to do a nice and thorough job knowing that everyone on the Internet might possibly be viewing their pages.

Phylum Mollusca, Class Cephalopoda
("Head-Foot" Mollusks)

GENERAL INFORMATION

The most advanced mollusks, the cephalopods, include squids, octopi, nautiluses and cuttlefish. Cephalopoda means head-foot (*cephalo-* means head, *-pod* means foot). This name refers to the fact that a cephalopod's head is attached to its foot, which is divided into arms. The eight flexible arms are equipped with round suction disks that are used to gather food and to manipulate objects. In addition to these arms, cuttlefish and squid also have two long, slender tentacles with suckers on the end. Nautiluses have many more arms (38 to 90) than other cephalopods. Their arms lack suckers but are made sticky by a mucus-like covering. All cephalopods are marine predators. Although they may have a radula, carnivorous mollusks such as octopi, squid, and certain sea slugs typically use sharp jaws to eat their prey. Like cone shells, some octopi produce poison to subdue their prey.

Like all mollusks, cephalopods have nephridia to remove nitrogenous wastes from the body. And like all aquatic mollusks, they use gills to respire. But their circulatory and nervous systems differ greatly from other mollusks.

The streamlined bodies of cephalopods permit rapid swimming. Cephalopods swim by *jet propulsion* by expelling a column of water from their mantle cavity. Some cephalopods, such as squid and octopi, discharge an inky fluid from the mantle in times of danger to distract predators and make an escape. The flow of blood through sinuses is not efficient enough for fast-moving octopi and squids. Therefore, unlike other mollusks, cephalopods have a *closed circulatory system,* in which blood always moves inside blood vessels.

Most animals have a closed circulatory system. This means that the walls of the heart and the blood vessels are continuously connected. Large blood vessels transport blood rapidly to and from tissues, and a system of smaller vessels, called *capillaries,* provide for a slower diffusion of blood at various tissue sites. [It should be noted that even a closed circulatory system is not completely sealed off. A small amount of fluid is constantly filtering out of the capillaries, and materials are continually passing between the capillaries and the surrounding tissues.]

Octopi and other tentacled mollusks are active and intelligent predators that have the most highly developed nervous systems of all members of their phylum. Because of their well-developed brain, these animals can remember things for long periods of time. The numerous complex sense organs they possess help them distinguish shapes by sight and texture by touch. The eye of an octopus, with its lens and focusing muscle, is remarkably similar to the eyes of humans and other vertebrates. Also like vertebrates, cephalopods have highly developed sense organs for balance. Octopi can be trained to perform different tasks in order to obtain a reward or avoid punishment. Because of these abilities, octopi are often studied by psychologists interested in studying how animals learn.

The chambered nautilus is so named because of its coiled shell with many chambers. By taking in and releasing gas from chambers in its shell, a nautilus can change its buoyancy. This means that it can live at any level of the ocean, from the surface down to 600 m. The spiral shell of the chambered nautilus has provided inspiration for many paintings and other works of art. One of the most famous contemporary artists to use the chambered nautilus in her work is Georgia O'Keeffe. A chambered nautilus is the subject of one of her well-known series of posters for the Santa Fe Music Festival.

One trend in cephalopod evolution that should be noted has been the reduction in the size of the shell. Most modern species have a small shell or no shell at all. For example, nautiluses (the most ancient of the cephalopods) have a large shell, squids and cuttlefish have a small, internal shell, and octopi have no shell.

ANATOMY AND PHYSIOLOGY

The name "cephalopoda" means "head-foot," for in these mollusks the foot, which is divided up into a number of arms, is closely associated with the head. As in gastropods, the shell of cephalopods has been reduced in varying degrees. The nautiluses have large, external, coiled shells; the squid have only a thin vestige of a shell embedded in the mantle; the octopi have virtually no shell at all.

A squid has many features that illustrate how it has adapted the molluscan body plan to an active, free-swimming life. A squid relies for protection not on a heavy shell but chiefly on its ability to leave the scene of danger in a hurry. The shell is vestigial and is represented by a *pen* with a texture like that of cellophane. (A *vestigial structure* is a remnant of a structure that was functional in an ancestral organism but is reduced in size and serves little or no function in a modern animal.) The *mantle* is thick and muscular and is the chief swimming organ. At the pointed posterior end the mantle surface is extended into a pair of triangular folds, the *lateral fins,* which act as stabilizers; they can be undulated to move the animal slowly and adjusted to steer. The mantle terminates in a free edge, the *collar,* which surrounds the neck between the head and visceral mass. The *siphon,* a conical muscular tube derived from part of the foot, projects beyond the collar on the underside of the head. When the mantle is relaxed, water enters the *mantle cavity* around the collar. When the mantle contracts, the collar is tightly sealed against the visceral mass and water is forced out through the siphon. When a squid is excited by the sight of prey, the mantle is contracted strongly, forcibly expelling a jet of water from the siphon. The tip of the siphon is bent backward, and the jet sends the squid quickly forward to seize its prey. If a squid is threatened, the siphon is directed forward, and the animal shoots backward like a torpedo; this is its usual behavior in escape. When attacked, a squid may emit a cloud of inky material from a special *ink sac* that opens into the siphon. Discharges from the anus, reproductive and excretory organs also exit through the siphon.

Besides the siphon, the eight *sucker-bearing tentacles* that surround the mouth are derived from the foot of a squid. When the animal is swimming, the tentacles are pressed together and aid in steering. Two more arms, longer and more slender than the tentacles, can be extended forward to seize the prey and draw it toward the mouth. There it is held firmly by the other tentacles, while two strong hard *jaws* in the mouth pierce the prey, biting out large pieces and swallowed so rapidly.

The active life of squid is made possible by more complex respiratory and circulatory systems than those of other mollusks. On the other hand, squid require well-aerated water and could not survive in the poorly oxygenated waters were many other mollusks live. A squid expends considerable energy in constant contractions and expansions of the muscular mantle that provide a vigorous circulation of water through the mantle cavity, in which lie two large *gills.* Unlike the open circulatory systems of other mollusks, *the circulatory system of squid is closed.* The blood flows within blood vessels. There are separate pumping mechanisms for blood going to the gills and that going out to the other organs. Deoxygenated blood enters two *gill hearts,* each of which pumps blood through one gill. This gives the blood a fresh impetus, so that it passes through the gills at higher speed and pressure. Freshly oxygenated blood from the gills enters a single *systemic heart,* from which it is pumped out again to the various organs.

The *nervous system* of a squid is highly complex and centralized — in sharp contrast with that of the slow-moving clams. Squid have a large *brain* that lies between the eyes. The eyes are large and can form images. They are remarkably like primitive human eyes.

Cuttles, or cuttlefishes, resemble squid in structure and habits. The shell, a plate embedded in the fleshy mantle, is the "cuttlebone" given to caged birds as a source of calcium.

An octopus has virtually no shell. Unlike the ten-armed squid, octopi have only eight arms and are mostly bottom-dwelling cephalopods. They move by pulling themselves over the rocks with their arms or swim short distances by forcibly expelling water from the siphon. In a sheltered place among the rocks the entrance to an octopus home may sometimes be identified by the accumulated litter of crab shells. The octopus breaks open the shells by a pair of horny jaws and the radula. As far as is known, the octopi are the most intelligent invertebrates; their ability to learn and to make subtle visual discriminations has been intensively studied. Octopus eggs are guarded and are kept clean and aerated by the mother. The eggs, like those of other cephalopods, contain a large amount of yolk and hatch directly into juveniles.

A few shy species of nautilus in the tropical Indo-Pacific are the only living cephalopods with an external shell, although shelled cephalopods were dominant members of ancient seas. Nautiluses appear to be active at night, using their 90 or so short tentacles to seize crabs and fishes; they rest during the day in coral crevices.

REFERENCES

http://is.dal.ca/~ceph/TCP/index.html
http://www.nrcc.utmb.edu/MODEL.html#Summary%20of%20the%20Class%20Cephalopoda
http://www.marinelab.sarasota.fl.us/OCTOPI.HTM

Rubric for Cyber Cephalopods on the Web

Cephalopod you are researching: (Common name and scientific name)

_____

Team Member Names:

Team member #1_____

Team member #2_____

Team member #3_____

Mechanics	No	Yes
Banner/header of text at top of the page displaying the title of your web site	0 pts.	2 pts.
Scientific name of your selected cephalopod in the correct *Genus species* format	0 pts.	2 pts.
Three (3) links to other Internet cephalopod web sites + a description of link		
• Link #1		
Works?	0 pts.	2 pts.
Description?	0 pts.	3 pts.
• Link #2		
Works?	0 pts.	2 pts.
Description?	0 pts.	3 pts.
• Link #3		
Works?	0 pts.	2 pts.
Description?	0 pts.	3 pts.
Three (3) images of your selected cephalopod		
• Image #1 of cephalopod displayed	0 pts.	6 pts.
• Image #2 of cephalopod displayed	0 pts.	6 pts.
• Image #3 of cephalopod displayed	0 pts.	6 pts.
Attractive background with coordinating font colors for link, alink, vlink (check *Netscape Composer* tutorial) (Right click on page properties to see these listed)	0 pts.	2 pts.

Mechanics Subtotal	_____pts.		
Content	Absent	Present but unclear	Clearly Explained
Three (3) survival strategies of your selected cephalopod			
• Survival strategy #1	0 pts.	3 pts.	6 pts.
• Survival strategy #2	0 pts.	3 pts.	6 pts.
• Survival strategy #3	0 pts.	3 pts.	6 pts.
Three (3) interesting facts about your selected cephalopod			
• Interesting fact #1	0 pts.	3 pts.	6 pts.
• Interesting fact #2	0 pts.	3 pts.	6 pts.
• Interesting fact #3	0 pts.	3 pts.	6 pts.
Seven (7) essential functions and descriptions for your selected cephalopod			
• Feeding/digestion	0 pts.	2 pts.	4 pts.
• Respiration	0 pts.	2 pts.	4 pts.
• Internal transport	0 pts.	2 pts.	4 pts.
• excretion	0 pts.	2 pts.	4 pts.
• response to stimuli	0 pts.	2 pts.	4 pts.
• movement	0 pts.	2 pts.	4 pts.
• reproduction	0 pts.	2 pts.	4 pts.
Content Subtotal	_____pts.		

Deductions	Point value	X number of occurrences	= lost points
Image and web page file names must contain all lowercase letters (including extensions .jpg, .html, .gif)	- 5		
Image and web page file names must not have any commas, dashes, or spaces	- 5		
Words must all be correctly spelled (- 5 for each incorrectly spelled word)	- 5		
Deductions Subtotal	_____pts.		
Comments:			
TOTAL SCORE	_____pts.		

Cyber Cephalopods on the Web
Creating Web Pages as a Research Product

PURPOSE

To create a professional-looking web page(s) that celebrates the life, survival strategies and seven essential functions of cephalopods (Class: Cephalopoda): squid, octopi and cuttlefish.

MATERIALS

computer with Internet access
Rubric for Cyber Cephalopods on the Web

Netscape Communicator/Composer software
Phylum Mollusca, Class Cephalopoda ("Head-Foot" Mollusks) readings

You will have 2 class periods to finish this web project. It will be due at the end of the 2nd class period. Your time will be limited to these class periods and you will have to work very cooperatively and quickly to finish. The project may be accomplished in one or multiple web pages. The entire project must be completed in class. No work may be done outside of class!

PROCEDURE

1. Your first task will be to begin researching a specific cephalopod (squid, octopus or cuttlefish) on the Internet with two other class members. Look at your student answer page and assign a student job to each of the three team members in your group.

2. Write your name next to your assigned job as either Team member # 1, # 2 or # 3, on the student answer page.

3. Go to the following website:
   The Cephalopod Page
   http://is.dal.ca/~ceph/TCP/index.html

4. Decide which cephalopod that your team will be researching. Let your teacher know which one you want to research and write the common name of that cephalopod and its scientific name in <u>Genus species</u> (underlined when it is written) format in the blank provided on your student answer page.

5. Start your research and gather content for your web page. You will find lots of appropriate content and images at *The Cephalopod Page* website. You will also find content from the reading: *Phylum Mollusca, Class Cephalopoda ("Head-Foot" Mollusks)*. These are both great places to start.

   Of course, you can also search for information through http://www.google.com or some other search engine but do not forget that you will have a limited amount of time to complete this website.

6. Once you have found all of the information and images for your web page you may begin constructing the page(s) using Netscape Composer. Please make sure that you first review the steps of creating a web page in the *Creating Web Pages Using Netscape Composer* activity before starting this project.

7. Each web project must include the following content and or images:
   - A banner of text at the top of the page displaying the title of your web site. Try to be creative.
   - The scientific name of your selected cephalopod, in the format of *Genus species* or <u>Genus species</u>
   - Attractive background color or image with complementary font colors (including alink, vlink and link colors. (Check the *Netscape Composer* tutorial for explanations of these.)
   - Three (3) links to other cephalopod web sites (with description of links) on the Internet
   - Three (3) images of your selected cephalopod
   - Three (3) survival strategies of your selected cephalopod
   - Three (3) interesting facts about your selected cephalopod
   - Seven (7) essential functions of your cephalopod, with descriptions:
        feeding/digestion
        respiration
        internal transport
        excretion
        response to stimuli
        movement
        reproduction

8. As each person on your team completes a specified task, check that task off on your student answer page by putting a ✓ in the blank next to each task. At the end of the research period, each team member's task list should be fully completed.

9. Once your web project is completed, review the *Rubric for Cyber Cephalopods on the Web Webpage* to determine whether or not you have met the requirements. Each person is to turn in their own student answer page once your website is ready to be scored.

Name _____

Period _____

Cyber Cephalopods on the Web
Creating Web Pages as a Research Product

TEAM MEMBERS' RESPONSIBILITIES

Each web project will be created and designed by a team of 3 people. The project may be accomplished in one or more web pages. The responsibilities of each team member are listed below.

You will have 2 class periods to finish this project. It will be due at the end of the 2nd class period.

Cephalopod you are researching: (Common Name and scientific name)

_____

Team member #1 _____
(name)

Place a ✓ in the blank to the left of the item when you or your team member has completed that item on the web project:

_____ 2 essential functions: respiration and internal transport
_____ 1 link to other cephalopod web site
_____ 1 survival strategy of your selected cephalopod
_____ 1 image of your selected cephalopod
_____ 1 interesting fact about your selected cephalopod
_____ title/header of your web page(s)

Team member #2 _____
(name)

Place a ✓ in the blank to the left of the item when you or your team member has completed that item on the web site:

_____ 3 essential functions: response to stimuli, movement, reproduction
_____ 1 link to other cephalopod web site
_____ 1 survival strategy of your selected cephalopod
_____ 1 image of your selected cephalopod
_____ 1 interesting fact about your selected cephalopod

Team member #3 _____
                    (name)

Place a ✓ in the blank to the left of the item when you or your team member has completed that item on the web site:

_____ 2 essential functions: excretion and feeding/digestion
_____ 1 link to other cephalopod web site
_____ 1 survival strategy of your selected cephalopod
_____ 1 image of your selected cephalopod
_____ 1 interesting fact about your selected cephalopod
_____ the scientific name of your selected cephalopod in the correct format

Snails Versus Humans
Comparing Relative Strength of Snails and Humans

OBJECTIVE
Students will compare the relative pulling strength of humans with that of land snails (*Helix aspersa*) by mathematically calculating and then graphing the percentage of body mass pulled.

LEVEL
Middle Grades: Life Science

NATIONAL STANDARDS
UCP.1, UCP.2, UCP.3, A.1, A.2, C.5, C.6

TEKS
6.1(C), 6.2(A), 6.2(B), 6.2(C), 6.2(D), 6.2(E), 6.4(A), 6.4(B), 6.6(B), 6.10(A), 6.10(C), 6.12(A)
7.1(A), 7.2(A), 7.2(B), 7.2(C), 7.2(D), 7.2(E), 7.4(A), 7.11(B)
8.2(A), 8.2(B), 8.2(C), 8.2(D), 8.2(E), 8.4(A), 8.11(A)
IPC 2(C), 2(D)

CONNECTIONS TO AP
AP Biology:
   III. Organisms and Populations: B. Structure and Function of Plants and Animals 1. Reproduction, growth, and development 2. Structural, physiological, and behavioral adaptations 3. Response to the environment

TIME FRAME
90 minutes

MATERIALS
(For a class of 28 working in groups of 3 for 4)

8 small plastic disposable cups	8 triple beam balances
8 small containers with nuts, screws, washers to be used as weights	8 20 cm pieces of dental floss
clear "shiny" cellophane tape	8 rubber bands
bathroom scale	150 kg of various weights
backpack	rope
shoulder or duffel bag	1 head of lettuce
metric ruler	masking tape

TEACHER NOTES
Snails Versus Humans mathematically compares the percent body mass pulled by a land snail, *Helix aspersa*, and the percent body mass pulled by a human (student).

Obtaining *Helix aspersa* land snails can be done one of two ways. If you live in Texas, Arizona, California, New Mexico, Utah, or Washington, you have the option of ordering the snails from an educational supply company such as Delta Education, http://www.delta-education.com. Teachers, residing in only these states are allowed by the U.S.D.A., with a federal permit, to obtain *Helix aspersa* land snails by mail. The snails are banned from being transported to all other states. If you live in one of the above listed states and end up ordering the snails from an educational supply company, you have to first obtain a federal permit from the U.S.D.A. (see the link below for the U.S.D.A. official page about *Helix aspersa* permits) as *Helix aspersa* are considered agricultural pests. The permit process takes about 6-8 weeks to complete so plan accordingly. The first link below in the reference section describes and explains all aspects of this process.

If you live in another state or if you cannot or do not want to purchase the snails, you can go out and collect *Helix aspersa* in a garden, around the bottoms of shrubs, or other cool dark locations in the natural world, as they are indeed common garden pests. Another way of obtaining snails is to offer extra credit to students for each healthy specimen. Figure 1 shows a typical *Helix aspersa* specimen.

If you live in a state not listed above, and you cannot find the snails in your area, the lab can be performed with another organism, *Odontotaenius disjunctus*, commonly known as the Bess beetle. The Bess beetles are around $30-40 for a set of 12 (also available at Delta Education) and will not breed in captivity. If you choose to do the lab with Bess beetles, see the link in the references section below, where there is a 32 page lab resource in a PDF format.

Obtaining a permit, collecting snails, or buying the beetles might seem, at first, like a convincing reason for you *not* to do this activity. However, witnessing the amazement and hearing the screams of excitement from the students performing this lab makes it worth whatever difficulty you encountered in securing the living materials. The level of student motivation and excitement in this lab is beyond anything that could be described in a written format. They simply love this lab.

When the snails arrive or after you collect them, place them in an aquarium and spray them with distilled water. To get maximum pulling power from the snails, do not feed the snails for a day or two before doing the experiment with students. If you are keeping them for a longer period of time until you actually perform the lab, the snails will be very content eating celery. They will also eat lettuce and most leaves or even grass.

Before students arrive on the day of the lab the snails need to be "awakened". Pour some distilled, slightly warm, water into a 500 mL or 1000 mL beaker. Drop the dormant snails into the water. After a minute or so, sometimes longer, the snails will begin to come out of their shells and will begin to move around. As each one awakens, remove it from the water and put it back into the aquarium. Do the same with all the other snails.

To assemble the snail's pulling apparatus, take a small (4 ounce) plastic disposable cup and place a rubber band around the mouth of the cup just under the lip. Cut off a piece of dental floss or string approximately 20 cm in length. To make a harness for the snail, tie each end of the dental floss to opposite points on the cup's rubber band. There should be 180° between the two attachment points.

Loop a piece of masking tape around the middle of the dental floss harness, sticky side to sticky side, so that the non sticky side is facing out on the top and the bottom. The looped piece of masking tape provides a surface to which you will adhere a small piece of clear tape. The small piece (3 or 4 cm) of clear tape should connect the masking tape tab to the shell of the snail.. Be sure to gently wipe the snail's shell with a paper towel before adhering the tape. The best results are obtained with the clear cellophane tape that has the "shiny" finish and not the satin finish clear tape. The "shiny" clear tape seems to adhere better to the snail's shell.

A labeled photograph of a snail and the snail pulling apparatus appears below.

You can keep track of individual snails and their usage by using a fine tipped permanent marker to label the dried shell with a unique number.

Snails are living creatures and do not always perform "on cue". It is always a good idea to have more snails on hand than you have lab groups, in all your classes.

Remind students that when the snail is pulling, they should not allow the snail to eat the lettuce (or whatever food item is being used to entice it to pull). If the snail feeds, it will become satiated and stop pulling. The students should keep the food just in front of the snail's head, perhaps even gently touching the food to the antennae. After the snail has pulled its maximum weight, it can be rewarded with the celery or other food item.

Figure 1: *Helix aspersa* specimen with attached pulling apparatus

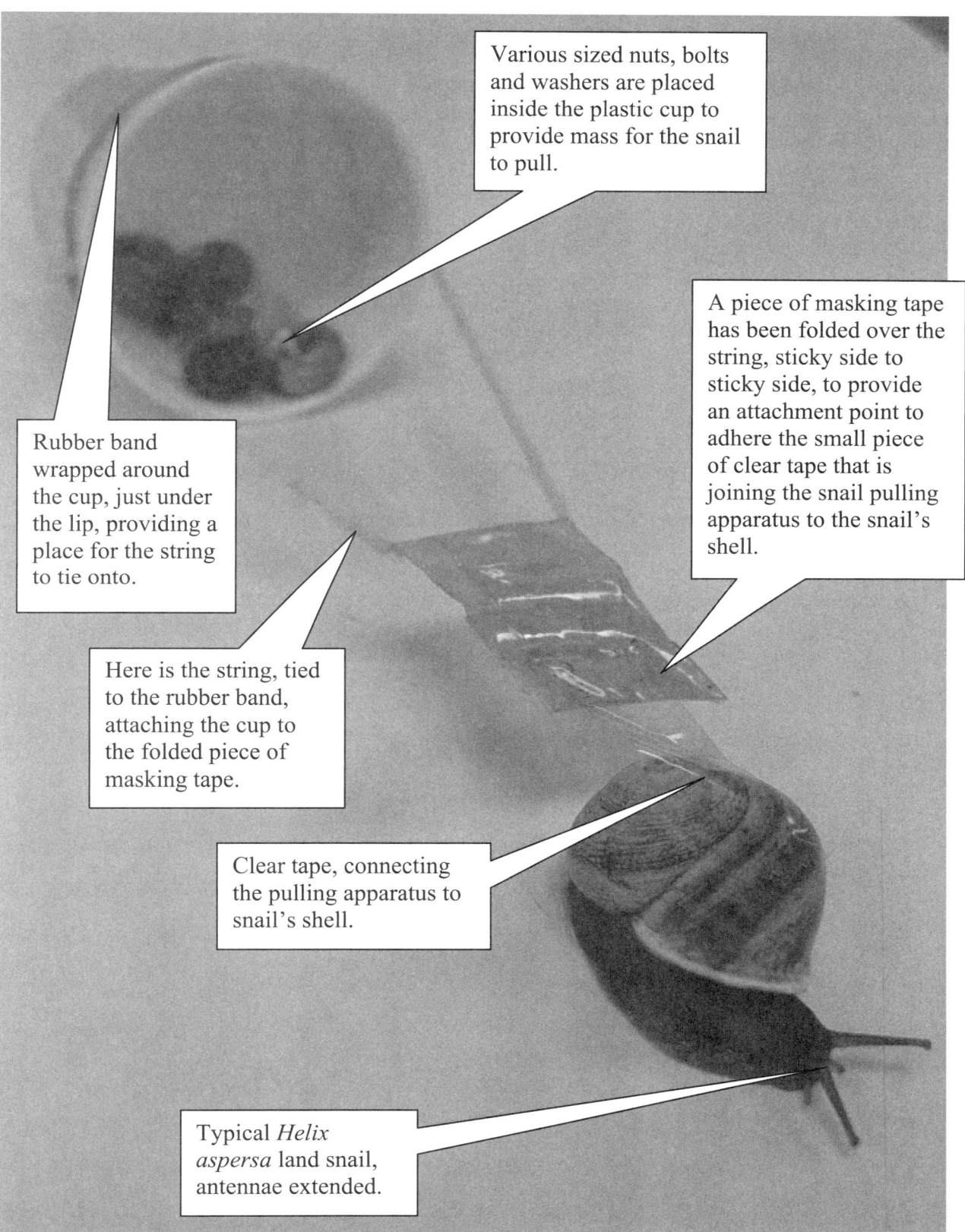

Various sized nuts, bolts and washers are placed inside the plastic cup to provide mass for the snail to pull.

A piece of masking tape has been folded over the string, sticky side to sticky side, to provide an attachment point to adhere the small piece of clear tape that is joining the snail pulling apparatus to the snail's shell.

Rubber band wrapped around the cup, just under the lip, providing a place for the string to tie onto.

Here is the string, tied to the rubber band, attaching the cup to the folded piece of masking tape.

Clear tape, connecting the pulling apparatus to snail's shell.

Typical *Helix aspersa* land snail, antennae extended.

TEACHER PAGES

To set up the pulling apparatus for the humans, you will need to get approximately 150 kg (330 pounds or so, in various sizes) of weights, typically available in the school weight room or the athletic department. A bathroom scale, a sturdy backpack, a 10 meter section of strong rope, and a shoulder or duffel bag with shoulder straps will also need to be collected and brought into your classroom. When its time to for students to pull the weights, take one group at a time into the hallway, where the weights and the human pulling apparatus are set up. Lay the shoulder/duffel bag on the floor with the rope looped through the shoulder loops. The shoulder/duffel bag should be quite sturdy and potentially able to withstand 150 kg of force. The rope is to be knotted firmly at each end to the hard frame of the backpack. You may find other ways to construct this apparatus based on what you have available.

The *Pulling Technician* student should put on the back pack and walk slowly away from the weights until the rope pulls taut. Remind the students to use the utmost care in pulling the weights and that at the first sign of serious personal strain they must stop pulling. Use your judgment for each student as how much weight to add but remember that the strongest students will be pulling only approximately 150% of their weight, maximum. Lay a meter stick on the floor in front of the student that is pulling the weight. If the student manages to smoothly, without jerking or straining, pull the weights for one meter then it counts towards the maximum amount of weight pulled.

While each group is out in the hallway, the *Zoo Keeper* must keep up with the snail. If left unattended, snails can easily crawl to the edge of the desk and fall off.

Make an overhead of each of the CLASS data tables from the student answer page and allow students to record their results once each group has finished doing the snail pull, the human pull and their calculations. Write in the snail and human values for each group in both the data tables. All students will copy the class data into their own data tables to do the bar graph in the ANALYSIS section.

If needed, refer to Foundation Lesson II, *Numbers in Science,* for practice with conversions such as pounds to kilograms. 1kg = 2.20 lbs.

REFERENCES

Helix aspersa land snail information page. Everything you always wanted to know about *Helix aspersa*.
http://lhsfoss.org/fossweb/teachers/materials/plantanimal/landsnails.html

USDA official page about snail permits as well as rules and regulations.
http://www.aphis.usda.gov/ppq/permits/plantpest/snails_slugs.html

32 page PDF lesson on Bess Beetles
http://www.delta-education.com/foss/bess_beetles.pdf

SAMPLE OBSERVATIONS

There are many observations that could be made regarding including the following:

- two tentacles on head
- uses tentacles to feel in front of itself
- moves on muscular foot
- withdraws into shell if touched
- produces a mucus trail
- eats lettuce
- has a coiled shell
- touches antenna to the lettuce

SAMPLE CLASS DATA

Snail-group number	Maximum Mass Pulled in grams	Snail Mass in grams	Percent of Body Mass Pulled
1	153.8	5.8	2651.74%
2	161.2	5.8	2779.31%
3	53.5	7.0	764.29%
4	87.5	8.0	1093.75%
5	150.7	5.7	2643.86%
6	52.3	5.0	1046.00%
7	22.3	1.1	2027.27%
8	65.9	7.1	928.17%
Average	93.4	5.69	1741.80%

Human-group number	Maximum Mass Pulled in kg	Human Mass in kg	Percent of Body Mass Pulled
Group #1	75.00	47.27	158.65%
Group #2	70.00	59.09	118.46%
Group #3	54.55	52.27	104.34%
Group #4	86.36	47.73	180.95%
Group #5	47.73	65.91	72.41%
Group #6	50.91	40.00	127.27%
Group #7	60.91	42.73	142.55%
Group #8	45.45	39.55	114.94%
Average	61.36	49.32	127.45%

SAMPLE ANALYSIS/BAR GRAPH

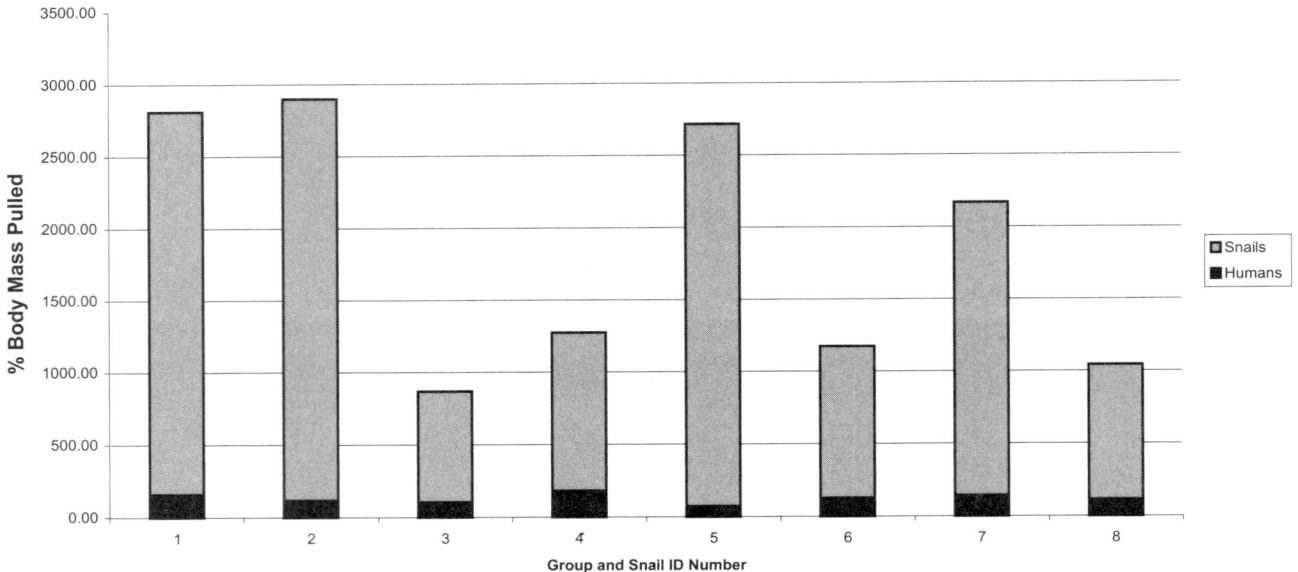

Comparison of % Body Mass Pulled: Humans and Snails

T E A C H E R   P A G E S

Snails Versus Humans
Comparing Relative Strength of Snails and Humans

PURPOSE
In this activity you will compare the relative pulling strength of humans with that of snails by mathematically calculating the percentage of body mass pulled.

MATERIALS

small plastic disposable cup	triple beam balance
small container with nuts, screws, washers to be used as weights	20 cm piece of dental floss
	rubber band
clear "shiny" cellophane tape	150 kg of various weights
bathroom scale	rope
backpack	lettuce
shoulder or duffel bag	metric ruler

Safety Alert
1. Wash your hands after handling the snail.
2. If you are the one pulling the weights, be extremely careful to not strain your back or any muscles in your body.

PROCEDURE
Part I: Snail Pulling Strength

1. Determine at your table who will do each of the jobs in this lab. The descriptions of each job are below.

   Weight Technician: Responsible for making all weight measurements on the balance.
   Zoo Keeper: Responsible for maintaining the safety and well-being of the snail at all times.
   Pulling Technician: Responsible for pulling the weight for the group, representing human pulling strength for your group.

   Each student is responsible for completing their own the student answer page.

   In the space provided on the student answer page, write each team member's name next to the assigned job title.

2. On the student answer page, under PROBLEM, copy the following statement: "Which organism — a human or a snail — is able to pull a higher percentage of its body mass?"

3. Discuss with your group what you think the answer to the problem question will be. In the HYPOTHESIS section on your student answer page, write *your* hypothesis according to what you think the answer to the problem question will be. It is acceptable to have different hypotheses within your group.

4. After your teacher brings the snail to your table, let it crawl around on your desk and observe how it responds to its environment. Use lettuce and water to assist you in making observations of its response to stimuli.

5. List five observations about the behavior of the land snail in the OBSERVATIONS section on your student answer page. The observations can be about its physical characteristics or its behavior.

6. Make certain your balance is calibrated by following your teacher's instructions. Use the balance to record the mass of the snail in grams (to the nearest 0.1g) in the SNAIL PULL DATA table on your student answer page.

7. Measure and record the mass of your snail pulling apparatus, i.e. plastic cup with dental floss and tape.

8. Record the number on the snail's shell in the space provided in your data table.

9. Now you are going to see how much mass the snail is able to pull. Take the snail pulling apparatus and gently attach it with clear transparent tape to the top of the snail's shell. *NOTE:* Make certain the shell is dry before you try to attach the tape. If necessary wipe it off with a small piece of paper towel. Add a couple of small weights to the plastic cup.

10. Place the snail in the center of your table and lay the ruler next to the cup. Use the celery to entice the snail to move forward. If the snail is able to drag the weighted cup at least one centimeter, add more weight to the cup. In order to determine the maximum amount of mass the snail is able to pull, continue adding weights until the snail is no longer able to drag the cup. When you have maximized your snail's pulling capacity remove the last weight you added (since it was NOT able to pull that mass) and take your cup and weights to the balance. The *Zoo Keeper* should now remove the tape and return the snail to the teacher.

11. Record the total mass of the objects the snail was able to pull in the SNAIL PULL DATA table.

12. What percentage of the snail's body mass was pulled? Use the following formula to determine the percent:

$$\text{Percent body mass pulled} = \frac{\text{Maximum Mass Pulled in Grams}}{\text{Snail Mass in Grams}} \times 100$$

Part II: Human Pulling Strength

1. Your teacher will help test the *Pulling Technician's* pulling ability. First, weigh the duffel bag, rope, and backpack frame that *Pulling Technician* will be wearing. Convert the *Pulling Technician's* mass to kg by using the following conversion factor:

   1 kilogram = 2.20 pounds

   Record the mass of the pulling apparatus, in kg, in the HUMAN PULL DATA table on the student answer page and remember it will count toward the total mass pulled.

2. Determine the mass of the *Pulling Technician* on the bathroom scale. Record the mass in the HUMAN PULL DATA table on the student answer page. Convert the *Pulling Technician's* mass to kg by using the following conversion factor:

   1 kilogram = 2.20 pounds

3. Have the *Pulling Technician* wear the backpack frame with the rope connected to the duffel bag. The teacher will begin with 90 pounds by placing two 45-pound weights on the duffel bag. CAUTION: the *Pulling Technician* should keep his/her back straight, lean over at a 45° angle and keep their knees bent! He or she must be able to pull the weights at least 1 meter without straining or jerking. Lay the meter stick on the floor so you can see how far the weights have been pulled. Add weights as needed until the *Pulling Technician* can no longer steadily drag the duffel bag.

4. Record the amount of weight pulled by the *Pulling Technician* in the HUMAN PULL DATA table. Be sure to add the weight of the pulling apparatus to the "Maximum Mass Pulled in kg".

5. What percentage of the human's body mass was pulled? Use the following formula to determine the percent:

   $$\text{Percent body mass pulled} = \frac{\text{Maximum Mass Pulled in Kilograms}}{\text{Human Mass in Kilograms}} \times 100$$

6. After each group has finished both parts of the lab, your teacher will write the class data for each group's snails and humans on an overhead or on the chalkboard. Copy this data into the *CLASS: HUMAN PULL DATA* table and the *CLASS: SNAIL PULL DATA* table.

7. Construct a bar graph comparing the snail's percent body mass pulled to human's percent body mass pulled on the graph paper found in the ANALYSIS section. Use the data from your class data tables to construct this graph. On the *x*-axis will be the identification number for each group and snail. On the *y*-axis, you will graph the % of body mass pulled. Make sure that your graph is done neatly. Shade the bars such that it will be easy to compare the snail data to the human data and make a key describing your scheme. Make sure that your axes are labeled and your graph has an appropriate title.

8. From your graph and your data, write a conclusion to your experiment. Your conclusion will answer your original problem question, "*Which organism — a human or a — is able to pull a higher percentage of its body mass?*" Also, in your conclusion, you should relate to your original hypothesis, whether the data from the class (and your group) support your hypothesis or not.

Name _____

Period _____

Snails Versus Humans
Comparing Relative Strength of Snails and Humans

LAB JOBS

Job Title	Person responsible for those job duties
Weight Technician:	
Zoo Keeper:	
Pulling Technician:	

PROBLEM

HYPOTHESIS

OBSERVATIONS

1.

2.

3.

4.

5.

DATA

GROUP: SNAIL PULL DATA

Snail number	Mass of Pulling Apparatus	Maximum Mass Pulled in Grams	Snail Mass in Grams	Percent of Body Mass Pulled

GROUP: HUMAN PULL DATA

Human-group number	Mass of Pulling Apparatus	Maximum Mass Pulled in kg	Human Mass in kg	Percent of Body Mass Pulled

CLASS: SNAIL PULL DATA

Snail-group number	Maximum Mass Pulled in Grams	Snail Mass in Grams	Percent of Body Mass Pulled
Average			

CLASS: HUMAN PULL DATA

Human-group number	Maximum Mass Pulled in kg	Human Mass in kg	Percent of Body Mass Pulled
Average			

ANALYSIS

CONCLUSION

Planaria Chainsaw Massacre
Observing, Dissecting and Regenerating a Planarian

OBJECTIVE

Students will observe a planarian's (Genus *Dugesia*) reactions to external stimuli, predict its ability to regenerate and observe the regeneration process after making incisions.

LEVEL

Middle Grades: Life Science

NATIONAL STANDARDS

UCP.1, UCP.2, UCP.5, A.1, A.2, C.5, C.6

TEKS

6.1(A), 6.2(A), 6.2(B), 6.2(C), 6.2(D), 6.2(E), 6.4(A), 6.10(A), 6.10(B), 6.10(C), 6.12(A), 6.12(B)
7.1(A), 7.2(A), 7.2(B), 7.2(C), 7.2(D), 7.2(E), 7.4(A), 7.9(B), 7.11(B)
8.1(A), 8.2(A), 8.2(B), 8.2(C), 8.2(D), 8.2(E), 8.4(A), 8.6(B), 8.11(A)
IPC 1(A), 2(A), 2(B), 2(C), 2(D)

CONNECTIONS TO AP

AP Biology:
    III. Organisms and Populations: B. Structure and Function of Plants and Animals 1. Reproduction, growth, and development 2. Structural, physiological, and behavioral adaptations 3. Response to the environment
AP Environmental Science:
    I. Interdependence of Earth's Systems: Fundamental Principles and Concepts E. The Biosphere 1. organisms: adaptations to their environments

TIME FRAME

90 minutes

MATERIALS

(For a class of 28, working in groups of 2 or 3)

9 planaria	30 Petri dishes
9 scalpels	9 hand lenses
9 toothpicks	purified spring water
9 index cards	9 beakers
9 eye droppers or plastic disposable transfer pipets	9 wax markers or permanent markers
9 stereoscopic microscope	paper towels
9 polystyrene or plastic containers for ice	

TEACHER NOTES

Planaria Chainsaw Massacre allows students to observe the behavior of a live planarian (Genus *Dugesia*), make predictions about its regenerative abilities, make prescriptive incisions and then observe the regenerative abilities of the planarian.

Planaria can be obtained in two ways. They can be collected in the wild or they can be bought from a scientific supply company.

If you would like to try to collect them yourself, you must have access to a clear, fast-running, rocky-bottom, freshwater creek or stream. Planaria thrive in clear running water. Locate fist sized and larger flat bottom rocks from the bottom of the creek. Wash the underside of the rocks into a large container such as a 5 gallon bucket. Another option is to use a disposable transfer pipette and suction off the individual planarian from the bottom of the rocks and then place them into a container filled with creek water. Planaria are often found on the underside of leaves, and other objects in the shallow water.

Another method of collection is to use bait — a cube of raw beef liver tied with a string and left to drift near submerged rocks for 15 or 20 minutes. Retrieve the liver and shake off any planaria adhering to the surface and place them into a container of creek water. One other suggested method of capturing planaria is to secure a nylon stocking, with a small pellet of canned pet food inside, along the bottom of a creek. Leave the stocking overnight in the water and by morning planaria will be crawling all over it.

Depending on the water source and your persistence, 30 to 40 planaria could be collected in an hour.

If collecting in a creek is not an option, planaria can be ordered from many science supply companies for approximately $5-10 per class set.

To ensure success in this experiment, there are some suggested maintenance procedures. After you bring the planaria back to your classroom, do not use creek water for any further step in your experimental set up or storage. Creek water has microorganisms that can infect the planaria once it has been cut open and the infected planaria will simply disintegrate! Do not use distilled water or tap water for storing planaria. Distilled water does not have any minerals and tap water has chlorine and other ions that can kill the planaria. Use only spring water for storage, set up and student procedures.

When making incisions always use fresh, sterile blades. Dirty scalpel blades can lead to the infection and death of the planaria.

If you are keeping the planaria for more than a week before performing the experiment, it is imperative that you replace the dirty water with fresh spring water. Remove the old water with a pipet, being careful not to accidentally dispose of your planaria. It is recommended that you do not feed the planaria as it can cause infection. If you are keeping the planaria for a very long period and you absolutely must feed them, planaria can be fed a very small piece of liver. However, you should replace the water daily with fresh spring water while feeding.

On the day of the lab, obtain plastic transfer pipets, spring water, as many Petri dishes (preferably sterile) as you have groups, your stock supply of planaria, and a supply of clean small pebbles, between

.5 cm and .7 cm. Pour a few mL of spring water into each of the Petri dishes and add a planaria into each one with the transfer pipet. Add one pebble to each dish, and label (see student instructions as to how this could be done) the lid in some way to designate the different groups. The set up should take less than 30 minutes.

Seeing a two-headed planaria regenerate in the Petri dish is an unforgettable experience for both teacher and students!

POSSIBLE RESULTS

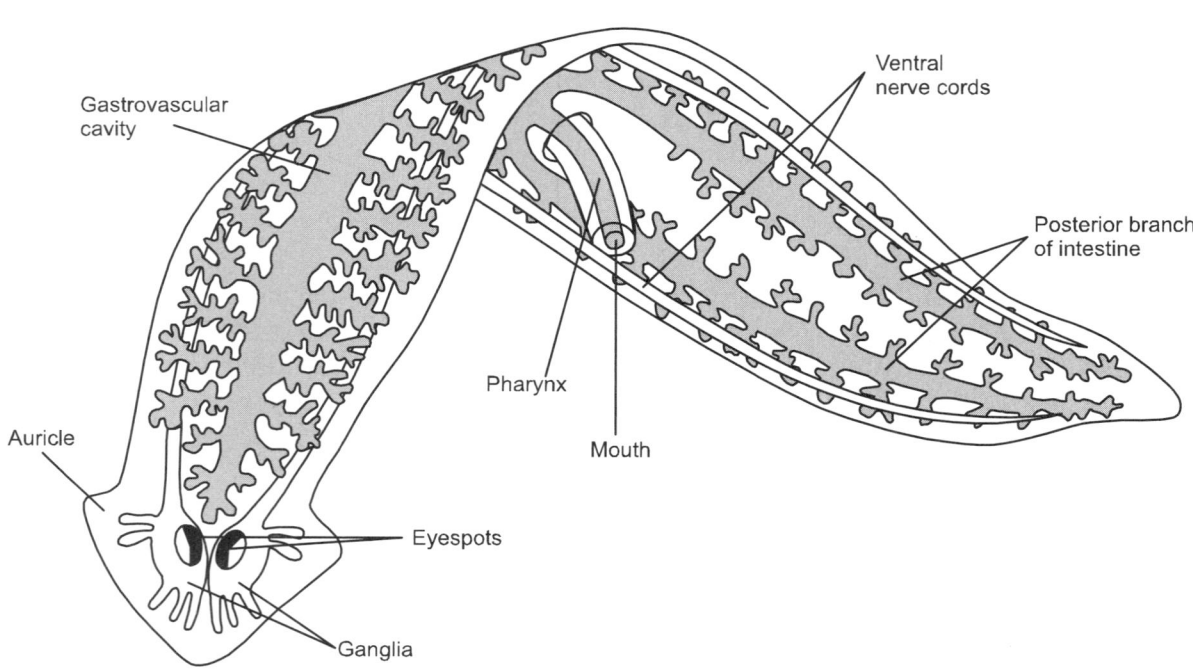

Planaria Anatomy

Diagrams Illustrating Various Incision Types Which
May Be Used to Demonstrate Regeneration in Planaria

Descriptions of Resultant Regenerated Planaria

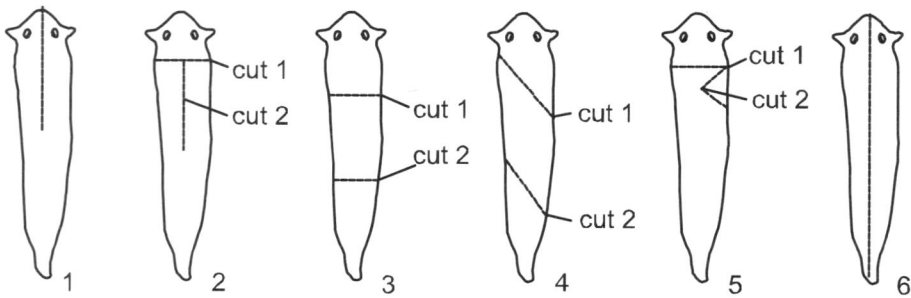

- Incision Type 1: The regenerated planaria typically has two heads arising from its body trunk. This is the most dramatic regeneration result for students, typically. (See the diagram below.)

- Incision Type 2: If students make careful enough incisions, such as in the diagram, two planaria will regenerate, one with two heads, regenerating from the posterior end and one from the anterior end.
- Incision Type 3: There should be 3 new planaria regenerated from the 3 body fragments.
- Incision Type 4: There should be 3 new planaria regenerated from the 3 body segments.
- Incision Type 5: students typically have problems making cut 2 and end up cutting the section just posterior to the head into two fragments. But a new planaria will ideally regenerate from all 3 fragments.
- Incision Type 6: Two whole planaria will regenerate from the two halves.

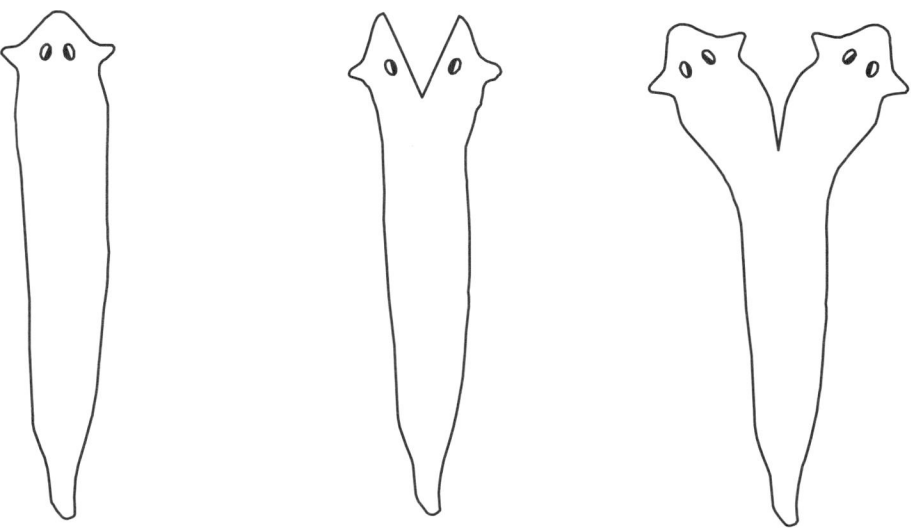

The above diagram shows the progressive regeneration, from left to right, of one intact head, to an incised split head and finally to the two-headed, regenerated planaria.

Regeneration takes anywhere from 3 days to two weeks. Blastema tissue should begin forming within 2 days as you begin your observations. It is recommended that you make observations every 2 days for about two weeks. The entire process of regeneration can then be observed by the students.

ANSWERS TO THE CONCLUSION QUESTIONS

1. Before you picked up the Petri dish, where was the planarian? [For example, was it under the rock, swimming in the middle, near the edge of the dish, etc.?]
   - Answers will vary, but if placed in a dish exposed to light, they immediately turn and move toward the darkest part of the dish, usually under the rock.

2. Describe the movement of the planarian.
   - A planarian moves in a characteristic slow, gliding fashion, with the head bending from side to side. They do not swim freely through the water but move only when in contact with a solid object or on the underside of the surface film.

3. How did the planarian react to the shaded area?
   - A planarian shows a preference for darkness.

4. Based on its behavior, what can you infer about the planarian's sensitivity to light?
   - Planarians avoid light and are generally found in dark places, under stones or leaves.

5. How does the planarian react to being touched in the posterior region?
   - Planarians may tend to ball up when touched in the posterior region.

6. How does the planarian react to being touched in the anterior region? Were the reactions to being touched different in any way? Explain.
   - Planarians respond quickly to touch in the anterior region by contracting their bodies into a ball. Their response to a touch on the head is much more rapid than a touch to the tail.

7. Which direction does the planarian face, up current (with water moving toward its head) or down current (with water moving in the same direction that the planarian is moving)?
   - Planarians react to water currents by facing upstream.

8. What is different about the planarian's body and physiology that allows it to regenerate lost body parts?
   - Regeneration, the ability to "re-grow" lost body parts, is possible because planarians have the ability to form a *blastema*, a dome-shaped accumulation of undeveloped and undifferentiated cells, just beneath the epidermis at the site of the wound. These undifferentiated cells will eventually grow and differentiate into the missing parts.

9. After you make an observation in a few days, how can you tell, by looking at your newly regenerating planarian that it has actually started regenerating?
   - The *blastema* and the newly regenerated parts of the planaria lack any pigmentation and are easily distinguishable from older body parts.

10. Why do you think that planaria evolved this amazing asexual way of reproducing?
    - It has been noted that natural separation and regeneration in planaria occurs more frequently when populations of planarians are low and less frequently in large populations. In such a condition, when the surrounding planaria population is low, there is less of a chance of finding a mate to breed with sexually. If sexual reproduction is not an option, then asexual reproduction, via regeneration is the next best thing.

REFERENCES

For information on *Dugesia* behavior and other interesting facts:.
http://www.ns.purchase.edu/biology/bio1560lab/planaria.htm

Good explanation of regeneration in *Dugesia*:
http://www.umanitoba.ca/faculties/science/biological_sciences/lab11/biolab11_4.html

Information for collecting *Dugesia* in the wild :
http://wardsci.com/pdf/Working_with_Planaria.pdf

Classics of Biology: Planaria
http://ebiomedia.com/gall/classics/Plan/plan_about.html

Planaria Chainsaw Massacre
Observing, Dissecting and Regenerating a Planarian

PURPOSE
In this activity you will observe a planarian's reactions to external stimuli, predict its ability to regenerate and observe the regeneration process after making incisions.

MATERIALS

planarian	Petri dish
scalpel	hand lens
toothpick	purified spring water
index card	beaker
eye dropper or plastic disposable transfer pipet	wax marker or permanent marker
stereoscopic microscope	paper towel
ice in plastic or polystyrene container	

Safety Alert
1. Use extreme caution when using the scalpel as it is razor sharp.
2. Be very careful not to spill any water when using the stereoscopic microscope.
3. Carefully wipe any moisture or water off the bottom of the Petri dish before attempting to view the planarian.

PROCEDURE
Step 1: Observations of planarian
1. Determine at your table who will perform each of the jobs listed in this lab. A description of each job follows. Each student is responsible for completing their own the student answer page. Write in all team members' names next to the job titles in the space provided on the student answer page.

Zoo Keeper: Handling, moving and transferring the planarian or the planarian dissected parts, using the pipet

Incision Technician: Make the incisions and keep up with the scalpel at all times so no one is injured

Observation Specialist: Primary operator of the stereoscopic microscope, making the primary observations of the planarian's behavior

2. The *Zoo Keeper* will obtain a planarian in a Petri dish from the teacher. (Tell the *Zoo Keeper* to observe the position of the planarian in the dish BEFORE picking it up!) Each dish contains a planarian, a rock, and spring water. The Petri dish is marked with a unique number such as 1-2. The first number on the dish stands for your class period and the second number stands for your group number. Write your group's unique number in the space provided on your student answer page.

3. Using the hand lens or stereoscopic microscope, the *Observation Specialist* should describe how the planarian moves. All group members should observe the planarian while it moves and describe and discuss its movements.

4. The *Observation Specialist* should remove the rock from the dish, place it on a paper towel and leave the planarian inside the dish. Place the Petri dish under the stereoscopic microscope and turn on the overhead light only. The *Observation Specialist* should use an index card to shade the half of the Petri dish that does not contain the planarian. Wait at least 3 or 4 minutes, being careful not to shake the table or otherwise disturb the planarian.

5. The *Zoo Keeper* should gently touch the posterior (tail) end of the planarian with a toothpick as the *Observation Specialist* carefully observes the planarian's reaction.

6. The *Zoo Keeper* should gently touch the anterior (head) end of the planarian with the toothpick as the *Observation Specialist* carefully observes the planarian's reaction.

7. The *Zoo Keeper* should gently stir the water with an eyedropper (or plastic transfer pipet) to set up a circular current inside the Petri dish.

8. The *Observation Specialist* should view the planarian using a dissecting scope or magnifying glass. Carefully allow all 3 members to view the planarian in the scope. Find the diagram of the planarian on your student answer page. Lines are drawn to 7 parts on the diagram. Identify those 7 from the list below and write the name in the blanks provided.

   eyespots –photoreceptors used to detect light
   pharynx – used to suck food into the gut
   mouth – food intake and output point
   auricle – ear-like extensions where chemical receptors are located
   anterior and posterior intestine – digestion occurs here
   cerebral ganglion – brain
   lateral nerve cord – nervous system

9. If your teacher instructs you to clean up, make sure that you replace the lid of the Petri dish, put the rock back into the Petri dish and return the planarian to your teacher for storage. If you are completing part 2 of the lab in the same period, go on to the next section.

10. Answer the conclusion questions on your student answer page.

Step 2: Dissection of planarian

Read this introductory section very carefully before making any of the incisions below. The *Incision Technician* will use a scalpel to make the cut that has been selected for the group's planarian. CAUTION: THE SCALPEL IS EXTREMELY SHARP! Each group will complete one of the incision types show below, as assigned by your teacher.

Diagrams Illustrating Various Incision Types Which
May Be Used to Demonstrate Regeneration in Planaria

Descriptions of Resultant Regenerated Planaria

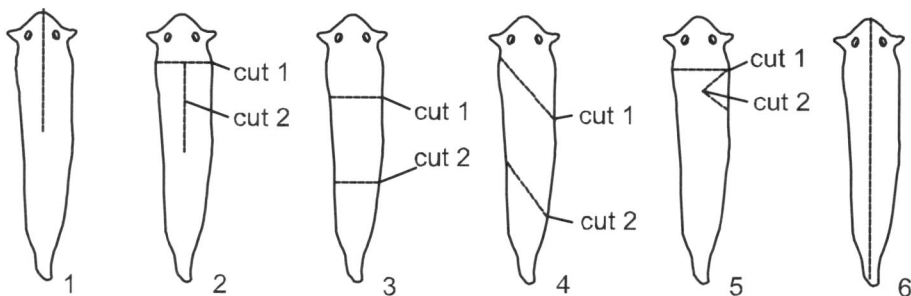

- Incision Type 1: Cut forward from the head end of the body to a point just short of the middle of the worm as shown in the diagram above. Place the dissected specimen into a clean Petri dish with spring water. (If your group is doing this incision, you need to collect **one** extra Petri dish from your teacher.)
- Incision Type 2: Remove the head as illustrated in the diagram at cut 1 and place it in a separate Petri dish of spring water. Then make cut 2 from the head end of the body to a point just in back of the mouth. Transfer the posterior end of the specimen to a clean Petri dish of spring water. (If your group is doing this incision, you need to collect **two** extra Petri dishes from your teacher.)
- Incision Type 3: The specimen will be cut into three pieces as illustrated in the diagram. Immediately after making cut 1, remove the head and place it in a separate clean Petri dish of spring water. After making cut 2, place each piece into a separate clean Petri dish of spring water. (If your group is doing this incision, you need to collect **three** extra Petri dishes from your teacher.)
- Incision Type 4: The specimen will be cut into three diagonal pieces as illustrated in the diagram. Immediately after making cut 1, remove the head and place it in a separate clean Petri dish of spring water. After making cut 2, place each piece into a separate clean Petri dish of spring water. (If your group is doing this incision, you need to collect **three** extra Petri dishes from your teacher.)
- Incision Type 5: This is a tricky one! Remove the head at cut 1 as illustrated in the diagram and place it in a separate Petri dish of spring water. Then cut a triangular piece out of the top right side of the worm as shown in cut 2. Place each piece into a separate clean Petri dish of spring water. (If your group is doing this incision, you need to collect **three** extra Petri dishes from your teacher.)
- Incision Type 6: Dissect the worm into two halves lengthwise as shown in diagram 6. Place each piece into a separate Petri dish of spring water. (If your group is doing this incision, you need to collect **two** extra Petri dishes from your teacher.)

PROCEDURE

1. Before proceeding, write a hypothesis on your student answer page to answer today's problem question: *What will happen to the planarian if incisions are made into it?*

2. *Zoo Keeper*: Using an eyedropper or transfer pipet place a few drops of spring water on the inside of the lid of the Petri dish.

3. *Zoo Keeper*: QUICKLY use the eyedropper or transfer pipet to transfer the planarian to the lid. (If you're not fast, the planarian or portions of it will adhere to the inside of the eyedropper and be very difficult to remove).

4. Allow the worm to extend fully on the Petri dish.

5. *Incision Technician*: Place the Petri dish containing the planarian on top of the container of ice, making sure that the entire bottom surface of the Petri dish is in contact with the ice. Allow the Petri dish to remain on the ice for approximately 1 to 2 minutes, until the planarian has slowed its movement.

6. *Incision Technician*: Remove the Petri dish from the ice, wipe off any excess water from the bottom of the Petri dish with a paper towel and place the Petri dish on the stage of the stereoscopic microscope.

7. *Incision Technician*: Using the magnification of the stereoscopic microscope, and with a steady, quick movement, make your incision(s) with the scalpel blade, according to the incision type assigned to your group and described above.

8. *Zoo Keeper*: After *Incision Technician* makes the cut(s), use the eyedropper or transfer pipet to place each piece of the planarian into separate Petri dishes, as outlined above. Make certain all Petri dishes have plenty of fresh spring water — planaria will die without it!

9. The *Observation Specialist* should label the lid of each Petri dish with a wax pencil or permanent marker. Two labels are required — one to indicate your class period and team number (that is, 7-1 would be seventh period, group 1) and the other to indicate the part of the planarian that you have placed in the dish (head, middle, tail, small chunk taken just posterior to auricle, etc.).

10. Once the dissection is complete and the pieces have been separated, the *Observation Specialist* should place each Petri dish under the stereoscopic microscope. Once the *Observation Specialist* has made the initial observation, let each group member observe each piece.

11. It is often difficult to make the exact cuts that you were instructed to make. It is important, therefore, that you record the actual cuts made on your planarian in the data section of your student answer page. Follow the directions on that page carefully, including predicting how each part of your planarian may (or may not) regenerate.

12. The *Zoo Keeper* should take all labeled Petri dishes to the designated area. Everyone should help in cleaning up, including wiping up all water from the table. The rock and other equipment should be returned to the area designated by your teacher.

13. In a few days, you will make your first observations of your dissected planaria. Carefully observe your planaria specimen in the stereoscopic microscope and follow any directions and instructions your teacher provides.

Name _____

Period _____

Planaria Chainsaw Massacre
Observing, Dissecting and Regenerating a Planarian

Group Number:

Lab Group Job	Student Name
Observation Specialist	
Incision Technician	
Zoo Keeper	

HYPOTHESIS

DATA AND OBSERVATIONS

Label the parts of the planarian on the diagram below using the terms from the list in step 8 of your lab procedure.

Planaria Anatomy

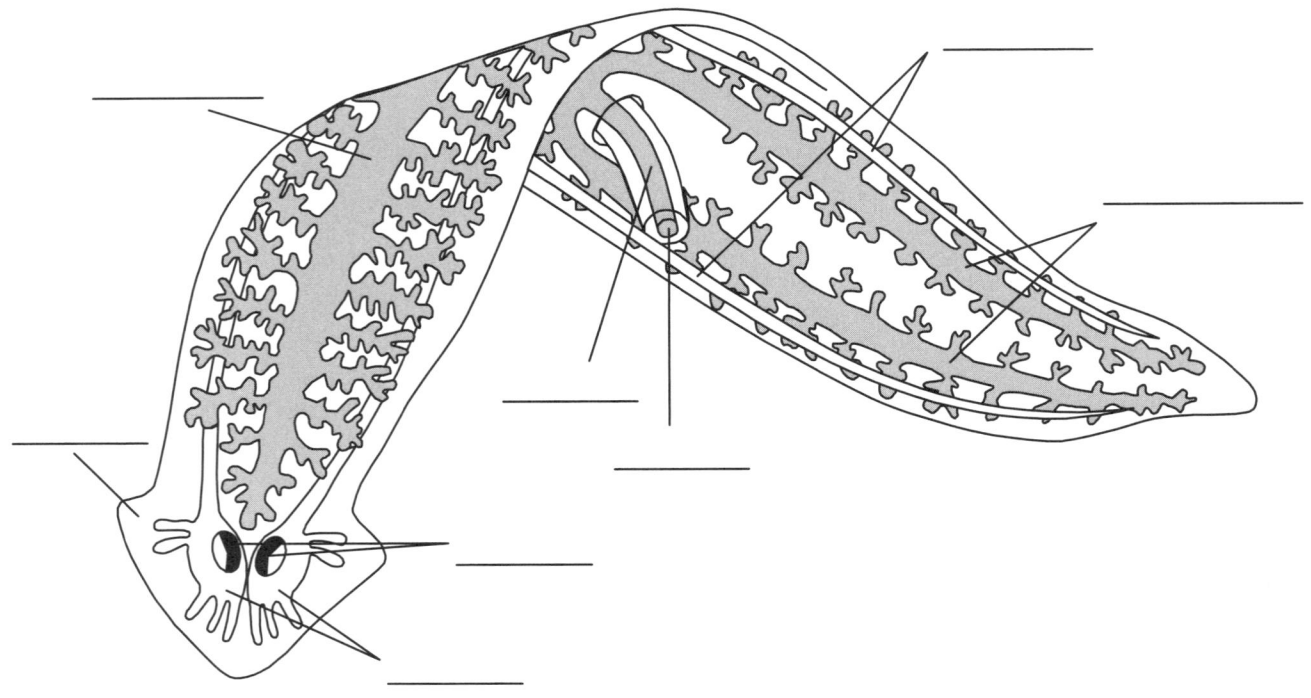

DATA AND OBSERVATIONS

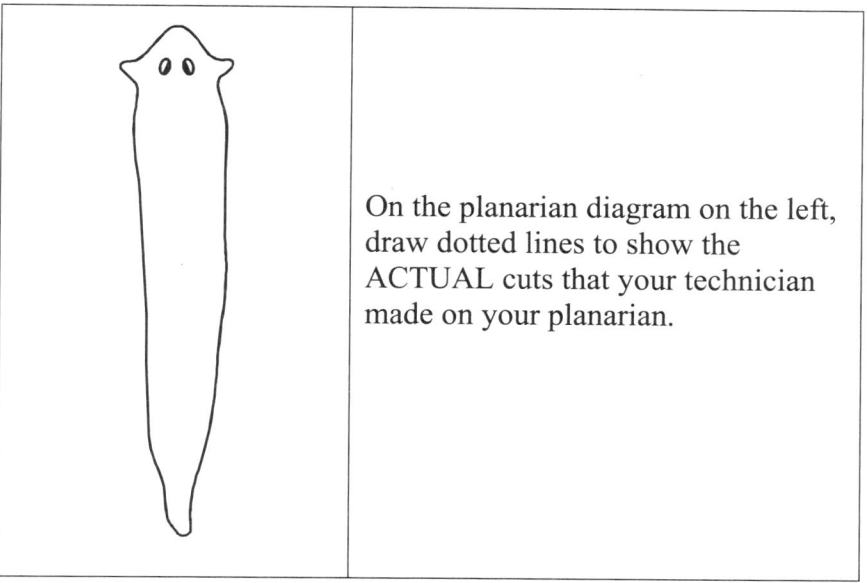

On the planarian diagram on the left, draw dotted lines to show the ACTUAL cuts that your technician made on your planarian.

In the table below, use the boxes under OBSERVED CUTS to describe where your cuts were made. (For example, head, middle, tail, left, right, etc…). Provide a diagram that shows the shape and markings of **each dissected planarian piece**.   Use the boxes under PREDICTED GROWTH to diagram what you think those body section pieces will look like in a few days. Some groups, depending on the cuts made, will not need to use all boxes.

OBSERVED CUTS

PREDICTED GROWTH

CONCLUSION QUESTIONS

1. Before you picked up the Petri dish, where was the planarian? [For example, was it under the rock, swimming in the middle, near the edge of the dish, etc.?]

2. Describe the movement of the planarian.

3. How did the planarian react to the shaded area?

4. Based on its behavior, what can you infer about the planarian's sensitivity to light?

5. How does the planarian react to being touched in the posterior region?

6. How does the planarian react to being touched in the anterior region? Were the reactions to being touched different in any way? Explain.

7. Which direction does the planarian face, up current (with water moving toward its head) or down current (with water moving in the same direction that the planarian is moving)?

8. What is different about the planarian's body and physiology that allows it to regenerate lost body parts?

9. After you make an observation in a few days, how can you tell, by looking at your newly regenerating planarian that it has actually started regenerating?

10. Why do you think that planaria evolved this amazing asexual way of reproducing?

Planaria Chainsaw Massacre
Observing, Dissecting and Regenerating a Planarian

BACKGROUND INFORMATION: PLANARIA

Classification: KINGDOM: Animalia PHYLUM: Platyhelminthes, CLASS: Turbellaria, FAMILY: Planariidae, GENUS: *Dugesia*

Planaria are free living flatworms, members of the Phylum Platyhelminthes ("flat worms"), the simplest animals with bilateral symmetry and that exhibit enough cephalization, or development of the anterior end, to have what we call a head. Because flatworms really are flat, the name of the phylum is quite appropriate.

Free-living flatworms like planaria, in addition to a ladder-type nervous system, have a pair of light-sensitive organs called ocelli, or eyespots. These eyespots do not see objects as our human eyes do; they simply detect whether the animal is in light or in darkness. Their nervous system allows them to gather information from their environment — data they use to locate food and to find dark hiding places beneath stones and logs during the day. By virtue of abundant sensory cells (that are specialized to receive the stimuli of touch, water currents and chemicals), specialized sense organs for light reception, and a centralized nervous system, planarians show a more varied behavior and much more rapid responses than many other invertebrates.

Planaria also have a fairly sophisticated excretory system. This lies in an internal layer of cells called mesoderm and consists of a network of fine tubules called flame cells that run the length of the animal on each side and open to the surface by tiny pores. The primary function of the flame cells is to remove excess water and liquid wastes from the tissues of the planarian. Many flame cells join together to form a network that empties through tiny pores in the animal's skin.

Diagram of Structure of Flame Cell in Planaria

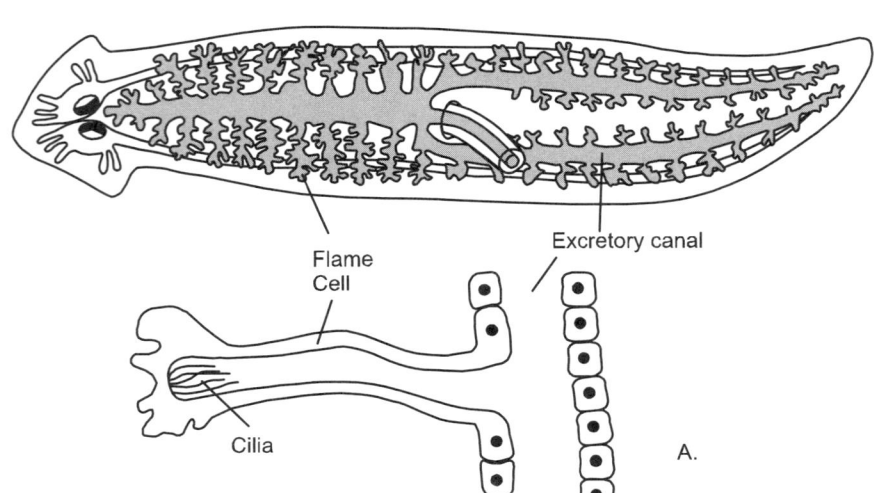

Planaria use two means of locomotion at once. Cilia on their epidermal cells help them glide through the water and over the bottom. Muscle cells controlled by the nervous system allow them to twist and turn so that they are able to react to environmental conditions rapidly.

Planarian reproduction can be either sexual or asexual. A planarian has a highly complicated reproductive system for sexual reproduction, in contrast to the simple clumps of eggs or sperm found among cells of other tissues in more primitive life forms such as sponges or jellyfish. Like other invertebrates planaria are hermaphrodites, forming both male and female reproductive organs in every individual. During sexual reproduction, two worms join in pairs and exchange sperm. Fertilization is internal, and the fertilized eggs are shed in capsules and hatch in a few weeks. Following the breeding season, the reproductive system degenerates and is regenerated anew at the beginning of the next sexual period.

As well as sexual reproduction, many members of the flat worm class, *Turbellaria*, have the amazing ability to regenerate. Regeneration is the ability to "re-grow" lost body parts. This is possible because planarians have the ability to form a *blastema*, a dome-shaped accumulation of undeveloped and undifferentiated cells, just beneath the epidermis at the site of the wound. These undifferentiated cells will eventually grow and differentiate into the missing parts. (The *blastema* and the newly regenerated parts of the planaria lack any pigmentation and are easily distinguishable from older body parts.)

In nature, planarian flatworms reproduce by taking advantage of their regeneration capabilities. When conditions are favorable, the organism will attach its tail-end to the ground and pull forward with its head-end until it tears itself in half. Each end will then regenerate its missing half. This is a form of *asexual* reproduction called *fission*. The two "new" planaria are clones of each other, both possessing identical genes. Recent studies have shown that natural separation and regeneration occurs more frequently when populations of planarians are low and less frequently under crowded conditions.

http://biocourse.bio.tamu.edu/course/zool344/Lab101A.PDF
http://www.upb.pitt.edu/scienceinmotion/biology_labs.htm

Shed Some Light on the Problem
Exploring the Relationship between Light and Photosynthesis

TEACHER PAGES

OBJECTIVE
Students will determine if light is necessary to initiate photosynthesis in the aquatic plant, *Elodea* in a scientific experiment complete with variables, controls and constants.

LEVEL
Middle grades: Life Science

NATIONAL STANDARDS
A.1, A.2, C.5

TEKS
6.1(A), 6.2(A), 6.2(B), 6.2(C), 6.2(D), 6.3(C), 6.4(A), 6.7(A)
7.1(A), 7.2(A), 7.2(B), 7.2(C), 7.2(D), 7.3(C), 7.4(A), 7.8(B)
8.1(A), 8.2(A), 8.2(B), 8.2(C), 8.2(D), 8.3(C), 8.4(A)
IPC 1(A), 2(A), 2(B), 2(C), 2(D)

CONNECTIONS TO AP
AP Biology:
  I. Molecules and Cells: A. Chemistry of Life 1. Water C. Cellular Energetics 3. Photosynthesis
  III. Organisms and Populations: B. Structure and Function of Plants and Animals 3. Response to the environment
AP Environmental Science:
  I. Interdependence of Earth's Systems: Fundamental Principles and Concepts (25%) A. The Flow of Energy 1. forms and quality of energy E. The Biosphere 3. ecosystems and change: biomass, energy transfer, succession

TIME FRAME
90 minutes

MATERIALS
(For a class of 28 working in groups of 4)

28 test tubes (16mm x 150mm) with stoppers	9 test tube racks
7 wax pencils	deionized water in 250 mL beaker
bromothymol blue solution (0.04%)	28 coffee straws
Elodea sample in deionized water	7 graduated cylinders (10 mL)
7 100 mL beakers (for bromothymol blue)	7 metric rulers
	28 rubber stoppers or corks

TEACHER NOTES

Shed Some Light on the Problem is a hands-on scientific inquiry to determine how light affects the process of photosynthesis. In addition to clarifying concepts of photosynthesis and providing an inquiry into how the presence of light affects photosynthesis, it reinforces the components of the scientific method. Students will need to have a good understanding about the steps of the scientific method, variables, constants, and controls. This lab is not an introduction to the concept of photosynthesis and a lesson or two on the process of photosynthesis should precede this lab. For an introduction to photosynthesis, there are many text books and Internet sources that you can use an an introduction. A quick search on http://www.google.com will present many results. You may also wish to include more information about the chemical reactions between carbon dioxide and water and/or the chemical reactions of photosynthesis. For your reference, photosynthesis is represented by the following balanced chemical reaction.

$$6CO_2 + 6H_2O + \text{light energy} \rightarrow C_6H_{12}O_6 + 6O_2$$

Also, the formation of carbonic acid from carbon dioxide in water is represented by the following balanced chemical reaction.

$$CO_2 + H_2O \rightarrow H_2CO_3$$

Best results for this lab will be obtained if the weather outside is sunny and warm. The more sunlight available, the more dramatic the results. A lamp with a "photoflood" bulb, 100-250 watts, can be substituted for the sunlight. These are available at most hardware stores.

Each group will need a test tube rack at their lab station. You will also need to provide a test tube rack in your "sunny" location and your "dark" location. Sunny windowsills and dark cabinets make good choices for this experiment. The intensity of your light source will affect the amount of time necessary to observe a color change. Typically the color change (from yellow to back to blue) will be evident after 30 minutes in a sunny location but the results will vary according to light intensity, season, etc....

It is a good idea to have students practice blowing through a coffee straw into a test tube with just water to get an idea of how hard to blow without splashing. A teacher demonstration of this step may also be in order.

Be sure to use either the rubber stoppers or corks on the test tubes as carbon dioxide will diffuse out of the water solution in an open test tube, possibly yielding false results.

Elodea is available at pet stores or aquarium supply stores for about $2.00-$3.00 a bunch. It is a good idea to obtain at least two bunches to insure an ample supply. Aqueous 0.04% bromothymol blue is widely available in most scientific supply catalogs for less than $10.00 a bottle.

REFERENCES:
http://chem.lapeer.org/bio1docs/PhotoLab.html
http://www.mbari.org/education/EARTH/Iron/Photosynthesis_Lab_1_.pdf
http://www.the-aps.org/education/k12curric/activities/pdfs/carswell.pdf

POSSIBLE ANSWERS TO THE CONCLUSION QUESTIONS AND SAMPLE DATA

1. What is the independent variable in this experiment?
   - Exposing the *Elodea* plants to light or not exposing the plants to light.

2. What is the dependent variable?
   - Whether or not the plants undergo photosynthesis and cause a color change from yellow back to blue.

3. Describe 2 constants in the experiment?
   - Same size *Elodea* sprig
   - Adding bromothymol blue to all test tubes.
   - Adding the same amount of water to the test tubes
   - Same amount of bromothymol blue

4. What is the purpose of setting up test tubes 2 and 4?
   - They serve as controls in the experiment. By definition, a control is a part of the experiment where the independent variable is not applied.

5. Why did we blow through the straw into each tube?
   - To add carbon dioxide into the solution.

6. Describe the color of the bromothymol blue solution in each test tube BEFORE you exhaled into it?
   - Blue

7. Explain what the color of the solution in the test tube tells us (the answer to number 6)?
   - The solution is not acidic and has no carbon dioxide.

8. Describe the color change in the bromothymol blue solution in each test tube AFTER you exhaled into it?
   - Yellow

9. Explain what this color change tells us (the answer to number 8)?
   - That carbon dioxide has been added to the water solution. Carbon dioxide in a water solution makes a weak acid, and therefore, in the presence of bromothymol blue, turns yellow in color.

10. Why did we use bromothymol blue in this experiment?
    - Bromothymol blue indicates the presence or absence of dissolved carbon dioxide in a solution, in the form of a weak acid, or not.

11. After the 30 minute testing time, which of the test tubes changed color?
    - Test tube number 1 is the only one with carbon dioxide, bromothymol blue, water, *Elodea* and light, and therefore, it is the only one that will change color.

12. Explain what caused the test tube(s) in the previous question to change colors?
    - *Elodea* took in the carbon dioxide during photosynthesis. Because it took in the carbon dioxide, the water was not as acidic, and therefore changed from yellow back to blue.

13. Justify why were *Elodea* springs placed in both test tubes 1 and 3?
    - Test tubes 1 and 3 had the same ingredients added: water, bromothymol blue, carbon dioxide, and *Elodea*, all necessary ingredients for photosynthesis. But only test tube 1 was placed in the light. This demonstrates that light is a necessary requirement to initiate *Elodea* photosynthesis. When photosynthesis begins, *Elodea* will begin using carbon dioxide and the color of the solution will turn from yellow to blue.

14. Describe the differences you observed between the *Elodea* in the light and the *Elodea* in the dark?
    - *Elodea* in the light undergoes photosynthesis and *Elodea* in the dark does not.

15. How do your results of your experiment demonstrate the requirements necessary for photosynthesis to occur?
    - All the necessary ingredients must be present for photosynthesis to occur: carbon dioxide, water, *Elodea*, and light. The experiment demonstrates, by the use of controls, that all of the ingredients must be present for photosynthesis to occur.

Test Tube	Color before Blowing	Color after Blowing	Color after 30 minutes
1	Blue	Yellow	Blue
2	Blue	Yellow	Yellow
3	Blue	Yellow	Yellow
4	Blue	Yellow	Yellow

Shed Some Light on the Problem
Exploring the Relationship between Light and Photosynthesis

All life on earth depends on photosynthesis. With the exception of the abyssal zone ecosystem, it is the most important energy-capturing process in all ecosystems. In photosynthesis, plants use the energy from sunlight to make food (glucose) by combining carbon dioxide and water. The glucose made by plants is used by the plants and animals that eat them as a source of energy.

When a plant undergoes photosynthesis, carbon dioxide is absorbed and oxygen is released. Humans and other non-photosynthetic organisms release carbon dioxide as a waste product. (Humans exhale carbon dioxide when they breathe.) When carbon dioxide is introduced into water, it dissolves to form a weak acid, carbonic acid. Bromothymol blue indicator, can indicate the presence of carbonic acid. When little or no carbonic acid is present in a solution, bromothymol blue will show a blue color. Depending upon the amount of carbon dioxide that is added to the solution, bromothymol blue will change to green or yellow. The more yellow the color the greater the concentration of carbonic acid present. As a result, scientists can use an acid-base indicator such as bromothymol blue to indicate the presence or absence of dissolved carbon dioxide. Today you will use bromothymol blue as an indicator to test whether plants need light to perform photosynthesis.

PURPOSE
In this activity you will determine if light is necessary is needed to initiate photosynthesis in the aquatic plant, *Elodea*. You will conduct a scientific study, paying close attention to controls, constants, and variables.

MATERIALS
4 test tubes (16mm x 150mm) with stoppers	test tube rack
wax pencils	deionized water in 250 mL beaker
bromothymol blue solution (0.04%)	4 coffee straws
Elodea sample in deionized water	graduated cylinder (10 mL)
100 mL beaker (for bromothymol blue)	metric ruler

Safety Alert
1. Wear safety goggles while other table members are blowing through the straw.
2. Do not inhale the bromothymol blue as you are gently blowing into the test tube.
3. Blow very gently into your test tube avoiding splashing so you do not get any of the dye on your clothes or hands.

PROCEDURE

1. Your teacher will assign each person at your table a number, 1-4. Each person will then be responsible for preparing one of the test tubes as described below. Each person is required to do complete their own student answer page.

2. In the space marked HYPOTHESIS on your student answer page, use your previous knowledge about photosynthesis to write a hypothesis answering the question, how does light affect Elodea's ability to undergo photosynthesis?

3. Label the test tubes 1-4 with the wax pencil.

4. Using your graduated cylinder, measure out and transfer 20.0 mL of deionized water to each test tube.

5. Using the graduated cylinder, transfer 2.0 mL of bromothymol blue solution to each test tube. Record the color of each test tube solution in the data table on your student answer page in the space labeled "Color before blowing".

6. Using a coffee straw, gently blow into the water/bromothymol blue solution in each test tube until the solution changes color completely. *Be extremely careful not to inhale the solution or to blow so hard that the mixture bubbles over the side of the test tube.*

7. Observe your test tubes and determine what color of the solution is in each test tube and record that color under the "Color after Blowing" column in the data table on your student answer page.

8. Persons 1 and 3, using your metric ruler, carefully measure and cut two 10.0 cm pieces of Elodea from the sample at your table.

9. Place one piece of Elodea into test tube 1, and one piece into test tube 3. Put the stoppers on all of the test tubes.

10. Place test tubes 1 and 2 in bright light (either under the grow lamp or in the direct sun) and test tubes 3 and 4 in the dark as your teacher instructs for 30 minutes.

11. While you are waiting, begin answering the conclusion questions on your student answer page.

12. After 30 minutes, observe the test tubes and record the color of each solution.

13. Complete the conclusion questions on your student answer page. Consult the introduction and your data table as needed to answer the questions.

Name _____

Period _____

Shed Some Light on the Problem
Exploring the Relationship between Light and Photosynthesis

HYPOTHESIS

DATA AND OBSERVATIONS

Test Tube	Color before Blowing	Color after Blowing	Color after 30 minutes
1			
2			
3			
4			

CONCLUSION QUESTIONS

1. What is the independent variable in this experiment?

2. What is the dependent variable?

3. Describe 2 constants in the experiment.

4. What is the purpose of setting up test tubes 2 and 4?

5. Why did we blow through the straw into each tube?

6. Describe the color of the bromothymol blue solution in each test tube BEFORE you exhaled into it?

7. Explain what the color of the solution in the test tube tells us (the answer to number 6)?

8. Describe the color change in the bromothymol blue solution in each test tube AFTER you exhaled into it?

9. Explain what this color change tells us (the answer to number 8)?

10. Why did we use bromothymol blue in this experiment?

11. After the 30 minute testing time, which of the test tubes changed color?

12. Why did the test tube(s) that changed color do so?

13. Justify why were *Elodea* springs placed in both test tubes 1 and 3?

14. Describe the differences you observed between the *Elodea* in the light and the *Elodea* in the dark?

15. How do your results of your experiment demonstrate the requirements necessary for photosynthesis to occur?

Antacid Analysis
Neutralizing Stomach Acid

TEACHER PAGES

OJECTIVE
Students will determine which brand of antacid neutralizes stomach acid the most effectively and is the most cost efficient.

LEVEL
Middle Grades: Life Science

NATIONAL STANDARDS
UCP.1, UCP.2, UCP.3
C.5

TEKS
6.1(A), 6.2(B), 6.2(C), 6.2(D), 6.2(E), 6.3(A), 6.3(B), 6.4(A)
7.1(A), 7.2(B), 7.2(C), 7.2(D), 7.2(E), 7.3(A), 7.3(B), 7.4(A), 7.9(A)
8.1(A), 8.2(B), 8.2(C), 8.2(D), 8.2(E), 8.3(A), 8.3(B), 8.4(A), 8.4(B), 8.9(A)
IPC 1(A), 2(A), 2(B), 2(C), 2(D)

CONNECTIONS TO AP
AP Biology:
   III. Organisms and Population B. Structure and Function of Plants and Animals 2. Structural, physiological, and behavioral adaptations
AP Chemistry:
   III. Reactions A. Reaction types 1. Acid-base reactions

TIME FRAME
45 minutes

MATERIALS
(For a class of 28 working in pairs)

various brands of antacids	7 pestle and mortars / one for each group of four
14 50 mL graduated cylinders	or waxed paper and a rolling pin or hammer
28 50 mL beakers	1.5 L of 0.1 M HCl solution
distilled water	14 glass stirring rods
28 thin stem pipets	42 (3 antacids x 14 groups) coffee filters
14 dropper bottles of phenolphthalein indicator solution	14 12-well microplates
	14 dropper bottles of 0.01 M NaOH solution

TEACHER NOTES

The set up for *Antacid Analysis* is based on 28 students working in pairs with four students at each table. Each pair of students will test three different brands of antacids.

If dropper bottles are not available for the indicator solution, you can substitute with test tubes and thin stem pipets. When the solutions are not in use, cover them with plastic wrap or Parafilm or a cork so evaporation does not occur.

If 28 beakers of the same size are not available; smaller or larger ones can be used. If beakers are not available at all polystyrene or plastic cups may be substituted.

Students will need the following information from the antacid boxes: tablets per dose, cost of package, names of active ingredients, and amount of active ingredient. Include this information on the board or overhead if antacid boxes are not available.

You might need to take the opportunity to explain what the "M" represents in the materials and introduction section of the lab. The students do not need to know how to calculate it, but they do need to understand that the "M" represents molarity which is a measure of the concentration of the acid. The higher the "M", the more concentrated the solution.

Preparation of solutions:
0.01 M HCl acid (1.0 Liter): Fill a liter container (the most accurate you have available, graduated cylinder, volumetric flask, etc) with about 500 mL of distilled water, then add 12.5 mL of concentrated (12M) HCl to the water. Bring the solution to a total volume of 1.0 L. Alternately, 0.1M HCl acid can be purchased from a laboratory supply company if you choose not to buy concentrated HCl.

0.01 M NaOH solution (500 mL): measure out 0.2 grams of solid NaOH, place in a 500 mL volumetric flask and add distilled water to the 500 mL mark.

Taking It Further:
Students can display their antacid data on poster board and present their findings to the class. As each group discusses their data, the students in class can be copying the data into a class table to better analyze the information. Once all of the data has been shared, students can return to their groups and discuss which antacid is the best at neutralizing acid and the most cost effective.

You may want to prepare an acid/base solution that has been neutralized so that students can see the color pink you wish for them to reach. They will need to know that when the solution turns a slight pink and remains that way, the solution has been neutralized.

PRELAB QUESTIONS

Use the antacid box, Internet, or other sources of information to answer the pre-lab questions:

1. What is the active ingredient in your antacid? Give the chemical name and chemical formula.
   - The most common substances found in antacids are hydroxides (such as magnesium hydroxide $Mg(OH)_2$ or aluminum hydroxide $Al(OH)_3$), calcium carbonate $CaCO_3$, and sodium bicarbonate $NaHCO_3$.

2. What is the job of an antacid?
   - Antacids are used to neutralize the excess acid in the stomach.

3. Which type of acid is predominantly found in your stomach?
   - HCl, hydrochloric acid

4. What is the approximate pH of the acid in your stomach?
   - pH 1

5. What is the job of the acid found in your stomach?
   - The job of stomach acid is to digest food and provide the proper pH for enzyme function.

6. What is the role of an indicator solution?
   - An indicator solution is used to determine when a change in pH has taken place. Phenolphthalein is often used in acid base titrations since it turns from colorless in acid solution to pink in basic solution. This allows you to visually determine when neutralization has occurred.

7. Define endpoint.
   - This is the point at which the neutralized solution begins to just turn pink.

8. Define neutralization reaction. In this definition include the two types of compounds produced.
   - A neutralization reaction is a reaction between an acid and a base to produce water and a salt.

9. What is an ionic compound?
   - A compound involving a positive and negative ion or a metal and nonmetal.

POSSIBLE ANSWERS TO THE CONCLUSION QUESTION AND SAMPLE DATA

Antacid Brand	# of drops	Tablets/dose	Cost/dose	Active Ingredients
Rolaids				$NaAl(OH)_2CO_3$
Walmart Brand				
Tums				$CaCO_3$

1. Which brand of antacid was the most effective at neutralizing the stomach acid? Explain your answer.
   - Answers will vary based on the types of antacids selected.
   - Explanation should address that the most effective antacid was the one whose solution required the LEAST additional drops of base (NaOH).

2. Which brand of antacid was the least effective at neutralizing the stomach acid? Explain your answer.
   - Answers will vary based on the types of antacids selected.

- Explanation should address that the least effective antacid was the one whose solution required the MOST additional drops of base (NaOH).

3. List at least two possible sources of error in your lab and explain how these sources of error would specifically alter your results.
   - Answers will vary but may include
     - o Inaccurate counting of drops. If the numbers of drops are not recorded accurately, then the most effective or least effective antacid reported could also change.
     - o Size of the drops used was not consistent. Again, this might have changed the most effective or least effective antacid.
     - o Failure to pulverize the antacid tablets or allow them to dissolve as completely as possible. If the tablets are not pulverized, all of the active ingredient may not be released to neutralize the acid. It may be discarded upon filtration.

4. Cindy was an assistant for you and your partner in the laboratory. You asked her to help with some of the antacid testing to speed up the collection of data. Everything was going great until she noticed that on her last set of tests she had accidentally placed 3 drops of indicator solution into the acid instead of 1 drop. Would this affect the results? Explain your answer.
   - Adding additional indicator solution will not affect the final results. Indicator is simply a solution used to determine when the neutralization reaction has been achieved; it is not part of the reaction.

5. Based on the results of your three antacids, which antacid should be the most cost effective? You will need to analyze the cost per dose and effectiveness per dose and determine which is the best brand to buy based on how well it works and how much it cost. Show all calculations below.
   - The students will need to analyze the cost, how much active ingredient per dose, and how well the antacid worked and create their own method for determining cost effectiveness.

Antacid Analysis
Neutralizing Stomach Acid

The human stomach regularly secretes acid to help digest food. Hydrochloric acid of 0.1 M is involved in this important process. Although acid plays an important role in our digestive system, it is also a corrosive substance. For this reason, your stomach has a mucosal lining which keeps the acid from coming in direct contact with the sensitive stomach tissue.

Sometimes excess acid builds up in the stomach or esophagus and causes painful indigestion or heartburn. If the acid gets too excessive, it can lead to a stomach ulcer or a hole in the stomach lining. Current research, however, indicates that many ulcers are caused by bacterial infections not acid build-up. Between 10 to 20 percent of Americans suffer from ulcers at some point in their lives.

For those who suffer from acid build-up, antacids are available. These medicines contain bases which remove excess acid through a neutralization reaction. A neutralization reaction is the reaction between an acid and a base to produce water and an ionic compound also known as a salt.

$$HCl + NaOH \rightarrow H_2O + NaCl$$

PURPOSE
You and your partner have been hired by the Food and Drug Administration to test various antacids and determine which one is the best at neutralizing stomach acid and the most cost effective. In order to accomplish this task, you will need to determine how much antacid (base) is needed to neutralize 0.1 M hydrochoric (HCl) acid. As a knowledgeable biologist, you know that an indicator must be used to determine when neutralization has occured. The indicator you have chosen is phenophthalein because it turns pink when neutralization has taken place. The point at which the solution turns pink is known as the end point.

MATERIALS
various brands of antacids
100 mL graduated cylinder
2 50 mL beakers
distilled water
1 thin stem pipet
1 dropper bottle of phenolphthalein indicator
   solution

1 pestle and mortar / one for each group of four
or waxed paper and rolling pin or hammer
0.1 M HCl solution
1 glass stirring rod
3 coffee filters
1 12 well microplate
1 dropper bottle of NaOH

Safety Alert
1. Wear safety goggles and aprons throughout the entire lab.
2. 0.1 M HCl is slightly toxic by ingestion. It is also corrosive so avoid tissue contact. If contact does occur, wash thoroughly and inform your teacher.
3. Phenophthalein is a tissue irritant. Wash thoroughly if contact occurs.
4. 0.01 NaOH is slightly toxic by ingestion and skin absorption. It is a skin irritant and causes eye burns. If contact does occur, was thoroughly and inform the teacher.

PROCEDURE

1. Use the antacid box, Internet, or other sources of information to answer the pre-lab questions on your student answer page.

2. Obtain three different brands of antacids from the supply table.

3. Based on any experiences you have had with antacids, hypothesize as to which antacid you believe will be the best at neutralizing stomach acid. Record your hypothesis in the space provided on your student answer page.

4. Use the graduated cylinder to measure 25.0 mL of the 0.1 M HCl. Pour the acid into a 50 mL beaker. Be sure to read the volume from the bottom of the meniscus.

5. Pulverize **1 dose** of an antacid. Make sure you read the antacid box to determine the number of tablets in one dose. Record the brand name of the antacid, dose amount, and cost in the data table. Pour the antacid into the beaker with the acid and stir thoroughly using a glass stirring rod. The antacid will neutralize only a portion of the acid. You will add NaOH to neutralize any remaining acid. The antacid that requires the least amount of NaOH neutralizes best.

6. Rinse the graduated cylinder with distilled water two times. Measure 10.0 mL of distilled water and pour it into the beaker with the acid and antacid. Use a stirring rod to mix the solution.

7. Place a coffee filter over the top of another beaker and secure it with your hand. Your partner will slowly pour the antacid solution through the coffee filter making sure not to spill any of the solution on your hand. If this does occur, wash your hands immediately. Dispose of as directed by your teacher.

8. Add one drop of indicator solution to the filtered solution.

9. Using a thin stem pipet add 10 drops of the acid solution into one of the wells on your microplate.

10. Using the dropper bottle labeled NaOH (sodium hydroxide), add one drop at a time to the acid solution in the microplate until the endpoint is reached. You will know when you have reached the endpoint because the solution will remain pink.

11. Record the number of NaOH drops needed to neutralize the acid solution in the data table.

12. Repeat steps 9-11 two more times in separate microplate wells.

13. Average the number of drops needed for the three trials. Record the average to two significant figures.

14. Repeat steps 4-13 with the other two antacids.

15. Once you have completed the lab, dispose of the materials as directed by your teacher. Make sure to clean the microplate with a test tube brush to rid the container of any solution residue.

Name _____

Period _____

Antacid Analysis
Neutralizing Stomach Acid

HYPOTHESIS

PRELAB QUESTIONS

1. What is the active ingredient in your antacid? Give the chemical name and formula.

2. What is the job of an antacid?

3. Which type of acid is predominantly found in your stomach?

4. What is the approximate pH of the acid in your stomach?

5. What is the job of the acid found in your stomach?

6. What is the role of an indicator solution?

7. Define endpoint.

8. Define neutralization reaction. In this definition include the two types of compounds produced.

9. What is an ionic compound?

DATA AND OBSERVATIONS

Antacid Brand	# of drops	Tablets/dose	Cost/dose	Active Ingredients

CONCLUSION QUESTIONS

1. Which brand of antacid was the most effective at neutralizing the stomach acid? Explain your answer.

2. Which brand of antacid was the least effective at neutralizing the stomach acid? Explain your answer.

3. List at least two possible sources of error in your lab and explain how these sources of error would specifically alter your results.
   a.

   b.

4. Cindy was an assistant for you and your partner in the laboratory. You asked her to help with some of the antacid testing to speed up the collection of data. Everything was going great until she noticed that on her last set of tests she had accidentally placed 3 drops of indicator solution into the acid instead of 1 drop. Would this affect the results? Explain your answer.

5. Based on the results of your three antacids, which antacid should be the most cost effective? You will need to analyze the cost per dose and effectiveness per dose and determine which is the best brand to buy based on how well it works and how much it cost. Show all calculations below.

21

A Fishy Tale
Observing the Circulatory System of a Goldfish with a Compound Light Microscope

OBJECTIVE

Students will compare and contrast the structure and function of an artery, vein, and capillary. Students will then observe blood flowing through arteries, veins, and capillaries in the tail of a live goldfish.

LEVEL

Middle Grades: Life Science

NATIONAL STANDARDS

UCP.1, UCP.2, UPC.5, A.1, A.2, C.1, C.5

TEKS

6.1(A), 6.2(A), 6.2(B), 6.2(C), 6.2(D), 6.2(E), 6.3(A), 6.3(C), 6.4(A), 6.5(B), 6.10(A), 6.10(B), 6.10(C), 6.12(B)

7.1(A), 7.2(A), 7.2(B), 7.2(C), 7.2(D), 7.2(E), 7.3(A), 7.3(C), 7.4(A), 7.4(B), 7.6(C), 7.9(A), 7.11(B)

8.1(A), 8.2(A), 8.2(B), 8.2(C), 8.2(D), 8.2(E), 8.3(A), 8.3(C), 8.4(A)

IPC: 1(A), 2(A), 2(B), 2(C), 2(D)

CONNECTIONS TO AP

AP Biology

    I. Molecules and Cells B. Cells 2. Membranes

    I. Molecules and Cells C. Cellular Energetics 2. Fermentation and cellular respiration

    III. Organisms and Populations B. Structure and Function of Plants and Animals 1. Reproduction, growth, and development 2. Structural, physiological, and behavioral adaptations

TIME FRAME

90 minutes

MATERIALS

(For a class of 28 working in pairs)

14 computers with internet access	14 beakers of aquarium water
14 disposable transfer 1 mL pipets	14 cotton balls
14 goldfish	14 compound light microscopes
14 sets of colored pencils	14 microscope cover slips
14 petri dishes	

TEACHER NOTES

A Fishy Tale provides students with the opportunity to explore the circulatory system using an Internet exercise and then the opportunity to directly observe and explore the circulatory system in a goldfish tail using a microscope. This lab could serve as an introduction to the circulatory system and typically would be completed within a body system unit.

The students are expected to make a biological diagram. The lesson, *Picturing Life: Making a Biological Diagram*, found elsewhere in this book, should be completed prior to performing this exercise.

It is absolutely imperative for you to practice the goldfish observation component of the lab before trying with students. It is highly recommended that you demonstrate the proper wrapping procedure and placement in the Petri dish for students.

The goldfish, commonly called "feeder goldfish", can be obtained from any pet store for a minimal cost, probably $0.10-0.30 a fish. The fish need to be kept in a tank of spring (or de-chlorinated) water with an aeration device (available at most pet stores for less than $10). If kept in non-aerated water, the fish will live only a few hours. If kept in aerated water, the fish will survive well over a week. It is recommended that you purchase more fish than you think you will need. Students, even acting very carefully and ethically, can kill the goldfish during the observation due to the intense heat of the microscope. (Of course, prior to any lab that uses live organisms, a discussion of your expectations regarding ethical and humane treatment of the animals is mandatory at the junior high/middle school level.)

Please refer to Figure 2 in the student procedure for the correct set up of the goldfish in the Petri dish. Also refer to Figure 1, a photograph taken through the microscope lens at 100 X magnification showing the correctly labeled parts of the fish tail. However, your best guide for learning the logistics of this lab is to place the goldfish in the Petri dish and directly observe the blood coursing through the vessels of the tail using your own microscope.

If necessary Parts I and II may be completed on separate days. Your students will really enjoy this lab. You will find that seeing the individual blood cells moving through the capillaries will not be easily forgotten by anyone, including you!

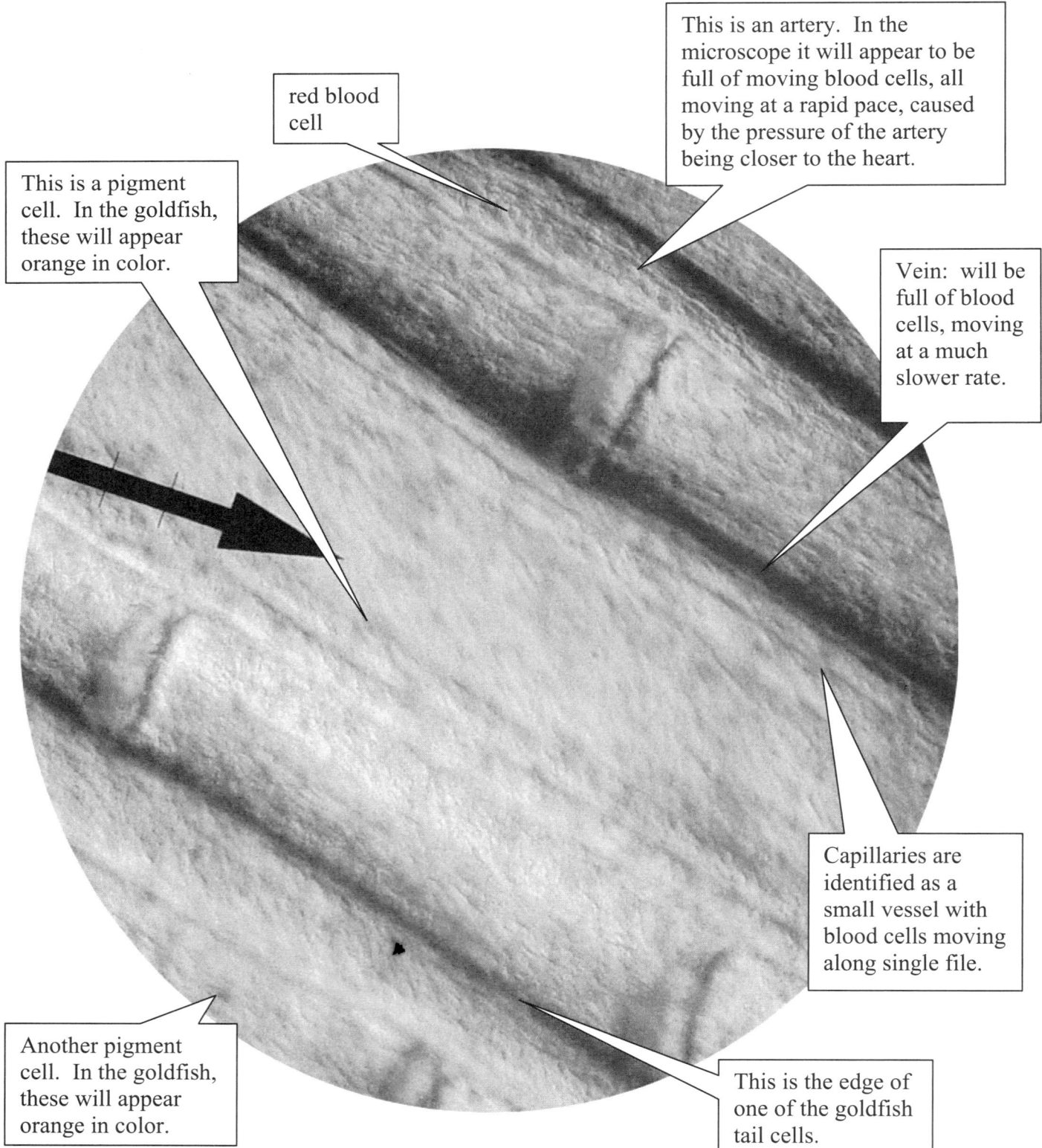

Figure 1: Photo taken through the microscope eyepiece lens, viewing the goldfish tail at 100X magnification.

POSSIBLE ANSWERS TO THE CONCLUSION QUESTIONS
Part I: Exploring the Circulatory System via the Internet

1. What is the responsibility of the cardiovascular and respiratory systems?
   - The cardiovascular and respiratory systems together are responsible for carrying oxygen from the air to the blood stream and expelling the waste product of carbon dioxide.

2. Why is blood such an essential fluid to the cardiovascular and respiratory systems?
   - Blood carries oxygen (to the cells) and carbon dioxide (from the cells) throughout the body, through the vessels from the heart.

3. How do arteries differ from veins?
   - Arteries carry blood filled with nutrients away from the heart to all parts of the body.
   - Arteries are thick-walled tubes.
   - Arteries contain a filling of muscle that absorbs the tremendous pressure wave of a heartbeat.
   - Veins deliver deoxygenated blood back to the heart.
   - Veins, unlike arteries, have thin, slack walls.

4. How does the function of capillaries fit into the cardiovascular system?
   - Capillaries join arteries to veins. Also, the blood's work is done in the capillaries when it gives up what the cells need (primarily O_2) and takes away the waste products (primarily CO_2) that they don't need.

5. How does the pressure of the blood in the veins compare to the pressure of the blood in the arteries?
   - Blood from the heart is under much greater pressure and moves much more quickly away from the heart. As deoxygenated blood moves back to the heart, it is under much less pressure and "oozes" much more slowly.

6. How many square miles do capillaries cover in our bodies?
   - Capillaries provide a total surface area of 1,000 square miles.

7. Why do you think there are so many square miles of capillaries?
   - Capillaries are the site of cellular respiration and gas exchange in the cells. To provide the oxygen and carbon dioxide exchange in so many cells requires a huge amount (and length) of capillary vessels.

8. How much more pressure does an artery have than a vein?
   - The pressure of an artery is roughly fifty times more than a vein.

9. Why do you think that the pressure in an artery is so much greater than in a vein?
   - The artery is taking blood directly from the heart, and as this system of tubes is closer to the heart, receives more pressure than the veins, which are taking blood back to the heart and lungs and have less pressure.

10. As the blood in the arteries approaches the capillaries, what gas is the blood rich in?
    - The blood, as it approaches the capillaries, is rich in oxygen.

11. In your own words, describe the flow of blood as it leaves the heart and moves through the various vessels until it returns to the heart. Include the following terms in your description: blood cells, heart, artery, vein, capillary, O_2, CO_2.
    - *Blood cells* flow from the *heart* in *arteries* through the body to capillaries, after dropping off O_2 and picking up CO_2 and returning to the *heart* through the *veins*.

Part II: Exploring the Circulatory System in the Fish's Tail
1. Explain how you can quickly tell the difference between the three kinds of vessels (capillary, vein and artery) in the tail of a fish.
    - Capillaries are much smaller and you can see the blood, individual cell by individual cell, moving through the capillary. Veins have much slower rate of blood movement. Arteries have much faster rate of blood movement.

2. Which type of vessel seemed to be most numerous in the tail of the fish? Explain why it is essential that there be more vessels of this kind?
    - Capillaries are found all over the tail. They are so numerous because the capillary is the primary site of cellular respiration.

3. In your observation of the fish's tail, in which blood vessels does the blood move the slowest? Why?
    - Veins have less pressure, being further removed from the heart, and cause the blood to move at a slower rate.

4. In your observation of the fish's tail, in which blood vessels does the blood move the fastest? Why?
    - Arteries have much more pressure, being closer to the heart, and cause the blood to move at a much higher rate.

5. Describe another way that you could tell the difference between the arteries and veins in the tail of the fish. (Hint: All the images that you see in the microscope are reversed!)
    - The direction of blood flow would be another way to observe the difference between arteries and veins. The blood in the arteries is coming from the anterior end of the fish, where the heart is in relation to the tail. The blood in the veins is traveling back to the heart and gills, from the posterior end toward the anterior. But all this is reversed in the compound light microscope due to optics. So the blood that, in the microscopic view, looks like it is coming at high pressure and velocity from the posterior end, from the tail! But it is really coming from the anterior end of the fish.

A Fishy Tale
Observing the Circulatory System of a Goldfish with a Compound Light Microscope

PURPOSE
In this activity you will compare and contrast the structure and function of an artery, vein, and capillary. You will then observe blood flowing through arteries, veins, and capillaries in the tail of a living fish.

MATERIALS
computer with Internet access	beaker of aquarium water
disposable transfer 1 mL pipet	cotton ball
goldfish	compound light microscope
Petri dish	microscope cover slip

Safety Alert
1. Always dry the bottom of the Petri dish completely before placing it onto the microscope.
2. Whenever you are not viewing the fish remember to turn off the microscope — too much exposure to the heat of the microscope light will kill the fish.

PROCEDURE
The lab is divided into in two parts. The first part of the experiment will be spent exploring a computer animation of the human body's circulatory system. The second part of the experiment is for you to directly observe vessels in the circulatory system of a goldfish using a microscope.

Part I: Exploring the Circulatory System on the Internet
1. Go to http://www.innerbody.com.

2. Click on "Human Anatomy Online"

3. Click on 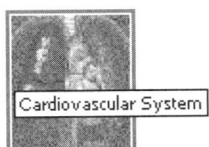 (the image for the Cardiovascular System link)

4. Scroll down below, the model of the human cardiovascular system.

5. Click on the link, Cardiovascular System Overview, and read the information that appears on the right side of the screen.

6. Click on the lower Cardiovascular System (simplified) link (there are two, one under the other) and read the information that appears on the right side of the screen.

7. Click the first Cardiovascular System (simplified) link and an overview image of the cardiovascular system will be displayed in the middle of the page.

8. Click on Animations at the bottom of the page.

9. Click the Capillary link (on the right side of the screen).

10. In the middle of the screen an animation will show the blood cells moving through the capillaries.

11. Draw, color and label a sketch of the capillary animation in the space provided on your student answer page. Show all the details of the animation, including the flow of oxygen, carbon dioxide and water. Use the chemical symbols O_2 to represent oxygen, CO_2 to represent carbon dioxide and H_2O to represent water and show how each of these are diffused in and out the capillary and the cell. Use all rules for biological diagrams when making your diagram. Title your drawing: *Model of Capillary in the Circulatory System*.

12. Answer the Part I questions on the student answer page using the animations and the text from the Human Anatomy Online web site.

Part II: Exploring the Circulatory System in the Fish's Tail

READ ALL DIRECTIONS CAREFULLY BEFORE PERFORMING THIS PROCEDURE!
Whenever you are not viewing the fish remember to turn off the microscope — too much exposure to the heat of the microscope light will kill the fish.

1. Dip the cotton ball into the aquarium water.

2. Gently pull the wet cotton ball and stretch it slightly into a flattened disk..

3. Lay the stretched and wet cotton ball into the Petri dish.

4. Use the transfer pipet to put 2 or 3 mL of aquarium water into the Petri dish.

5. Your teacher will come to your table to bring your group a fish. Try to handle the goldfish as little as possible because touching it removes its protective mucus coating.

6. Gently lay the fish onto your cotton in the Petri dish.

7. Wrap the wet cotton around the body, head, and gill area of fish so that just the tail is visible. Make sure that the head of the fish is covered completely with the cotton. The fish will move very little when secured in the cotton.

8. Place a microscope cover slip over just the tail fin, in the Petri dish. See Figure 2, showing a properly positioned fish in the Petri dish.

9. Carefully wipe any water off the bottom of the Petri dish before placing it on the microscope stage.

10. Place the Petri dish on the stage of the microscope. Adjust the diaphragm settings to allow a LOW amount of light to pass through the specimen. Too much heat generated by the light will kill your fish.

11. Turn the nosepiece until the low power objective (4X) is positioned over the tail. Focus on a thin section of the tail using the low power objective.

12. Use the coarse adjustment knob to bring your specimen into focus.

13. Carefully turn the nosepiece to the medium power objective (10X). Focus using the fine adjustment knob. Again refer to Figure 2 to make sure you have the Petri dish, fish, cotton ball and cover slip set up correctly.

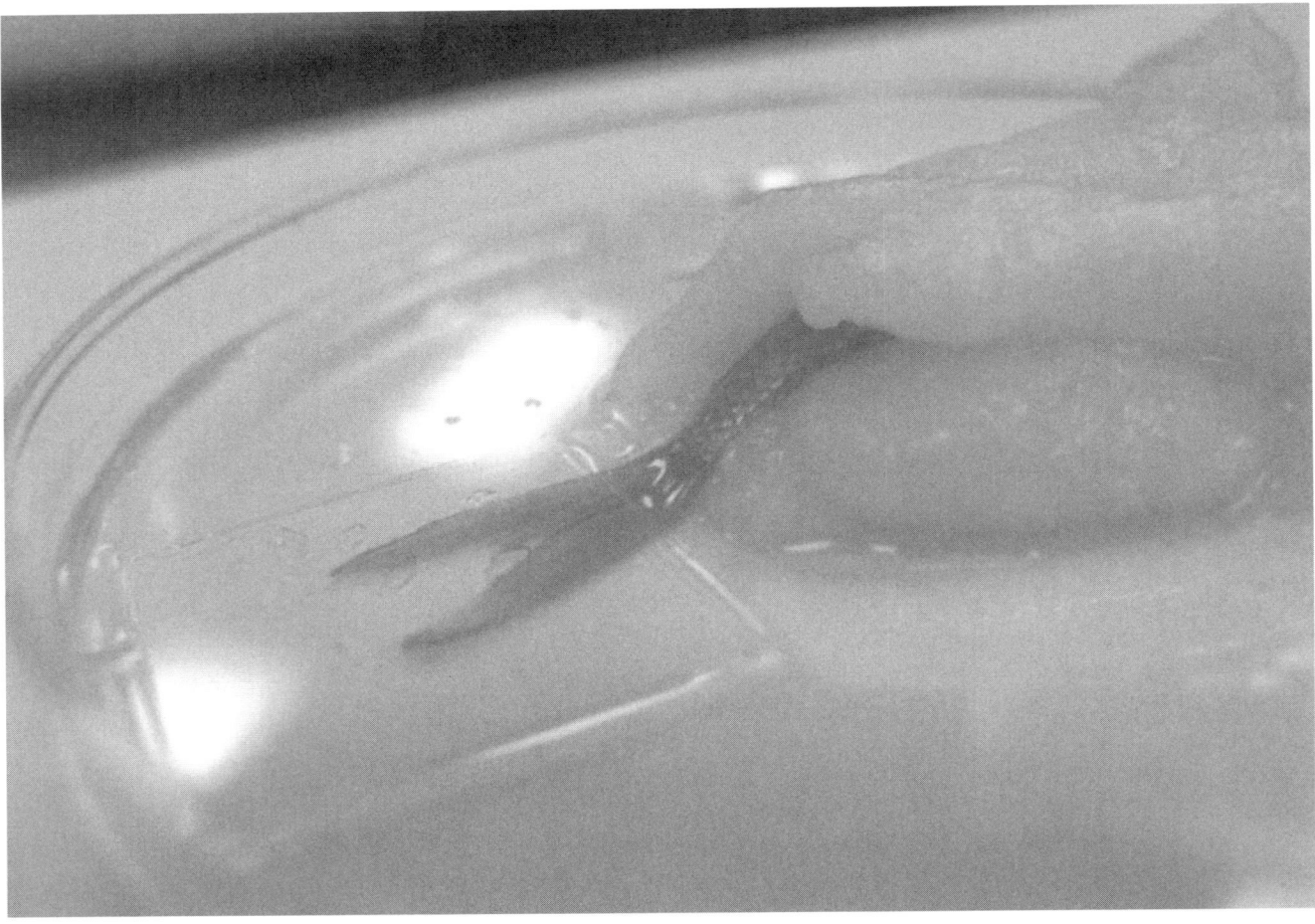

Figure 2: This photo shows a fish properly positioned in the Petri dish, securely wrapped in the cotton, and ready to be observed with the compound light microscope. Be sure there is plenty of aquarium water saturating the cotton and the cover slip over the tail.

14. Closely observe the blood flowing through the blood vessels. Determine which vessels are arteries, which are veins, and which are capillaries. Make sure that when you or your partner are not directly observing the fish tail that you turn OFF the light to avoid overheating the fish. Work carefully but quickly to make sure that your fish survives this procedure.

15. Use colored pencils to draw a section of the tail as it appears under 100X magnification (or medium power) in the space provided on the student answer page. Follow all rules for creating a biological diagram and indicate labels for: an artery, a vein, and a capillary. Be sure to add as much realism and detail as you can. The orange dots that you see in the tail are pigment cells and not blood cells.

16. As soon as you and your partner have observed and made diagrams of the capillaries/arteries/veins, return the fish to the aquarium water.

17. Using your knowledge about the circulatory system from the website and your observations in the microscope, answer the Part II conclusion questions on your student answer page.

Name _____

Period _____

A Fishy Tale
Observing the Circulatory System of a Goldfish with a Compound Light Microscope

BIOLOGICAL DIAGRAM-PART I

BIOLOGICAL DIAGRAM-PART II

CONCLUSION QUESTIONS

Part I: Exploring the Circulatory System on the Internet

1. What is the responsibility of the cardiovascular and respiratory systems?

2. Why is blood such an essential fluid to the cardiovascular and respiratory systems?

3. How do arteries differ from veins?

4. How does the function of capillaries fit into the cardiovascular system?

5. How does the pressure of the blood in the veins compare to the pressure of the blood in the arteries?

6. How many square miles do capillaries cover in our bodies?

7. Why do you think there are so many square miles of capillaries?

8. How much more pressure does an artery have than a vein (a %)?

9. Why do you think that the pressure in an artery is so much greater than in a vein?

10. As the blood in the arteries approaches the capillaries, what gas is the blood rich in?

11. In your own words, describe the flow of blood as it leaves the heart and moves through the various vessels until it returns to the heart. Include the following terms in your description: blood cells, heart, artery, vein, capillary, O_2, CO_2.

Part II: Exploring the Circulatory System in the Fish's Tail!

1. Explain how you can quickly tell the difference between the three kinds of vessels (capillary, vein and artery) in the tail of a fish.

2. Which type of vessel seemed to be most numerous in the tail of the fish? Explain why it is essential that there be more vessels of this kind?

3. In your observation of the fish's tail, in which blood vessels does the blood move the slowest? Why?

4. In your observation of the fish's tail, in which blood vessels does the blood move the fastest? Why?

5. Describe another way that you could tell the difference between the arteries and veins in the tail of the fish. (Hint: All the images that you see in the microscope are reversed!)

How Does Your Heart Rate?
Examining the Effect of Exercise on Heart Rate

OBJECTIVE
Students will investigate the effect of exercise on heart rate. They will graph their data and determine if there is a correlation between exercise and increased heart rate.

LEVEL
Middle Grades: Life Science

NATIONAL STANDARDS
UCP.1, UCP.2, UPC.5, A.1, A.2, C.1, C.5

TEKS
6.1(A), 6.2(A), 6.2(B), 6.2(C), 6.2(D), 6.2(E), 6.3(A), 6.3(C), 6.4(A), 6.5(B)
7.1(A), 7.2(A), 7.2(B), 7.2(C), 7.2(D), 7.2(E), 7.3(A), 7.3(C), 7.4(A), 7.4(B)
8.1(A), 8.2(A), 8.2(B), 8.2(C), 8.2(D), 8.2(E), 8.3(A), 8.3(C), 8.4(A)
IPC: 1(A), 2(A), 2(B), 2(C), 2(D)

CONNECTIONS TO AP
III. Organisms and Populations B. Structure and Function of Plants and Animals 1. Reproduction, growth, and development 2. Structural, physiological, and behavioral adaptations

TIME FRAME
45 minutes

MATERIALS
(For a class set of 28 working individually)

28 graphing calculator   28 timing devices with seconds

TEACHER NOTES
Prior to the day of this exercise, students should be told to wear comfortable clothing and tennis shoes to class. You might suggest girls avoid wearing a skirt or dress.

If possible, demonstrate to students what you mean by "jogging" in place. They tend to go over board with the jogging and then the data is extremely skewed. Discuss controlling this portion of the lab so that the data will be more uniform.

You may want to use an LCD panel and go through the graphing portion with the students if this is their first time to use graphing calculators.

Use the chalkboard or overhead to provide a data table for students to record their average heart rate during this exercise. This data table will need to be divided into girls versus boys.

Take the time to practice taking pulses. Students have a difficult time finding the pulse in their necks. If you prefer, use the wrist pulse.

ANSWERS TO CONCLUSION QUESTIONS AND SAMPLE DATA

Seconds	Heartbeat/10 seconds
30	16
60	25
90	30
120	35
150	40
180	45
R =	.993

POSSIBLE ANSWERS TO THE CONCLUSION QUESTIONS

1. What is the numerical value for your y-intercept (b)_____, slope (a) _____.
   - y intercept = 12.33 and the slope = .186

2. What happened to your heart rate as the time increased?
   - It increased

3. Would time versus heart rate be a direct or indirect relationship? Explain your answer.
   - A direct relationship. The more you run, the more your heart rate increases.

4. Another name for a linear regression is a "line of best fit". The correlation coefficient (r) gives an indication of how well your points fit on the mathematical line. When r = 1 or –1 there is a perfect fit and all data points will lie on the mathematical line. Using your correlation coefficient (r), how well does your data fit your line? Explain your answer.
   - Answers will vary. However, for the most part students should see a steady increase in their heart rate. Any value close to r = 1.00 would be considered a good fit and therefore indicating a positive correlation between exercise and increasing heart rate.

5. Look at the class data (boys vs. girls), is there a difference in the average heart rate? If so, explain. If not, explain.
   - Usually there is a difference between the boys versus the girls. Most of the time the boys had slower heart rates. This can lead to discussion on how many of the boys were in sports versus the girls.

6. According to the American Council of Exercise (ACE), a fit heart will beat less times per minute than an unhealthy heart. Explain.
   - A healthier heart can beat less per minute because it is able to pump more blood with less time. In other words, the heart muscle is healthy and can do the job more efficiently per beat.

7. List at least one variable that was not controlled in our experiment and how this could affect the accuracy of the results.
   - There could be many different answers here: style of jogging, different clocks used, etc… If one person was jogging at a much higher pace than the rest of the class, their heart rate will be exceptionally high skewing the class data as well as the boys versus girls data.

How Does Your Heart Rate?
Examining the Effect of Exercise on Heart Rate

We often hear in the news, "eat healthy and exercise more". The reason everyone should exercise is not necessarily to lose weight, but to strengthen their hearts. The heart is a muscle and needs to be exercised just like the skeletal muscles in our bodies.

PURPOSE

In this activity you will determine the effects of exercise on heart rate.

MATERIALS

graphing calculator

timing device with seconds
writing utensil

Safety Alert
1. Watch your heart rate carefully. If you feel dizzy or short of breath at any time, please stop the experiment.

PROCEDURE

1. In the space marked HYPOTHESIS on your student answer page, write a hypothesis statement in if-then format that explores the effect of jogging on your heart rate.

2. Each member of the pair should collect their own data. Complete the lab with one partner and then continue through the steps repeat the procedure with the second partner.

3. Practice locating your pulse by placing your index and middle fingers over the carotid artery in the neck. This artery should be located just under the jawbone on either side of the throat. When you can reliably locate your pulse you are ready to begin data collection. A 10 second pulse rate will be determined every 30 seconds over a 3 minute period of jogging in place.

4. When you are ready, note the time on a clock or watch and jog in place for 30 seconds. After 30 seconds, immediately find your pulse on your neck. Watch the clock and count how many times you feel your pulse beat in 10 seconds. Your partner can watch the clock if it is too difficult for you to count and watch the clock.

5. Record your data in the data table on your student answer page.

6. Once your pulse has been record, resume jogging for another 30 seconds. After this interval stop and record your 10 second pulse rate. Continue this procedure until you have jogged a total of 180 seconds (3 minutes).

7. Partner #2 should repeat steps 2-6.

8. When both partners have completed their trials, record your information on the class data table as your teacher instructs.

GRAPHING CALCULATOR PROCEDURE

This portion of the lab will be done individually since you will each have different data.

1. Press the [STAT] key on your calculator. Go to EDIT. If there is information in the lists, it must be cleared. If there is no information in your list then skip step 10 and proceed to step 11.

2. To clear the list, arrow up until your cursor is on the name of the list (L_1, L_2, …). Press [CLEAR] and then [ENTER]. Do this for both L_1 and L_2.

3. Enter your number of seconds under the $L_{.(}$. (30, 60, etc…)

4. Enter your 10 second pulse in L_2. (Each partner should have different values).

5. To set up your graphs, press [2nd][Y=] to access the STATPLOT features. Arrow down to choice 4 and turn off all plots, press [ENTER]. Return to the STATPLOT menu with [2nd][Y=], position your cursor over Plot 1 and press [ENTER]. Arrow down until the cursor is on ON and press [ENTER]. Move down a row. Select the first choice, ⌐⋅⋅, press [ENTER]. Be sure that Xlist is L_1 and Ylist is L_2. Arrow to the next row and choose a point protector style.

6. Before continuing, make sure that all Y = functions have been cleared. Press [Y=] and clear any functions which may have been previously entered there. To do this, position your cursor atop the = sign and press [CLEAR].

7. To automatically scale the window and axes to fit your data, the ZOOMSTAT feature can be used. Press [ZOOM] and arrow down to ZOOMSTAT (#9). Take a look at your graph, does it look linear in nature?

8. To investigate how well your data fits a linear function (y=ax+b), let your calculator perform a linear regression and examine the correlation coefficient, r. Press [STAT] and arrow over to CALC. Arrow down to #5 LinReg and press [ENTER]. This will paste the LinReg function on your screen. To execute the function, press [ENTER] again. Your calculator should now produce a line of best fit in the form of y = ax + b for the data which you have in your lists.

9. Record your correlation coefficient (r) in the data table. If your calculator did not display a value for "r", you will need to turn this feature on and repeat step 15. To turn on this function go to the calculator's catalog ([2nd][0]) arrow down to DiagnosticsOn and press [ENTER] then [ENTER] again. After turning the DiagnosicsOn, repeat step 14 to obtain the r value for your data table.

Name _____

Period _____

How Does Your Heart Rate?
Examining the Effect of Exercise on Heart Rate

HYPOTHESIS

DATA AND OBSERVATIONS

Seconds	Heartbeat/10 seconds
30	
60	
90	
120	
150	
180	
R =	

CONCLUSION QUESTIONS

1. What is the numerical value for your y-intercept (b)_____, slope (a) _____.

2. What happened to your heart rate as the time increased?

3. Would time versus heart rate be a direct or indirect relationship? Explain your answer.

4. Another name for a linear regression is a "line of best fit". The correlation coefficient (r) gives an indication of how well your points fit on the mathematical line. When r = 1 or –1 there is a perfect fit and all data points will lie on the mathematical line. Using your correlation coefficient (r), how well does your data fit your line? Explain your answer.

5. Look at the class data (boys vs. girls), is there a difference in the average heart rate? If so, explain. If not, explain.

6. According to the American Council of Exercise (ACE), a fit heart will beat less times per minute than an unhealthy heart. Explain.

7. List at least one variable that was not controlled in our experiment and how this could affect the accuracy of the results.

Human Computer
Simulating the Internal Operations of a Computer

OBJECTIVE
The students will be able to name the six main function areas inside a computer and simulate the internal operations of a computer completing a problem. In addition the students will be able to compare and contrast the difference between human and computer parts that perform input, output, process, and storage functions.

LEVEL
Middle Grades: Life Science

NATIONAL STANDARDS
E.1, E.2

TEKS
6.3(C)
7.3(C)
8.3(C)

TIME FRAME
60 - 75 minutes

MATERIALS
For a class of 28 working in groups of 7 (1 demo and 2 timed runs per group)

computer opened to show CPU	1 stop watch
2 black markers	28 sample memory chips (optional)
28 exercise rules	28 exercise scripts
8 red 5 x 8 cards	2 blue 5 x 8 cards
20 yellow 5 x 8 cards	printer paper

TEACHER NOTES
Prepare the 5 x 8 cards ahead of time. Construct 1 set of red cards numbered 2 through 9. Construct 1 set of blue cards marked with x and +. The yellow cards stay blank.

Prepare the Scripts ahead of time by labeling sets for each team. (For example, six scripts could have an "A" at the top, 6 have a "B" at the top, and so on.)

Suggested teaching procedures

Give the following introduction:

Q: How has technology affected your life?
A: Cars, planes, televisions, music, computers, movies, etc.

Q: How have computers affected your life?
A: Makes writing papers faster and more accurate, makes math calculations faster and more accurate, provides access to the Internet, computer games, etc.

Use the open computer to identify the components as they are discussed. There are many parts of a computer that must work together to make the computer operate properly. These memory chips are samples of some of the integrated circuits that make computers function. **(Pass the memory chips around so that the students can look at them as you go on.)** Today we'll be imitating the functions of a computer. These functions work together, courtesy of the motherboard which connects the various parts.

Input: Information must somehow get into the computer and then the information must be translated into digital form. Input comes from various devices, such as the keyboard, mouse, scanner, microphone, camera, and graphic tablet.

Process: The computer needs to do something with (**process**) what it has received. The CPU, or central processing unit, is the main processing unit of a computer. It coordinates all of the actions of the machine like carrying out instructions, performing calculations, and interacting with all the components used to operate the computer. More importantly, the CPU handles the **fetch**, **decode**, and **execute** steps of the computer system. To better understand the efficiency of a computer's processing system, you will become the parts of the computer and perform these functions.

Storage: Computers have two types of storage: temporary and long-term.

- **Hard drives, CD-ROMs, floppy disks**, and **magnetic optical disks** are examples of long-term storage devices that keep information whether the computer is on or off.
- **ROM** or *Read Only Memory* holds important information that the computer needs each time it runs.
- **RAM** or *Random Access Memory* is a type of temporary storage because it stores information as you use it but it is constantly being erased and rewritten as you open and close files.

Output: When the computer is finally ready to display the information that it has been processing, it **outputs** information by using tools like a monitor, printer, and speakers. The user needs to be able to retrieve the information of the result of the instructions given the computer.

EXERCISE INTRODUCTION

Suggested script: We are going to walk through the sequence that takes place in the computer when you give it a command. We will begin by identifying the various functions on the chalkboard. Then I will call on 6 of you to come up front and stand under one of the terms. Each of you will be simulating the function of that part of the computer.

1. Print the following functions on the board, leaving ample room for a student to stand under the function label: USER, I/O, BUS, CPU, MEMORY, PRINTER.

2. Call 6 students to the board, each to stand under one of the terms.

3. Distribute the Rules and Script to each person so that everyone can follow along during the demonstration.

4. Place the RED numbered cards and the BLUE function cards in front of the user.

5. Place the YELLOW cards and one marker in front of the CPU.

6. Place the printer paper and one marker in front of the printer.

7. Go over the rules so that the students understand what is allowed.

8. Walk the demonstration group through the activity sequence as they perform their roles.

Suggested script:

"One of the most important aspects of a computer is its speed. So we're going to time the group to see how fast they can perform the function. The rest of you will be divided into groups so you, too, can become a human computer and be timed as you go through the functions. We will be using a stopwatch to time your teams, and you'll be competing for the following titles:

- 1st Place Nanosecond Team
- 2nd Place Microsecond Team
- 3rd Place Millisecond Team"

Point out to the students that the time is not restarted if they have to restart the process. The time includes the entire time it takes to obtain the correct answer.

"I will call each team up by letters and let you walk through once for practice, then I will time you. If time permits, you may get two timed runs."

Write the times for the teams on the board so they can see their scores.

Note: If one or more of the groups has more than 6 students, consider having the "left over" students participate in a second timed run for the teams with more than 6.

HUMAN COMPUTER EXERCISE RULES

1. No talking while the computer is running.

2. If the rules are not followed, the computer must be turned OFF and then ON again and the process started over.

3. If an incorrect answer is given, the computer must be turned OFF and then ON again and the process started over.

4. USER must stay under his or her name.
   USER selects a problem either multiplying or adding two numbers between 2 & 9; e.g., 2 x 9.
   USER can receive a card only from I/O or printer.
   USER can give a card only to I/O.

5. I/O must stay under his or her name.
   I/O can receive a card only from USER or BUS.
   I/O can give a card only to USER or BUS.

6. BUS must do whatever I/O or CPU says and then return to his or her name.
   BUS can receive a card only from I/O, CPU, or MEMORY.
   BUS can give a card to everyone except USER.

7. CPU must stay under his or her name.
   CPU can receive a card only from BUS.
   CPU can give a card only to BUS.

8. MEMORY must stay under his or her name.
   MEMORY can receive a card only from BUS.
   MEMORY can give a card only to BUS.

9. PRINTER must stay under his or her name.
   PRINTER can receive a card only from BUS.
   PRINTER must rewrite on printer paper whatever information BUS gives and wait for USER to
      come pick it up.

HUMAN COMPUTER EXERCISE SCRIPT
(Note: Micron, see reference, has a .pdf version that is an excellent format for printing.)

BEGIN:
1. EVERYBODY in position.
2. EVERYBODY on their knees, except USER: COMPUTER is OFF.

TURN ON:
3. USER selects a problem and writes it on board; *e.g.*, 2 x 9.
4. USER touches I/O's head to turn computer ON.
5. EVERYBODY stands up.

NUMBER:
6. USER picks the first RED (2) card and gives it to I/O.
7. I/O gives the RED card to BUS.
8. BUS gives the RED card to CPU.
9. CPU gives the RED card back to BUS.
10. BUS runs to MEMORY and gives MEMORY the RED card.
11. BUS returns.

FUNCTION:
12. USER picks a BLUE (x) card and gives it to I/O.
13. I/O gives the BLUE card to BUS.
14. BUS gives the BLUE card to CPU.
15. CPU keeps the BLUE card.

NUMBER:
16. USER picks another RED (9) card and gives it to I/O.
17. I/O gives the RED card to BUS.
18. BUS gives the RED card to CPU.
19. CPU keeps the RED card.
20. BUS runs to MEMORY and gets the first RED card from MEMORY.
21. BUS returns and gives CPU the RED card.

ANSWER:
22. CPU looks at both RED cards and the BLUE card and writes the answer (18) on a YELLOW card.
23. CPU gives the YELLOW card to BUS.
24. BUS gives the YELLOW card to MEMORY and returns.
25. CPU writes the answer (18) on another YELLOW card and gives it to BUS.

PRINTOUT:
26. BUS gives the YELLOW card to the PRINTER and returns.
27. USER goes to the PRINTER.
28. PRINTER looks at the YELLOW card and prints the answer (18) on the printer paper.
29. PRINTER gives USER the printer paper with the answer on it.
30. USER returns and writes the answer on the board.

Announce which teams won 1st, 2nd, and 3rd places and discuss the recorded times when the teams have completed their runs. Ask how many knew that all of these processes take place each time the computer solves the most simple problems.

CONCLUDING ACTIVITY

Instruct the students to fill in the table on the student answer pages to show how the elements of a computer relate to the thought processes of a person. Tell them to think of both a computer and a person as information processing machines and to identify the four components of an information-processing system for both in order to complete the table.

POSSIBLE ANSWERS TO THE CONCLUSION QUESTIONS AND SAMPLE DATA

	Computer	Person
Input done with	keyboard, mouse, scanner, microphone, camera	eyes, ears, feeling, taste
Storage done with	hard drive, CD-ROMs, memory chips & boards	brain, notes, computer, etc.
Information processing done with	CPU	calculator, computer, brain, friend
Output done with	monitor, printer, speakers, etc.	pencil & paper, typewriter, telephone, speakers, chalkboard

CONCLUSION QUESTIONS

1. Where do you think technology will take us?
   - To Mars
   - To totally automated cars, houses, work places
   - To producing enough food and water that everyone on earth has what they need
   - To more new inventions and useful applications than ever before

2. What new directions do you think technology will go?
   - There will be more and more computer capability available to everyone so that each person can study and learn about anything he or she would like and listen to her or his favorite music while learning.
   - There will be all kinds of wireless applications so that we know what's happening anywhere we like—where our parents are, what the children we're babysitting are doing while we grab a snack or watch a TV program, to visit with our grandparents or best friend who's moved away, etc.
   - Note: There could be as many ideas as there are students.

3. What doesn't your computer do now that you think it will in the future?
   - Have a wireless connection to the Internet
   - Act as a communications (telephone, pager, etc.) center whether user is on-line or off-line
   - Write papers and reports from garbled notes
   - Guide blind people safely

REFERENCES:

http://micron.com/K-12/lessonplans/humancomp

Destination Digital Brochure and http://www.destinationdigital.org

P. Van Zant, *Microchip Fabrication*, Fourth Edition, McGraw Hill Publishing, 2000, glossary

Name _____

Period _____

Human Computer
Simulating the Internal Operations of a Computer

There are many parts of a computer that must work together to make the computer work. Integrated circuits, or chips, are the products inside computers that make computers work. When you look on any circuit board from inside a computer, you will see many integrated circuits mounted on the boards. The industry that designs, creates and produces the integrated circuits used in computers, as well as in most of the electronic products we use every day, is the *semiconductor industry*. Boards connect the integrated circuits with other devices, such as capacitors, resistors, etc., to perform the designed functions. Today you will be simulating one of the functions of a computer. These functions work together, courtesy of the motherboard which connects the various parts:

Input: Information must somehow get into the computer and then the information must be translated into digital form. Input comes from various devices, such as the keyboard, mouse, scanner, microphone, camera, and graphic tablet.

Process: The computer needs to do something with (**process**) what it has received. The central processing unit, or *CPU*, is the main processing unit of a computer. It coordinates all of the actions of the machine like carrying out instructions, performing calculations, and interacting with all the components used to operate the computer. More importantly, the CPU handles the **fetch**, **decode**, and **execute** steps of the computer system. To better understand the efficiency of a computer's processing system, you will become the parts of the computer and perform these functions.

Storage: Computers have two types of storage: temporary and long-term.

- **Hard drives, CD-ROMs, floppy disks**, and **magnetic optical disks** are examples of long-term storage devices that keep information whether the computer is on or off.
- **ROM** or *Read Only Memory* holds important information that the computer needs each time it runs.
- **RAM** or *Random Access Memory* is a type of temporary storage because it stores information as you use it but it is constantly being erased and rewritten as you open and close files.

Output: When the computer is finally ready to display the information that it has been processing, it **outputs** information by using tools like a monitor, printer, and speakers. The user needs to be able to retrieve the information of the result of the instructions given the computer.

Semiconductor fabrication is one of the names for the mass production of integrated circuits. Fabrication includes all of the processes that can be done in wafer form. If you want to know more about the semiconductor industry, refer to www.destinationdigital.org.

GLOSSARY
Bus: The data transport function in computers
Chip (or microchip): An individual integrated circuit built in a tiny, layered rectangle or square on a silicon wafer. There may be as many as hundreds of these chips on a single wafer.

CPU: Central processing unit in computers

Discrete Devices: Individual electronic elements, such as transistors, capacitors, resistors, *etc.*, that have a single electrical function

Integrated Circuit: An electronic circuit containing as many as millions of microscopic transistors that work together to perform specific functions. All elements of the circuit are fabricated and interconnected in and on a single chip of semiconducting material.

I/O: The input and output functions for computers

Memory: The information storage capability for computers

RAM (Random Access Memory): A type of temporary storage that stores information as it is used but is constantly being erased and rewritten as files are opened and closed

ROM (Read Only Memory): Storage capability that holds important information that the computer needs each time it runs

Semiconductor: A material that can be an electrical conductor or insulator. Silicon is the most common semiconductor used to manufacture integrated circuits.

PURPOSE

You will be able to name the six main function areas inside a computer and will simulate the internal operations of a computer completing a problem. In addition, you will be able to compare and contrast the difference between human and computer parts that perform input, output, process and storage functions.

MATERIALS

computer opened to show CPU	1 stop watch
2 black markers	memory chips (optional)
exercise rules for each team member	exercise scripts for each team member
set of 8 red 5 x 8 cards numbered 2-9	set of 2 blue 5 x 8 cards marked with x and +
1 yellow 5 x 8 card	printer paper

PROCEDURE

1. Read and follow the exercise rules and script when your teacher instructs you to do so.

2. Carefully observe or participate in the demonstration run.

3. Perform the function you are assigned as quickly and accurately as possible. (Remember that the least time taken to complete the process wins 1st place, but an incorrect answer causes the entire process to be restarted while the time continues to count.)

4. Fill in the table below showing how the elements of a computer relate to the thought processes of a person.

	Computer	Person
Input done with		
Storage done with		
Information processing done with		
Storage done with		
Output done with		

CONCLUSION QUESTIONS

1. Where do you think technology will take us?

2. What new directions do you think technology will go?

3. What doesn't your computer do now that you think it will in the future?

Molecular Motion: Are You Current on Convection?
Experimenting with Heat and Molecules

OBJECTIVE
Students will be able to describe the effect that heat has on the speed of water molecules and then connect that learning to density and convection.

LEVEL
Middle Grades: Earth Science

NATIONAL STANDARDS
UCP.2, UCP.3, A.1, A.2, B.5, B.6, D.1

TEKS
6.2(A), 6.2(B), 6.2(C), 6.2(D), 6.2(E), 6.3(C), 6.4(A), 6.4(B), 6.5(C), 6.8(B)
7.2(A), 7.2(B), 7.2(C), 7.2(D), 7.3(C), 7.4(A), 7.4(B), 7.5(C)
8.2(A), 8.2(B), 8.2(C), 8.2(D), 8.3(C), 8.4(A), 8.4(B), 8.5(C), 8.10(A)
IPC 2(A), 2(B), 2(C), 2(D), 3(A), 6(B)

CONNECTIONS TO AP
AP Environmental Science:
I. Interdependence of Earth's Systems: Fundamental Principles and Concepts A. The Flow of Energy

TIME FRAME
60 minutes

MATERIALS
(For a class of 28 working in pairs)

14 400 mL beakers	hot plate or tea kettle
ice, small cooler full	small bucket or bowl for ice
blue food coloring	electric heating pad
10 gallon aquarium	28 pairs of safety goggles

TEACHER NOTES
Molecular Motion: Are You Current on Convection is an introductory activity that explains how adding heat changes the speed of molecules. This is a fundamental concept for students and serves as the basis for explaining what happens in the atmosphere as a result of solar heating. Weather, winds, ocean currents and hurricanes all start with this basic concept.

Part I of the lesson requires ice and hot water. One small cooler of ice should be plenty for one class. An electric teapot makes heating and transporting the water from table to table very easy and safe. It is also possible to use a hot plate and pot or large beaker to heat the water. Have students wear goggles while you distribute the hot water. Any color of food coloring will work, but yellow is the least effective. Red, blue and green are all very visible.

Part II of the lesson will demonstrate the movement of food coloring through a tank of water that has different temperature regions. The demonstration needs an aquarium with water set on top of both a heat source and a cooling source. A dual-temperature base can be created by using a stack of books and a wide bowl that are the same height. A hot pad laid over the stack of books creates the heated side and a bowl of ice creates the cooling side. Place the aquarium atop this base and fill it with some water. Set this up just prior to the demonstration.

When it is time for the demonstration, ask the students where they want you to drop the food coloring. Again, red, blue and green work well. Add several drops of the food coloring and allow students to observe what happens. A perfect little convection cell should develop after a minute or so.

Once students have completed and handed in their answer pages, review the concepts of heat (energy), density, volume, and convection.

POSSIBLE ANSWERS TO THE CONCLUSION QUESTION AND SAMPLE DATA
Part I

Observations: Draw both beakers and write a description of each one below.

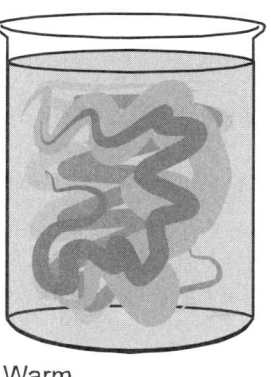
Warm

Cold

1. In which beaker did the food coloring spread out faster?
   - The food coloring spread out much faster in the beaker containing hot water.

2. What caused the molecules to move faster? How could you tell which beaker had faster molecules?
   - The hot water had more energy than the cold water. The heating of the water sped up the water molecules. The food coloring acted as an indicator and showed us the behavior of the water molecules moving.

3. How does temperature change the speed of molecules?
   - Increasing the temperature of the water increases the speed of the molecules. Decreasing the temperature of water decreases the speed of molecules.

Part II

1. Define "heat" in terms of molecular motion.
   - As you add thermal energy and heat up the water, you increase temperature and increase the average molecular motion.

2. If the temperature of a liquid is raised, will the molecules within the sample collide with more or less force?
   - As the temperature of a liquid rises, its molecules will collide with more force because they are moving faster and have more energy.

3. If the temperature of a liquid is raised, predict whether the liquid sample will tend to take up more or less space?
   - Answers will vary, but most students will say that as the temperature of a liquid rises it will take up more space since it expands. Don't miss this opportunity to combat a common misconception — expansion occurs due to an increased distance *between* molecules, not that the molecules themselves become enlarged.

4. If the mass of a substance stays the same and its volume increases (see previous question), how is the density of a substance affected?
   - Density is the ratio of mass to volume, so if volume goes up density will go down.

5. Which floats better....low density substances or high density substances?
   - The lower the density the better something will float. One way to test if an object is less dense than water is to toss it into water and see if it floats. If it floats then the object is less dense than water.

6. Based on your answers to the questions above, predict the movement of water (if any) in the unevenly heated container drawn below.
   - Drawings of predictions will vary.

Heat Source   Ice

7. Draw what actually happened during the demonstration.

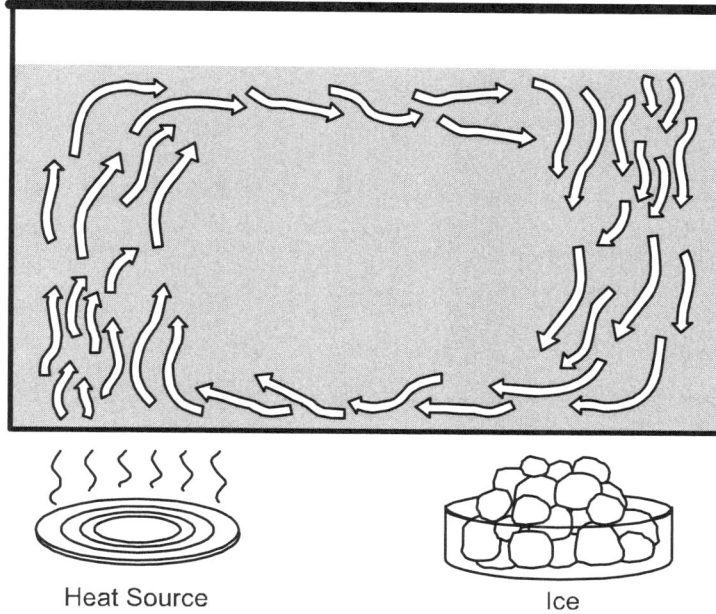

Heat Source   Ice

Molecular Motion: Are You Current on Convection?
Experimenting with Heat and Molecules

Convection is the transport of energy due to density differences. As a liquid or gas is heated it expands and becomes less dense and therefore lighter. If a cooler material with greater density is above the warmer layer of fluid the warmer fluid material will be displaced and "rise" through the cooler material to the surface. The rising material will dissipate its heat (energy) into the surrounding environment, become more dense as it cools, and will sink to start the process over.

PURPOSE
In this activity you will explore the effect heat energy has on the speed of water molecules.

MATERIALS
> 2 400 mL beakers
> ice
> blue, red or green food coloring
> safety goggles

PROCEDURE
Part I
1. Make sure you and your group members are wearing safety goggles.

2. Obtain two 400 mL beakers and fill them with 200 mL of tap water.

3. In one beaker, add 100 mL of ice. In the second beaker <u>your teacher</u> will add 100 mL of hot water. Do not touch the beaker containing hot water.

4. Allow both beakers to settle for one minute.

5. While the beakers are settling, predict which beaker will have the most molecular movement. Write your answer under HYPOTHESIS on your student answer page.

6. After waiting one minute, add one drop of food coloring to each beaker. Observe each beaker and draw what you see in the OBSERVATIONS section of your student answer page.

7. Answer the conclusion questions for Part I.

Part II

1. Answer the first five conclusion questions from Part II based on what you observed in Part I.

2. For question number 6, make a prediction about what will happen in the described demonstration. Draw your prediction on the student answer page.

3. Watch the demonstration. Draw what you saw happen, and describe it in words below your drawing.

Name _____

Period _____

Molecular Motion: Are You Current on Convection?
Experimenting with Heat and Molecules

HYPOTHESIS

OBSERVATIONS

Draw both beakers and write a description of each one below your drawing.

HOT H_2O COLD H_2O

CONCLUSION QUESTIONS

Part I

1. In which beaker did the food coloring spread out faster?

2. What caused the molecules to move faster? How could you tell which beaker had faster molecules?

3. How does temperature change the speed of molecules?

Part II

1. Define "heat" in terms of molecular motion.

2. If the temperature of a liquid is raised, will its molecules collide with more or less force?

3. If the temperature of a liquid is raised, will it tend to take up more or less space?

4. If the mass of a substance remains the same and its volume increases (see previous question), how is the density of a substance affected?

5. Which floats better....low density substances or high density substances?

6. Based on your answers to the questions above, predict the movement of water (if any) in the unevenly heated container drawn below.

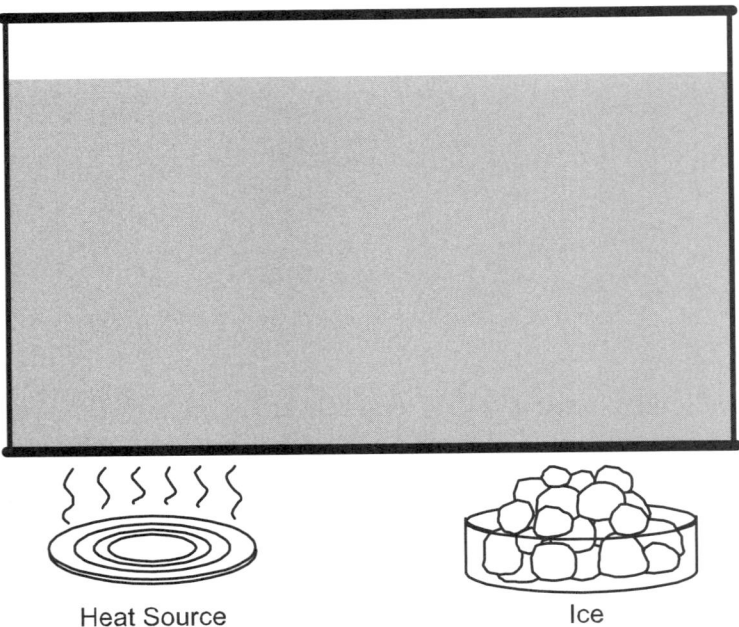

Heat Source   Ice

7. Watch the demonstration. Draw what actually happened to the water molecules during the demonstration and describe what happened in words below your drawing.

Heat Source   Ice

Write Your Notes and Ideas Here!

Evaporation and Condensation
Investigating Energy Transfer During a Phase Change

OBJECTIVE
Students will be able to explain that evaporation uses energy and that condensation releases energy.

LEVEL
Middle Grades: Earth Science

NATIONAL STANDARDS
UCP.2, UCP.3, B.2, B.3, D.2

TEKS
6.1(A), 6.2(A), 6.2(B), 6.2(C), 6.2(D), 6.2(E), 6.4(A), 6.8(A), 6.9(B), 6.14(C)
7.1(A), 7.2(A), 7.2(B), 7.2(C), 7.2(D), 7.2(E), 7.4(A), 7.14(C)
8.1(A), 8.2(A), 8.2(B), 8.2(C), 8.2(D), 8.2(E), 8.4(A), 8.9(A), 8.9(C), 8.10(A), 8.10(B), 8.12(B)
IPC 1(A), 2(A), 2(C), 2(D), 8(A), 8(E), 9(A), 9(B), 9(C)

CONNECTIONS TO AP
AP Chemistry:
    II States of Matter, A. Gases, B. Liquids and solids

TIME FRAME
50 minutes

MATERIALS
(For a class of 28 working in groups of 4)

    7 alcohol swabs
    7 200 mL beakers
    7 Celsius thermometers (alcohol)
    ice for beakers
    7 mini desk fans (or 1 large fan)

TEACHER NOTES
The Evaporation and Condensation Lab allows students to experiment on a small scale the same energy interactions that happen in the atmosphere. It may be taught while discussing weather, atmosphere and climate, or during phase change discussions. As a closure to the lab, draw the phase change diagram on the board and have students come up and explain what happened in the lab using the diagram. Explain for students the idea of latent heat.

Students must understand the relationship between energy and the processes of evaporation and condensation. Evaporation moves the stored solar energy in the ocean to the atmosphere, and then condensation releases that energy into the atmosphere. Storm systems like hurricanes need huge amounts of energy and this cycle provides the mechanism for moving energy in the earth's system.

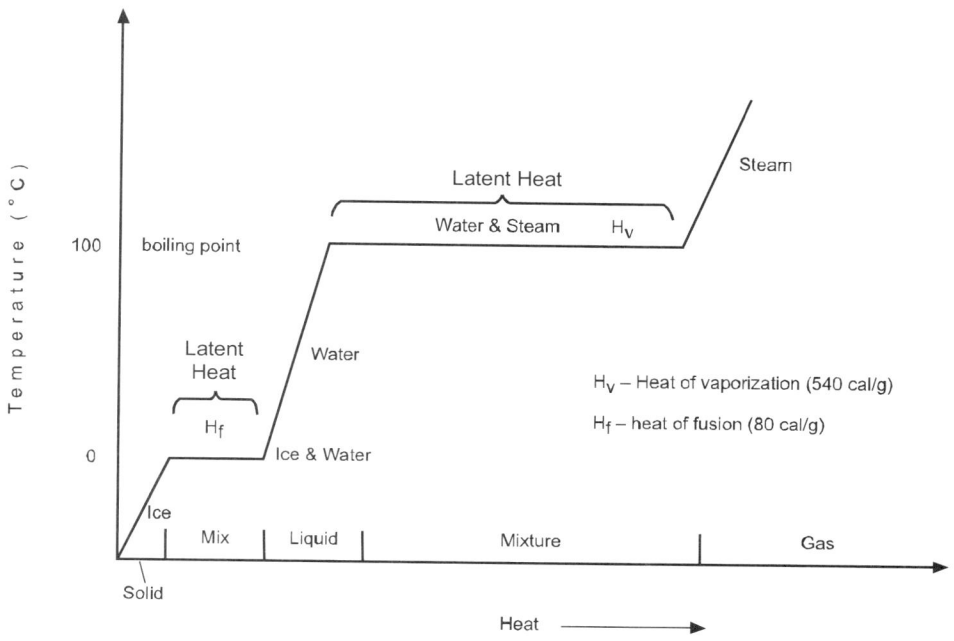

For more information, the following web sites have good discussions of latent heat and weather: http://www.usatoday.com/weather/wlatent.html

POSSIBLE ANSWERS TO THE CONCLUSION QUESTIONS AND SAMPLE DATA

Part I: Evaporation

1. Did the alcohol absorb or release energy when it evaporated?
   - The hand felt cold as the alcohol evaporated. Evaporation absorbed energy from the hand resulting in a lower hand temperature and a cool sensation.
   - Use this as a link to life science and discuss with students how the human body keeps cool by perspiring.

2. Name two variables that affect the rate of evaporation.
   - Temperature and wind speed affect evaporation.
   - Holding a hand in front of a fan allowed the alcohol to evaporate faster and thus the hand felt much cooler sooner.
   - Evaporation rate is increased with higher temperatures (think of puddles on a warm summer day vs. puddles on a cool winter day).

3. In what way do winds affect the rate of evaporation of ocean water?
   - Wind increases the rate of evaporation of ocean water, and therefore increases the rate of cooling.

4. State two reasons why bodies of water heat up more slowly than land masses when both surfaces receive an equal amount of sunlight.
   - Ocean water is constantly evaporating which causes cooling.
   - The specific heat capacity of ocean water is higher (it takes more energy to heat a cubic centimeter of water than it takes to heat up a cubic centimeter of land).

Part II: Condensation

1. What happens to a gas (water vapor) when it is cooled?
   - Water condenses into a liquid when water vapor (gas) is cooled. The pressure of an enclosed gas also decreases since the molecules slow down.

2. If the temperature at which the water begins to condense is called the dew point, what is the dew point in the classroom today?
   - Answers will vary. The temperature could range from 25 - 30 degrees Celsius.

3. If the air in the room was more humid, do you think that the dew point would be higher or lower?
   - If the air was more humid (more saturated with water vapor) the dew point temperature would be higher because it would be easier for water vapor to condense.

4. As humid air rises and adiabatic cooling occurs, what may happen to the water vapor in the air?
   - As humid air rises and is cooled water vapor condenses.

5. If condensation is the true opposite of evaporation, how does condensation affect the temperature of the air from which the water is condensing?
   - Evaporation cools the air by absorbing energy to speed up the molecules and change them from a liquid to a gas. Condensation actually warms the air as energy is released to the atmosphere as the molecules change between gas and liquid.

TEACHER PAGES

Evaporation and Condensation
Investigating Energy Transfer During a Phase Change

Imagine being outdoors on a hot day with low humidity. As you exercise you begin to sweat, but the sweat quickly disappears. What is happening? Where does the sweat go? How do you feel as the sweat disappears?

PURPOSE
In this activity you will discover what happens to energy during evaporation and condensation.

MATERIALS
    1 alcohol swab
    1 200 mL beaker
    1 Celsius thermometer (alcohol)
    ice for beakers
    1 mini desk fan (or 1 large fan in front of room)

PROCEDURE
Part I: Evaporation
1. Write the answer to the following question as your hypothesis for Part I on your student answer page:
   - How do you think that evaporation will affect the temperature of the surface from which the liquid is evaporating?

2. Wipe an alcohol pad across the back of your hand and the hands of each person in the group. Report how your hand feels in the observations section on your student answer page.

3. Repeat step 2, but this time, hold the surface of your hand in front of the mini fan on your desk. Report your observations and be sure to compare to the non-fan trial. Did the air passing over your hand in trial 2 increase or decrease the speed of evaporation?

Part II: Condensation
1. Write the answer to the following question as your hypothesis for Part II on your student answer page.
   - If the air surrounding a beaker of water is cooled, what do you think will happen to the water vapor in the air?

2. Fill the beaker with 100 mL of tap water and add 50 mL ice.

3. Watch the outside of the beaker for 60 seconds.

4. Record your observations.

5. Empty your beaker and dry the outside of the beaker.

6. Again fill a beaker with 100 mL of water.

7. Insert the thermometer into the water and record the temperature in the section for observations on your student answer page.

8. Leave the thermometer in the beaker add 50 mL of ice to the beaker.

9. Record the temperature of the water in the beaker when water vapor begins to condense on the outside of the beaker.

Evaporation and Condensation
Investigating Energy Transfer During a Phase Change

HYPOTHESIS

Part I: Evaporation

Part II: Condensation

OBSERVATIONS

Part I: Evaporation

Part II: Condensation

Initial temperature of water _____ °C

Temperature when you saw condensation _____ °C

CONCLUSION QUESTIONS

Part I: Evaporation

1. Does alcohol absorb or release energy when it evaporates?

2. Name two variables that affect the rate of evaporation.

3. In what way does wind affect the evaporation of ocean water?

4. State two reasons why bodies of water heat up more slowly than land masses when both surfaces receive an equal amount of sunlight.

Part II: Condensation

1. What happens to a gas (water vapor) when it is cooled?

2. If the temperature at which the water begins to condense is called the dew point, what is the dew point in the classroom today?

3. If the air in the room was more humid, do you think that the dew point would be higher or lower? Justify your answer.

4. As humid air rises and cooling occurs, what may happen to the water vapor in the air?

5. If condensation is the true opposite of evaporation, how does condensation affect the temperature of the air from which the water is condensing?

The Watched Pot Never Boils…Or Does It?
Exploring Heating and Phase Changes

OBJECTIVE
Students will make close observations about the effect of heat on a beaker of water.

LEVEL
Middle Grades: Earth Science

NATIONAL STANDARDS
UCP.3, A.1, A.2, B.6, D.1, D.2

TEKS
6.4(A), 6.9(B)
7.2(B), 7.4(A)
8.2(B), 8.4(A), 8.4(B), 8.10(A)
IPC.1(A), 2(B), 3(A), 6(H)

CONNECTIONS TO AP
AP Chemistry:
    II. States of Matter B. Liquids and solids 3. Changes of state

TIME FRAME
45 minutes

MATERIALS
(For a class of 28 working in pairs.)
    14 beakers (600 mL to 800 mL)   14 test tubes 16 mm x 125 mm
    14 hot plates   14 alcohol thermometers*
    14 thermometer holders (Flinn AP5408)   beaker tongs, or hot pads.

*Data collection devices and temperature probes may be used in place of thermometers.

TEACHER NOTES
Even though boiling seems like a simple process, it is often misunderstood even by AP science students. This activity asks the students to carefully observe a beaker of water as it heats and comes to a boil. They should record their observations and then use their observations to answer the questions about heating and boiling.

Safety Alert

1. Students and teacher should wear safety goggles.
2. Students will be dealing with hot plates and boiling water. Students should not remove the beaker of boiling water from the hot plate. The teacher should use beaker tongs or a hot pad to remove the beaker for each group.

POSSIBLE ANSWERS TO THE CONCLUSION QUESTION AND SAMPLE DATA

Temperature (° C)	Observations	Time
20		
25		
30	Small bubbles start to form on sides of beaker, thermometer, test tube, etc.	
35		
40		
45		
50	Small gas bubbles increase	
55		
60		
65		
70	Large bubbles start to form on bottom of the beaker where it is hotter; however they collapse before rising to the surface of the water.	
75		
80	Small gas bubbles should be gone from the surfaces. Larger bubbles forming on bottom.	
85		
90		
95	Test tube begins to collect gas and may start to float.	
100	Full boil. Test tube full of gas and bobbing around.	
95	Water has returned to the test tube except for a small amount of gas at the tip.	
90	If the tap water contains calcium ions, the water may be cloudy and when the thermometer is removed it may be coated with white powder.	

TEACHER PAGES

1. At what temperature did you first notice little bubbles on the sides of the beaker?
   - Around 30° C.

2. At what temperature do these little bubbles disappear? What do you think is inside the little bubbles?
   - The bubbles contain the gases that were dissolved in the water. This would include oxygen, nitrogen, and some carbon dioxide. [Connect to biology and how fish "breathe"]
   - The bubbles disappear around 80° C.

3. At what temperature do the large bubbles start forming on the bottom of the beaker?
   - This will vary depending upon the hot plates and the amount of water used.

4. What happens to the bubbles as they try to come to the top of the beaker? Suggest a possible explanation.
   - They collapse before they get to the top.
   - The water at the top of the beaker is cooler than the water at the bottom close to the hot plate.

5. At what temperature do the large bubbles finally reach the top of the beaker? What do you think is inside the large bubbles?
   - The bubbles finally reach the top around 100° C. This may vary slightly due to changes in barometric pressure and the fact that thermometers are not calibrated.
   - The bubbles contain water vapor. Students should be forced to recognize that the bubbles do NOT contain air.

6. What did you observe about the test tube? What do you think is inside the test tube?
   - The test tube is starting to float around.
   - The test tube is filling with water vapor. (Be sure that the students do not think that it is filled with air.)

7. What happens to the test tube after the heat is removed?
   - The vapor in the test tube condenses and the water refills the tube.

8. Does the water in your beaker look any different now than it did when you started to heat it? If so, suggest a reason.
   - If the water in your region is hard, it may be cloudy. If calcium ions are present in the water, the white cloudiness is probably due to the formation of calcium carbonate.

9. Is there any gas in the test tube? What kind of gas do you think it is?
   - There should be a small amount of gas in the test due to the collection of the gases which were dissolved in the water at the beginning of the lab.
   - The students should realize that this gas is due to the air that was originally dissolved in the water.

10. Define or identify the following:
    A. boiling
       • The point where the liquid rapidly turns into vapor with bubbles large enough and frequent enough to disturb the surface of the water.
    B. out-gassing
       • Gases coming out of solution due to the increase in temperature.
    C. boiling point of water in Celsius
       • 100° C
    D. freezing point of water in Celsius
       • 0° C

TEACHER PAGES

The Watched Pot Never Boils...Or Does It?
Exploring Heating and Phase Changes

The simple process of boiling water is probably one you have seen many time. But have you ever really observed what happens when water boils?" In this lab, you will make careful observations as the water heats and comes to a boil. You will record your observations in a data table and use those observations to answer the conclusion questions.

PURPOSE
You will heat a beaker of water and make observations every 5.0° C. The observations will be used to answer the questions about heating and boiling.

MATERIALS
    1 beaker (600 mL to 800 mL)
    hot plate
    1 thermometer support

    1 test tube
    1 alcohol thermometer or data collection device with
        temperature probe as your teacher instructs
    tap water

Safety Alert
1. Wear your safety goggles.
2. You will be dealing with hot plates and boiling water. Do not attempt to remove the beaker from the hot plate by yourself. Allow your teacher to move it.

PROCEDURE
1. Fill a large beaker with water.

2. Fill a test tube with water. Put your thumb over the top of the test tube and invert it into the beaker. Try to invert the test tube so that there are no bubbles in the test tube.

3. Use a utility clamp to secure a thermometer inside the beaker . Do not let the thermometer touch the bottom of the beaker.

4. Place the beaker on the hot plate and turn the hot plate on high. Note the time you begin and make observations every 5° Celsius. Record your observations in the data table on your student answer page. Be specific in your observations. What do you see? Where do you see it?

5. Continue heating the beaker until the water has been in a full rapid boil for 2 minutes. Raise your hand and wait until your teacher comes to remove the beaker from the hot plate.

6. Continue making observations every 5° Celsius until the temperature falls to 90° C.

7. Clean up your lab area as your teacher instructs.

Name _____

Period _____

The Watched Pot Never Boils…Or Does It?
Exploring Heating and Phase Changes

DATA AND OBSERVATIONS

Temperature (^{o}C)	Observations	Time
20		
25		
30		
35		
40		
45		
50		
55		
60		
65		
70		
75		
80		
85		
90		
95		
100		
95		
90		

CONCLUSION QUESTIONS

1. At what temperature did you first notice little bubbles on the sides of the beaker?

2. At what temperature do these little bubbles disappear? What do you think is inside the little bubbles?

3. At what temperature do the large bubbles start forming on the bottom of the beaker?

4. What happens to the bubbles as they try to come to the top of the beaker? Suggest a possible explanation.

5. At what temperature do the large bubbles finally reach the top of the beaker? What do you think is inside the large bubbles?

6. What did you observe about the test tube? What do you think is inside the test tube?

7. What happens to the test tube after the heat is removed?

8. Does the water in your beaker look any different now than it did when you started to heat it? If so, suggest a reason.

9. Is there any gas in the test tube? What kind of gas do you think it is?

10. Define or identify the following:
    A. Boiling

    B. out-gassing

    C. boiling point of water in Celsius

    D. freezing point of water in Celsius

Relative Humidity
Measuring the Amount of Water Vapor in the Air Versus the Total Amount the Air Can Hold

OBJECTIVE
Students will use a psychrometer to measure the relative humidity of 5 places around campus. Students will be able to explain what relative humidity is and why it varies inside vs. outside.

LEVEL
Middle grades: Earth Science

NATIONAL STANDARDS
UCP.2, UCP.3, B.2, B.3, D.2

TEKS
6.2(B), 6.2(C), 6.2(D), 6.2(E), 6.14(B)
7.2(A), 7.2(B), 7.2(C), 7.2(D)
8.2(A), 8.2(B), 8.2(C), 8.2(D)
IPC 1(A), 2(A), 2(C), 2(D), 8(A), 8(E), 9(A), 9(B), 9(C)

CONNECTIONS TO AP
AP Chemistry:
    II States of Matter, A. Gases, B. Liquids and solids
AP Physics B:
    II Fluid Mechanics and Thermal Physics B. Temperature and Heat

TIME FRAME
50 minutes

MATERIALS
(For a class of 28 working in pairs/groups of 4)

    7 psychrometers (with socks and rubber bands)
    distilled water
    7 eye droppers or pipets

TEACHER NOTES
Relative Humidity is a lab activity that has students measure the relative humidity's of several locations using an instrument called a psychrometer. Psychrometers can be purchased from most science equipment supply houses for around $15 a piece.

A psychrometer is an instrument used to measure relative humidity. Psychrometers have two thermometers connected on a handle so the thermometers can be spun to simulate wind. One thermometer is covered with a small knit "sweater" that is then wet and when spun it cools as the water evaporates off the sweater. The knit sweaters and rubber bands are included when you order the psychrometers.

Relative humidity is a measure of the amount of water vapor in the air versus the amount of water vapor the air can possibly hold, multiplied by 100 and written as a percentage. Humidity alone is a simple measure of the amount of water vapor in the air and is recorded in grams per cubic meter. Scientists use relative humidity to compare different locations that have different temperatures. Warm air can hold more moisture than cold air, so the humidity might be higher than cooler air, and yet their relative humidity might be the same.

If relative humidity were 100% then you would expect to see some type of precipitation (snow, rain, drizzle, hail). The air is considered saturated at 100% relative humidity.

POSSIBLE ANSWERS TO THE CONCLUSION QUESTIONS AND SAMPLE DATA

Location	Dry bulb temperature (^{0}C)	Wet bulb temperature (^{0}C)	Relative humidity %
Classroom	20*	16	67
Outside shade	23	22	92
Outside sun	25	23	84
Main office	18	13	57
Locker room	23	22	92

* because the relative humidity chart is in whole numbers temperatures need to be rounded to the nearest whole number.

1. Why is there a difference between the temperature of the wet bulb and the temperature of the dry bulb? What would it mean if the temperatures were the same?
   - The wet bulb experiences evaporation of the water from the cloth on the thermometer, which in turn cooled the temperature reading of the wet bulb thermometer. The greater the difference between the dry bulb and the wet bulb readings, the easier it was for water to evaporate and the lower the relative humidity.
   - If the temperatures were the same then no evaporation would have occurred and the relative humidity would be 100%….it would be raining.

2. What location measured the lowest humidity? Why do you think it was lowest?
   - Answers will vary, but the lowest relative humidity is probably in the principal's office.
   - She/He has the best air conditioning unit, which removes the most moisture from the air, causing the lowest relative humidity measurement. Low relative humidity makes humans feel very comfortable because it allows sweat to evaporate off our bodies.

TEACHER PAGES

3. Explain why sweat is a good, if sometimes stinky thing.
   - Sweat allows the body to regulate its temperature and maintain homeostasis. Humans sweat in response to some stress. When sweat evaporates it uses energy, which it takes from your body. The moisture (sweat) on the surface of your body evaporates and cools you off. In the process of evaporation liquid particles gain energy and move into the gaseous phase. As the particles escape the surface they take energy with them, thus, we feel cooler.

4. In central Texas we have air conditioning units that remove humidity from the air while at the same time produce cool air. Why do we feel more comfortable if humidity is lower?
   - The lower the humidity the easier it is for sweat to evaporate off our skin and cool us off.

Relative Humidity
Measuring the Amount of Water Vapor in the Air Versus the Total Amount the Air Can Hold

A psychrometer is an instrument used to measure relative humidity. Relative humidity is a measure of the amount of water vapor in the air versus the amount of water vapor the air can possibly hold, multiplied by 100 and written as a percentage. Humidity alone is a simple measure of the amount of water vapor in the air recorded in grams per cubic meter. Scientists use relative humidity because you can use it to compare places of different temperatures. Warm air can hold more moisture than cold air, so the humidity might be higher than cooler air, and yet their relative humidity might be the same.

If relative humidity were 100% then you would expect to see some type of precipitation (snow, rain, drizzle, hail). The air is considered saturated at 100% relative humidity.

PURPOSE
To use a psychrometer to measure wet bulb/dry bulb temperatures and be able to calculate relative humidity from the results.

MATERIALS
        psychrometer (with sock and rubber band)
        distilled water
        eye dropper or pipet

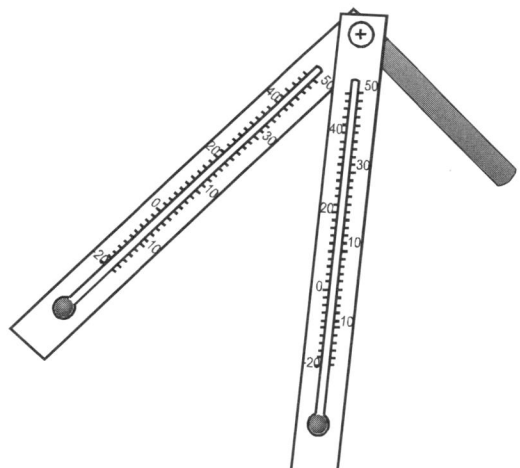

Safety Alert
1. Thermometers are made of glass. Spin the psychrometer with care not to hit anything.

PROCEDURE

1. Read the locations you will be measuring relative humidity on your student answer page and make a hypothesis on your student answer page as to which location will have the highest relative humidity.

2. Place the cloth "sock" on one of the thermometers of your psychrometer. Use a rubber band to hold it in place.

3. With an eyedropper or a small pipet, use distilled water to wet the cloth jacket that covers one of the thermometers on the psychrometer.

4. Spin the psychrometer for one minute. Be sure to hold the pychrometer level in your hand so that the thermometers swing at a 90° angle from the floor. Read the temperature of the wet bulb and the dry bulb and record them on your student answer sheet.

5. Place your thumb gently over the wet bulb and warm the thermometer until the two thermometers again read the same temperature. Proceed to the next location and repeat steps one and two.

6. Repeat steps 3-5 of the procedure for each location you are testing.

7. To calculate relative humidity, subtract the dry bulb temperature from the wet bulb temperature. Use this temperature difference and the dry bulb temperature to find the relative humidity from the chart found on the student answer page. Record your results in the data table on the student answer page.

Name _____

Period _____

Relative Humidity
Measuring the Amount of Water Vapor in the Air Versus the Total Amount the Air Can Hold

HYPOTHESIS

DATA AND OBSERVATIONS

Location	Dry bulb temperature (°C)	Wet bulb temperature (°C)	Relative humidity
Classroom			
Outside shade			
Outside sun			
Main office			
Locker room			

CONCLUSION QUESTIONS

1. Why is there a difference between the temperature of the wet bulb and the temperature of the dry bulb? What would it mean if the temperatures were the same?

2. What location measured the lowest humidity? Why do you think it was lowest?

3. Explain why sweat is a good, if sometimes stinky thing.

4. In central Texas we have air conditioning units that remove humidity from the air at the same time as putting out cool air. Why do we feel more comfortable if humidity is low?

Relative Humidity Table (in percent)
Dry Bulb Minus Wet Bulb

°C	1	2	3	4	5	6	7	8	9	10
10	88	77	66	55	44	34	24	15	6	
11	89	78	67	56	46	36	27	18	9	
12	89	78	68	58	48	39	29	21	12	
13	89	79	69	59	50	41	32	22	15	7
14	90	79	70	60	51	42	34	25	18	10
15	90	81	71	61	53	44	36	27	20	13
16	90	81	71	63	54	46	38	30	23	16
17	90	81	72	64	55	47	40	32	25	18
18	91	82	73	65	57	49	41	34	27	20
19	91	82	74	65	58	50	43	36	29	22
20	91	83	74	67	59	53	46	39	32	26
21	91	83	75	67	60	53	46	39	32	26
22	91	83	76	68	61	54	47	40	34	28
23	92	84	76	69	63	55	48	42	36	30
24	92	84	77	69	63	56	49	43	37	31
25	92	84	77	70	63	57	50	44	39	33

(Left axis label: **Dry Bulb Temperature °C**)

Emission Possible
Student Research Project on Greenhouse Gases and Global Warming

OBJECTIVE
Students will be able to describe the problems associated with society's impact on increased production of certain greenhouse gases that have been cited as being responsible for trapping additional heat in the earth's atmosphere.

LEVEL
Middle grades: Earth Science

NATIONAL STANDARDS
UPC.2, UPC3, A.1, A.2, B.3, D.2, F.4, F.5

TEKS
6.3(A), 6.3(D)
7.3(A), 7.3(D), 7.14(A), 7.14 (B), 7.14 (C)
8.3(A), 8.3 (D), 8.5(A), 8.11(A), 8.12(C)
IPC 1(A), 2(A), 2(C), 2(D), 8(A), 8(E), 9(A), 9(B), 9(C)

CONNECTIONS TO AP
AP Environmental Science:
    V. Global Changes and Their Consequences A. First-order Effects 1. atmosphere; B. Higher-order
    Interactions 1. atmosphere

TIME FRAME
120 minutes in class, 1 week of out of class homework

MATERIALS
(For a class of 28 working in groups of 4)

    7 computers
    Microsoft PowerPoint or HyperStudio
    computer projector

TEACHER NOTES:
Since the beginning of the industrial revolution, atmospheric concentrations of carbon dioxide have increased nearly 30%, methane concentrations have more than doubled, and nitrous oxide concentrations have risen by about 15%. These increases have enhanced the heat-trapping capability of the earth's atmosphere. Sulfate aerosols, a common air pollutant, cool the atmosphere by reflecting light back into space; however, sulfates are short-lived in the atmosphere and vary regionally.

Why are greenhouse gas concentrations increasing? Scientists generally believe that the combustion of fossil fuels and other human activities are the primary reason for the increased concentration of carbon dioxide. Plant respiration and the decomposition of organic matter release more than 10 times the CO_2 released by human activities; but these releases have generally been in balance during the centuries leading up to the industrial revolution with carbon dioxide absorbed by terrestrial vegetation and the oceans.

What has changed in the last few hundred years is the additional release of carbon dioxide by human activities. Fossil fuels burned to run cars and trucks, heat homes and businesses, and power factories are responsible for about 98% of U.S. carbon dioxide emissions, 24% of methane emissions, and 18% of nitrous oxide emissions. Increased agriculture, deforestation, landfills, industrial production, and mining also contribute a significant share of emissions. In 1997, the United States emitted about one-fifth of total global greenhouse gases.

Estimating future emissions is difficult, because it depends on demographic, economic, technological, policy, and institutional developments. Several emissions scenarios have been developed based on differing projections of these underlying factors. For example, by 2100, in the absence of emissions control policies, carbon dioxide concentrations are projected to be 30-150% higher than today's levels

Global mean surface temperatures have increased 0.5-1.0°F since the late 19th century. The 20th century's 10 warmest years all occurred in the last 15 years of the century. Of these, 1998 was the warmest year on record. The snow cover in the Northern Hemisphere and floating ice in the Arctic Ocean have decreased. Globally, sea level has risen 4-8 inches over the past century. Worldwide precipitation over land has increased by about one percent. The frequency of extreme rainfall events has increased throughout much of the United States.

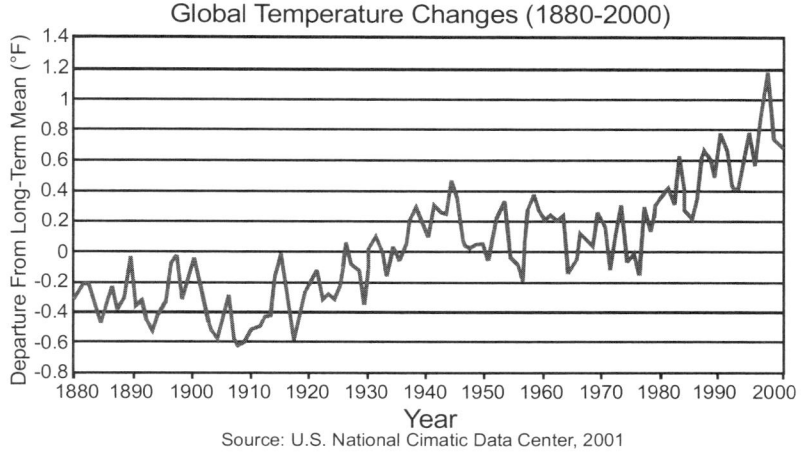

Source: U.S. National Cimatic Data Center, 2001

Increasing concentrations of greenhouse gases are likely to accelerate the rate of climate change. Scientists expect that the average global surface temperature could rise 1-4.5°F (0.6-2.5°C) in the next fifty years, and 2.2-10°F (1.4-5.8°C) in the next century, with significant regional variation. Evaporation will increase as the climate warms, which will increase average global precipitation. Soil moisture is likely to decline in many regions, and intense rainstorms are likely to become more frequent. Sea level is likely to rise two feet along most of the U.S. coast.

Calculations of climate change for specific areas are much less reliable than global ones, and it is unclear whether regional climate will become more variable.

Since 1979, scientists have generally agreed that a doubling of atmospheric carbon dioxide increases the earth's average surface temperature by 3-8°F (1.5-4.5°C). More recent studies have suggested that the warming is likely to occur more rapidly over land than the open seas. Moreover, the warming in temperatures tends to lag behind the increase in greenhouse gases. At first, the cooler oceans will tend to absorb much of the additional heat and thereby decrease the warming of the atmosphere. Only when the ocean comes into equilibrium with the higher level of CO_2 will the full warming occur.

The Intergovernmental Panel on Climate Change estimates that the concentration of CO_2 will double from pre-industrial levels by the mid- to late 21st century. Currently, the panel projects a global average warming of 1.0-4.5°F (0.6-2.5°C) in the next fifty years and 2.5 to 10.4°F (1.4 to 5.8°C) by the year 2100, compared with the global average temperature in 1990. The wide range in projected temperatures is due to varying assumptions about future trends in greenhouse gas emissions and sulfate aerosols.

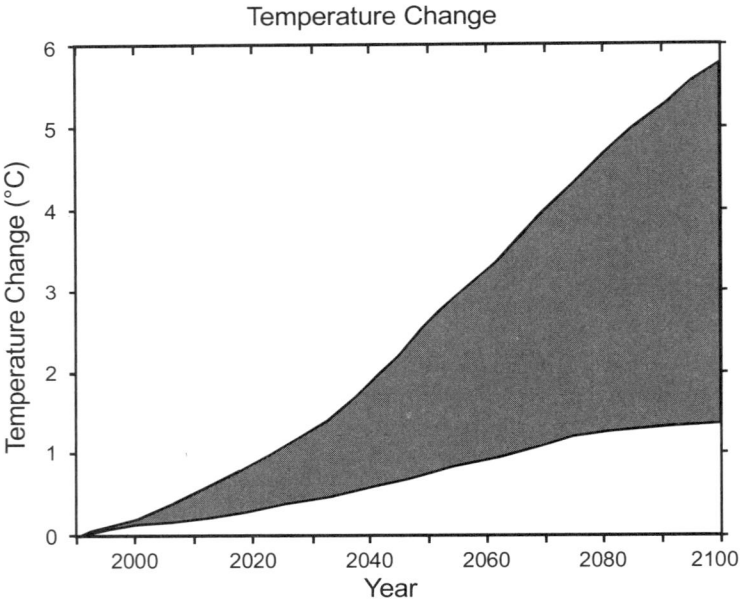

This background information is from the Environmental protection Agency

For more information see:
http://yosemite.epa.gov/oar/globalwarming.nsf/content/index.html

SAMPLE DATA

Amount of Gas in Atmosphere

Gas	1900	1950	2000	2050
Carbon dioxide	280 ppm	330 ppm	380 ppm	540-970 ppb
methane	750 ppb	1250 ppb	1750 ppb	1570-3730 ppb
Nitrous oxides	270 ppb	278 ppb	290 ppb	354-460 ppb
CFC's	0	40 ppt	618 ppt	650 ppt

ppm = parts per million, ppb = parts per billion, ppt = parts per trillion

Average Global Temperature Change (°C)

1900	1950	2000	2050	2100
−°.2	+°.2	+°.8	+°2.5	+5.6

Emission Possible
Student Research Project on Greenhouse Gases and Global Warming

PURPOSE
In this activity you will analyze the problems associated with human impact on increased production of certain greenhouse gases responsible for trapping additional heat in the earth's atmosphere.

MATERIALS
    7 computers
    Microsoft PowerPoint or HyperStudio
    computer projector

PROCEDURE
The President of the United States has requested a briefing on the effects of global warming and an outlook for the future. Your team has been named as a special consultant to the Secretary of the Interior and has been called upon to prepare this brief and present it to the President. For your presentation, you must address:

1. Background facts addressing global warming and the effect of greenhouse gases. You should also assess the current conditions on earth. Your background should include:
   - the natural atmospheric warming that occurs (*group*)
   - definition of the problem (*group*)
   - background about the sources, uses, and atmospheric effects of one of the greenhouse gases (*individual*)

   You will cover the first two points as a group. For the third point, each member of your team will research one of the greenhouse gases, CH_4 (methane), CO_2 (carbon dioxide), N_2O (nitrous oxide), or CFCs (chlorofluorocarbons), for the background section of the report. **Each of you must have at least 5 sources for your background, listed by section in the bibliography.** *Only one of these may be an encyclopedia. Be sure to properly cite internet sources.*

2. Projections for conditions on the earth resulting from several different scenarios of either continued or reduced usage of greenhouse gases by industry and the public. (*group*)

   Use the computer to research projected levels of your gases over the next 50 years. Data tables are provided for you on the student answer page. Then using this data, write a brief summary of your findings and predict future outcomes for our earth under various conditions. Use graphs to show the overall increase or decrease in gas levels, temperature, and water levels.

3. Recommendation for action. (*group*)

   Taking into account the degradation of the environment versus the dependence of our society on petroleum products, CFCs, and other materials contributing to the greenhouse effect, propose a plan of action for the president.

What your group must produce:

Contract — Each member of the group agrees to their part of the project and to specific due dates. The contract is to be typed and signed by all members. The contract will include times for computer use, paper due dates for individual papers that will enable the group to be successful in writing the common sections. The more specific you are in the contract, the more likely your group will be successful.

Background research — Your group background research should answer all of the following questions:
- What is global warming and what might be the projected consequences?
- What is the greenhouse effect?
- What is the current United State's position on global warming?
- Describe current and past international agreements dealing with the atmosphere.

Proposal for Action — Your group should develop recommendations to give the president describing ways to remediate global warming. There should be a recommendation for each gas and an overall recommendation.

The Final Project — The final project will consist of both a written and oral presentation of your findings.

A group written brief — Your brief should contain the following sections in order:
1. Background section
   group general background (~1-2 pg)
   individual gas papers, each with bibliography (~2-3 pgs w/o bib. each)
2. Future gas predictions
   written summary of your findings (~1-2 pg)
   graphs showing your findings
3. Proposal for action
   specific solutions to the problem that address **all** of the gases (~1-2 pg)

These written documents must be typed, double-spaced, 12 pt professional font with 1-inch margins. Include a cover page listing the names of each group member and the gas each researched. Page numbers and a table of contents are also recommended.

An oral MS PowerPoint presentation—whereby your group will present its findings to the President.

In this presentation, your group will make a PowerPoint presentation **with no more than 13 slides**:
- title slide
- general background
- 1 to 2 slides per individual gas (total 3-8)
- proposal for action
- closing slide

What you as an individual need to produce:

A 2-3 page paper on the greenhouse gas that you have chosen. This paper must include sources of the gas, uses of it, and the methods by which it enters the atmosphere. Your paper must include a bibliography of at least 5 sources, only one of which can be an encyclopedia. Your paper must follow the guidelines for the group paper as stated above (i.e. must be typed, double spaced, 12 pt professional font, 1 inch margins, introduction, conclusion, thesis, proper citations: in short, it must be a professional piece of work). Most of the research and writing of your paper will be done outside of class.

Grading will be as follows:

1. Daily grades on expected completed items
2. An individual grade on your section of the background part paper
3. A group grade on your group's written brief, research, and graphs
4. A grade on your group's PowerPoint presentation based on both individual and group achievement
5. An individual grade based on peer evaluation of your cooperation and work in the group

Sample Calendar:

10/2-10/3 - Project assigned
10/2-10/3 - Contracts signed
10/6-10/7 - Simulations and graphs complete (end of period)
10/11   - Rough draft of contract work due/ in class edit
           - "Slide basics" completed by end of period
10/18   - Final group briefs due at the beginning of the class period, presentations begin

Sample bibliography:

Works Cited

Atchley, John A. and Susan, Heeger. *When Will We Laugh Again?* New York: Norton, 1990

Brumberg, John Jacobs. "The Origins of Anorexia Nervosa." *Harper's Magazine*, August 1993: 31-32

Goldman, Nechama. "Fatal Obsession." Reprint from *JERUSALEM POST* ,19 Mar. 1993. 64. SIRS: Medical Science, 1993. Ed.

Webb, Adrienne. "The Starvation of Demons." *McLeans*, 2 May 1994: 50.

Fredrickson, B.L. (200, March 7). "Cultivating Positive Emotions to Optimize Health and Well-being." *Prevention and Treatment,* 3, Article 0001a. Retrieved November 20, 2003, from http://journals.apa.org/prevention/volume3/pre0030001a.html

Britannica Junior Encyclopedia, Volume 2, NY, USA: Encyclopedia Britannica Inc., 1970, page 121

Name _____

Period _____

Emission Possible
Student Research Project on Greenhouse Gases and Global Warming

ANALYSIS

Amount of gas in atmosphere

Gas	1900	1950	2000	2050
Carbon dioxide				
methane				
Nitrous oxides				
CFC's				

Average Global Temperature Change (°C)

1900	1950	2000	2050	2100

RUBRICS

Oral Presentation Assessment

	Excellent	Good	Average	Below Average	Poor	Not Present
General introduction: introductions, definition of problem, natural greenhouse effect, history of global warming **Individual Gases**: sources, association with global warming, other facts	20	18	15	12	8	0
_____	30	25	22	18	12	0
_____	30	25	22	18	12	0
_____	30	25	22	18	12	0
_____	30	25	22	18	12	0
Climate Simulation Overview: thoroughness, interpretation of data, organized presentation	20	18	15	12	8	0
Proposal for Action: covers all gasses, reasonable, original, supportable	20	18	15	12	8	0
Overall Presentation: volume, use of language, stance, eye contact, not reading from slide						
_____	10	8	7	6	4	0
_____	10	8	7	6	4	0
_____	10	8	7	6	4	0
_____	10	8	7	6	4	0

TOTAL SCORE:

_____   _____

_____   _____

_____   _____

_____   _____

Emission Impossible
Individual Written Grade

Name: _____

Gas: _____

	Score (out of 5)	Weight	Total Points
Scientific Content (accurate logical reasoning, info based on contemporary research, facts relevant to topic)		5x	
Organization (logical flow, transitions used to connect ideas, concise)		3x	
Thesis (motivated by a question or problem, provides a clear map of paper)		2x	
Introduction and Conclusion (grabs attention, suggests further research, conclusion rephrases thesis)		3x	
Works Cited (conforms to proper format)		3x	
Mechanics (1 inch margins, 12 point consistent font, typed, double-spaced, proper length)		4x	
Total score			

Emission Impossible
Peer Critique

Name of evaluator _____

Class period _____

Instructions: Every member of your group should complete one of these forms. Be sure to clearly list all members of your group including yourself. Take time and think about each person, including yourself, and truthfully decide the amount of effort each devoted to this project. For instance, if someone came to class unprepared, they may receive only a 2 in preparation yet still earn a 5 in cooperation. This is difficult! Be fair! Complete each section and write a total for each person. Please keep this form private. Hand the form into your teacher after completion.

Group members names	Cooperation	Preparation for class	Work in class	Total points
	5 4 3 2 1	10 8 6 4 2	10 8 6 4 2	
	5 4 3 2 1	10 8 6 4 2	10 8 6 4 2	
	5 4 3 2 1	10 8 6 4 2	10 8 6 4 2	
	5 4 3 2 1	10 8 6 4 2	10 8 6 4 2	

Emission Impossible
Group Written Grade

Group members: _____

_____

	Excellent	Good	Average	Below Average	Poor	Not Present
Introduction/background	20	18	15	12	8	0
Climate simulation summary including graphs	20	18	15	12	8	0
Proposal for action	20	18	15	12	8	0
Climate simulation sheets	10	8	7	6	4	0
Gas papers included	10	8	7	6	4	0
Format/appearance/ mechanics	20	18	15	12	8	0

Group Written Grade Total _____

Greenhouse Effect
Investigating Global Warming

OBJECTIVE

Students will design three different environments, including a control group. They will identify which environment results in the greatest temperature change. Using the temperature probes and the computer graphing software, data will be collected and analyzed for each environment modeled in the experiment. At the end of the experiment, students will be able to define the greenhouse effect and predict future changes in our atmosphere.

LEVEL

Middle Grades: Earth Science

NATIONAL STANDARDS

UPC.2, UPC3, A.1, A.2, B.3, D.2, F.4, F.5

TEKS

6.1(A), 6.2(A), 6.2(B), 6.2(C), 6.2(D), 6.2(E), 6.3(A)
7.1(A), 7.2(A), 7.2(B), 7.2(C), 7.2(D), 7.2(E), 7.14(C)
8.1(A), 8.2(A), 8.2(B), 8.2(C), 8.2(D), 8.2(E), 8.6(C), 8.9(A), 8.9(C), 8.12(C), 8.14(C)
IPC 1(A), 2(A), 2(C), 2(D), 8(A), 8(E), 9(A), 9(B), 9(C)

CONNECTIONS TO AP

AP Environmental Science:
  IV. Environmental Quality A. Air/Water/Soil 1. major pollutants
  V. Global Changes and Their Consequences A. First-order Effects 1. atmosphere B. Higher-order Interactions 1. atmosphere

TIME FRAME

50 minutes

MATERIALS

(For a class of 28 working in groups of four)

7 Lab Pro's	1 large bag of soil
7 computers w/Logger Pro software	28 600 mL beakers
21 temperature probes	21 rulers
7 lamps with 100 watt bulbs	plastic wrap
tape	200 grams baking soda
850 mL of vinegar	

TEACHER NOTES

The Greenhouse Effect deals with global warming as a result of changes in the composition of the atmosphere. The lab may be done when studying the atmosphere, energy sources or environmental changes.

The students will be using the materials provided to create three different environments and measure the changes in temperature for each environment when under a heat lamp for 15 minutes. At the beginning of class, question students regarding their knowledge of the earth's atmosphere. After brainstorming information, describe the three beakers the students will be testing during the experiment. The uncovered beaker acts as a control, the covered beaker represents the earth with its atmospheric blanket, and the CO_2 beaker represents an atmosphere with high levels of CO_2. Students will investigate how changing the composition of the atmosphere might change the heat trapping ability of the atmosphere.

POSSIBLE ANSWERS TO THE CONCLUSION QUESTIONS AND SAMPLE DATA

Data Table 1			
	Beaker 1	Beaker 2	Beaker 3
Time (minutes)	Probe 1 (Celsius) Control Group	Probe 2 (Celsius) W/out Gas Added	Probe 3 (Celsius) W/ Gas Added
0-minute Temp.	27	27	27
1-minute Temp.	27	27	27
2-minute Temp.	28	28	28
3-minute Temp.	28	28	29
4-minute Temp.	29	29	29
5-minute Temp.	29	30	30
6-minute Temp.	30	30	30
7-minute Temp.	29	30	30
8-minute Temp.	29	29	30
9-minute Temp.	29	29	29
10-minute Temp.	28	29	29
11-minute Temp.	28	30	30
12-minute Temp.	29	31	31
13-minute Temp.	30	31	31
14-minute Temp.	30	32	32
15-minute Temp.	30	32	33

Data Table 2		
Time (minutes)	Temperature difference between beaker 1 and beaker 2	Temperature difference between beaker 1 and beaker 3
0-minute Temp.	0	0
1-minute Temp.	0	0
2-minute Temp.	0	0
3-minute Temp.	0	1
4-minute Temp.	0	0
5-minute Temp.	1	1
6-minute Temp.	0	0
7-minute Temp.	1	1
8-minute Temp.	0	1
9-minute Temp.	0	0
10-minute Temp.	1	1
11-minute Temp.	2	2
12-minute Temp.	1	2
13-minute Temp.	1	1
14-minute Temp.	2	2
15-minute Temp.	2	3

ANALYSIS

1. In the spaces provided in Data Table 2, subtract to find the temperature differences.
   * on chart

CONCLUSION QUESTIONS

1. During periods when the lamp was on, did the covered beakers warm faster or slower than the control? Did the covered beakers (beakers 2 and 3) have about the same temperature or different temperatures throughout the experiment?
   * The covered beakers heated faster. The covered beakers were not the same. The beaker with carbon dioxide had a higher temperature.

2. Give a possible explanation for your answers in question one.
   • The covered beakers let heat in, but did not let heat out. Carbon dioxide retains more heat than regular air.

3. What important greenhouse gas did the air in beaker 3 contain?
   • Carbon dioxide

4. During the periods when the lamp was off, did the uncovered beaker cool faster or slower than the covered beakers? Justify your answer.
   • The control beaker (uncovered) cooled off more quickly because it was uncovered and lost heat faster.

5. Explain why a closed automobile heats up in the sun.
   • A closed automobile in the sun allows heat to come into the car but then traps the heat and doesn't let the heat get out. A closed car acts like a greenhouse when parked in the sun.

6. Design an experiment to test the ability of methane to trap heat.
   • Repeat the experiment as done here, but replace carbon dioxide in the third beaker with methane. Heat with a lamp and measure the change in temperature in the three beakers.

Greenhouse Effect
Investigating Global Warming

The earth is surrounded by a layer of gases which help to retain heat and act like a greenhouse. Greenhouses allow gardeners to grow plants in cold weather. Radiation from the sun passes through the glass and experiences a change in its wavelength. The new wavelength radiation is unable to pass back through the glass and is trapped inside the greenhouse. As a result the temperature of the air inside the greenhouse is increased. This, along with the lack of mixing between the inside and outside air, keeps the greenhouse consistently warm.

Similarly the gases in our atmosphere trap heat. The main components of our atmosphere are N_2, O_2, CO_2, H_2O and Ar.

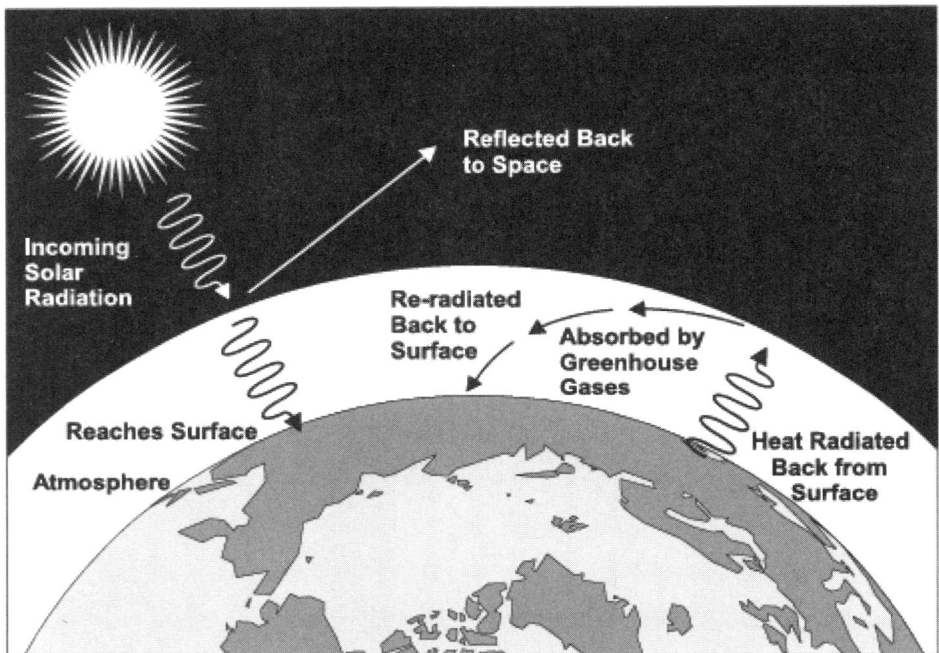

Fig. 1

In this experiment you will have three beakers to model different environments. The first beaker will be filled with soil, remain uncovered, and serve as our control. The second beaker will contain soil and have a plastic cover, representing the earth with its atmospheric layer. The third beaker will contain soil, a plastic cover and CO_2, which is a gas that has been increasing in our atmosphere over the last 100 years.

PURPOSE
You will analyze temperature data from the three beakers, draw conclusions and make predictions from the data.

MATERIALS

1 Lab Pro and computer
3 temperature probes
1 lamp with 100-watt bulb
4 600 mL beakers
15 grams of baking soda

3 rulers
tape
soil
plastic wrap
20 mL of vinegar

Safety Alert
1. Students should avoid touching the heat lamp.
2. Students should wear goggles when mixing the baking soda and vinegar together.

PROCEDURE

1. Make a hypothesis about which beaker will retain the most heat.

2. Plug the temperature probes into channels 1, 2 and 3 of your **Vernier** computer interface.

3. Tape each temperature probe to a ruler as shown in Figure 2. The probe tips should each be 3 cm from the ruler ends and the tape should not cover the probe tips.

4. Prepare the computer for data collection by opening **Logger Pro** on your computer. The computer should automatically detect the temperature probes that are connected in each channel. A data collection window should open. Check with your teacher if the computer does not detect the probes.

5. Obtain four beakers and prepare three of them for data collection.

6. Place a layer of soil 1 cm deep in each beaker.

7. Place the temperature probes into the beakers as shown in Figure 2.

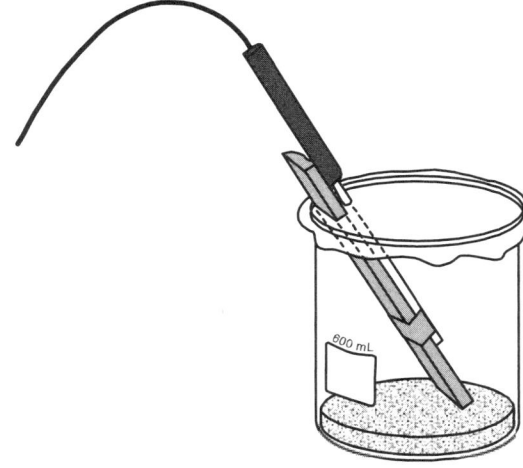

Fig. 2

8. Cover the top of beakers 2 and 3 tightly with plastic wrap. Remove any excess plastic wrap covering the sides of the beaker. Beaker 1 should be open to air (NO PLASTIC WRAP) and is the control. Beakers 2 and 3 represent your covered greenhouses.

9. In a **separate clean beaker**, combine **15 grams of baking soda and 20 mL of vinegar**. The mixture should immediately produce a gas, as demonstrated in the following equation.
$CH_3COOH_{(aq)} + NaHCO_{3 (s)} \rightarrow NaCH_3COO_{(aq)} + H_2O_{(l)} + CO_{2 (g)}$

10. Remove the plastic covering from beaker 3 and pour the gas **slowly** into the beaker. Be careful and **do not allow any liquid to be transferred**. After all the gas has been successfully poured into beaker 3, immediately cover it with plastic wrap.

11. Position a light bulb the same distance from all three beakers, about 7 cm above the tabletop and the same distance from all three temperature probe tips.

12. Click ▶Collect to begin data collection. Turn on the lamp.

13. Monitor the time in the meter window. When 5 minutes have passed, **turn off** the lamp. Data will continue to be collected.

14. At the 10-minute mark, **turn the lamp back on**. Data collection will stop after 15 minutes.

15. When data collection stops, turn the lamp off and remove the temperature probes from the beakers.

16. Turn on the **EXAMINE** feature by clicking the **EXAMINE** button, ▦ on the toolbar.

17. Move the cursor to the 0-minute mark on the graph. Use the **EXAMINE BOX** to determine the temperatures in beakers 1, 2, and 3, and record them in the data table.

18. Use the same method to determine the temperatures at the 1-, 2-, 3- minute, etc. marks and record them in the data table.

19. Print copies of the graph as directed by your teacher.

20. Choose **STORE LATEST RUN** from the **DATA** menu.

Name _____

Period _____

Greenhouse Effect
Investigating Global Warming

HYPOTHESIS

DATA AND OBSERVATIONS

	Beaker 1	Beaker 2	Beaker 3
Time (minutes)	Probe 1 (Celsius) Control Group	Probe 2 (Celsius) W/out Gas Added	Probe 3 (Celsius) W/ Gas Added
0-minute Temp.			
1-minute Temp.			
2-minute Temp.			
3-minute Temp.			
4-minute Temp.			
5-minute Temp.			
6-minute Temp.			
7-minute Temp.			
8-minute Temp.			
9-minute Temp.			
10-minute Temp.			
11-minute Temp.			
12-minute Temp.			
13-minute Temp.			
14-minute Temp.			
15-minute Temp.			

Data Table 1

	Data Table 2	
Time (minutes)	Temperature difference between beaker 1 and beaker 2	Temperature difference between beaker 1 and beaker 3
0-minute Temp.		
1-minute Temp.		
2-minute Temp.		
3-minute Temp.		
4-minute Temp.		
5-minute Temp.		
6-minute Temp.		
7-minute Temp.		
8-minute Temp.		
9-minute Temp.		
10-minute Temp.		
11-minute Temp.		
12-minute Temp.		
13-minute Temp.		
14-minute Temp.		
15-minute Temp.		

ANALYSIS

Printed Graph

1. In the spaces provided in the Data Table 2, subtract to find the temperature differences.

CONCLUSION QUESTIONS

1. During periods when the lamp was on, did the covered beakers warm faster or slower than the control? Did the covered beakers (beakers 2 and 3) have about the same temperature or different temperatures throughout the experiment?

2. Give a possible explanation for your answers in question one.

3. What important greenhouse gas did the air in beaker 3 contain?

4. During the periods when the lamp was off, did the uncovered beaker cool faster or slower than the covered beakers? Justify your answer.

5. Explain why a closed automobile heats up in the sun.

6. Draw a sketch and describe an experiment to test the ability of methane to trap heat.

Are You Meeting the Kyoto Protocol
Calculating Your Carbon Dioxide Footprint

OBJECTIVE
Students will calculate their yearly CO_2 emissions and compare their results with the amounts listed in the international Kyoto Agreement on greenhouse gases.

LEVEL
Middle Grades: Earth Science

NATIONAL STANDARDS
UPC.2, UPC3, A.1, A.2, B.3, D.2, F.4, F.5

TEKS
6.2(C), 6.2(D), 6.2(E)
7.2(C), 7.2(D), 6.2(E)
8.2(C), 8.2(D), 8.2(E), 8.6(C), 8.12(C), 8.14(C)
IPC 1(A), 2(A), 2(C), 2(D), 8(A), 8(E), 9(A), 9(B), 9(C)

CONNECTIONS TO AP
AP Environmental Science:
    IV. Environmental Quality A. Air/Water/Soil 1. major pollutants
    V. Global Changes and Their Consequences A. First-order Effects 1. atmosphere B. Higher-order
    Interactions 1. atmosphere

TIME FRAME
45 minutes

MATERIALS
Calculators, Global Warming Wheel Card

TEACHER NOTES
Calculation Your Carbon Dioxide Footprint has students explore how human activity has resulted in changing the composition of the atmosphere. The lab may be done when studying the atmosphere, energy sources or environmental changes.

Environmental scientists call the amount of anything you produce into the environment your "footprint". In this lab students will be specifically calculating the amount of carbon dioxide they are emitting into the atmosphere and therefore calculating their CO_2 footprint. Other footprint labs might have students calculate their solid waste footprint, waste water footprint, methane footprint or even a combination of all categories just called their "environmental footprint".

TEACHER PAGES

The Kyoto Protocol was developed by the United Nations Framework Convention on Climate Change and was adopted by the United Nations in December of 1997. The protocol brought an international focus to limiting the emission of greenhouse gases and specifically focused on carbon dioxide, methane, and nitrous oxides. While 84 nations, including the U.S., have signed the treaty, not all nations have ratified it, which would require nations to enforce it. The Kyoto Protocol has different recommendations for developed vs. developing countries and only recommends limiting emissions for the 38 nations with the highest emissions. Thirty-one of the high-emission nations have ratified it and are making efforts to reduce their emissions. The U.S. is one of the seven countries that has signed the protocol but not put any limits in place. Even for countries complying with the Kyoto Protocol, it is not a binding contract, and countries police themselves.

A copy of the Protocol is available at: http://unfccc.int/resource/docs/convkp/kpeng.html

Prior to doing this activity in class, give your students the homework assignment to talk to their parents and research their household amounts of heating cost (natural gas or heating oil), electricity cost, miles traveled and recycling (the listed categories of carbon dioxide production from the student answer sheet). The students will bring this data to class to calculate their carbon dioxide emissions.

Each group will need a Global Warming Wheel Card.
http://yosemite.epa.gov/oar/globalwarming.nsf/uniquekeylookup/shsu5bwjq7/$file/wheelcard.pdf

You will need to print these and put them together prior to class.

TEACHER PAGES

Global Warming —

What's Your

Score?

In the United States, a typical household of two people generates approximately 60,000 pounds of carbon dioxide (CO_2) emissions every year from household activities and personal transportation.

Emmissions Source

Pounds of Co_2 Emitted Per Year

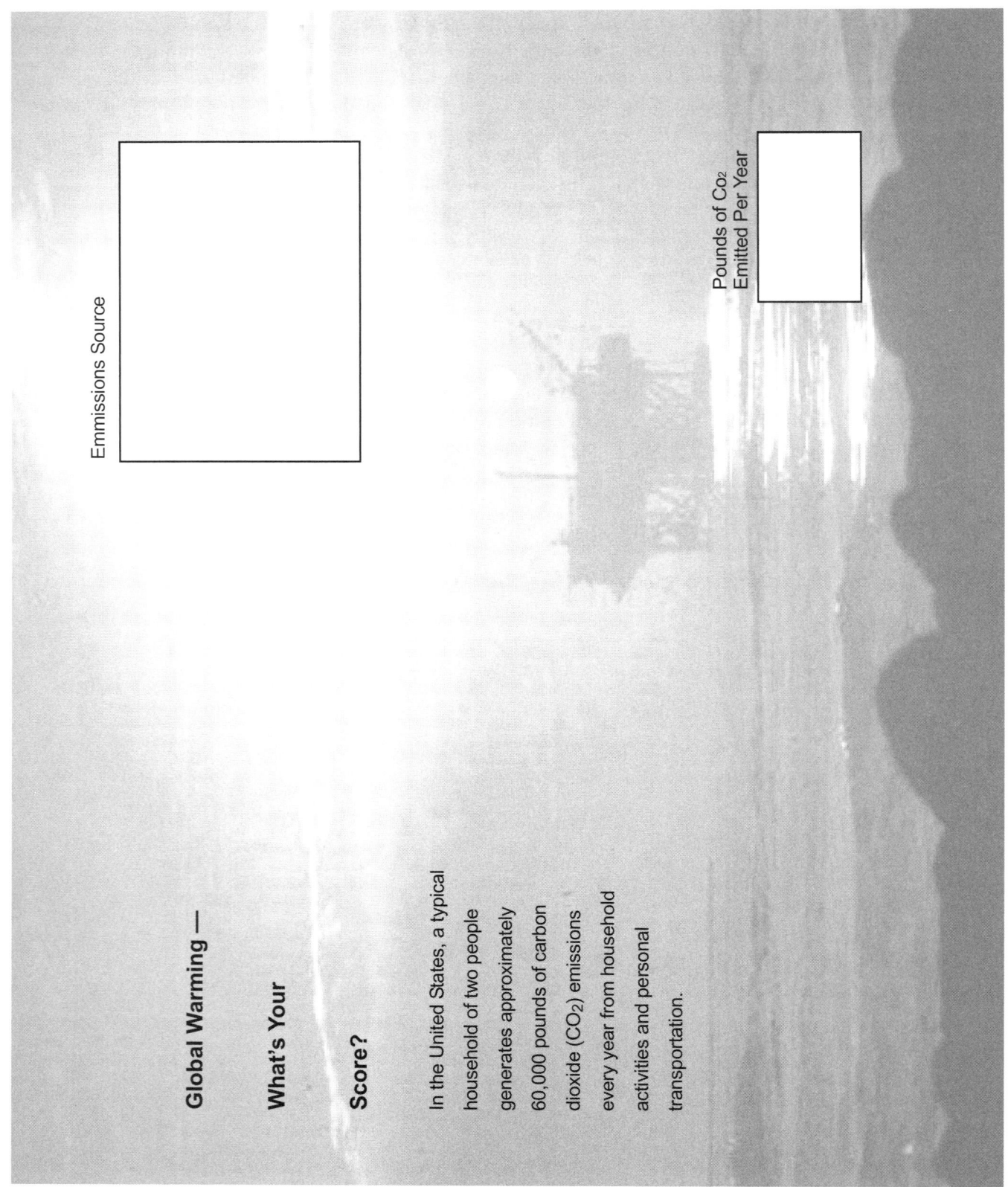

Transportation

24,000 lbs./year
12,000 lbs./year
6,000 lbs./year

On average, how many miles does
your household put on your car(s)
per week? Pick closest amount.

- 200 miles (10,400 miles/year)
- 400 miles (20,000 miles/year)
- 600 miles (31,200 miles/year)

Home Heating

9,800 lbs./year
19,700 lbs./year
29,500 lbs./year

Which of the following scenarios
best describes your household's
waste disposal practices?

- We don't recycle at all.
- We recycle about half of all
  recyclable materials (paper,
  plastics, glass).
- We recycle everything we can.

Global Warming —
What's Your Score?

On average, how much does your
household spend on electricity
each month? Pick closest amount.

- $40
- $80
- $120

11,800 lbs./year
23,600 lbs./year
35,400 lbs./year

Waste Disposal

- $25
- $50
- $100

On average, how much does your
household spend on natural gas
or fuel oil each month?
Pick closest amount.

6,300 lbs./year
3,900 lbs./year
1,500 lbs./year

Electricity Use

Global Warming — What Can You Do?

Electricity Use

- Recycle half of your potentially recyclable materials (paper, plastics, glass).
- Reduce your waste generation by 10 percent.

100 lbs./year
1,700 lbs./year
1,500 lbs./year

Waste Disposal

1,500 lbs./year
1,700 lbs./year
100 lbs./year

- Buy a car that averages 10 miles per gallon more than your present vehicle.
- Avoid driving 25 miles a week by waling, biking, carpooling, or taking mass transit instead.

Home Heating

- Replace 10 60-watt incandescent light bulbs with 10 13-watt compact flourescent bulbs.
- Replace old refrigerator with a new Energy Star® model.
- Turn up your central air conditioner's thermostat by 2°F in summer

7,000 lbs./year
1,500 lbs./year

Transportation

- Replace your old furnace or boiler with an Energy Star® model.
- Replace single-glazed windows with Energy Star® windows.
- Turn down your thermostat by 10° Fahrenheit each night in winter.

2,400 lbs./year
1,200 lbs./year

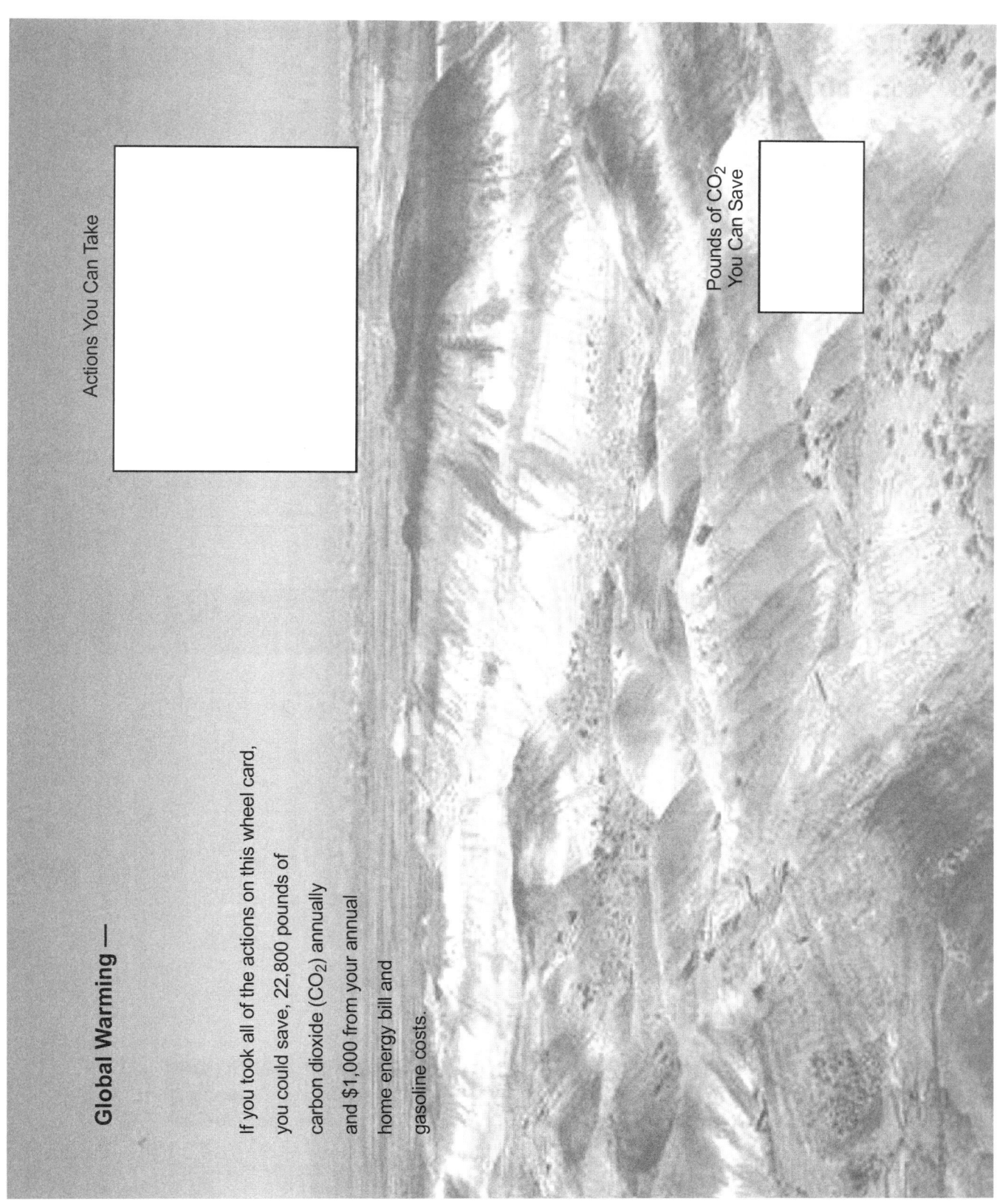

Global Warming —

If you took all of the actions on this wheel card, you could save, 22,800 pounds of carbon dioxide (CO_2) annually and $1,000 from your annual home energy bill and gasoline costs.

Actions You Can Take

Pounds of CO_2 You Can Save

POSSIBLE ANSWERS TO THE CONCLUSION QUESTION AND SAMPLE DATA

Data – What's Your Score

Category	Your Estimated Household Amount	Pounds of CO_2 Emitted Per Year
Waste Disposal	recycle half	3,900
Transportation	400 miles	23,600
Home Heating/Water Heating	$25	6,000
Electricity Use	$120	29,500

Analysis – What's Your Score

Yearly Household Total (lbs. of CO_2)	Number of People in Household	Your Individual Total (lbs. of CO_2)
63,000	3	21,000

Data – What Can You Do?

Category	Actions You Can Take	Pounds of CO_2 Saved Per Year
Waste Disposal	Reduce waste by 10%	1,200
Transportation	Improve mileage by 10 miles per gallon and cut driving 25 miles per week	8,500
Home Heating/Water Heating	Turn thermostat down 10 degrees Fahrenheit each night	250
Electricity Use	Replace 60-watt incandescent bulbs with 13-watt fluorescent bulbs	1,500

Analysis – What Can you Do?

Total Saved (lbs. of CO_2)	Number of People in Household	Your Individual Amount Saved (lbs. of CO_2)
11,450	3	3,817

CONCLUSION QUESTIONS

1. How does your individual total compare to the total carbon dioxide emissions recommended by the Kyoto Protocol?
   - Answers will vary.
   - For the sample data, the total was twice the recommended amount.

2. What category produced the most CO_2 in your household? List ways you could reduce this total.
   - Answers will vary. Transportation was by far the largest category. Students might suggest taking the bus, walking, carpooling or arranging to take fewer trips by car. We also could buy a car with better gas mileage or a hybrid that gets great gas mileage.

3. How much were you able to reduce your emissions by making the changes recommended by *What Can You Do?*
   - In the sample data, almost 4,000 pounds or 2 tons. This is still above the Kyoto recommended amounts, but that is a great improvement.

4. Individuals in the developing countries produce much less CO_2 than individuals in the United States. Why is this true?
   - They use less energy because they have fewer cars and less money for expensive house heating and cooling systems.

5. If the answer to question number four is true, why is it still important for developing countries to work on reducing emissions?
   - Population is growing more rapidly in developing countries and the more people a country has, the more emissions they produce. Also, developing countries have more forest to clear for agriculture.

6. The United States has signed the Kyoto Protocol, but does not follow or enforce the agreement. What recommendations would you make to the government about this treaty? Should it be modified for the United States?
   - Answers will vary. The U.S. government has concerns that the Kyoto Protocol would adversely effect the economy. There are many aspects to reducing emissions that would not effect the economy, especially those focused on individuals making changes in their day to day energy use. It is not necessary to modify the protocol, as the amounts are **goals** to be met by countries and there is no enforcement arm. Having a low amount as a goal is a good thing. The Kyoto Protocol is already modified for the developed countries.

TEACHER PAGES

7. In your opinion, is there any hope for something like the Kyoto Protocol, or will countries of the world never be able to reach such accords?

- Yes, the Montreal Protocol, which was drafted to address the destruction of ozone in Earth's atmosphere by CFC's, has been very successful in reducing the use of CFC's. In contrast to the Kyoto Protocol, the Montreal Protocol was ratified and enforced by the ratifying nations, leading to a reduction in CFC releases to the atmosphere. Projections are that this will lead to a reduction of the hole in the ozone layer. So, while the Montreal Protocol provides a reason for optimism in regard to question 7, there are some key differences between the two issues. Students should compare the relative impacts on a nation's economic infrastructure between 1) replacing one manufactured chemical compound used for aerosol products and refrigerants with another manufactured chemical compound (i.e., result of Montreal protocol) and 2) replacing one form of energy (e.g., fossil fuels) with another (e.g., electric/hybrid vehicles, hydrogen cells, solar, wind and other alternative energies).

TEACHER PAGES

Are You Meeting the Kyoto Protocol
Calculating Your Carbon Dioxide Footprint

The Kyoto Protocol is an international agreement written by the United Nations Framework Convention on Climate Change to limit or reduce the emission of greenhouse gases. Carbon dioxide is one of the gases which is addressed by this Protocol. The recommended amount of carbon dioxide emissions per person is 5.5 tons per year (1 ton = 2000 pounds). How much carbon dioxide do you produce a year? Can you reduce that number?

PURPOSE
Students will caclulate the amount of carbon dioxide they contribute to the atmosphere every year.

MATERIALS
calculators

PROCEDURE
Part I – What's Your Score?
1. Make a hypothesis about the amount of CO_2 you produce a year.

2. Use the global warming wheel and the information you gathered for homework to fill in the data table on the student answer page.

3. In the Part I analysis section of your student answer page calculate your family's total pounds of CO_2 emitted. To adjust this number to reflect the amount of CO_2 you as an individual produce, divide the total emissions by the number of people in your household. Record this value in the appropriate box on the analysis table.

Part II – What Can You Do?
1. Use the global warming wheel to complete the Part II data table.

2. Calculate your family's total CO_2 savings and record it in the Part II analysis table. Divide by the number of people who live in your family to get your individual savings.

3. Use the data you collected and the information on this page to answer the conclusion questions on your student answer page.

Name _____

Period _____

Are You Meeting the Kyoto Protocol
Calculating Your Carbon Dioxide Footprint

HYPOTHESIS

Part I – What's Your Score?

Data – What's Your Score

Category	Your Estimated Household Amount	Pounds of CO_2 Emitted Per Year
Waste Disposal (percentage of waste you recycle)	%	
Transportation (miles per week)	miles	
Home Heating/Water Heating ($ per month)	$	
Electricity Use ($ per month)	$	

Analysis – What's Your Score

Yearly Household Total (lbs. of CO_2)	Number of People in Household	Your Individual Total (lbs. of CO_2)

Part II – What Can You Do?

Data – What Can You Do?

Category	Actions You Can Take	Pounds of CO_2 Saved Per Year
Waste Disposal		
Transportation		
Home Heating/Water Heating		
Electricity Use		

Analysis – What Can You Do?

total Saved (lbs of CO_2)	Number of people in household	your individual Amount Saved (lbs. of CO_2)

CONCLUSION QUESTIONS

1. How does your yearly household total compare to the total carbon dioxide emissions as recommended by the Kyoto Protocol?

2. What category produced the most CO_2 in your household? List ways you could reduce this total.

3. How much were you able to reduce your emissions by making the changes recommended by *What Can You Do?*

4. Individuals in the developing countries produce much less CO_2 than individuals in the United States. Why is this true?

5. If the answer to question number four is true, why is it still important for developing countries to work on reducing emissions?

6. The United States has signed the Kyoto Protocol, but does not follow or enforce the agreement. What recommendations would you make to the government about this treaty? Should it be modified for the United States?

7. In your opinion, is there any hope for something like the Kyoto Protocol, or will countries of the world never be able to reach such accords?

Acid Rain Drops Keep Falling on My Head
Investigating Acid Rain

OBJECTIVE

Students will be able to describe how CO_2, NO_2, and SO_2 gases cause acid rain by forming the following acids in the atmosphere:

- Carbonic acid, H_2CO_3
- Nitrous acid, HNO_2
- Nitric acid, HNO_3
- Sulfurous acid, H_2SO_3

LEVEL

Middle Grades: Earth Science

NATIONAL STANDARDS

UPC.2, UPC3, A.1, A.2, B.3, D.2, F.4, F.5

TEKS

6.1(A), 6.2(A), 6.2(B), 6.2(C), 6.2(D), 6.2(E), 6.3(A)
7.1(A), 7.2(A), 7.2(B), 7.2(C), 7.2(D), 7.2(E), 7.14(C)
8.1(A), 8.2(A), 8.2(B), 8.2(C), 8.2(D), 8.2(E), 8.9(A), 8.9(C), 8.14(C)
IPC 1(A), 2(A), 2(C), 2(D), 8(A), 8(E), 9(A), 9(B), 9(C)

CONNECTIONS TO AP

AP Chemistry:
   III. Reactions A. Reaction types 1. Acid-base reactions
AP Environmental Science:
   IV. Environmental Quality A. Air/Water/Soil 1. major pollutants
   V. Global Changes and Their Consequences A. First-order Effects 1. atmosphere B. Higher-order
   Interactions 1. atmosphere

TIME FRAME

60 minutes

MATERIALS

(For a class of 28 working in groups of 4)

2 L distilled water	14 thin-stem Beral pipet with 1.0 M HCl
50 grams solid potassium nitrite, KNO_2(or $NaNO_2$)	14 100 mL beakers
50 grams solid sodium bicarbonate, $NaHCO_3$ (baking soda)	14 24 well-plates
50 grams solid sodium hydrogen sufite, or sodium bisufite, $NaHSO_3$	30 mL universal indicator
	tap water for cleanup
42 thin-stem Beral pipets stretched to make a small opening	42 thin-stem Beral pipets cut to have a large opening
	28 pairs safety goggles

TEACHER NOTES

Investigating Acid Rain allows students to see how different gases added to the atmosphere might change the pH of rain. It may be taught while studying the atmosphere, environmental issues or even when teaching acid-base reactions and chemical changes.

Students are using thin-stem Beral pipets. They are extremely cheap and easy to use. To perform this lab students must make 4 pipets with a narrower stem in order to insert one pipet into another. You can easily do this by holding the pipet in the palm of you hand and gently stretching the stem of the pipet until it stretches into a uniform narrow diameter. Cut the stem of the pipet so that there is 3 cm of narrow stem. Each group will need three small opening pipets for gas collectors and one for HCl. To make the wide opening pipets, cut off the stem at the junction to the bulb. This should make it easier to draw up small amounts of the solids. Each group will need 3 wide opening pipets.

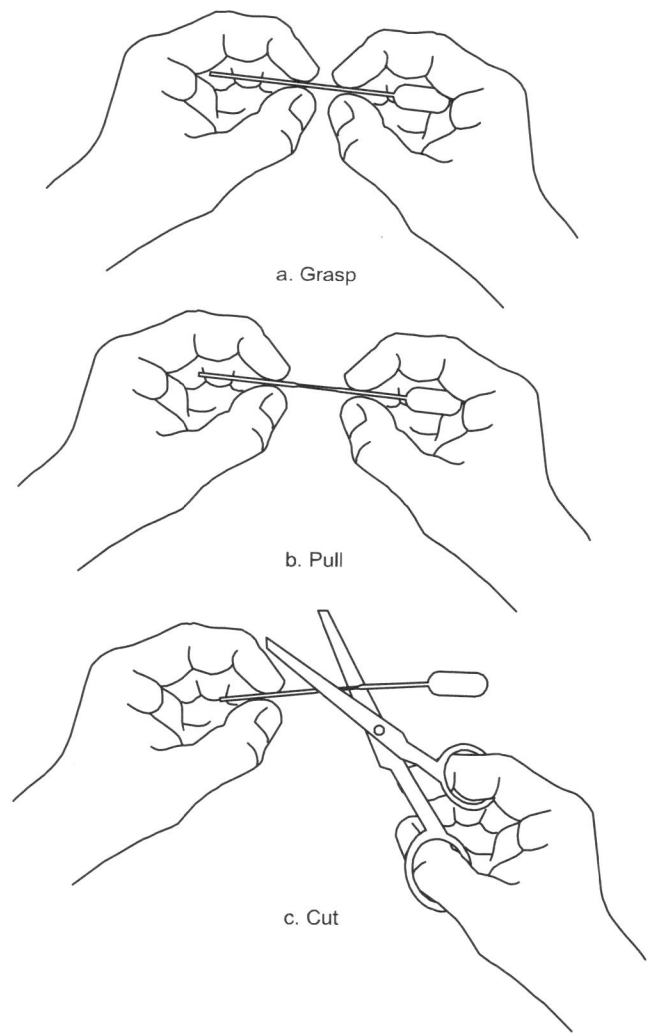

a. Grasp

b. Pull

c. Cut

In your pre-lab discussion be sure to show students how to use the color of the universal indicator solution to determine the pH of a solution. Also, review the ranges of the pH scale. A pH less than 7 is considered acidic and a pH greater than 7 is considered basic.

Safety Alert
1. Students must wear goggles at all times.
2. Potassium nitrite is a strong oxidizer and a fire and explosion risk if heated.
3. Sodium bisulfite is a severe skin irritant as an aqueous solution and is moderately toxic.

POSSIBLE ANSWERS TO THE CONCLUSION QUESTIONS AND SAMPLE DATA

Data Table					
Gas	Initial pH	After 10 bubbles	After 10 more bubbles	Final pH	ΔpH
CO_2	7	7	6	6	1
NO_2	7	6	5	3-4	3-4
SO_2	7	6	5	3-4	3-4

1. For each of the three gases, calculate the change in pH (ΔpH), by subtracting the final pH from the initial pH. Record these values.
   - Recorded on data table above.

2. In the experiment, which gas caused the smallest change in pH?
   - Carbon dioxide showed the least change.

3. Which gas (or gases) caused the largest change of pH?
   - Usually sulfur dioxide shows the greatest change, but some results may be similar between nitrogen dioxide and sulfur dioxide.

4. Coal from western states such as Montana and Wyoming is known to have a lower percentage of sulfur impurities than coal found in the eastern United States. How would burning low sulfur coal lower the level of acidity in rainfall? Use specific information about gases and acids to answer the question.
   - Burning low sulfur coal would release less sulfur dioxide and therefore cause less sulfurous acid to form in the atmosphere. (SO_2 acts as condensation nuclei for water molecules to form sulfurous acid.)

5. High temperatures in the automobile engine cause nitrogen and oxygen gases from the air to combine to form nitrogen oxides. What two acids in acid rain result from the nitrogen oxides in automobile exhaust?
   - Nitrous acid and nitric acid both result from nitrogen oxides produced by cars.

6. Which gas would produce acid rain from air that is unpolluted?
   - Carbon dioxide is a natural byproduct of respiration and when found in the atmosphere acts as a condensation nucleus to form carbonic acid.

7. Why is acid rain more of a problem in the northeastern U.S. and Canada then in central Texas?
   - The bedrock and soils in the northeastern United States and Canada tend to be slightly acidic, due to the presence of granite. Central Texas has a limestone bedrock with very basic soils, which remediate the pH of acid rain.

Acid Rain Drops Keep Falling on My Head
Investigating Acid Rain

Carbonic acid is produced when carbon dioxide gas dissolves in rain droplets of unpolluted air:

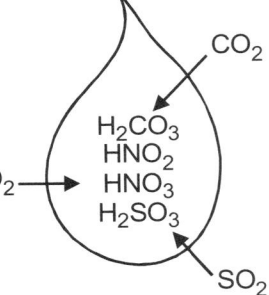

$$(1)\ CO_{2(g)} + H_2O_{(l)} \rightarrow H_2CO_{3(aq)}$$

Nitrous acid and nitric acid result from a common air pollutant, nitrogen dioxide (NO_2). Most nitrogen dioxide in our atmosphere is produced from automobile exhaust. Nitrogen dioxide gas dissolves in rain drops and forms nitrous and nitric acid:

$$(2)\ 2NO_{2(g)} + H_2O_{(l)} \rightarrow HNO_{2(aq)} + HNO_{3(aq)}$$

Sulfurous acid is produced from another air pollutant, sulfur dioxide (SO_2). Most sulfur dioxide gas in the atmosphere results from the burning of coal containing sulfur impurities. Sulfur dioxide dissolves in rain drops and forms sulfurous acid:

$$(3)\ SO_{2(g)} + H_2O_{(l)} \rightarrow H_2SO_{3(aq)}$$

In the procedure outlined below, you will first produce and collect the three gases listed above by reacting the solids with HCl. After collecting the gas produced, you will then bubble the gases though water, producing the acids found in acid rain. Using universal indicator, which changes colors to indicate pH, you will monitor the pH change of the water sample.

PURPOSE
In this activity you will describe the pH changes caused when the gases CO_2, SO_2 and NO_2 act as condensation nuclei in the atmosphere.

MATERIALS
100 mL distilled water
solid potassium nitrite, KNO_2
solid sodium bicarbonate, $NaHCO_3$ (baking soda)
solid sodium hydrogen sufite, or sodium bisufite, $NaHSO_3$
3 thin-stem Beral pipets stretched to make a small opening
1 thin-stem Beral pipet with 1.0 M HCl

100 mL beaker
24 well-plate
2 mL universal indicator
tap water for cleanup
3 thin-stem Beral pipets cut to have a large opening
safety goggles

Safety Alert

1. Students must wear goggles at all times. Potassium nitrite is a strong oxidizer and a fire and explosion risk if heated. Sodium bisulfite is a severe skin irritant as an aqueous solution and moderately toxic.

PROCEDURE

1. Read the background information and make a hypothesis about which gas will cause the greatest change in pH. Record your hypothesis on your student answer page.

2. Obtain and wear safety goggles.

3. Obtain your lab set up consisting of one well plate containing 3 small stem and 3 wide opening Beral pipets. The large opening ones should be named for each solid: $NaNO_2$, $NaHCO_3$ and $NaHSO_3$. The small stem pipets should be named for the gases: CO_2, NO_2 and SO_2. Always set the Beral pipets with the stems pointing upward in your well plate.

4. Take your large opening pipets to the dispensing area and obtain the solid substances. Squeeze the bulb of the pipet to expel all of the air, hold the opening in the solid and release the bulb. Some of the solid will be drawn into your pipet. Do this several times for each solid until you have enough solid to fill the curved end of the bulb. (See figure 1) *Caution, avoid inhaling dust from the solids.*

Fig. 1

5. Obtain a small stem Beral pipet with 1.0 M HCl from your teacher. **Caution: HCl is a strong acid. Gently hold the pipet with the stem pointing up, so that the HCl drops do not escape.** One at a time, insert the narrow stem of the HCl pipet into the large opening of the pipet containing the solid. Gently squeeze the HCl pipet to release about 20 drops of HCl into the solid. When finished, remove the HCl pipet and gently swirl reaction pipet to mix them together. **Leave the pipets open end up in your well plate.** (The gases you have produced are denser than the air in the classroom and will remain in the bulb of the pipet.)

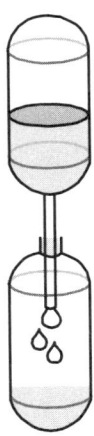

6. Obtain the small stem pipet labeled CO_2. Squeeze the air out of the bulb and collect the gas from the reaction pipets by inserting the small stem pipets into the reaction pipet bulb. Slowly release the bulb so that the gas is drawn into the small stem pipet. Place the CO_2 pipet, stem up, in your well plate. Repeat this procedure for the remaining two gases.

7. Fill three wells of your well plate half-way with distilled water. Add ten drops of universal indicator solution to each water well. Comparing the color of the solution to the color chart, determine the initial pH of your water. Record the pH in your data table.

8. Using the CO_2 gas pipet, insert the tip of the pipet into one of the wells filled with water and indicator solution. Slowly bubble the gas through the water, counting the bubbles as they come out of the pipet. Observe the color change and record the corresponding pH after every ten bubbles.

9. Repeat step 8 for the remaining two gases.

10. Dispose of your pipets and clean up your lab area as instructed by your teacher.

Name _____

Period _____

Acid Rain Drops Keep Falling on My Head
Investigating Acid Rain

HYPOTHESIS

DATA AND OBSERVATIONS

Data Table					
Gas	Initial pH	After 10 bubbles	After 10 more bubbles	Final pH	ΔpH
CO_2					
NO_2					
SO_2					

CONCLUSION QUESTIONS

1. For each of the three gases, calculate the change in pH (ΔpH), by subtracting the final pH from the initial pH. Record these values.

2. In the experiment, which gas caused the smallest change in pH?

3. Which gas (or gases) caused the largest change in pH?

4. Coal from western states such as Montana and Wyoming is known to have a lower percentage of sulfur impurities than coal found in the eastern United States. How would burning low sulfur coal lower the level of acidity in rainfall? Use specific information about gases and acids to answer the question.

5. High temperatures in the automobile engine cause atmospheric nitrogen and oxygen gases to combine to form nitrogen oxides. What two acids in acid rain result from the nitrogen oxides in automobile exhaust?

6. Which gas would produce acid rain from air that is unpolluted?

7. Why is acid rain more of a problem in the northeastern U.S. and Canada than in central Texas?

Investigating Porosity and Permeability
Aquifer Study

OBJECTIVE

Students will test the porosity and permeability of several different substances and from the results draw conclusions about which substances make good aquifers and which make good confining beds, or aquitards.

LEVEL

Middle Grades: Earth Science

NATIONAL STANDARDS

UPC.2, UPC.3, A.1, A.2, D.2, F.3

TEKS

6.2(B), 6.2(C), 6.2(D), 6.2(E), 6.14(B)
7.2(A), 7.2(B), 7.2(C), 7.2(D), 8.2(A)
8.2(B), 8.2(C), 8.2(D), 8.14(A)
IPC 1(A), 2(A), 2(B), 2(C), 2(D), 3(A)

CONNECTIONS TO AP

AP Environmental Science:
   I. Interdependence of Earth's Systems: Fundamental Principles and Concepts B. The Cycling of Matter 1. water
   II. Renewable and Nonrenewable Resources: Distribution, Ownership, Use, Degradation A. Water 1. fresh

TIME FRAME

50 minutes

MATERIALS

(For a class of 28 working in groups of 4)

1 10 lb bag of sand	7 rubber bands
1 10 lb bag of clay	7 ring stands
1 10 lb bag of gravel	7 stop watches
1 10 lb bag course sand	1 or 0.5 liter carbonated beverage bottles with cap
7 100 mL graduated cylinders	several pairs of old nylon stockings

TEACHER NOTES

The sand and gravel can be purchased at a garden store. Regular art clay can be used for the clay sample. The pantyhose tend to break easily so have extra on hand and replace as needed. All of the empty bottles need to be the same size, but either 1 liter or 0.5 liter bottles will work, depending on the size of the rings you have for your ring stands.

After getting 7 same sized bottles the teacher should cut the bottom of the bottles off so they are open and easy to fill. The cap end of the bottom stays untouched so that the lid can be taken off and put back on during the lab. Use a permanent marker to draw a line around each bottle 10 centimeters from the cap end.

The teacher needs to measure and draw a black line to the same level on each of the bottles so that each bottle will contain the exact same amount of material to be tested. For a 0.5 liter bottle measure from the bottom of the neck

POSSIBLE ANSWERS TO THE CONCLUSION QUESTIONS AND SAMPLE DATA

Answers will vary widely with the provided materials. A range of answers is given below.

Sample	Porosity (mL)	Flow time (min)	Recovered water (mL)	Permeability (mL/min)
Gravel	45 - 65	2 - 5	190 - 200	100 - 40
Coarse Sand	40 - 60	2 - 10	150 - 190	95 - 15
Fine Sand	30 - 55	8 - 11	145 - 185	23 - 13
Clay	0	0	0	0

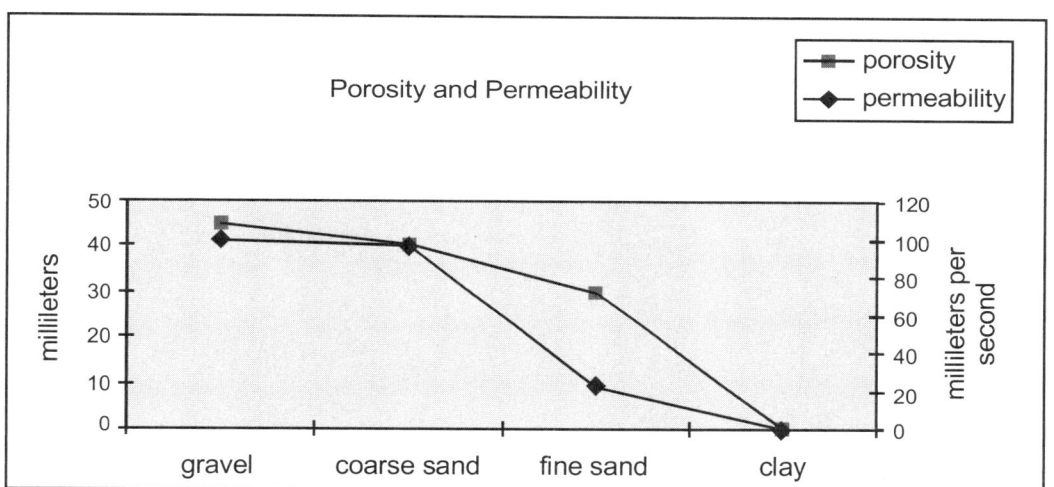

1. Which sample held the most water? Which held the least?
   - The gravel held the most water, clay held very little or no water.

2. Describe the amount of available pore space in the material that held more water?
   - The material that held the most water has the most porosity, or rather has the most space between particles to hold water.

3. Which sample was the most permeable? Which was the least permeable?
   - Gravel was the most permeable. Clay was the least permeable.

4. Why would gravel be a poor filter for water? Which material might make a good filter?
   - Gravel allowed the water to pass through too quickly to filter the water. Sand would be a much better filter.

5. Why would clay act as a confining layer (aquitard) for an aquifer?
   • Clay does not allow water to pass through and therefore is an excellent confining layer.

6. Aquifers must be permeable. List the samples which would be good aquifers.
   • Gravel and both types of sand would be good aquifers because they are both porous and permeable.

7. Discuss the truth of statement "What goes in, must come out." How does your data support or refute this statement.
   • The volume of water that was recovered was less than the volume of water that was added. Some of the water remained in the material clinging to the particles. Therefore, only *some* of what goes in comes out!

Investigating Porosity and Permeability
Aquifer Study

Porosity and permeability are related terms used to describe any rock or loose sediment. Both of these properties are essential to the formation of an aquifer.

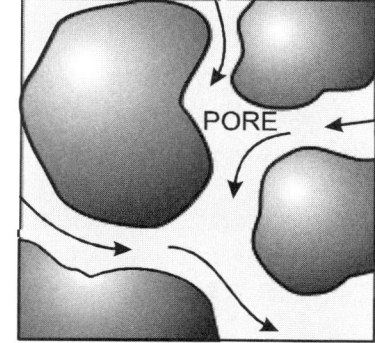

Specifically, *porosity* of a rock is a measure of its ability to hold a fluid. *Permeability* is a measure of the amount of fluid able to flow through a rock.

An *aquifer* is a geologic formation that will yield water to a well in sufficient quantities to make the production of water from this formation feasible for beneficial use. It contains permeable layers of underground rock or sand that hold or transmit groundwater below the water table. *Artesian aquifers* are confined on the top and bottom by non-permeable rock layers, sometimes called *aquitards*, which put the aquifer under pressure and allow springs and wells to flow without being pumped.

PURPOSE
Students will be able to describe porosity and permeability in sand, gravel, and clay.

MATERIALS
fine sand	rubber bands
clay	ring stand
gravel	stop watch
course sand	1 or .5 liter carbonated beverage bottle with cap
100 mL graduated cylinder	pantyhose

PROCEDURE

1. In the space marked HYPOTHESIS on your student answer sheet, write a hypothesis addressing which of the four materials (fine sand, course sand, gravel, and clay) will be the most porous and which will be the most permeable.

2. Cut a 20 cm square of pantyhose and fold it in half. Remove the cap from the bottle, place the pantyhose over the opening of the bottle, and secure with a rubber band.

3. Replace the cap and firmly attach it so that water will not leak out of the bottle. See the picture. Pre-test your bottle with some water to ensure that it does not leak.

4. Empty the water from the bottle, and proceed to the next step.

5. Fill the bottle to the black line with the sample material that will be tested.

6. To measure the porosity we will determine the volume of water it takes to just saturate the test material. Fill the graduated cylinder to 100 mL. Slowly add water to the sample material until water begins to pool at the surface of the sample. If you have added too much water, carefully pour off the excess back into the graduated cylinder.

7. Record the volume of water in mL that was added to the sample in the data table in the Porosity column on your student answer page. Then, record this value for your class data as your teacher instructs.

8. To measure the permeability we will determine the amount of time in takes for 200 mL of water to flow through the sample material. Add the remaining water in your graduated cylinder to the sample container. Add any additional water to the sample so that the TOTAL VOLUME of the water added is 200 milliliters. For example, if you added 60 milliliters to the sample in the porosity test, you will have to add 140 milliliters more to the container.

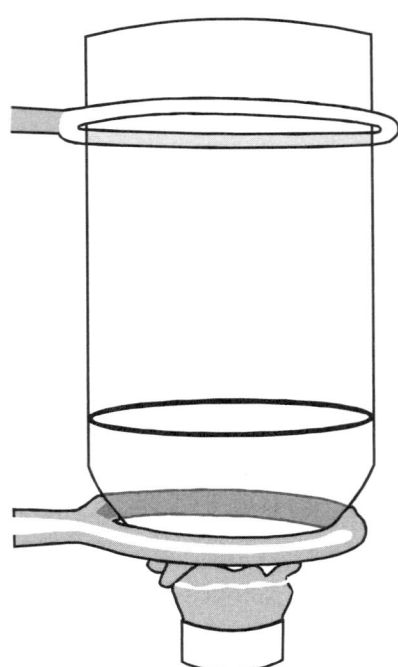

9. Position a 250 mL beaker under the neck of the bottle to catch the run-off water. Simultaneously, start the stopwatch and remove the cap from the bottle. Stop the timer when the dripping water from the bottle has a frequency of one drop every 10 seconds.

10. Record the time in the data table on your student answer page in the Flow time column. Then, record this value for your class data as your teacher instructs.

11. Using the run-off water in your beaker, measure the amount of water that passed through your sample with a graduated cylinder. Record this value in the data table on your student answer page in the Recovered water column. Then record this value for your class data as your teacher instructs.

12. Using the volume of recovered water and the elapsed time, calculate the porosity in mL/min for each sample. Record these values in your data table in the Permeability column and with the class data as your teacher instructs.

Name _____

Period _____

Investigating Porosity and Permeability
Aquifer Study

HYPOTHESIS

DATA AND OBSERVATIONS

Sample	Porosity (mL)	Flow time (min)	Recovered water (mL)	Permeability (mL/min)
Gravel				
Coarse Sand				
Fine Sand				
Clay				

ANALYSIS

Make one graph that shows both porosity and permeability using the two *y*-axes to show the data.

CONCLUSION QUESTIONS

1. Which sample held the most water? Which held the least?

2. What can you assume about the material that held more water?

3. Which sample was the most permeable? Which was the least/not permeable?

4. Why would gravel be a poor filter for water? Which material might make a good filter?

5. Why would clay act as a confining layer for an aquifer?

6. Aquifers must be permeable. List the samples which would be good aquifers.

7. Discuss the truth of statement "What goes in, must come out." How does your data support or refute this statement.

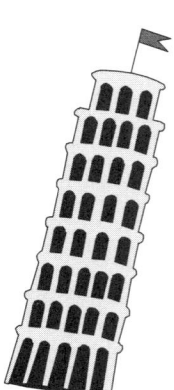

That Sinking Feeling
Investigating Land Subsidence

OBJECTIVE
Students will examine maps and data from areas experiencing subsidence. Students
will then calculate the rate of subsidence and evaluate issues surrounding land subsidence,
which results from pumping groundwater from unconsolidated rock.

LEVEL
Middle Grades: Earth Science

NATIONAL STANDARDS
UCP.2. UCP.3, D.2, F.3, F.5

TEKS
6.2(C), 6.2(D), 6.2(E), 6.3(A), 6.4(B), 6.6(C), 6.14(B)
7.2(C), 7.2(D), 7.2(E), 7.4(B), 7.14(C)
8.2(C), 8.2(D), 8.2(E), 8.4(B), 8.14(A)
IPC: 2(C), 2(D)

CONNECTIONS TO AP
AP Environmental Science:
    I. Interdependence of Earth's Systems: Fundamental Principles and Concepts B. The Cycling of
    Matter 1. water
    III. Renewable and Nonrenewable Resources: Distribution, Ownership, Use, Degradation A. Water
    1. fresh

TIME FRAME
50 minutes

MATERIALS
    calculators

TEACHER NOTES
The activity *Land Subsidence* has students review graphs, maps and data to draw conclusions and make
predictions about what happens when water is removed from unconsolidated rocks. The activity should
follow the Porosity and Permeability lab, also in this guide. The student pages have an explanation of
what causes subsidence and where it has been a concern.

If students have not had experience reading maps, especially contour maps, spend a few minutes at the
beginning of the activity explaining how to read a contour map. In this case the contours represent lines
of equal subsidence or equal sinking. So, the line with the number 6 on it indicates that everywhere that
line passes through has sunk 6 feet. Space between the lines fall somewhere between the numbers on
the contour lines.

Students also may need s refresher course on reading charts and graphs. Make an overhead of the charts, maps and graphs and do a quick questioning of students about what each chart, map or graph shows. This may not be necessary, or may not take a long time depending on what experiences your students have already had with these things.

POSSIBLE ANSWERS TO THE CONCLUSION QUESTIONS AND SAMPLE DATA

1. What cities have suffered the greatest subsidence in the Santa Clara Valley?
   - Santa Clara and San Jose both experienced subsidence greater than 8 feet.

2. Using the data in Table 1, what was the total subsidence in San Jose from 1934 to 1967?
   - The subsidence that occurred between 1934 and 1967 was 8.1 feet.

3. What was the average annual rate of subsidence for this same period of time?
   - The average annual rate of subsidence between 1934 and 1967 was .25 feet per year.

4. Describe the problems that people in the Santa Clara Valley may have had because of subsidence.
   - People in the Santa Clara Valley who lived near the bay may have experienced flooding. Houses in the areas with the greatest amount of subsidence, like Santa Clara and San Jose, would have had foundation problems (the leaning houses of Santa Clara, like the Leaning Tower of Pisa). Roads and sidewalks would have cracked and broken as the subsidence happened.

5. Would you expect subsidence to occur in the consolidated rocks shown on the map? Explain why or why not.
   - Consolidated rocks have no pore space and therefore they would not be expected to subside.

6. Examine the hydrograph (Figure 3). What years were the wells artesian (flowing without being pumped)?
   - The wells would have been artesian (flowing without being pumped) from around 1938-1944.

7. What is the explanation for the minor fluctuations in the hydrograph (Figure 3)?
   - Minor fluctuations were probably caused by seasonal rains combined with decreases in the amount pumped for use.

8. Why did the subsidence stop in 1967? What might that mean for cities like Houston and Las Vegas?
   - The Santa Clara Valley started importing water to pump back into their aquifer to slow down and stop subsidence. Houston and Las Vegas could try similar methods, but they will not be cheap in areas where water is already scarce like Las Vegas.

T E A C H E R   P A G E S

That Sinking Feeling
Investigating Groundwater Subsidence

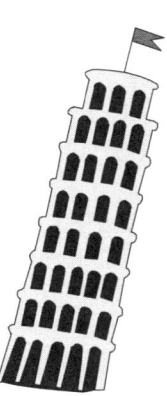

Groundwater removal from unconsolidated sediments may result in sediment compaction as pore spaces collapse when water is withdrawn. This can cause subsidence, or sinking, of the ground surface. In figure 33.1 below you see an artesian well flowing from a sand aquifer. As the water leaves the sand and empties the pore space, the sand compacts.

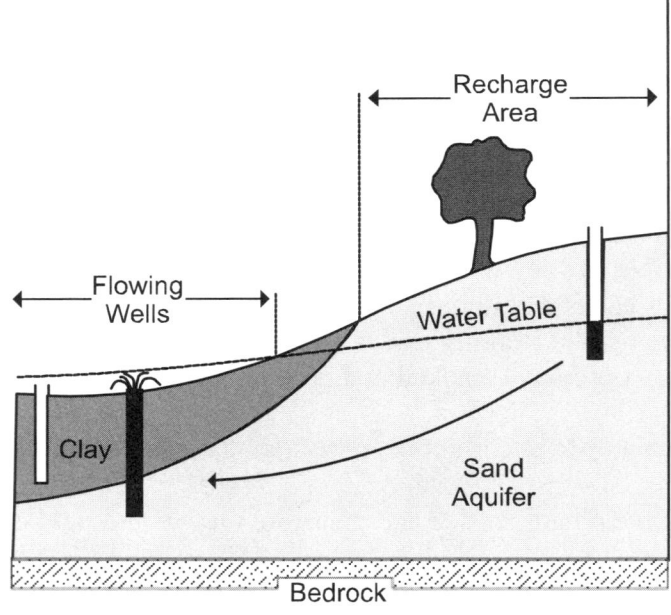

Fig. 1

Land subsidence because of withdrawal of ground water is a serious problem. Obviously, in coastal areas subsidence can mean that the land surface drops below sea level. In the Santa Clara Valley of California, some 17 square miles are now below the highest tide level in San Francisco Bay and must be protected by earthwork damns. Subsidence in cities like Houston causes serious flooding problems, especially during heavy rain events like hurricanes. Other centers of subsidence include Mexico City, Tokyo, and Las Vegas. With increasing withdrawal of groundwater and more intensive use of the land surface, the problem of subsidence may be expected to become more widespread.

Humans are usually the culprits in causing land subsidence. When we pump out large amounts of water for cities or for agriculture we can cause dramatic sinking.

The most famous case of land subsidence is in Pisa, Italy. The Leaning Tower of Pisa was built on soft, river valley sediments. Since the tower was built in 1174 AD, it has been sinking an average of 1 millimeter per year. In order to arrest the sinking and prevent the Leaning Tower from toppling over, modern engineers have worked to develop a plan of injecting cement into deep cores below the tower to firm up its footing. Interestingly, the Italian government has decided not to straighten the Leaning Tower, but to just prevent it from any further sinking.

Other cities have instituted plans to prevent land subsidence. Santa Clara, California has limited the amount of pumping from the water table aquifer to prevent further compaction and imports water to inject back into the aquifer. Arizona and Texas have both established new groundwater policies that address the problem of subsidence.

The map below shows the sinking in the Santa Clara Valley from 1934-1967.

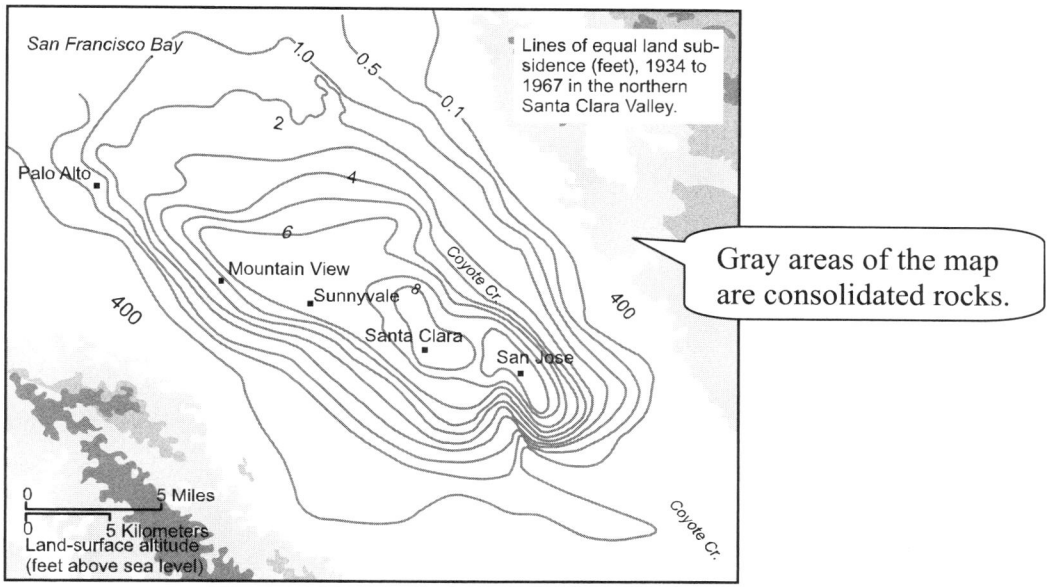

Fig. 2

Table 1 shows data from a USGS data point located in San Jose. The total sinking from 1912 to 1967 was 12.7 feet. Recharge measures were enacted in 1967 to arrest any further subsidence.

Table 1 Subsidence at bench mark P7 in San Jose

Year	Subsidence (feet)
1912	0.0
1920	0.3
1934	4.6
1935	5.0
1936	5.0
1937	5.2
1940	5.5
1948	5.8
1955	8.0
1960	9.0
1963	11.1
1967	12.7

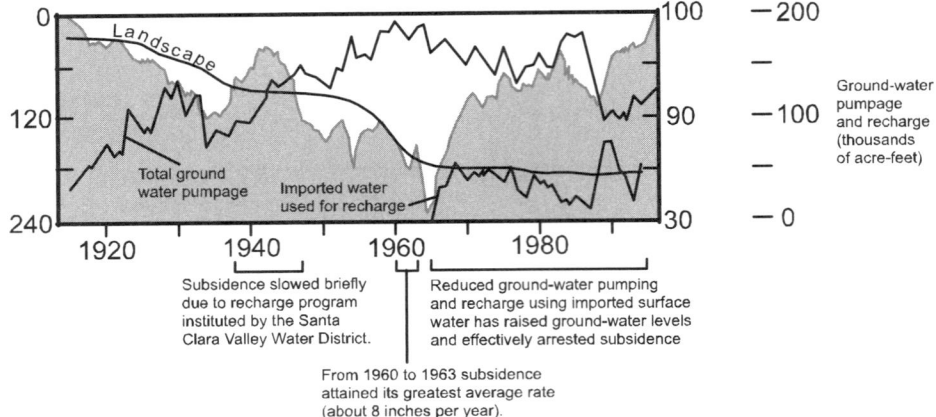

Land subsidence was a result of intensive ground-water pumping and the subsequent drop in water levels. Once pumping was stabilized by the introduction of imported surface water, subsidence was arrested.

Fig. 3 Hydrograph

PURPOSE

This exercise is intended to familiarize you with the subsidence, or sinking, of the land surface, owing to removal of underground water.

MATERIALS

calculator

PROCEDURE

1. Read the introductory information on this page.

2. Use the maps and data tables to answer questions on the student answer page.

Name _____

Period _____

That Sinking Feeling
Investigating Groundwater Subsidence

CONCLUSION QUESTIONS

1. What cities have suffered the greatest subsidence in the Santa Clara Valley?

2. Using the data in Table 1, what was the total subsidence in San Jose from 1934 to 1967?

3. What was the average annual rate of subsidence for this same period of time?

4. Describe the problems that people in the Santa Clara Valley may have had because of subsidence.

5. Would you expect subsidence to occur in the consolidated rocks shown on the map? Explain why or why not.

6. Examine the hydrograph (Figure 3). What years were the wells artesian (flowing without being pumped)?

7. What is the explanation for the minor fluctuations in the hydrograph (Figure 3)?

8. Why did the subsidence stop in the Santa Clara Valley in 1967? What might that mean for cities like Houston and Las Vegas?

How Wet Is Our Planet?
Modeling Fresh Water Resources on Earth
Water, Water Everywhere, But Not a Drop to Drink

OBJECTIVE

Students will be able to describe the amount and the distribution of water on the earth in the oceans, rivers, lakes, groundwater, icecaps, and the atmosphere.

LEVEL

Middle School: Earth Science

NATIONAL STANDARDS

UCP.2, D.2, F.3, F.4, F.5

TEKS

6.1(B), 6.2(D), 6.3(D), 6.14(B)
7.1(B), 7.2(D), 7.3(D), 7.14(C)
8.1(B), 8.2(D), 8.3(D), 8.12(C), 8.14(C)
IPC 1(A), 2(A), 2(B), 2(C), 2(D), 3(A)

CONNECTIONS TO AP

AP Environmental Science:
    I. Interdependence of Earth's Systems: Fundamental Principles and Concepts B. The Cycling of Matter 1. water
    II. Renewable and Nonrenewable Resources: Distribution, Ownership, Use, Degradation A. Water 1. fresh

TIME FRAME

45 minutes

MATERIALS

(For a class of 28 working in pairs)

1 five gallon tank	14 400 mL beakers
14 10 mL graduated syringes	14 calculators

TEACHER NOTES

How Wet Is Our Planet? is a striking way to demonstrate to students how little of the water on earth is actually readily available for human consumption. Because most of the earth's water is undrinkable and because of the steady increase in human use of natural waters, the earth is in crisis. It is not just overuse that threatens this resource. Pollution has ruined much of the water people depend on. Mineral fertilizers, pesticides and herbicides have seeped into surface and sub-surface waters contaminating them beyond human consumption and disrupting delicate ecosystems. Will there be enough water to accommodate the needs of future generations?

For more information and activities about fresh water:
http://www.thewaterpage.com/waterbasics.htm#resource

To start the activity, place the 5-gallon tank in the front or center of the room, or where it is easily accessible to all of the students.

Use a globe or a picture of earth from space and engage students with a discussion of what the earth looks like from space. Point out oceans, icecaps and large rivers and lakes. Identify what categories of water are fresh (rivers, atmosphere, glaciers/icecaps, groundwater) and which are salt (oceans, inland seas). Discuss that the percentages on the chart do not add up to 100% because of the water in living things, including human bodies.

POSSIBLE ANSWERS TO THE CONCLUSION QUESTION AND SAMPLE DATA

Source	Percentage	Portion of the 5 gallon tank in mL (18,927 milliliters in 5 gallons)
Oceans*	97.2%	18,397.04
Icecaps/glaciers	2.0%	378.54
Groundwater	0.62%	117.35
Fresh water lakes	0.009%	1.70
Inland seas	0.008%	1.51
Atmosphere	0.001%	.19
Rivers	0.0001%	.02
Total	99.8381%	18,896.36

*example: Oceans = 18,927 x .972 = 18,397.04

ANALYSIS QUESTIONS
1. Total the volumes of fresh water sources to calculate the amount of fresh water in milliliters in the five-gallon container.
   * 378.54 + 117.35 + 1.70 + .19 + .019 = 497.80 mL

2. In milliliters, how much of all of earth's water is represented by just lakes and rivers?
   * 1.70 + .02 = 1.72mL

3. In milliliters, how much of all of earth's water is represented by just the river water?
   * .02 mL

T
E
A
C
H
E
R

P
A
G
E
S

CONCLUSION QUESTIONS

1. What total percentage of water on the earth's surface is fresh water?
   - 2.63% of the earth's water is fresh.
   - (2.0% + 0.62% + 0.009% + 0.001% + 0.0001% = 2.63%)

2. What percentage of the above fresh water is readily available for consumption? (i.e. you could go to it with a cup, scoop some up and drink it.)
   - Only .091% of the earth's water is readily available for consumption.
   - (.009 + .0001 = .091%)

3. Where is most of the fresh water on the earth located?
   - Most of the earth's water is concentrated at the earth's poles in icecaps and glaciers and is unavailable for human use.

4. What types of pollution might affect the fresh water available to humans?
   - Water is vulnerable to both point pollution from industry and non-point pollution. Non-point pollution may include runoff of lawn fertilizers and pesticides and runoff from streets such as trash, oil and chemicals.

5. Why is fresh water so important?
   - All life is dependent upon water for survival. Humans can live without food for weeks, but without water we will die in a few days. We also use water everyday for cleaning, sewage, cooking, and dozens of other activities. Wildlife needs water too, and as water becomes more scarce people and animals are competing for the clean water available.

6. What will happen as our supply of clean fresh water disappears?
   - As water becomes more scarce it becomes more expensive to use. As areas are threatened with drought people are asked to conserve water. Many states and even countries are fighting over who gets the water that is still available. These disputes will become more serious as water disappears. For instance, Texas and Mexico have divided up all the water from the Rio Grande River for human and agricultural use. The Rio Grande no longer reaches the Gulf of Mexico because the water is all gone before it arrives there.

How Wet Is Our Planet?
Water, Water Everywhere, But Not a Drop to Drink

Consider the surface of the earth. Why do you think the earth is referred to as the water planet? When viewed from space the earth looks blue because of all of the water on the surface of the planet. But, because most of the earth's water is undrinkable and because of the steady increase in human use of natural waters, the earth is in crisis. It is not just overuse that threatens this resource. Pollution has ruined much of the water people depend on. Mineral fertilizers, pesticides and herbicides have seeped into surface and sub-surface waters contaminating them beyond human consumption and disrupting delicate ecosystems. Will there be enough water to accommodate the needs of future generations?

Somewhere between two-thirds and three-fourths of the earth's surface is covered with water. It breaks down into categories as follows:

Percentages of Water on Earth

Oceans	97.2%
Icecaps/glaciers	2.0%
Groundwater	0.62%
Fresh water lakes	0.009%
Inland seas	0.008%
Atmosphere	0.001%
Rivers	0.0001%
Total	99.8381%

MATERIALS

five gallon tank filled with water
1 mL graduated syringes

1 400 mL beaker
1 calculator

PROCEDURE

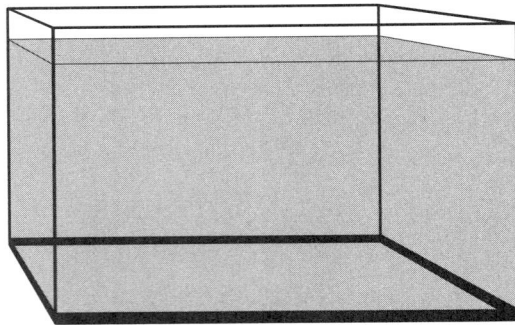

1. In order to fully comprehend the availability of our water resources, imagine that the five-gallon tank of water in the front of the room represents all of the water available to us on this planet.

2. Complete the table on the student answer page by calculating the volumes of water in milliliters for each category listed. You will need to know that there are 18,927 milliliters in one gallon.

3. Using your graduated syringe, carefully remove all the water that would be fresh water from the 5-gallon tank and place it into your beaker.

4. Answer the ANALYSIS and CONCLUSION questions on your student answer page. You may use additional resources including the Internet to answer the questions.

Name _____

Period _____

How Wet Is Our Planet?
Water, Water Everywhere, But Not a Drop to Drink

ANALYSIS

Source	Percentage	Portion of the 5 gallon tank in mL (18,927 milliliters in 5 gallons)
Oceans*	97.2%	
Icecaps/glaciers	2.0%	
Groundwater	0.62%	
Fresh water lakes	0.009%	
Inland seas	0.008%	
Atmosphere	0.001%	
Rivers	0.0001%	
Total	99.8381%	

1. Total the volumes of fresh water sources to calculate the amount of fresh water in milliliters in the five-gallon tank.

2. In milliliters, how much of all of earth's water is represented by just lakes and rivers?

3. In milliliters, how much of all of earth's water is represented by just the river water?

CONCLUSION QUESTIONS

1. What total percentage of water on the earth's surface is fresh water?

2. What percentage of the above fresh water is readily available for consumption? (i.e. you could obtain it with a cup, scoop some up and drink it.)

3. Where is most of the fresh water on earth located?

4. What types of pollution might affect the fresh water available to humans?

5. Why is fresh water so important?

6. What will happen as our supply of clean fresh water disappears?

Is Dilution the Solution to Pollution?
Washing Away Our Mess

TEACHER PAGES

OBJECTIVE
Students will dilute ink or dye until it is no longer visible. Students will be able to explain solute, solvent and solution and work with different concentration levels including parts per million (ppm), parts per billion (ppb) and parts per trillion (ppt).

LEVEL
Middle Grades: Earth Science

NATIONAL STANDARDS
UPC.3, A.1, D.2, F3, F.4, F.5, F.6

TEKS
6.1(B), 6.2(D), 6.3(D), 6.14(B)
7.1(B), 7.2(D), 7.3(D), 7.14(C)
8.1(B), 8.2(D), 8.3(D), 8.12(C), 8.14(C)
IPC 1(A), 2(A), 2(B), 2(C), 2(D), 3(A), 9(A), 9(D)

CONNECTIONS TO AP
AP Environmental Science:
   Scientific Analysis, Interdependence of Earth's Systems: Fundamental Principles and Concept, Renewable and Nonrenewable Resources

TIME FRAME
50 minutes

MATERIALS
(For a class of 28 working in groups of 4)

   7 microchem plates (24 well)
   14 small beral pipets
   Rit$^{©}$ dye (liquid) clothing dye
   paper towels
   14 100 mL beakers

TEACHER NOTES
Many students don't give much thought to what happens when they flush the toilet or rinse something down the drain, but new studies are showing that even small amounts of some toxins can ultimately be very dangerous. This lab uses an ink of dye to simulate a source of pollution and has students dilute it in 10% increments until the dye can no longer be seen. Students then complete dilutions until they reach one part per billion. They are then able to compare parts per billion with parts per million.

In addition to teaching about pollution this lab is an excellent introduction to the words solute, solvent and solution. At the conclusion of the lab, students work through several dilution problems to familiarize themselves with solving science word problems using math skills.

POSSIBLE ANSWERS TO THE CONCLUSION QUESTION AND SAMPLE DATA

Dilution Data Chart

Final Concentration = Source Concentration x Dilution Factor

Well #	Color	Source Concentration	Dilution Factor	Final Concentration
1	dye color, black	10% or 1/10	--------	1/10
2	black	1/10	1/10	1/100
3	gray	1/100	1/10	1/1,000
4	light gray	1/1,000	1/10	1/10.000
5	very very light	1/10.000	1/10	1/100,000
6	clear	1/100,000	1/10	1/1,000,000 - **ppm**
7	clear	1/1,000,000	1/10	1/10,000,000
8	clear	1/10,000,000	1/10	1/100,000,000
9	clear	1/100,000,000	1/10	1/1,000,000,000 - **ppb**

CONCLUSION QUESTIONS

1. A 10% solution of dye and water contains ____1____ part pigment for every ____10____ parts of solution.

2. Well #2 will be ____10____ times more dilute than well #1.

3. In well #2, there will be ____1____ parts dye for every ____100____ parts solution.

4. What is the number of the well in which the solution first appeared to be colorless?
   - 5 or 6

5. What is the concentration in that well?
   - answers will vary; typically 1/100,000 or 1/1,000,000

6. Do you think that there is any dye left in this well even though it is colorless? Explain.
   - Yes there is dye left, but it is in such small amounts it is impossible to see. It has one part of dye left compared to the 1,000,000 or 100,000 total.

7. On your data table label the well which has a concentration of one part per million with **ppm** and the well with a concentration of one part per billion with **ppb**. Which is the greater concentration, 1ppm or 1ppb?
   - See data table.
   - 1 ppm is a greater concentration than 1ppb because it has one part of solute compared to a smaller amount of total solution.

8. You are walking along a beautiful creek that runs through a park. The water looks crystal clear and you are very thirsty. Should you take a drink? Explain your answer using what you observed during this lab.
   - It would not be a good idea to drink from the creek because even though it might not look like there was anything dangerous in the water it could easily be there without being visible. Drinking from the creek could make you very sick.

9. Name one helpful chemical that is put into your drinking water by the water company. Why is this chemical used?
   - Fluorine is added for protecting teeth and chlorine is added for killing bacteria.

10. What can you as an individual do to make certain that harmful chemicals do not pollute your local drinking water?
    - Student may suggest being careful about what they do with toxic chemicals around the house. They should never dispose of them down the kitchen drain, storm drain, or street to run-off. They may also list recycling batteries, oil and other pollutants.

11. Name one benefit of allowing companies to get rid of their chemical waste by diluting it into lakes and streams. What are the disadvantages to this practice?
    - Dumping into lakes and streams is very cheap for the company but in the long term we all pay for the problems this may cause.
    - The disadvantages include fish kills, dramatic decreases in microorganisms and increases in bacteria.
    - Dumping even un-polluted water that is a different temperature than the water it is dumped into can cause serious ecological problems.

12. How would you make 50 grams of a 10% by mass solution of saltwater? List the amount of solute, solvent and solution you would need.
    - 5 grams salt (solute), 45 grams water (solvent) to make 50 grams of solution

13. How would you produce a 10% by weight solution of dye using powdered pigment and water?
    - 10 grams dye (solute), 90 grams water (solvent) for 100 grams total solution

Is Dilution the Solution to Pollution?
Washing Away Our Mess

Have you ever thought about what happens to the water that goes down your drain? All of this water eventually ends up back in our rivers, creeks, lakes and streams and, finally, our oceans. Many industrial waste products have been dumped into our water system because it was a cheap form of disposal and people thought there was enough water on the planet to dilute the waste until it disappeared. Scientists are now discovering that even extremely small amounts (one part per trillion!) of some substances like mercury can be dangerous. Recent health warnings about eating cold water fish which collect mercury in their tissues is a constant reminder of what happens to waste after it "disappears".

Water is an excellent solvent, which means it dissolves other substances easily. In this lab you will be using Rit$^{©}$ dye as your **solute** (the material you need to dissolve) and water as your **solvent** (your dilution agent). Since Rit$^{©}$ dye is already a solid dissolved in water, you are starting the experiment with a 10% solution.

The procedure you will be using is called a serial dilution because you will be making a series of more and more diluted solutions each from a sample of the previous solution. You will be making dilutions of 1 part colored solution: 9 parts water. Therefore you will be decreasing the concentration of dye by $1/10^{th}$ with each dilution. For example, if you start out with a sample of ink that is 100% ink and use 1 drop of ink to 9 drops of water, your new solution will be $100\% \times \frac{1}{10} = 10\%$. If you take one drop of the 10% solution and combine it with 9 drops of water your new solution will be $10\% \times \frac{1}{10} = 1\%$. For each successive dilution that you make the new concentration can be found by multiplying the source concentration by the dilution factor (1/10). It will be helpful to write answers less than one as fractions such as 1/100 or 1/1000.

PURPOSE
In this activity you will demonstrate the meanings of the terms dilution and concentration, and observe how pollution can affect the water that you drink.

MATERIALS
  1 microchem plate
  2 small beral pipets
  Rit$^{©}$ dye or ink
  paper towels
  2 100 mL beakers

PROCEDURE

1. In the space marked HYPOTHESIS on your student answer page, make a hypothesis about how many dilutions you will have to complete for the dye color to disappear.

2. Fill a 100 mL beaker with tap water to use when performing the dilutions. Fill the second beaker with water to use for rinsing your pipet between dilutions. Label your beakers so as not get the two beakers confused or you might contaminate your sample.

3. Your teacher will put 10 drops of the dye into well #1. The dye is a 10% solution. Using your eyedropper, take one drop of dye from well #1 and place it in to well #2. Rinse the eyedropper thoroughly in the "rinse" beaker after you have finished this step.

4. Now add 9 drops of water from the clean water beaker to well #2 to make the total 10 drops. Carefully swirl the contents of the microplate well.

5. Record the color of the solution in the data table. Calculate the final concentration for each well by multiplying the source concentration (the concentration of the well from which you removed one drop) by the dilution factor (1/10). Record this value on the student answer page.

6. Repeat steps 3 through 5 until you reach a dilution of one part per billion. Each well receives one drop from the preceding well and 9 drops of fresh water. The source concentration is always the same as the final concentration of the prior well.

7. Clean up your lab area and answer the conclusion questions.

Name _____

Period _____

Is Dilution the Solution to Pollution?
Washing Away Our Mess

HYPOTHESIS

_____

DATA AND OBSERVATIONS

DILUTIONS: Final Concentration = Source Concentration x Dilution Factor

Well #	Color		Source Concentration	Dilution Factor	Final Concentration
1					
2					
3					
4					
5					
6					
7					
8					
9					

CONCLUSION QUESTIONS

1. A 10% solution of dye and water contains _____ parts pigment for every _____ parts of solution.

2. Well #2 will be _____ times more dilute than well #1.

3. In well #2, there will be _____ parts dye for every _____ parts solution.

4. What is the number of the well in which the solution first appeared to be colorless?

5. What is the concentration in that well?

6. Do you think that there is any dye left in this well even though it is colorless? Explain.

7. On your data table, label the well which has a concentration of one part per million with **ppm** and the well with a concentration of one part per billion with **ppb**. Which is the greater concentration, 1ppm or 1ppb?

8. You are walking along a beautiful creek that runs through a park. The water looks crystal clear and you are very thirsty. Should you take a drink? Explain your answer using what you observed during this lab.

9. Name one helpful chemical that is put into your drinking water by the water company. Why is this chemical used?

10. What can you as an individual do to make certain that harmful chemicals do not pollute your local drinking water?

11. Name one benefit to allowing companies to get rid of their chemical waste by diluting it into lakes and streams. What is the disadvantage to this practice?

12. How would you make 50 grams of a 10% by mass solution of saltwater? List the amount of solute, solvent and solution you would need.

13. How would you produce a 10% by weight solution of dye using powdered pigment and water? List the amount of solute, solvent and solution you would need.

A Bugs Life
Judging Water Quality Based on Macroinvertebrates

OBJECTIVE
Students will understand how macroinvertebrates are used in determining the quality of water in lakes, streams, creeks, and any body of freshwater.

LEVEL
Middle Grades: Earth Science

NATIONAL STANDARDS
UCP.2, UCP.3, A.1, A.2, C.4, F.4, F.5, F.6

TEKS
6.1(B), 6.2(D), 6.3(D), 6.14(B)
7.1(B), 7.2(D), 7.3(D), 7.14(C)
8.1(B), 8.2(D), 8.3(D), 8.12(C), 8.14(C)
IPC 1(A), 2(A), 2(B), 2(C), 2(D), 3(A), 9(A), 9(D)

CONNECTIONS TO AP
AP Environmental Science:
> I. Interdependence of Earth's Systems: Fundamental Principles and Concepts B. The Cycling of Matter 1. water
> II. Renewable and Nonrenewable Resources: Distribution, Ownership, Use, Degradation A. Water 1. fresh

TIME FRAME
60 minutes

MATERIALS
(For a class of 28 working in groups of 4)

> Map of Aquaville with sites identified
> 7 pre-prepared bags containing organisms plates

TEACHER NOTES
A Bugs Life: Judging Water Quality Based on Macroinvertebrates has students use pictures of macroinvertebrate samples to evaluate water quality at several sites from the fictitious town Aquaville.

Benthic macroinvertebrates are invertebrates (animals without a backbone) that live on the bottom of streams during all or part of their life cycle. "Benthic" means bottom dwelling, and "macro" indicates that benthic macroinvertebrates are large enough to be seen with the naked eye. While many macroinvertebrates can be seen with the naked eye, most macroinvertibrates are small. Although benthic macroinvertebrates often go unnoticed because of their size and habitat, they are an extremely

TEACHER PAGES

important part of river ecosystems, and serve as a link in the food web between decomposing leaves and algae, and fish and other vertebrates.

Many benthic macroinvertebrates are the larval forms of flying insects, such as mayflies, stoneflies, and caddisflies. Some are small animals like crayfish. Many can travel between water and moist terrestrial environments, such as fresh water snails.

Fish, birds, turtles, salamanders, newts, and frogs all feed on macroinvertebrates during different stages of their life.

Several characteristics of benthic macroinvertebrates that make them useful indicators species.
- their populations fluctuate depending on physical and chemical changes in their habitat
- they include a great diversity of species
- they do not move far during their time in the stream, and can not move out of degraded areas
- they are easily collected in streams and rivers
- they provide a large number of species that occur in every conceivable aquatic habitat

To prepare for this lab you must produce seven sample packets for the students to rotate. These bags will represent the organisms found at each site labeled on the map. A photocopy master page containing the organisms for each site are provided on the following pages. Make one copy of each page on cardstock. Laminating them before you cut them out will increase their life span.

Once the students have read their student instruction pages, give one packet bag to each group of four students. Allow them five minutes to survey their site bag and then have the groups pass their bag to the next group. Rotate through all 7 bags.

Map of Aquaville

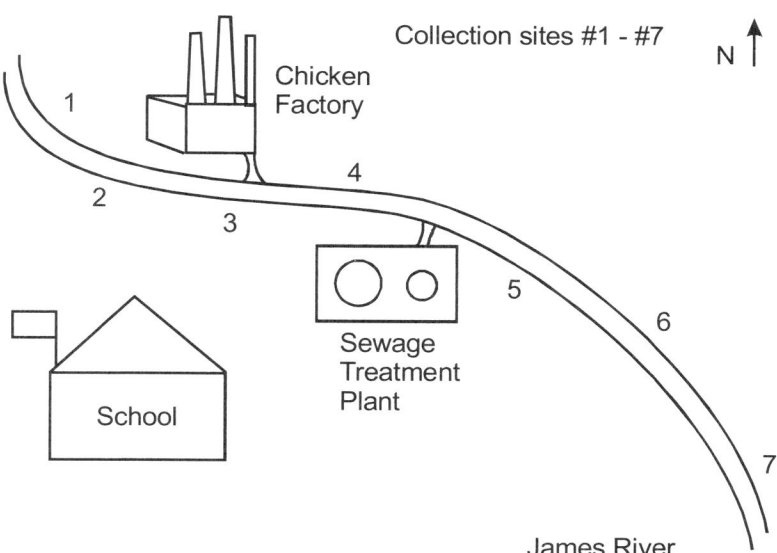

Packet for Site 1

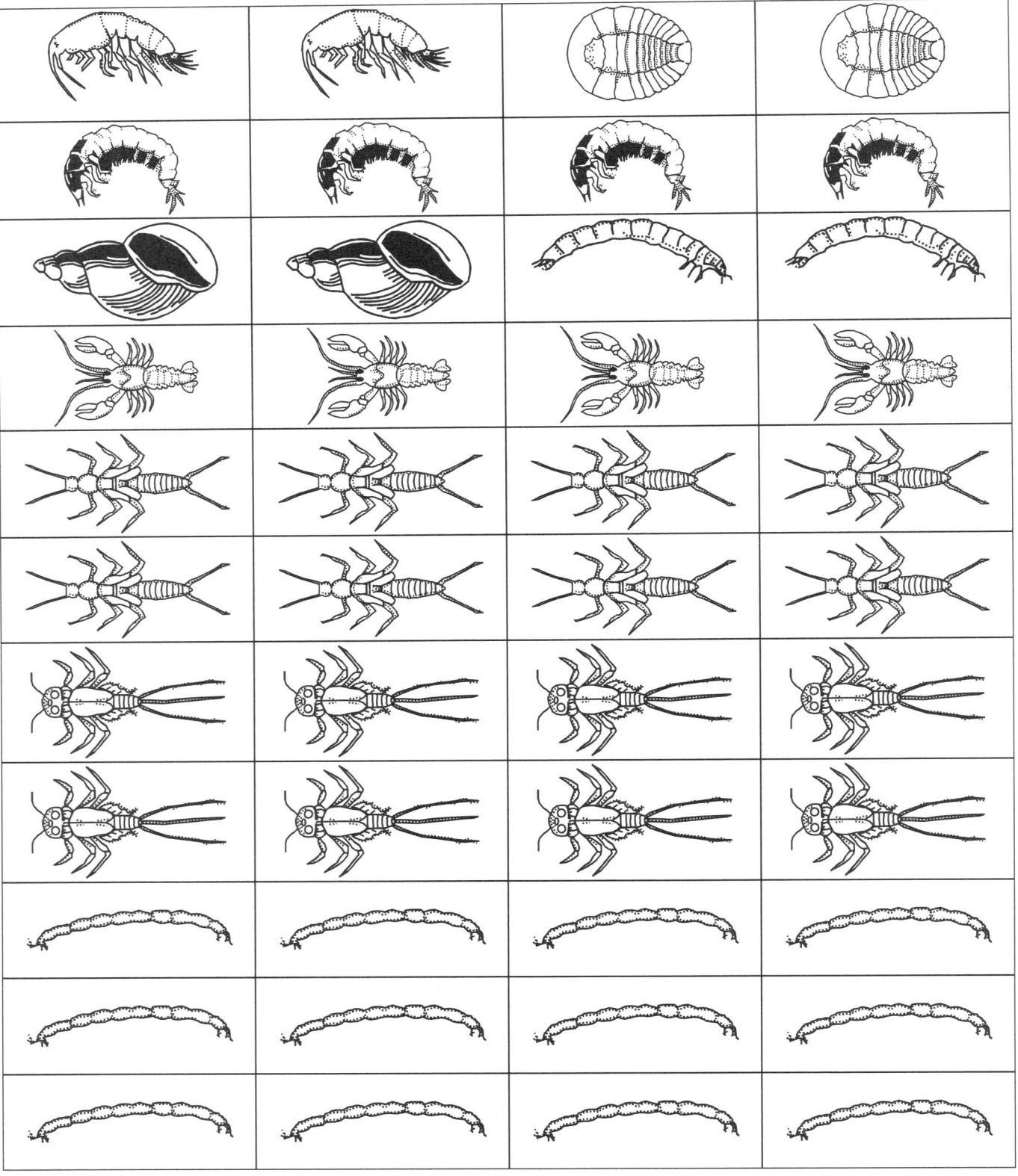

Packet for Site 2

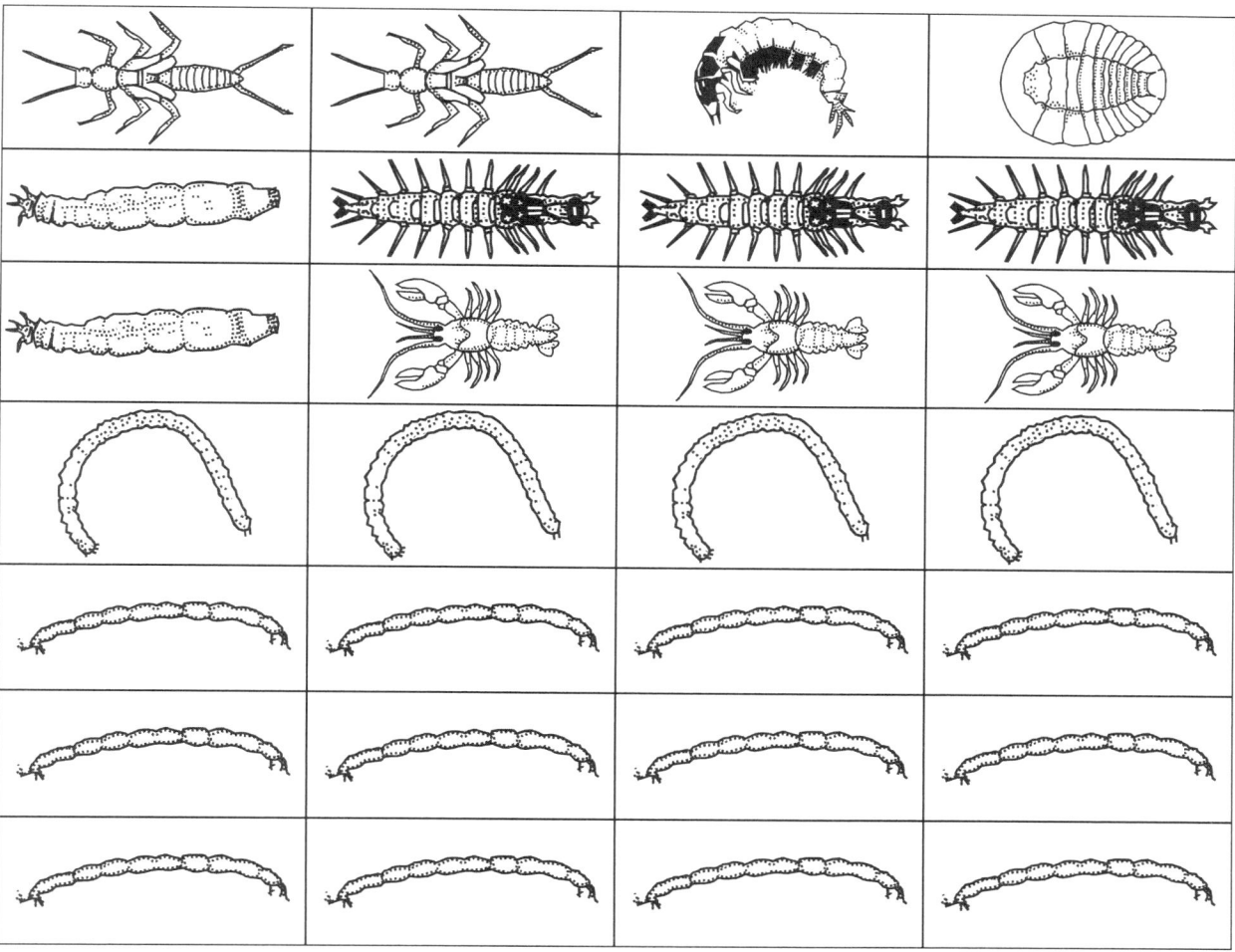

Packet for Site 3

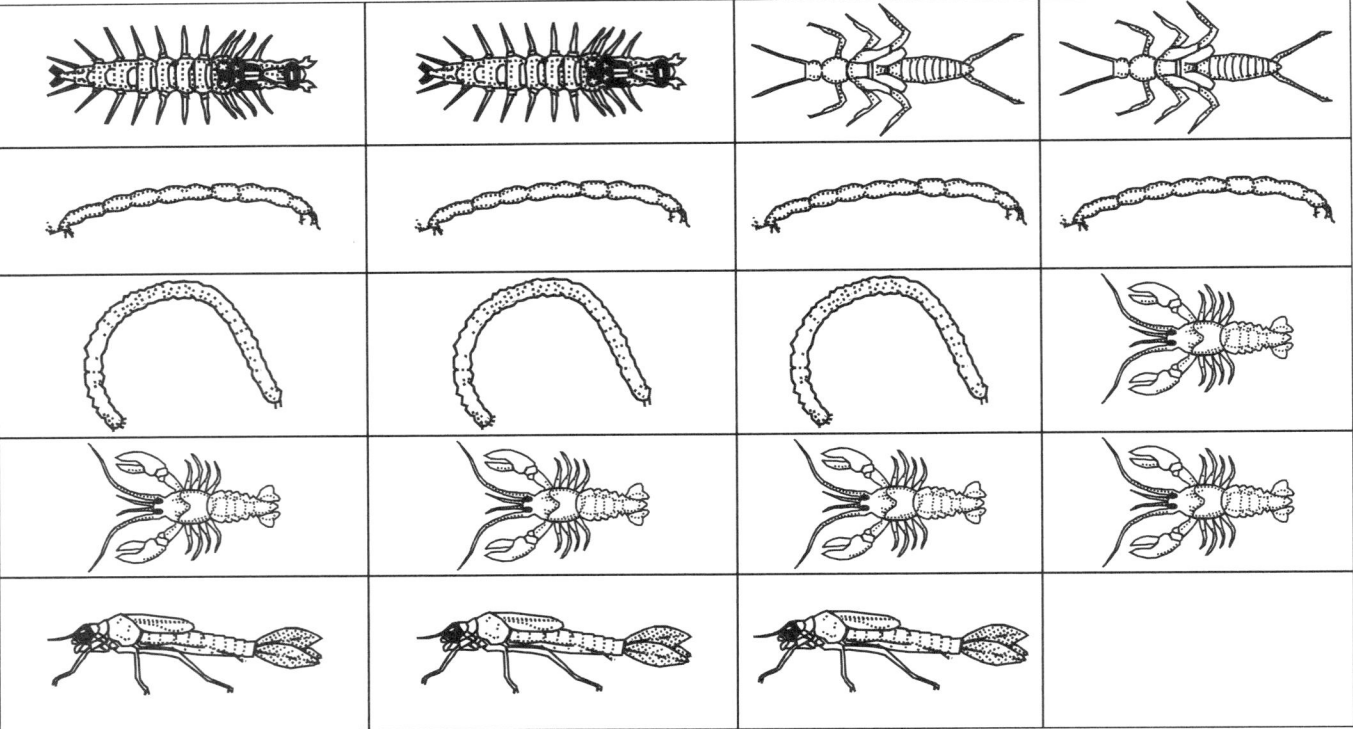

Packet for Site 4

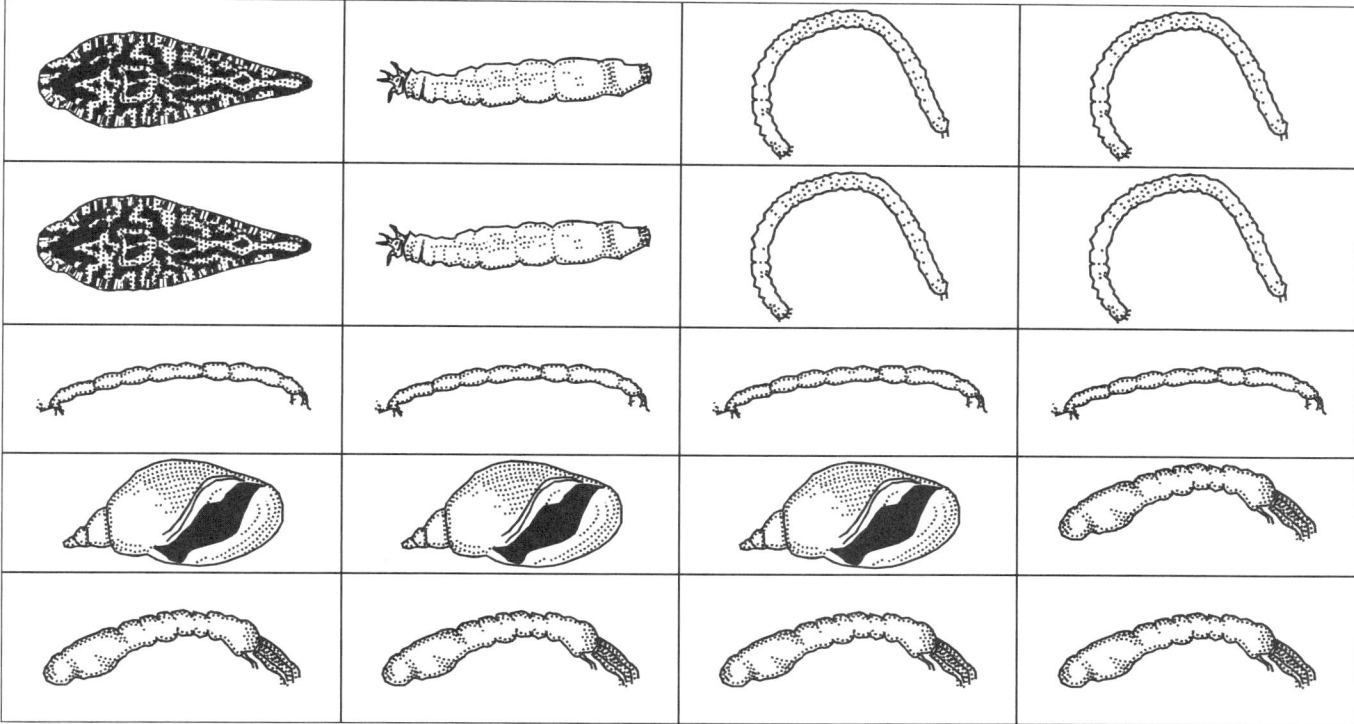

Sample Stream 5

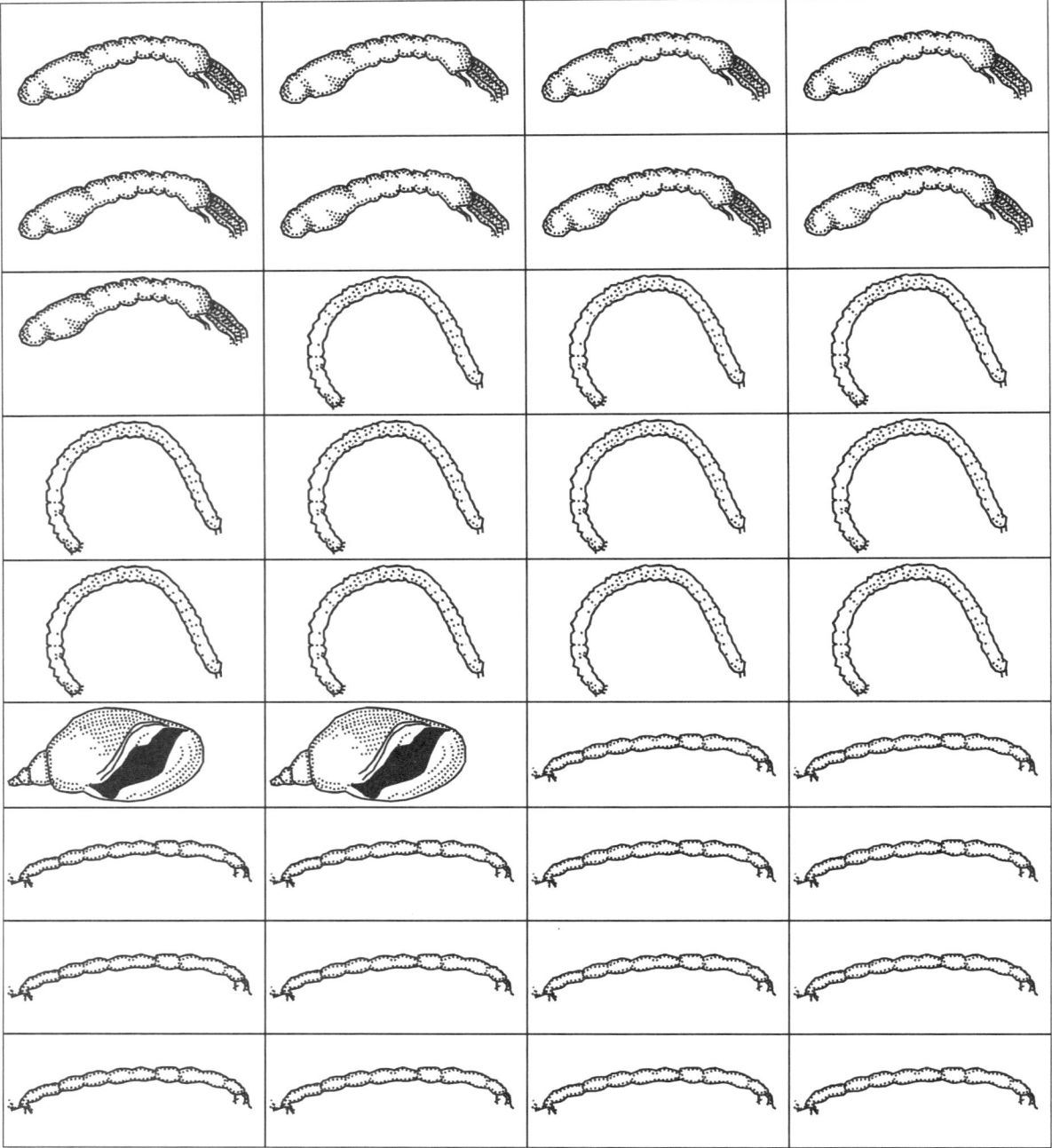

Sample Stream 6

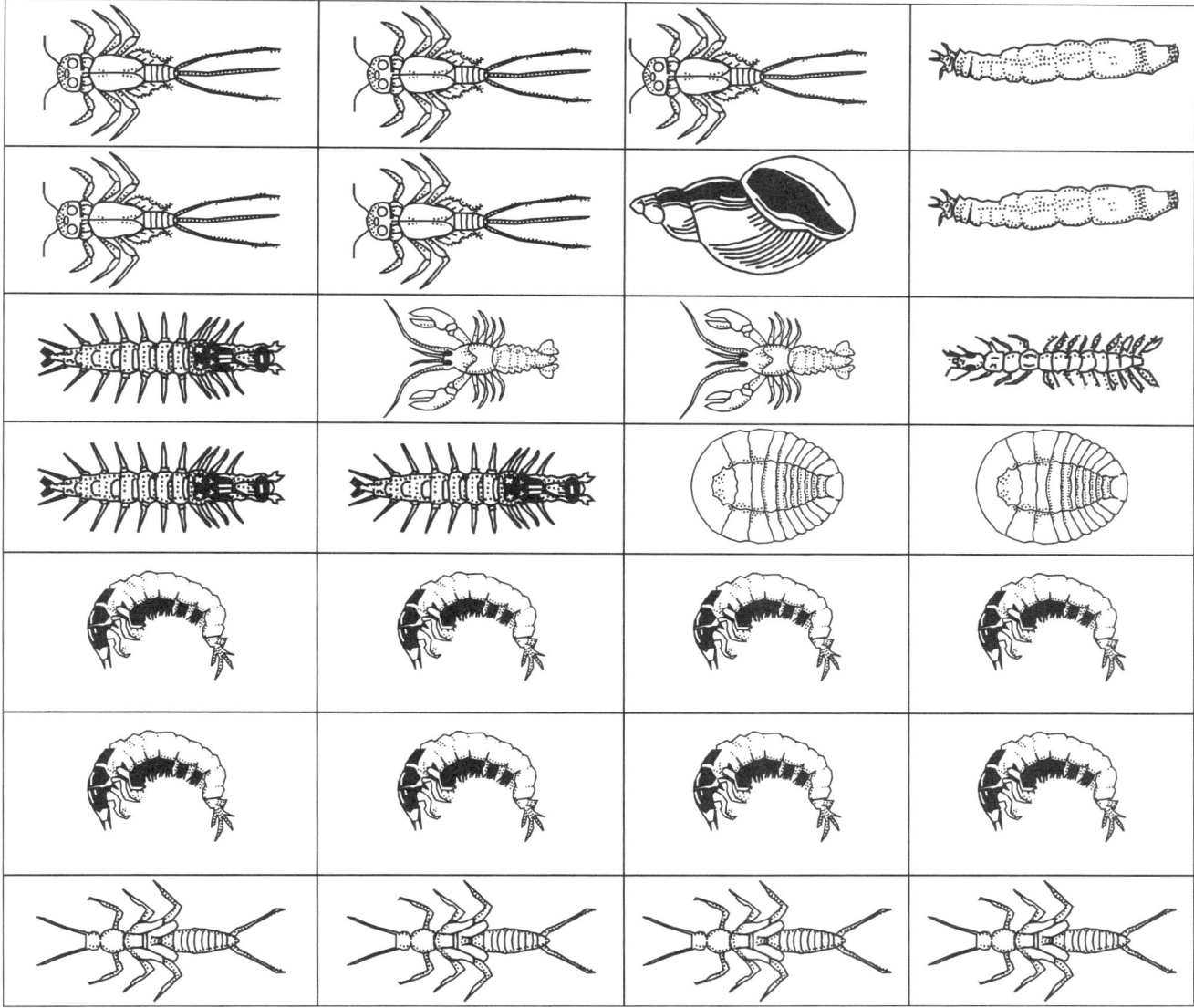

Sample Stream 7

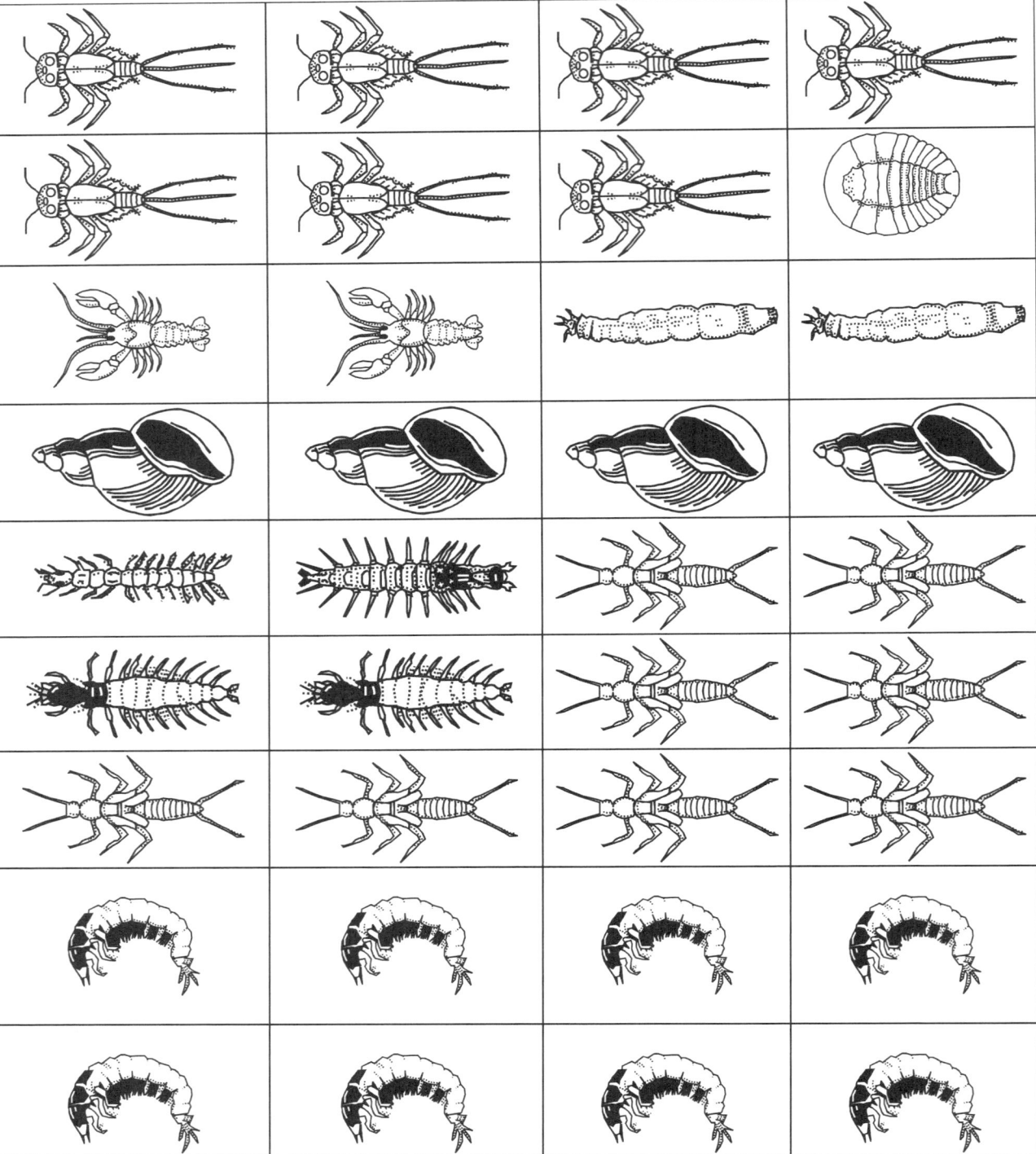

POSSIBLE ANSWERS TO THE CONCLUSION QUESTIONS AND SAMPLE DATA

Site 1 Data Sheet Key

Sensitive	Somewhat Sensitive	Tolerant
✓ _4_ caddisfly larvae	✓ _2_ riffle beetle larvae	___ aquatic worms
___ hellgrammite	___ clams	___ blackfly larvae
✓ _8_ mayfly larvae	___ crane fly larvae	___ leeches
✓ _2_ gilled snails	✓ _4_ crayfish	✓ _9_ midge fly larvae
___ riffle beetle adult	___ damselfly larvae	___ lunged snails
✓ _8_ Stonefly larvae	___ dragonfly larvae	
✓ _2_ water penny larvae	___ scuds	
	✓ _2_ sowbugs	
	___ fishfly larvae	
	___ alderfly larvae	
	___ watersnipe larvae	
	___ whirligig beetle larvae	
Boxes checked x 3 = ___15___ index value	Boxes checked x 2 = ___6___ index value	Boxes checked x 1 = ___1___ index value
TOTAL INDEX VALUE (SUM OF ALL CATAGORIES) ___22___	X Excellent (>22) Good (17-22)	Fair (11-16) Poor (<11)

Site 2 Data Sheet Key

Sensitive	Somewhat Sensitive	Tolerant
✓ **1** caddisfly larvae	___ beetle larvae	✓ **4** aquatic worms
___ hellgrammite	___ clams	___ blackfly larvae
___ mayfly larvae	✓ **2** crane fly larvae	___ leeches
___ gilled snails	✓ **3** crayfish	✓ **9** midge fly larvae
___ riffle beetle adult	___ damselfly larvae	___ lunged snails
✓ **2** Stonefly larvae	___ dragonfly larvae	
✓ **1** water penny larvae	___ scuds	
	___ sowbugs	
	✓ **3** fishfly larvae	
	___ alderfly larvae	
	___ watersnipe larvae	
	___ whirligig beetle larvae	
Boxes checked x 3 = __9__ index value	Boxes checked x 2 = __6__ index value	Boxes checked x 1 = __2__ index value
TOTAL INDEX VALUE (SUM OF ALL CATAGORIES) __17__	**Excellent (>22)** **X Good (17-22)**	**Fair (11-16)** **Poor (<11)**

Site 3 Data Sheet Key

Sensitive	Somewhat Sensitive	Tolerant
___ caddisfly larvae	___ beetle larvae	✓__4__ aquatic worms
___ hellgrammite	___ clams	___ blackfly larvae
___ mayfly larvae	___ crane fly larvae	___ leeches
___ gilled snails	✓__5__ crayfish	✓__4__ midge fly larvae
___ riffle beetle adult	✓__3__ damselfly larvae	___ lunged snails
✓__2__ Stonefly larvae	___ dragonfly larvae	
___ water penny larvae	___ scuds	
	___ sowbugs	
	✓__2__ fishfly larvae	
	___ alderfly larvae	
	___ watersnipe larvae	
	___ whirligig beetle larvae	
Boxes checked x 3 = __3__ index value	Boxes checked x 2 = __6__ index value	Boxes checked x 1 = __2__ index value
TOTAL INDEX VALUE (SUM OF ALL CATAGORIES) __11__	**Excellent (>22)** **Good (17-22)**	**X Fair (11-16)** **Poor (<11)**

Site 4 Data Sheet Key

Sensitive	Somewhat Sensitive	Tolerant
___ caddisfly larvae	___ beetle larvae	✓ __4__ aquatic worms
___ hellgrammite	___ clams	✓ __14__ blackfly larvae
___ mayfly larvae	✓ __2__ crane fly larvae	✓ __2__ leeches
___ gilled snails	___ crayfish	✓ __4__ midge fly larvae
___ riffle beetle adult	___ damselfly larvae	✓ __3__ lunged snails
___ Stonefly larvae	___ dragonfly larvae	
___ water penny larvae	___ scuds	
	___ sowbugs	
	___ fishfly larvae	
	___ alderfly larvae	
	___ watersnipe larvae	
	___ whirligig beetle larvae	
Boxes checked x 3 = _____ index value	Boxes checked x 2 = __4__ index value	Boxes checked x 1 = __5__ index value
TOTAL INDEX VALUE (SUM OF ALL CATAGORIES) __9__	**Excellent (>22)** **Good (17-22)**	**Fair (11-16)** **X Poor (<11)**

Site 5 Data Sheet Key

Sensitive	Somewhat Sensitive	Tolerant
___ caddisfly larvae	___ beetle larvae	✓ 11 aquatic worms
___ hellgrammite	___ clams	✓ 9 blackfly larvae
___ mayfly larvae	___ crane fly larvae	___ leeches
___ gilled snails	___ crayfish	✓ 14 midge fly larvae
___ riffle beetle adult	___ damselfly larvae	✓ 2 lunged snails
___ Stonefly larvae	___ dragonfly larvae	
___ water penny larvae	___ scuds	
	___ sowbugs	
	___ fishfly larvae	
	___ alderfly larvae	
	___ watersnipe larvae	
	___ whirligig beetle larvae	
Boxes checked x 3 = ____ index value	Boxes checked x 2 = ____ index value	Boxes checked x 1 = __4__ index value
TOTAL INDEX VALUE (SUM OF ALL CATAGORIES) __4__	Excellent (>22) Good (17-22)	Fair (11-16) X Poor (<11)

Site 6 Data Sheet Key

Sensitive	Somewhat Sensitive	Tolerant
✓ _8_ caddisfly larvae	___ beetle larvae	___ aquatic worms
___ hellgrammite	___ clams	___ blackfly larvae
✓ _5_ mayfly larvae	✓ _2_ crane fly larvae	___ leeches
✓ _1_ gilled snails	✓ _2_ crayfish	___ midge fly larvae
___ riffle beetle adult	___ damselfly larvae	___ lunged snails
✓ _4_ Stonefly larvae	___ dragonfly larvae	
✓ _2_ water penny larvae	___ scuds	
	___ sowbugs	
	✓ _3_ fishfly larvae	
	___ alderfly larvae	
	___ watersnipe larvae	
	___ whirligig beetle larvae	
Boxes checked x 3 = __15__ index value	Boxes checked x 2 = __6__ index value	Boxes checked x 1 = ____ index value
TOTAL INDEX VALUE (SUM OF ALL CATAGORIES) __21__	**Excellent (>22)** **X Good (17-22)**	**Fair (11-16)** **Poor (<11)**

TEACHER PAGES

Site 7 Data Sheet Key

Sensitive	Somewhat Sensitive	Tolerant
✓ __8__ caddisfly larvae	___ beetle larvae	___ aquatic worms
✓ __1__ hellgrammite	___ clams	___ blackfly larvae
✓ __7__ mayfly larvae	✓ __2__ crane fly larvae	___ leeches
✓ __4__ gilled snails	✓ __2__ crayfish	___ midge fly larvae
___ riffle beetle adult	___ damselfly larvae	___ lunged snails
✓ __6__ Stonefly larvae	___ dragonfly larvae	
__1__ water penny larvae	___ scuds	
	___ sowbugs	
	___ fishfly larvae	
	___ alderfly larvae	
	___ watersnipe larvae	
	___ whirligig beetle larvae	
Boxes checked x 3 = __18__ index value	Boxes checked x 2 = __4__ index value	Boxes checked x 1 = ___ index value
TOTAL INDEX VALUE (SUM OF ALL 3 INDEXES) __22__	**X Excellent (>22)** **Good (17-22)**	**Fair (11-16)** **Poor (<11)**

James River Water Quality Analysis		
Site	**Score (total points)**	**Water Quality**
1	11	Excellent
2	17	Good
3	11	Fair
4	9	Poor
5	4	Poor
6	21	Good
7	22	Excellent

CONCLUSION QUESTION

1. What does it mean for a macroinvertebrate to be pollution sensitive? What does it mean for a macroinvertebrate to be pollution tolerant?
   - Pollution sensitive macroinvertebrates cannot live if any pollution is present. Pollution tolerant macroinvertebrates can live even in polluted water.

2. Could a pollution tolerant macroinvertebrate live in clean water? Why?
   - Yes, pollution tolerant organisms can live in all kinds of water.

3. Which site (number and location) of the James River is the most polluted?
   - Site 5 was the most polluted with an index value of 4.

4. What are some possible causes of the pollution of the James River?
   - The chicken plant and the sewage treatment plant could be a source of pollution, as well as the general population in the area.

5. Can you be positive that the pollution is coming from these causes and not from a location upstream? Why or why not?
   - You can be fairly certain of the pollution sources because the sites up river from the pollution sources are cleaner.

6. Why could the Chicken Plant be a source of pollution? Give a point source and nonpoint source example.
   - Processing chicken produces waste products such as blood. Any of this that makes its way into the river is considered point source pollution. In addition, there is probably a lot of traffic in and out of the plant by trucks. These trucks and cars provide oil, gas and trash as non-point runoff pollution.

7. Which is better indication of the water quality, the number of organisms (for example 12 crayfish are found at one site) or the type of organisms (water pennies are found with lots of pouch snails and leeches) found at a site? Why?
   • It is not the number of organisms that is important, but rather the number of different species that are found. The number of organisms found might simply be a factor of collection time, whereas the number of different species able to live in the area gives a clearer picture of the pollution level.

8. In addition to taking macroinvertebrate specimens as a water quality test Mayor Pops has requested chemical analysis of the water. Why is this a good idea?
   • Chemical testing is a good idea because it can provide specifics about what is polluting the water. Knowing what chemicals are present may tell you what is causing the pollution as well as what type of mediation is recommended.

TEACHER PAGES

Macroinvertebrates That Are Sensitive to Pollution

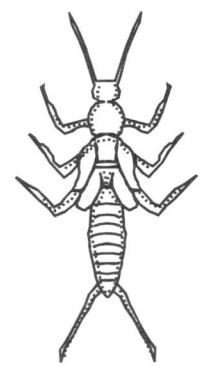
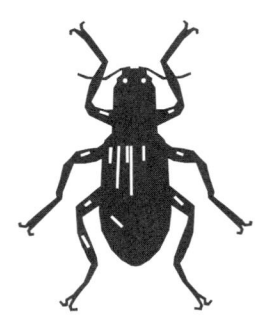
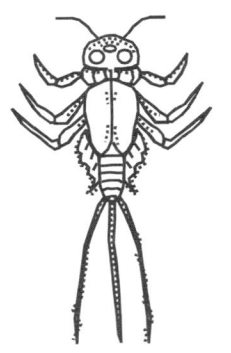
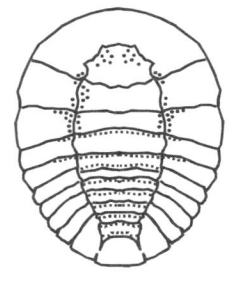

<u>Stonefly</u>   <u>Riffle Beetle Adult</u>   <u>Mayfly</u>   <u>Water Penny</u>

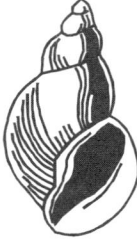

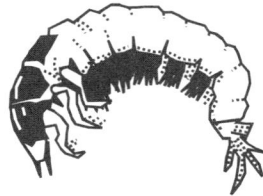
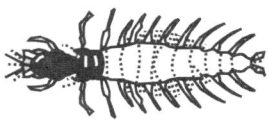

<u>Gilled Snail</u>   <u>Planarian</u>   <u>Caddisfly</u>   <u>Hellgrammite</u>

Macroinvertebrates That Are Somewhat Sensitive to Pollution

Crayfish

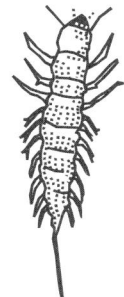

Alderfly

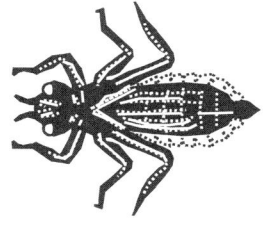

Dragonfly

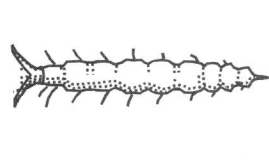

Watersnipe Fly

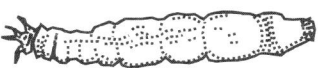

Crane Fly

Riffle Beetle Larva

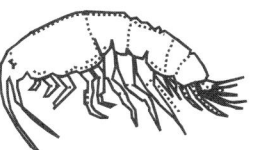

Scud

Whirligig Beetle Larva

Damselfly

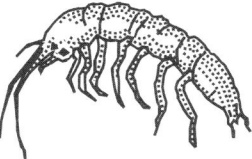

Sowbug

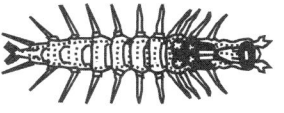

Fishfly

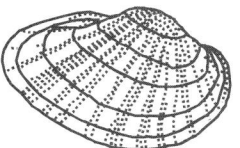

Clam or Mussel

Macroinvertebrates That Are Tolerant of Pollution

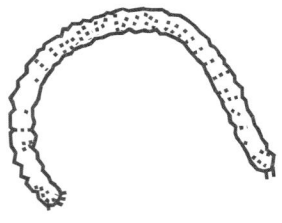

Aquatic Worm

Lunged Snail

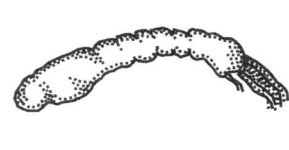

Black Fly

Leech

Midge Fly

A Bugs Life
Judging Water Quality Based on Macroinvertebrates

There is a whole world of life in rivers and lakes. Some of the tiny animals living in the water are *benthic*, meaning they live in the bottom of the body of water. Some are called *macroinvertebrates* because they are large and have no backbone. The most common of macroinvertebrates include insects, clams, snails, crayfish, and worms. Some live their whole lives in the water, others leave the water as adults to feed and reproduce.

Macroinvertebrates are important as food to all the creatures living in the water. Some are considered by scientists to be *indicator species* and are a way of telling whether or not a river or lake is polluted. In rivers, macroinvertebrates live attached to rocks and plants where there is fast flowing water. They are good indicators of water quality because they do not move around and are easy to collect. The moving water gives them food and oxygen. If the water is polluted, there is less food and oxygen for the aquatic macroinvertebrates. Some types of macroinvertebrates are harmed and even killed by the presence of pollutants in the water. Those that are killed by the pollutants are said to be pollution sensitive. Those that can survive in polluted water are said to be pollution tolerant. If the water contains *pollution sensitive* macroinvertebrates in it, then it is a good indication that the water is clean enough and of high enough quality for these sensitive individuals to survive. If there are mostly *pollution tolerant* macroinvertebrates, in the water this may indicate that the water is polluted because those types of species are able to survive the water conditions.

Water pollution can occur in two different ways, as point source pollution and as non-point source pollution. *Point source pollution* occurs at a specific place such as a leaky barrel of pesticide or a pipe discharging sewage. *Nonpoint source pollution* occurs over a large area and its cause cannot be pin pointed to a specific or easily identified place. Pollution such as runoff from a parking or pesticides on a lawn washed into a sewer system would be considered nonpoint source pollution. The severity of both point source and nonpoint pollution can be judged by determining the types of macroinvertebrates that are found in the body of water.

PURPOSE
In this activity you will explore how macroinvertebrates are used to determine the quality of water in lakes, streams, creeks, and any body of freshwater.

MATERIALS
    Map of Aquaville with sites identified
    7 pre-prepared bags containing organisms plates

PROCEDURE

In this activity your team of water quality specialists has been chosen by Mayor Sam "Pops" Waterson to analyze macroinvertebrate specimens taken from the James River in the nearby town of Aquaville. Seven different sites have been chosen and specimens were collected from each site. Using the "specimens" provided, the map of the James River and the town of Aquaville, and the Bug Picking data sheet attached, determine the water quality of the James River at your specific area.

1. Obtain the first sample bag from your teacher.

2. Organize the macroorganisms according to species.

3. Use the provided identification sheets to identify each species and count the number of each organism present at your site.

4. Record your site data on the appropriate data sheet provided in your student answer pages.

5. Calculate the water quality of the site by multiplying the number of species found by the species value, as given on your organism ID sheets.

6. When your teacher indicates that time is up, rotate your packet to the next group.

7. After collecting data for all seven sites, record your results on the James River data table and answer the conclusion questions.

Map of Aquaville

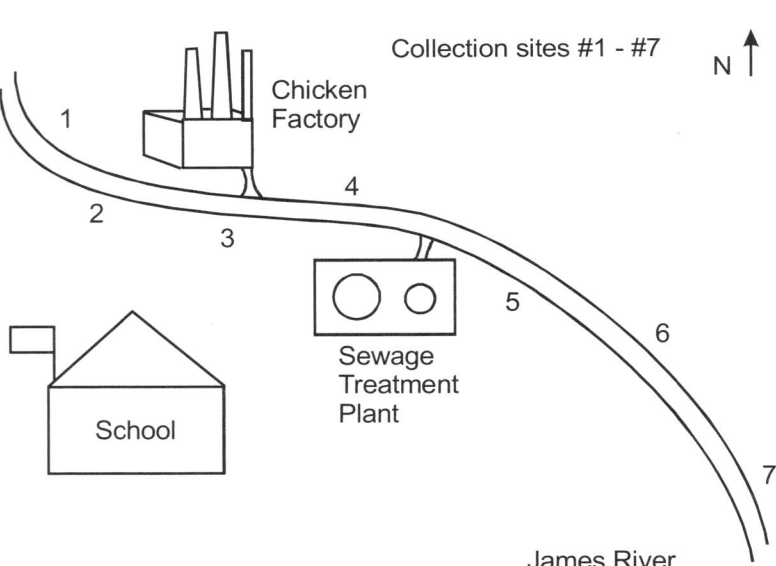

Name _____

Period _____

A Bugs Life
Judging Water Quality Based on Macroinvertebrates

ANALYSIS

Site 1 Data Sheet

Sensitive	Somewhat Sensitive	Tolerant
____ caddisfly larvae	____ beetle larvae	____ aquatic worms
____ hellgrammite	____ clams	____ blackfly larvae
____ mayfly larvae	____ crane fly larvae	____ leeches
____ gilled snails	____ crayfish	____ midge fly larvae
____ riffle beetle adult	____ damselfly larvae	____ lunged snails
____ Stonefly larvae	____ dragonfly larvae	
____ water penny larvae	____ scuds	
	____ sowbugs	
	____ fishfly larvae	
	____ alderfly larvae	
	____ watersnipe larvae	
	____ whirligig beetle larvae	
Boxes checked x 3 = _____ index value	Boxes checked x 2 = _____ index value	Boxes checked x 1 = _____ index value
TOTAL INDEX VALUE (SUM OF ALL CATAGORIES) _____	Excellent (>22) X Good (17-22)	Fair (11-16) Poor (<11)

Site 2 Data Sheet

Sensitive	Somewhat Sensitive	Tolerant
___ caddisfly larvae	___ beetle larvae	___ aquatic worms
___ hellgrammite	___ clams	___ blackfly larvae
___ mayfly larvae	___ crane fly larvae	___ leeches
___ gilled snails	___ crayfish	___ midge fly larvae
___ riffle beetle adult	___ damselfly larvae	___ lunged snails
___ Stonefly larvae	___ dragonfly larvae	
___ water penny larvae	___ scuds	
	___ sowbugs	
	___ fishfly larvae	
	___ alderfly larvae	
	___ watersnipe larvae	
	___ whirligig beetle larvae	
Boxes checked x 3 = _____ index value	Boxes checked x 2 = _____ index value	Boxes checked x 1 = _____ index value
TOTAL INDEX VALUE (SUM OF ALL CATAGORIES) _____	Excellent (>22) X Good (17-22)	Fair (11-16) Poor (<11)

Site 3 Data Sheet

Sensitive	Somewhat Sensitive	Tolerant
___ caddisfly larvae	___ beetle larvae	___ aquatic worms
___ hellgrammite	___ clams	___ blackfly larvae
___ mayfly larvae	___ crane fly larvae	___ leeches
___ gilled snails	___ crayfish	___ midge fly larvae
___ riffle beetle adult	___ damselfly larvae	___ lunged snails
___ Stonefly larvae	___ dragonfly larvae	
___ water penny larvae	___ scuds	
	___ sowbugs	
	___ fishfly larvae	
	___ alderfly larvae	
	___ watersnipe larvae	
	___ whirligig beetle larvae	
Boxes checked x 3 = _____ index value	Boxes checked x 2 = _____ index value	Boxes checked x 1 = _____ index value
TOTAL INDEX VALUE (SUM OF ALL CATAGORIES) _____	**Excellent (>22)** **X Good (17-22)**	**Fair (11-16)** **Poor (<11)**

Site 4 Data Sheet

Sensitive	Somewhat Sensitive	Tolerant
___ caddisfly larvae	___ beetle larvae	___ aquatic worms
___ hellgrammite	___ clams	___ blackfly larvae
___ mayfly larvae	___ crane fly larvae	___ leeches
___ gilled snails	___ crayfish	___ midge fly larvae
___ riffle beetle adult	___ damselfly larvae	___ lunged snails
___ Stonefly larvae	___ dragonfly larvae	
___ water penny larvae	___ scuds	
	___ sowbugs	
	___ fishfly larvae	
	___ alderfly larvae	
	___ watersnipe larvae	
	___ whirligig beetle larvae	
Boxes checked x 3 = _____ index value	Boxes checked x 2 = _____ index value	Boxes checked x 1 = _____ index value
TOTAL INDEX VALUE (SUM OF ALL CATAGORIES) _____	Excellent (>22) X Good (17-22)	Fair (11-16) Poor (<11)

Site 5 Data Sheet

Sensitive	Somewhat Sensitive	Tolerant
___ caddisfly larvae	___ beetle larvae	___ aquatic worms
___ hellgrammite	___ clams	___ blackfly larvae
___ mayfly larvae	___ crane fly larvae	___ leeches
___ gilled snails	___ crayfish	___ midge fly larvae
___ riffle beetle adult	___ damselfly larvae	___ lunged snails
___ Stonefly larvae	___ dragonfly larvae	
___ water penny larvae	___ scuds	
	___ sowbugs	
	___ fishfly larvae	
	___ alderfly larvae	
	___ watersnipe larvae	
	___ whirligig beetle larvae	
Boxes checked x 3 = _____ index value	Boxes checked x 2 = _____ index value	Boxes checked x 1 = _____ index value
TOTAL INDEX VALUE (SUM OF ALL CATAGORIES) _____	Excellent (>22) X Good (17-22)	Fair (11-16) Poor (<11)

Site 6 Data Sheet

Sensitive	Somewhat Sensitive	Tolerant
___ caddisfly larvae	___ beetle larvae	___ aquatic worms
___ hellgrammite	___ clams	___ blackfly larvae
___ mayfly larvae	___ crane fly larvae	___ leeches
___ gilled snails	___ crayfish	___ midge fly larvae
___ riffle beetle adult	___ damselfly larvae	___ lunged snails
___ Stonefly larvae	___ dragonfly larvae	
___ water penny larvae	___ scuds	
	___ sowbugs	
	___ fishfly larvae	
	___ alderfly larvae	
	___ watersnipe larvae	
	___ whirligig beetle larvae	
Boxes checked x 3 = _____ index value	Boxes checked x 2 = _____ index value	Boxes checked x 1 = _____ index value
TOTAL INDEX VALUE (SUM OF ALL CATAGORIES) _____	Excellent (>22) X Good (17-22)	Fair (11-16) Poor (<11)

Site 7 Data Sheet

Sensitive	Somewhat Sensitive	Tolerant
___ caddisfly larvae	___ beetle larvae	___ aquatic worms
___ hellgrammite	___ clams	___ blackfly larvae
___ mayfly larvae	___ crane fly larvae	___ leeches
___ gilled snails	___ crayfish	___ midge fly larvae
___ riffle beetle adult	___ damselfly larvae	___ lunged snails
___ Stonefly larvae	___ dragonfly larvae	
___ water penny larvae	___ scuds	
	___ sowbugs	
	___ fishfly larvae	
	___ alderfly larvae	
	___ watersnipe larvae	
	___ whirligig beetle larvae	
Boxes checked x 3 = _____ index value	Boxes checked x 2 = _____ index value	Boxes checked x 1 = _____ index value
TOTAL INDEX VALUE (SUM OF ALL 3 INDEXES) _____	Excellent (>22) X Good (17-22)	Fair (11-16) Poor (<11)

James River Water Quality Analysis		
Site	Score (total points)	Water Quality
1		
2		
3		
4		
5		
6		
7		

CONCLUSION QUESTIONS

1. What does it mean for a macroinvertebrate to be pollution sensitive? What does it mean for a macroinvertebrate to be pollution tolerant?

2. Could a pollution tolerant macroinvertebrate live in clean water? Why?

3. Which site (number and location) of the James River is the most polluted?

4. What are some possible causes of the pollution of the James River?

5. Can you be positive that the pollution is coming from this site and not from a location upstream? Why or why not?

6. Why could the Chicken Factory be a source of pollution? Give a point source and nonpoint source example.

7. Which is better indication of the water quality, the number of organisms (for example 12 crayfish are found at one site) or the type of organisms (water pennies are found with lots of pouch snails and leeches) found at a site? Why?

8. In addition to taking macroinvertebrate specimens as a water quality test Mayor Pops has requested chemical analysis of the water. Why is this a good idea?

Sonar Seas
Mapping the Ocean Floor

OBJECTIVE
Students will use a motion detector to measure distances and map a simulated ocean floor.

LEVEL
Middle Grades: Earth Science

NATIONAL STANDARDS
UCP.2, UCP.3, A.1, A.2, E.2, F.3, F.6, G.1, G.2

TEKS
6.2(B), 6.2(C), 6.2(D), 6.2(E), 6.3(C)
7.2(B), 7.2(C), 7.2(D), 7.2(E), 7.3(C)
8.2(B), 7.2(C), 7.2(D), 7.2(E), 8.3(C), 8.14(A)
IPC1(A), 2(A), 2(B), 2(C), 2(D), 3(A)

CONNECTIONS TO AP
AP Physics:
    IV. Waves and Optics A. Wave motion (including sound) 1. Properties of traveling waves
    C. Geometric Optics 1. Reflection and refraction

TIME FRAME
45 minutes

MATERIALS
(For a class of 28 working in groups of 4)

7 meter sticks	masking tape
7 Power Macintosh or Windows PC	21 cardboard boxes, various sizes
7 computer interface boxes	Logger*Pro* software
7 motion detector probes	

TEACHER NOTES
Sonar Seas: Mapping the Ocean Floor is an excellent introductory activity for students to become familiar with the sound probes and using the computer program called LoggerPro. This lab works equally well using graphing calculators with the probes. *Foundation Lesson VII: Introduction to Data Collection Devices* shows how to use the graphing calculators with probes.

Students will use motion detectors to measure the distance from the lab table to the top of a succession of boxes. Students must realize that the graph that appears on the computer is an inverse of the shape of the "ocean floor" they are mapping. It helps if students are allowed to use the motion detector probes to explore several different sample ocean floors by rearranging the boxes and repeating the procedure.

An easy way to manage the use of probes with the computer is to have the software, LoggerPro, on the computer and have a link to the software on the desktop. If the computer interface and probe is set up when the students arrive to class they should just be able to click on the LoggerPro icon. When the program opens the computer will automatically sense the motion detector and open the appropriate lab. Having this set up prior to class will reduce the amount of in class troubleshooting.

POSSIBLE ANSWERS TO THE CONCLUSION QUESTIONS AND SAMPLE DATA

DATA AND OBSERVATIONS

Sketch a graph representing the first box you mapped here:

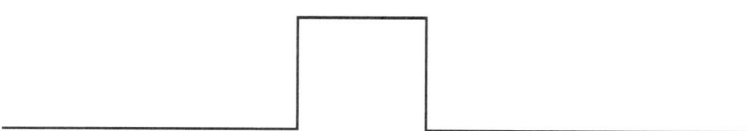

Sketch a graph representing the second "ocean floor" you created here:

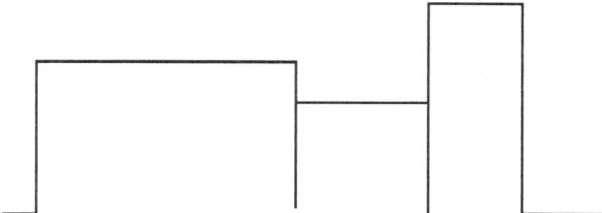

ANALYSIS

1. What was the mean of the three boxes?
   - Answers will vary, but usually range between 20 cm and 60 cm.

2. How does the graph the sonar reported to the computer differ from your drawing above? Explain why it is different.
   - The graph on the computer is an inverted picture of the actual shape of the boxes. This happens because the closer to the probe that the box is, the faster the sound bounces back.

CONCLUSION QUESTIONS

1. Explain how sonar works to map the ocean floor.
   - Sonar sends out a sound and then by measuring the amount of time it takes for the sound to hit something and bounce back, the distance can be calculated. The longer it takes for the sound to bounce back, the deeper the ocean floor is.

2. List three reasons it is important to have an accurate map of the ocean floor.
   - shipping
   - fishing
   - mining

3. What factors make it difficult to measure the ocean floor directly?
   - The ocean floor is difficult to map because we cannot see it. There is a lack of light and parts are so deep that we cannot physically go deep enough to map visually. Many instruments have difficulty surviving the pressure in the very deep parts of the ocean.

4. What animals use sonar or echo location? Why do these animals have to depend on sonar to survive? *(allow students to use resources to answer this question…textbook, Internet, encyclopedia)*
   - Bats use echo location because they are active at night and there is no light to see. Many ocean animals like dolphins and whales use echo location because they cannot see long distances in the ocean where they live.

Sonar Seas
Mapping the Ocean Floor

Oceanographers, marine geologists, and archeologists use echo sounders to investigate objects below the surfaces of bodies of water. An echo sounder consists of a transducer that sends out and receives sound waves. A signal is sent out and bounces back from a submerged surface. Scientists use the speed of sound in water and the time it takes for the signal to bounce back to calculate the depth of the object. The deeper the object, the longer it takes for the sound to return. A map of the ocean floor is made by sending out a series of "pings" in a grid pattern and recording the depths.

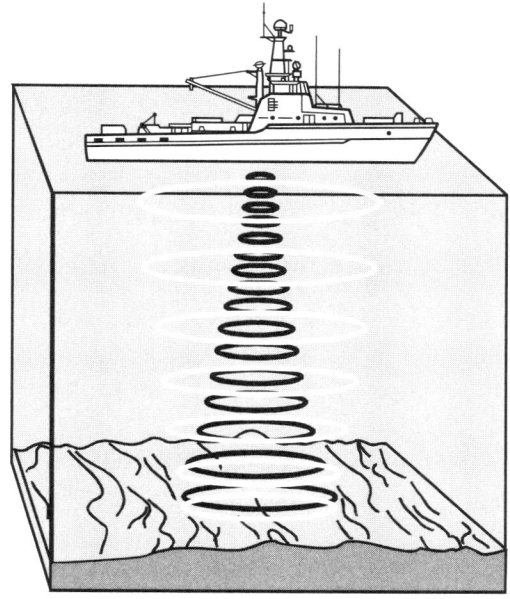

Maps of the ocean floor are used for many things, including mining, fishing and sailing. The most famous use of sonar in recent years was probably the sonar discovery of the *Titanic*, which was located deep on the bottom of the Atlantic Ocean.

Sonar, which is short for *so*und *na*vigation *r*anging, is the name given to this echo sounding system. It was invented during World War I to detect submarines. The motion detector you will use in lab works in the same way.

PURPOSE
In this activity you will use sonar to measure model ocean floors in the classroom and use a computer-generated graph to interpret ocean depths.

MATERIALS
meter stick

computer

computer interface box

motion detector probe

masking tape

3 cardboard boxes, various sizes

LoggerPro software

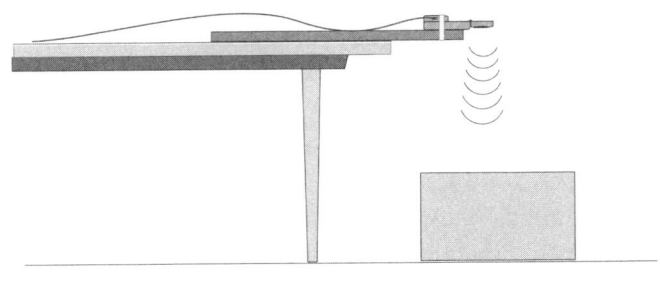

Fig. 1

PROCEDURE

1. Prepare the motion detector for data collection by taping your motion detector to a meter stick as shown in figure 1. Make sure the round screen of the ,motion detector is not covered and is pointed downward.

2. Connect the motion detector into Dig/Sonic 1 of the computer interface device.

3. Place the cardboard box on the floor underneath the motion detector, again as shown in figure 1. *Note:* The motion detector must be at least 40 cm from the top of the box.

4. Line up the motion detector so that when it is moved along the table edge it will pass over the box.

5. Turn on the computer. Prepare the computer for data collection by opening the LoggerPro software. The computer should automatically open the lab using the motion detector. f.

6. Zero the sensor. Position the motion detector on the table pointing down toward the box then click on ⬚ Zero ⬚ .

7. Move the meter stick to position the motion detector to the left of the box. Click ▶Collect to begin data collection. When you hear clicking, slowly slide the board across the tabletop so that the motion detector passes over and past the box.

8. After the motion detector has passed over all the boxes and data collection stops, use the mouse to select a flat portion of the graph that represents the box.

9. Click on the Statistics button, ⬚. Record the Mean (average) height of the box in meters.

10. Sketch and label your graph in the Data and Observation section on your student answer page.

11. Set up three boxes in a new orientation to imitate an ocean floor. The tallest box must be at least 40 cm from the motion detector.

12. Repeat Steps 4-7.

13. Find all three distances by clicking on the Statistics button, ⬚. Record the mean (average) height of each box in meters as the answer to question #1 in the Analysis section on the Student Answer Page.

14. Sketch and label the new graph

Name _____

Period _____

Sonar Seas
Mapping the Ocean Floor

DATA AND OBSERVATIONS

Sketch a graph representing the first box you mapped here:

Sketch a graph representing the second "ocean floor" you created here:

ANALYSIS

1. What was the mean of the three boxes?

2. How does the graph the sonar reported to the computer differ from your drawing above? Explain why it is different.

CONCLUSION QUESTIONS

1. Explain how sonar works to map the ocean floor.

2. List three reasons it is important to have an accurate map of the ocean floor.

3. What factors make it difficult to measure the ocean floor directly?

4. What animals use sonar or echo location? Why do these animals have to depend on sonar to survive?

Slope of the Beach
Determining the Slope of a Line

OBJECTIVE

Students often have difficulty connecting the concept of slope that is used in mathematics and science classes with the idea of slope that they see all around them. This exercise is designed to help them see that wherever it is found, a slope is simply "rise over run."

LEVEL

Middle grades: Earth Science

NATIONAL STANDARDS

UCP.1, UCP.3, A.1, A.2

TEKS

6.2(D), 6.4(A), 6.4(B)
7.2(E), 7.4(A), 7.4(B)
8.2(E), 8.4(A), 8.4(B)
IPC.1, IPC.2(B)

CONNECTIONS TO AP

AP Chemistry, AP Physics and AP Calculus all use the concept of "slope" and mathematical modeling to determine relations and functions.

TIME FRAME

30-45 minutes

MATERIALS

(For a class of 28 students working in groups of 4)

7 meter sticks	masking tape/clear tape
7 protractors	string
scissors	7 plastic coffee stirrers (straws may be used, but they are not as effective.)
7 washers	
tape measures at least 5 meters in length.	7 poles approximately 5 feet long, marked off in decimeter intervals.

TEACHER NOTES

Depending upon time constraints, you may have the students set up this lab, or you may do it before the lab. You will make a simple protractor sextant which will be taped to the top of the meter stick to be used both as a level and as a sighting device. Using clear tape, tape the coffee stirrer to the straight edge of the protractor. Thread a 20 cm length of string through the hole of the protractor and tie a washer to the other end of the string. This acts as a plumb line to keep the apparatus level. Tape the entire sextant assembly to the meter stick so that the coffee stirrer is level with the top of the meter stick. Any number of things may be used for the long poles. If you do not have anything around the house, go to a local home improvement store and purchase the least expensive strips of lumber you can find. It needs to be at least 6 feet long. Mark the pole in 10 cm intervals. Put labels at each mark in meters; i.e. 1.1 m, 1.2 m, etc.

Each group will need the long pole, the meter stick with attached sextant, a long tape measure, a student recording sheet and a pencil. They will need to find a "slope". If they can go outside, it will be no problem. If they cannot go out, handicapped ramps within a building work well.

POSSIBLE ANSWERS TO THE CONCLUSION QUESTIONS AND SAMPLE DATA

Student answers will vary. The data below is a sample of possible data. The answers provided for the conclusion questions will be based on this set of data.

The reading on the long pole is _____1.25 m_____. Be sure you record the measurement in meters.

Subtract 1.0 meter from the reading on the long pole to get the "rise" of your slope.

The rise is ____0.25 m____.

The distance between the poles is ___6.46 m___. Again, be sure that your measurement is in meters. This is the "run".

The value of the slope is ____0.039____.

1. Some highways are marked with % of grade. This is the slope x 100%. If a highway is marked with a 5% grade, how many meters of altitude is gained when you travel 1.0 km?

- $0.05 = \dfrac{\text{rise}}{1000\,\text{m}} = 50\,\text{m}$

- Note: This can be an easy problem if the run is considered to be the distance from the start point rather than the length of the road. Most students will make this assumption. It is up to the teacher whether to introduce the more difficult concept of the length of the road.

2. What are the units for your slope in this lab?
- This slope does not have units since the units in the numerator and denominator will cancel out.
- Point out that not all slopes are without units.

3. Pretend that the long pole used in your lab was really located at the edge of the water at the beach. If the tide rose 35 cm, how far up the beach would the water come? Use the slope that you calculated in your lab.

- $0.039 = \dfrac{0.35\,\text{m}}{x} = 9.0\,\text{m}$

Note: Students with limited math skills may profit from drawing a diagram to scale and measuring it.

Slope of the Beach
Determining the Slope of a Line

During the next several years you will be graphing many sets of data. All of your science classes will ask you to measure, collect data, and graph your results. One purpose for organizing data in a graphical form is to find relationships between variables. When you do this, it is often possible to determine the equation that will relate the variables. An essential portion of a graph is the slope of the line. You have probably learned in your math class that slope is "rise over run". In this activity, we will look at the slopes found around your school. They, too, are "rise over run".

PURPOSE
In this activity you will measure the elevation and distance between two points and calculate the slope.

MATERIALS
1 meter stick with protractor sextant attached   one long pole marked off in 10 cm intervals
tape measure   recording sheet and pencil

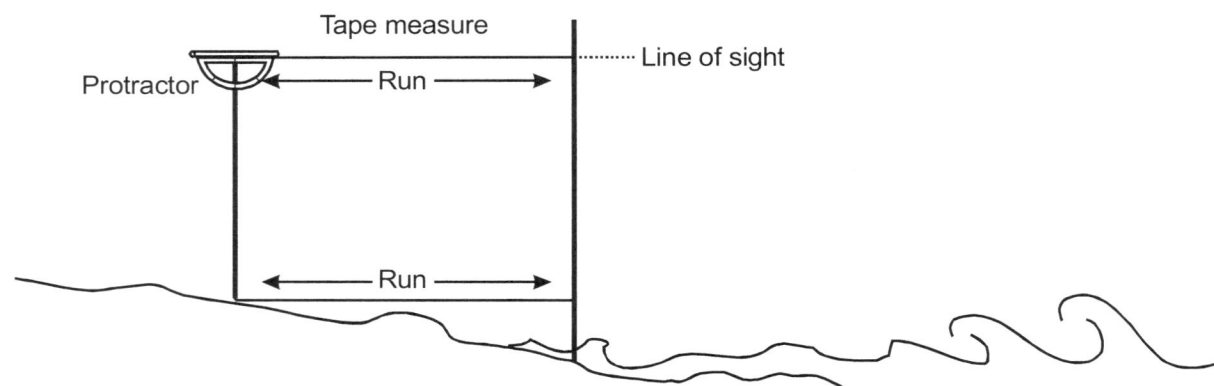

PROCEDURE
1. Locate a slope to measure. Your teacher will direct you to areas to work.

2. Place the long pole at the bottom of the slope. Use the meter stick with the plumb bob on it next to the pole and use it to get the long pole vertical. One to two team members should stay with the long pole.

3. Walk 5 to 10 paces uphill from the long pole carrying the meter stick with you. Place the meter stick on the ground. Use the plumb bob to level it. Point the straw at the long pole.

4. Have a team member located at the long pole place a finger at the 1.1 m mark and move his/her finger up and down the pole until the finger can be seen through the straw. Record the mark

5. Place one end of the tape measure on the top of the meter stick and stretch it to the 1.0 m mark on the long pole. Make an attempt to get the tape measure as straight as possible. Record the distance between the meter stick and the long pole.

6. Using the data collected, answer the questions assigned.

Name _____

Period _____

Slope of the Beach
Determining the Slope of a Line

DATA AND OBSERVATIONS

The reading on the long pole is _____. Be sure you record the measurement in meters.

Subtract 1.0 meter from the reading on the long pole to get the "rise" of your slope.

The rise is _____.

The distance between the poles is _____. Again, be sure that your measurement is in meters. This is the "run".

ANALYSIS

The value of the slope is _____. Hint: Slope $= \dfrac{\text{rise}}{\text{run}}$.

CONCLUSION QUESTIONS

1. Some highways are marked with % of grade. This is the slope x 100%. If a highway is marked with a 5% grade, how many meters of altitude is gained when you travel 1.0 km?

2. What are the units for your slope in this lab?

3. Pretend that the long pole used in your lab was really located at the edge of the water at the beach. If the tide rose 35 cm, how far up the beach would the water come? Use the slope that you calculated in your lab.

Write Your Notes and Ideas Here!

Staying Out of Hot Water
Modeling a Thermocline

OBJECTIVE

Students will be able to identify and model the thermocline, and describe this zone of rapid temperature change in the ocean.

LEVEL

Middle Grades: Earth Science

NATIONAL STANDARDS

UCP.2, UCP.3, UCP.4, A.1, B.6, D.2

TEKS

6.2(A), 6.2(B), 6.2(C), 6.2(D), 6.2(E), 6.3(C), 6.4(A), 6.4(B), 6.5(C)
7.2(A), 7.2(B), 7.2(C), 7.2(D), 7.3(C), 7.4(A), 7.4(B), 7.5(C), 7.12(A)
8.2(A), 8.2(B), 8.2(C), 8.2(D), 8.3(C), 8.4(A), 8.4(B), 8.5(C), 8.10(B)
IPC 1(A), 2(A), 2(B), 2(C), 2(D), 3(A)

CONNECTIONS TO AP

AP Environmental Science:

    I Interdependence of Earth's Systems: Fundamental Principles and Concepts A. Flow of Energy
    III Renewable and Nonrenewable Resources: Distribution, Ownership, Use, Degradation A. Water
    2. Oceans

TIME FRAME

60 minutes

MATERIALS

(For a class of 28 working in groups of 4)

42 thermometers	1 roll clear plastic tape
7 heat lamps	7 timing devices (seconds)
7 clear plastic overhead sheet (not copier ones)	colored pencils
7 permanent markers	7 3-liter bottles, cut off below the neck
7 metric rulers	

TEACHER NOTES

Modeling a Thermocline has students model the zone of rapid temperature change in the ocean. This zone is called the thermocline and it separates the surface waters warmed by the sun and the deep cold waters of the ocean.

Prior to performing the lab collect 3-liter bottles and cut off the tops with a box cutter (see photo). The protocol requires a wide opening for the thermometers to fit inside and to ensure that the heat lamp hits the entire surface of the water. Very large glass beakers (800 mL) can be used as well.

If you have limited thermometers in your classroom to perform this activity as a lab you can prepare a demonstration model in the front of the room and choose several students to carry out the procedure and relay the data to the class. If done as a classroom demonstration it is important to have research materials available for non-participating students so that they may be completing the lab questions during the data collection time.

Two line graphs are required for this lab. On the first graph students use different colored lines to plot temperature versus time for all of the depths investigated. The second line graph shows the temperatures for each depth at the 20 minute mark.

POSSIBLE ANSWERS TO THE CONCLUSION QUESTION AND SAMPLE DATA

Data Table: Temperatures At Different Depths						
Time (min)	2cm	4cm	6cm	8cm	10cm	12cm
2	21	21	21	21	20	20
4	21	21	21	20	20	20
6	21	21	21	21	20	20
8	22	22	22	22	20	20
10	23	22	22	22	20	20
12	24	23	22	22	20	20
14	24	23	22	22	20	20
16	25	24	22	23	20	20
18	25	24	23	23	20	20
20	25	24	23	23	21	20

Graph I

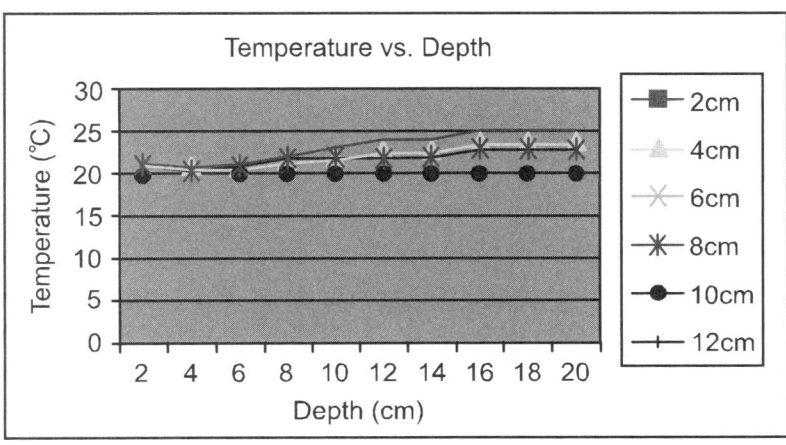

Graph II

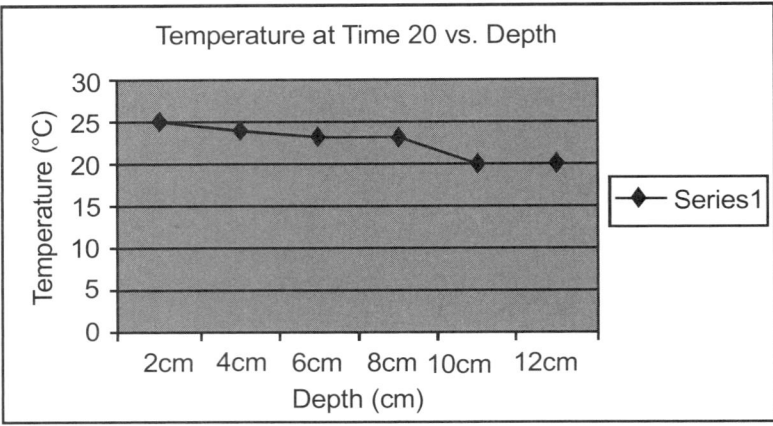

1. Which depth was the greatest range? What was the range in °C?
   - In this model the greatest temperature range was at 2 cm deep.
   - The range was from 21°C - 25°C.

2. Which depth was the smallest range? What was the range in °C?
   - The smallest range in this model was at 12 cm because there was no temperature change.
   - There was no range—the temperature remained constant.

3. Did you see a range in temperatures as depth increases? Why or why not?
   - No, as the depth increased the range dropped to zero.
   - A likely reason for this decrease in range is that not much heat was reaching deep into the container.

4. What is a thermocline?
   - A thermocline is the zone of rapid temperature change that separates the warm surface waters from the deep cold water.

5. What is the average depth of a thermocline in the ocean? What is the average temperature range of a thermocline in the ocean?
   - The average depth is somewhere between 200 and 1000 meters.
   - The average temperature range for an ocean thermocline is from 22 °C - 4°C.

6. In what vertical ocean life zone does the thermocline exist? Describe characteristics of that life zone?
   - The vertical zone of the ocean that contains the thermocline is called the twilight zone because there is only a small amount of light that reaches it.
   - The twilight zone has some neritic creatures (swimmers), but there is not enough light for phytoplankton to manufacture food, and so it has much less diversity of life compared to the epipelagicor sunlight zone.

Staying Out of Hot Water
Modeling a Thermocline

A thermocline is the layer of water between the warmer surface zone and the colder deep zone of any thermally stratified body of water. In the thermocline, temperature and oxygen concentration drop precipitously with depth.

In the ocean, the surface waters are warmed by the sun. Phyto-plankton use the sunlight to photosynthesize and manufacture food, and as a result the surface zone is full of organisms that depend on the phyto-plankton to survive. Below the surface zone though, there is very little light and not much mixing of the water. At this point there is a rapid temperature change between the warm surface waters and the cold deep water. This area is the thermocline.

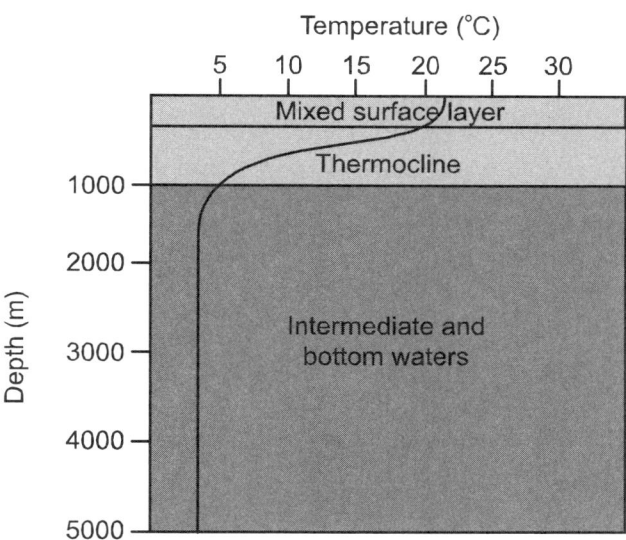

PURPOSE
In this activity you will model a thermocline and predict where this layer would occur in the ocean .

MATERIALS
6 thermometers	clear plastic tape
1 heat lamps	1 timing device (seconds)
1 clear plastic overhead sheet (not copier ones)	colored pencils
1 permanent markers	metric ruler

Safety Alert
1. Do not touch the heat lamp.

PROCEDURE

1. Predict which depth will have the greatest range of temperatures. Write your prediction under hypothesis on the student answer page.

2. Arrange 6 thermometers on the plastic sheet ten centimeters apart from each other, and in a descending order, each one 2 centimeters deeper than the one before.

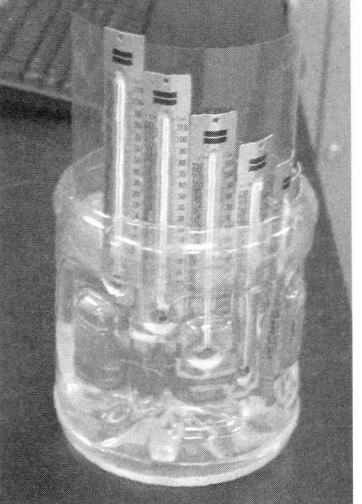

3. Fill the 3-liter bottle with water until the first thermometer is 2 cm below the surface.

4. Place the bottle under the heat lamp. Be sure that there is 20 centimeters between the top of the water and the heat lamp.

5. Turn the heat lamp on and start your timing device.

6. Take temperature readings every two minutes for each thermometer. Record your answer in the data table on the student answer page.

7. Use your collected data to complete the analysis graphs and the conclusion questions.

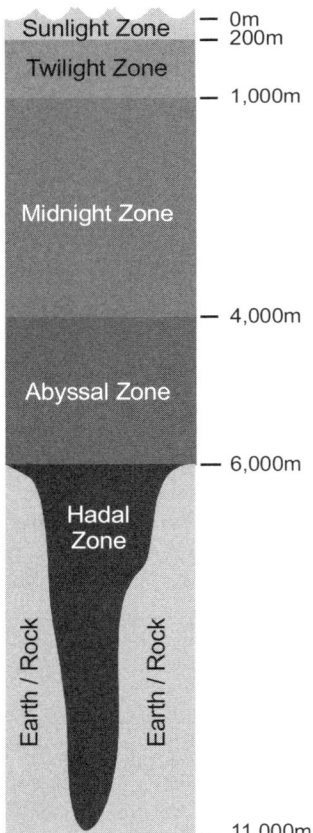

Name _____

Period _____

Staying Out of Hot Water
Modeling a Thermocline

HYPOTHESIS

DATA AND OBSERVATIONS

Data Table: Temperatures at Different Depths						
Time (min)	2cm	4cm	6cm	8cm	10cm	12cm
2						
4						
6						
8						
10						
12						
14						
16						
18						
20						

ANALYSIS

1. Make a line graph of the temperature versus time for each of the 6 depths measured. Use a different colored pencil for each different depth. You will have six lines on the one graph.

2. Make a second line graph plotting the data for final temperature (*y*-axis) versus depth (*x*-axis) at the end of the 20 minute period.

CONCLUSION QUESTIONS

1. Which depth had the greatest range of temperatures? What was the range in °C?

2. Which depth had the smallest range of temperatures? What was the range in °C?

3. Did you see a range in temperatures as depth increases? Why or why not?

4. What is a thermocline?

5. What is the average depth range of a thermocline in the ocean? What is the average temperature of a thermocline?

6. In what vertical ocean life zone does the thermocline exist? Describe characteristics of that life zone?

Underwater Avalanche: Turbidity Currents
Investigating Density Currents

OBJECTIVE

Students will investigate how changes in density affect the rate at which slurry moves down a slope and what effect changes in slope have on the rate of movement. Students will apply what they learn to ocean floor situations.

Fig. 1

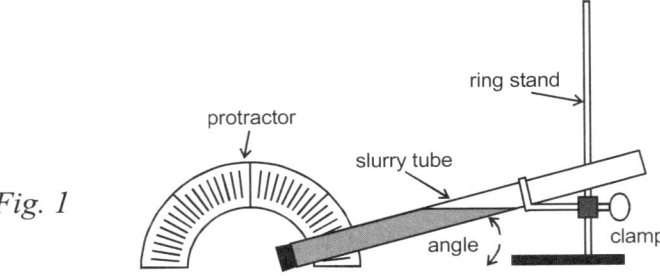

LEVEL

Middle Grades: Earth Science

NATIONAL STANDARDS

UCP.2, UCP.3, UPC.5, B.4, D.2

TEKS

6.1(A), 6.2(A), 6.2(B), 6.2(C), 6.2(D), 6.2(E), 6.3(A), 6.3(C), 6.4(A), 6.4(B), 6.6(A), 6.6(B), 6.6(C)
7.1(A), 7.2(A), 7.2(B), 7.2(C), 7.2(D), 7.2(E), 7.3(A), 7.3(C), 7.4(A), 7.4(B)
8.1(A), 8.2(A), 8.2(B), 8.2(C), 8.2(D), 8.2(E), 8.3(A), 8.3(C), 8.4(A), 8.4(B), 8.10(B)
IPC: 1(A), 2(A), 2(B), 2(C), 2(D), 3(A), 4(A), 7(A)

CONNECTIONS TO AP

AP Physics
    I. Newtonian Mechanics A. Kinematics
AP Chemistry
    III. Reactions D. Kinetics

TIME FRAME

100 minutes

MATERIALS

(For a class of 28 working in groups of 4)

7 sets of slurry samples, labeled #1-#4
7 ring stands and clamps
7 protractors
7 test tube racks
4 600-mL beakers
potting clay

7 long plastic tubes with stoppers
tap water
35 150mm standard test tubes
7 timers or stop watches
masking tape
pencil or marking pen

TEACHER NOTES

The Turbidity Current Lab demonstrates a density current in the ocean and shows students how the low angle slope of the continental shelf helps to increase the velocity of the sediment laden water. This lab may be taught while teaching about oceans or depositional processes.

Prior to doing the lab you must pre-mix slurries of four different densities and measure the speed of each as they move down a tube. To prepare the four slurries, mix 1/4 pound of clay with one gallon of tap water. Mix together until the clay is suspended in the water. Label the empty beakers #1, #2, #3 and #4. Fill the beakers labeled 1, 2, 3, and 4 with 400, 200, 100, and 50 mL respectively. Add tap water so that the total volume in each beaker is 600 mL. (For <u>each</u> class of 28 you will need 42 samples of slurry 1, or about 600 mL. You will need less of the other three samples, around 200-300 mL.)

SAMPLE DATA

Part I

Data Table 1									
Slurry #1	5 sec	10 sec	15 sec	20 sec	25 sec	30 sec	35 sec	40 sec	
distance (cm)	14	27	39	51	62	73	83	95	average velocity
velocity (cm/sec)	2.8	2.6	2.4	2.4	2.2	2.2	2.0	2.4	**2.375**
Slurry #2	5 sec	10 sec	15 sec	20 sec	25 sec	30 sec	35 sec	40 sec	
distance (cm)	21	29.5	33.5	37.5	43.5	50.5	56	62	average velocity
velocity (cm/sec)	4.2	1.7	0.8	0.8	1.2	1.4	1.1	1.2	**1.55**
Slurry #3	5 sec	10 sec	15 sec	20 sec	25 sec	30 sec	35 sec	40 sec	
distance (cm)	13	19	24	27	30.5	32	34.5	36	average velocity
velocity (cm/sec)	2.6	1.2	1.0	0.6	0.7	0.3	0.5	0.3	**0.90**
Slurry #4	5 sec	10 sec	15 sec	20 sec	25 sec	30 sec	35 sec	40 sec	
distance (cm)	18	20	22	23	25	28	30	32	average velocity
velocity (cm/sec)	3.6	0.4	0.4	0.2	0.4	0.6	0.4	0.4	**0.8**

Part II

Data Table 2		
Slope	Travel Time (sec)	Velocity (cm/sec)
15°	40	2.35
30°	41	2.31
45°	42	1.9
60°	47	1.71
75°	54	1.49
90°	65	1.23

Graph for Part I

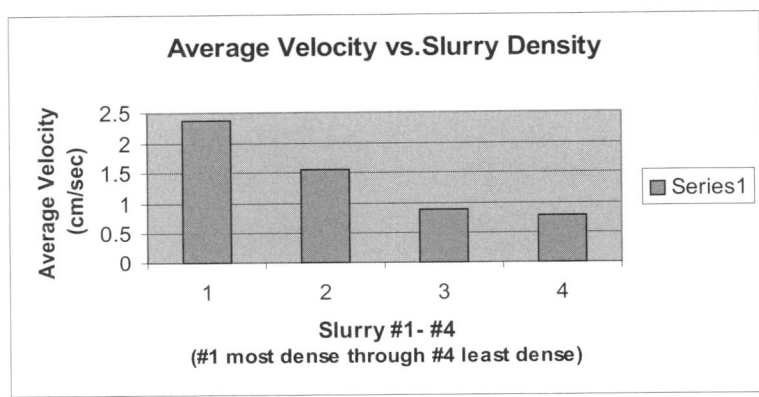

Graph for Part II

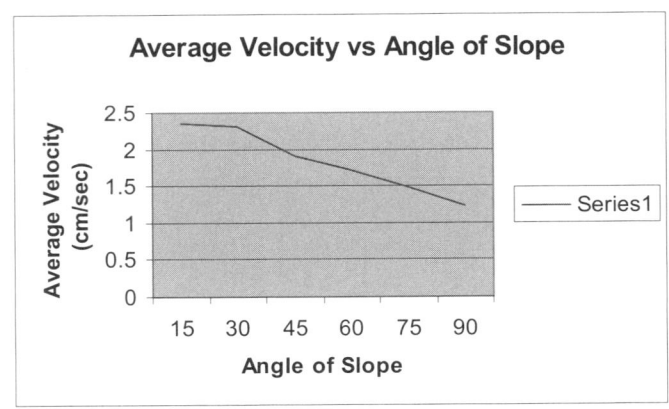

CONCLUSION QUESTIONS

1. From graph #1, what is the relationship between average velocity of a slurry and its density?
   - As the slurry density increases the velocity of the slurry increases.

2. Describe any velocity changes that you may have observed in each of the slurries as they traveled down the tube. What factors may have caused these changes?
   - The slurry slowed down as it moved down the tube and mixed with the water. he mixing with water decreased the density and caused it to slow down.

3. From the graph for Part II, describe the effect changes in slope had on the velocity of slurry #1. Did the outcome match your hypothesis? Explain why or why not.
   - The lower the angle of the slope, the faster the movement of the slurry.
   - Answers will vary depending on student hypotheses. Many students may have hypothesized that the greater the slope the faster the movement.

4. Your experiment differed from real density currents in one very important way. Real density currents (turbidity currents) move over the ocean floor collecting loose sediment as they move, thereby increasing in density. Did your samples increase in density as they moved down the tube? Explain your answer.
   - No, samples did not increase in density.
   - Samples decreased in density as they moved down the tube because they mixed with the fresh water and became less dense as they moved down the tube.

5. If the density of a turbidity current increases as it continues along its path, predict what happens to its velocity.
   - In the ocean, as a turbidity current moves down the continental slope and increases in density the velocity also increases.

6. Use what you have learned about turbidity currents to explain why continental sediment can be found far from continental shorelines.
   - On a continental shelf, as a turbidity current rolls across the low degree slope, it picks up more sediment, increasing density and therefore increasing velocity. As it leaves the continental shelf it is moving very fast and shoots over the continental slope. It continues to pick up additional sediment as it moves down the slope and finally it collapses onto the abyssal plain. The low angle of the abyssal plain can allow the sediment to travel long distances before coming to a stop.

Underwater Avalanche: Turbidity Currents
Investigating Density Currents

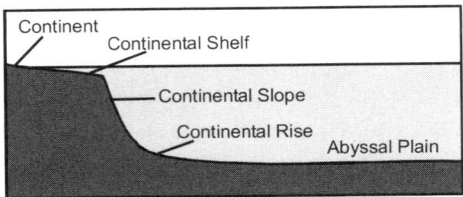

Model of the Ocean Floor

Turbidity currents are undersea flows of muddy sediments that are often triggered by earthquakes. They can also be caused by flooding events in which rivers carry high amounts of sediment into the ocean and by underwater landslides or slumps. Turbidity currents generally occur on the continental slope and rise, areas too deep to be affected by surface waves and tidal currents. As the continental slope flattens at its base into the continental rise and the abyssal plains of the deep sea, the turbidity flows slowly and the sediments settle into graded beds of sand, silt and mud called "turbidites."

PURPOSE
To investigate how slurry density affects the rate at which slurry moves down a slope and also what effect changes in slope have on the rate at which slurry moves down a slope.

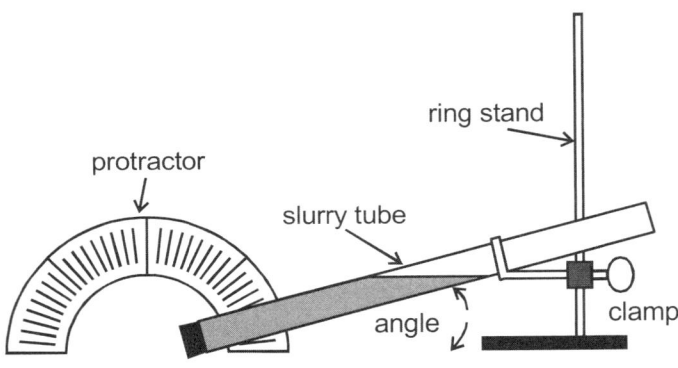

MATERIALS
slurry samples #1,#2, #3, #4   long plastic tube with stopper
ring stand and clamp   tap water
protractor   4 150 mm test tubes
test tube rack   timer or stop watch
pencil or marking pen

Safety Alert
1. Keep your lab table area clean and **dry**.

PROCEDURE

Part I – Slurry Flow Velocity vs. Slurry Density

1. In the space marked HYPOTHESIS on your student answer page, make a hypothesis as to which slurry will move fastest down the tube. (Slurry #1 is the most dense and #4 is the least dense.)

2. Apply a piece of masking tape down the vertical length of the long plastic tube so that it may be used for marking traveling distances. Set the long plastic tube in the ring stand as shown in Figure 1. Use your protractor to set the tube so that it forms a slope at a 15° angle with the tabletop (Figure 2).

3. Fill the long plastic tube with fresh water to within 10 cm of the top of the tube.

4. Label your test tubes #1-#4. Obtain samples of each slurry from your teacher. Stir each sample <u>before</u> pouring into your test tube.

5. Place your thumb over the end of test tube #1 and shake to mix. Quickly pour slurry #1 into the plastic tube of water. Use a pencil to mark the distance the slurry travels in 5 second intervals. Mark directly on your masking tape.

6. Measure the distances and record them in your data table on the student answer page.

7. Empty the long tube into the sink and refill the long tube with fresh water. For the next trial you may use the same piece of masking tape by either changing colors of marker for each trial or by labeling your marks for each trial prior to starting the next trial.

8. Repeat steps 1-7 for slurry #2, #3 and #4. Record the results in the data table for Part I.

Part II – Slurry vs. Slope

1. In the space marked HYPOTHESIS on your student answer page, make a hypothesis as to which slope angle will cause the slurry to move fastest down the tube.

2. Use your data from Part I for the 15° category. (You do not have to repeat this trial.)

3. Rinse the test tubes used in part I and remove their labels. Only slurry #1 will be used for this part of the experiment.

4. Refill the long tube with fresh water. Readjust the tube so that it forms a slope of 30°.

5. Obtain five more samples of slurry #1. Shake the slurry test tube and again add the slurry to the long tube. Measure the total time (in seconds) that it takes for the slurry to reach the end of the long tube. Record your measurements in data Table 2 on the student answer page.

6. Empty the long tube into the sink and refill the long tube with fresh water.

7. Repeat the procedure for 45°, 60°, 75°, and 90° angles. Record your data in Table 2 on the student answer page.

Name _____

Period _____

Underwater Avalanche: Turbidity Currents
Investigating Density Currents

HYPOTHESIS: PART I _____

HYPOTHESIS: PART II _____

DATA AND OBSERVATIONS

Part I

Data Table 1									
Slurry #1	5 sec	10 sec	15 sec	20 sec	25 sec	30 sec	35 sec	40 sec	
distance (cm)									average velocity
velocity (cm/sec)									
Slurry #2	5 sec	10 sec	15 sec	20 sec	25 sec	30 sec	35 sec	40 sec	
distance (cm)									average velocity
velocity (cm/sec)									
Slurry #3	5 sec	10 sec	15 sec	20 sec	25 sec	30 sec	35 sec	40 sec	
distance (cm)									average velocity
velocity (cm/sec)									
Slurry #4	5 sec	10 sec	15 sec	20 sec	25 sec	30 sec	35 sec	40 sec	
distance (cm)									average velocity
velocity (cm/sec)									

Part II

Data Table 2		
Slope	Travel Time (sec) (for entire length of plastic tube: 95 cm)	Velocity (cm/sec)
15°		
30°		
45°		
60°		
75°		
90°		

ANALYSIS

Part I

1. Calculate the velocities for each interval. Record them in the data table for Part I.

2. Draw a bar graph comparing the average velocities of the slurry and the slurry density, using your data in Part I. Average velocity should be your *y*-axis and the four slurry samples should be on your *x*-axis.

Part II

1. Calculate the velocity for each angle of slope by dividing the centimeters traveled by the time it took in seconds. Record your answers in your data table.

2. Draw a line graph comparing the velocity of the slurry's vs. the angle of the slope of the tube. Average velocity should be your *y*-axis and the angle of the slope of the plastic tube should be on your *x*-axis.

Graph for Part I

Graph for Part II

CONCLUSION QUESTIONS

1. From the graph for Part I, what is the relationship between average velocity of a slurry and its density?

2. Describe any velocity changes that you may have observed in each of the slurries as they traveled down the tube. What factors may have caused these changes?

3. From the graph for Part II, describe the effect changes in slope had on the velocity of slurry #1. Did the outcome match your hypothesis? Explain why or why not.

4. Your experiment differed from real density currents in one very important way. Real density currents (turbidity currents) move over the ocean floor collecting loose sediment as they move, thereby increasing in density. Did your samples increase in density as they moved down the tube? Explain your answer.

5. If the density of a turbidity current increases as it continues along its path, predict what happens to its velocity.

6. Use what you've learned about turbidity currents to explain why continental sediment can be found far from continental shorelines.

Dynamic Earth
Discovering Plate Tectonics

OBJECTIVE

Students will research earthquakes, volcanoes, plate boundaries and earth's history to put together a map and a presentation explaining plate tectonics.

LEVEL

Middle Grades: Earth Science

NATIONAL STANDARDS

UCP.2, UCP.3, A.1, D.3, D.2, F.5, G.3

TEKS

6.2(D), 6.2(E), 6.3(C), 6.3(E), 6.6(C)
7.2(D), 7.2(E), 7.3(C), 7.3(E), 7.5(A), 7.8(A), 7.14(A)
8.2(D), 8.2(E), 8.3(C), 8.3(E), 8.4(B), 8.12(A), 8.14(A)
IPC 1(A), 2(A), 2(B), 2(C), 2(D), 3(A)

CONNECTIONS TO AP

AP Environmental Science:
   I. Interdependence of Earth's Systems: Fundamental Principles and Concepts C. The Solid Earth
   2. Earth dynamics: plate tectonics, volcanism, the rock cycle

TIME FRAME

225 minutes

MATERIALS

(For a class of 28 working in groups of 4)

   7 meter sticks
   earth map overhead
   1 roll butcher paper
   colored pencils

TEACHER NOTES

The *Dynamic Earth* activity is a cooperative learning exercise that has students use research to discover the underlying causes for earthquakes, volcanoes and continental drift. Information gathered by each group will be plotted on a small map and small maps will be combined to make one large class map. Setting up a map contest between classes gives the students some investment in doing a good job.

Each lab group of four students is assigned a different portion of the earth. You can assign groups by using the longitudinal lines to divide the globe into portions. For instance, if you have six goups, divide the map every 60° of longitude so that the map ends up being divided into six pieces. Each group will then be responsible for a specific 60°, say from 0° to 60°W. The provided map can be copied onto a transparency and used for class discussion and region assignments.

Within the four member group each student has an assigned job; seismologist, volcanologist, paleogeographer and geologist. The provided rubrics outline the conceptual and procedural expectations for both the individuals and groups. Although not required, the seismologists (or other scientists) from different groups may wish to convene and discuss their research together.

Day one of the *Dynamic Earth* activity should be spent assigning roles and carefully reviewing job assignments. If there is time, students may start tracing their part of the map. Each student group will be responsible for generating a map of their section, however, since the maps will be re-combined to form the class map, agreements must be made as to what symbols, colors, and even brand of colored pencils will be used. To trace the map, project it on the chalk board or in the hallway so that the entire world map takes up the size of about one bulletin board. Tape up strips of white butcher paper to project map onto for tracing. No political divisions need to be drawn on the map. Students only need to trace the outlines of the continents and lines of latitude and longitude in their regions.

Research can be done on the Internet and from books. To facilitate research in the classroom you can check out relevant books from the library and have them available in your classroom. If access to the Internet is not available during class, you can print out lists of recent earthquakes and volcanoes for students to use in their map plots.

At the beginning of each class period lead a short discussion about the mechanics of the project. Allow students to share resources and information. Suggest methods for research and transferring information to the class map.

When the projects are completed students must present their information in groups. The presentation should be made in PowerPoint, but should also include an additional multimedia source such as photos, video clips, models or a guest speaker. Their presentations should explain the underlying causes to the events they are describing. The provided rubric outlines more specific criteria for the presentation.

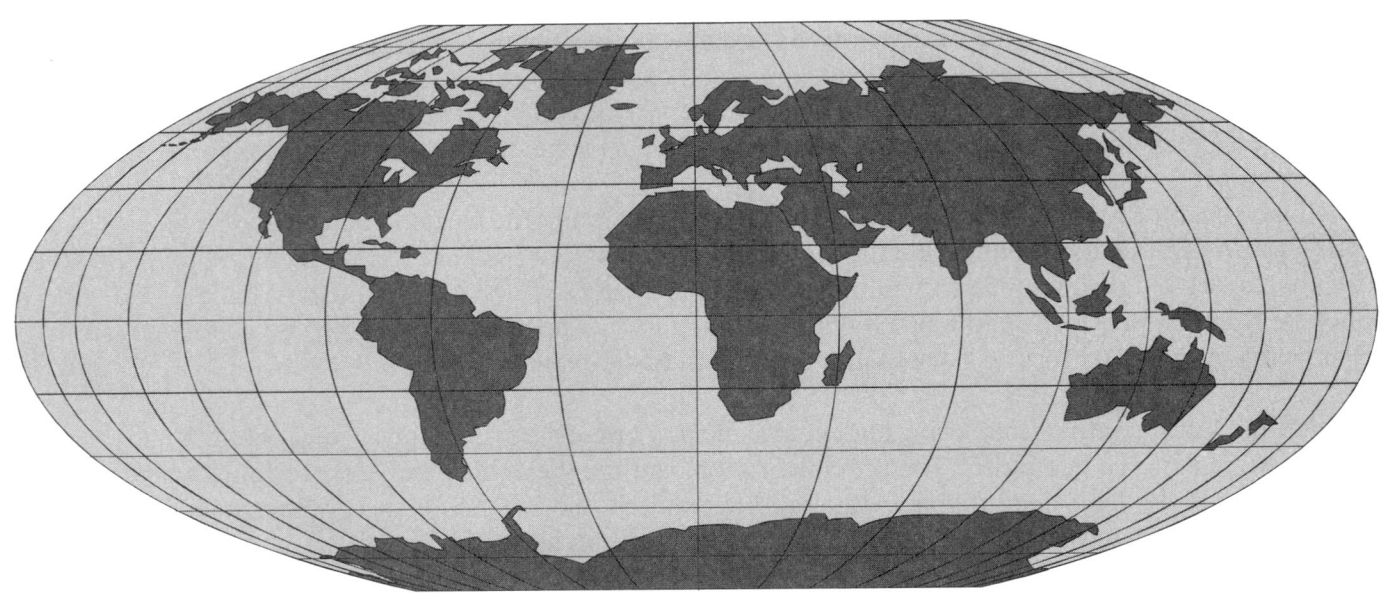

Dynamic Earth
Discovering Plate Tectonics

PURPOSE

In this activity you will identify, illustrate and explain plate tectonics.

MATERIALS

7 meter sticks
earth overhead
1 roll butcher paper
colored pencils

PROCEDURE

As a class you will be creating a large world map that shows information about significant geological events and formations. You will be divided into lab groups of four to research specific areas of the Earth's surface. Within your team you will have a seismologist, volcanologist, paleogeographer and geologist whose individual job assignments will vary, but each team will be responsible for preparing one map and giving a multimedia presentation explaining their region. Your individual group maps will be combined to make a class map. As such, it is important for teams to interact with one another during the research and development phase so that the map is coordinated and consistent.

Maps and presentations will include the following information:
- Land masses
- Plate boundaries showing direction of movement
- Oceans, seas, and large lakes
- Volcanoes
- Earthquake zones
- Longitude and latitude
- Mountains on land masses
- Mountains, ridges and trenches in the oceans

Your group presentation must use MS PowerPoint and one other multimedia source. You may choose to use videos, laser disc pictures, slides, books, movies, posters, computer links or guest speakers as you one additional multimedia source. Your presentation should be a minimum of 10 minutes long and you should be prepared to answer questions.

The responsibilities for each member of the team are outlined in the rubrics that follow. Since the quality of the class map depends on the quality of each piece, it is important that each job be completed as assigned. Cooperation within groups and within scientist teams will be necessary, but each group may only work on their own region, including labeling and coloring. The grading rubric attached shows the relative point values for each task within each job description. In addition, work grades will be assigned for in class work ethic. The final maps will be judged against the maps from the other classes, with a prize for the class with the best map.

Name _____

Period _____

Dynamic Earth
Discovering Plate Tectonics

Dynamic Earth Role Definitions:

Title	Research Activity	Map Activity	Presentation Contribution
Seismologist	1. Historical quakes in the region – famous or major w/damage and death statistics, the major geologic formations or faults that caused it. 2. Current activity – faults in the region, current quakes in region (last six months). 3. Future trouble spots.	1. Plot on map all historical and current quakes according to scale. 2. Head cartographer-responsible for tracing, whole map integration, senior colorist, symbol keeper.	1. Explain the general mechanics of a quake. 2. Detailed presentation of at least one historic quake or major fault movement (type of movement, magnitude and results) 3. Likelihood of future activity and why.
Volcanologist	1. Historical volcanoes in region – famous or major w/damage & death statistics, major geologic formations that caused it, type of volcano, etc. 2. Current activity- volcanology in region (last six months). 3. Future trouble spots.	1. Plot all historic and current volcanoes by type 2. Draw any major geologic features resulting from volcanic activity.	1. Explain the general mechanics of a volcano 2. Presentation of at least one major volcanic incident in the region (time, results, causes etc.). 3. Likelihood of future activity and why.
Geologist	1. Mountain ranges, ocean trenches and ridges, rift valleys and other geologic features. 2. Geologic causes for each (age and current status of orogenesis in region).	1. Label all features from research on map.	1. Description of mountains, trenches, etc. 2. Geologic processes that produced each feature 3. Future changes.
Paleo-geographer	1. Tectonic plates in the region. 2. Trace movements from Pangaea to present. 3. Types and location of plate boundary interactions demonstrated.	1. Draw tectonic plates. 2. Indicate the direction and relative rate of plate movement for each. 3. Indicate plate boundary interaction types.	1. Discuss how and when each modern plate came to its current location and current motion 2. Discuss plate boundary interactions in region and results (island arcs, orogenesis, sea floor spreading, etc.). 3. Predict future movement.

	Not Done	Poor	Below Average	Average	Good	Excellent
Introduction:						
Names, roles, region	0	8	12	15	18	20
Seismologist:						
Mechanics	0	5	6	7.5	8	10
Recent	0	5	6	7.5	8	10
1 Historic	0	5	6	7.5	8	10
Speaker pts	0	5	6	7.5	8	10
Multimedia	0	5	6	7.5	8	10
Q&A	0	5	6	7.5	8	10
Multimedia (as a group):						
Effort	0	6	7	8	9	10
PowerPoint:						
Slideshow (see other rubric)	0	6	7	8	9	10

	Not Done	Poor	Below Average	Average	Good	Excellent
Introduction:						
Names, roles, region	0	8	12	15	18	20
Volcanologist:						
Mechanics	0	5	6	7.5	8	10
Incident	0	5	6	7.5	8	10
Future	0	5	6	7.5	8	10
Speaker pts	0	5	6	7.5	8	10
Multimedia	0	5	6	7.5	8	10
Q&A	0	5	6	7.5	8	10
Multimedia (as a group):						
Effort	0	6	7	8	9	10
PowerPoint:						
Slideshow (see other rubric)	0	6	7	8	9	10

	Not Done	Poor	Below Average	Average	Good	Excellent
Introduction: Names, roles, region	0	8	12	15	18	20
Paleogeographer: _____						
Pos & Move	0	5	6	7.5	8	10
Plate bound	0	5	6	7.5	8	10
Future move	0	5	6	7.5	8	10
Speaker pts	0	5	6	7.5	8	10
Multimedia	0	5	6	7.5	8	10
Q&A	0	5	6	7.5	8	10
Multimedia (as a group): Effort	0	6	7	8	9	10
PowerPoint: Slideshow (see other rubric)	0	6	7	8	9	10

	Not Done	Poor	Below Average	Average	Good	Excellent
Introduction: Names, roles, region	0	8	12	15	18	20
Geologist: _____						
Feature Desc	0	5	6	7.5	8	10
Processes	0	5	6	7.5	8	10
Future Change	0	5	6	7.5	8	10
Speaker pts	0	5	6	7.5	8	10
Multimedia	0	5	6	7.5	8	10
Q&A	0	5	6	7.5	8	10
Multimedia (as a group): Effort	0	6	7	8	9	10
PowerPoint: Slideshow (see other rubric)	0	6	7	8	9	10

MAP RUBRIC

	Not Done	Poor	Below Average	Average	Good	Excellent
Overall Map:						
Key	0	1	2	3	4	5
Accuracy	0	1	2	3	4	5
Appearance	0	1	2	3	4	5
Integration	0	1	2	3	4	5
Seismologist:	_____					
Use of symbols	0	8	12	15	18	20
Accuracy	0	8	12	15	18	20
Appearance	0	8	12	15	18	20
Integration	0	8	12	15	18	20
Completeness	0	8	12	15	18	20
Volcanologist:	_____					
Use of symbols	0	8	12	15	18	20
Accuracy	0	8	12	15	18	20
Appearance	0	8	12	15	18	20
Integration	0	8	12	15	18	20
Completeness	0	8	12	15	18	20
Paleogeographer:	_____					
Use of symbols	0	8	12	15	18	20
Accuracy	0	8	12	15	18	20
Appearance	0	8	12	15	18	20
Integration	0	8	12	15	18	20
Completeness	0	8	12	15	18	20
Geologist:	_____					
Use of symbols	0	8	12	15	18	20
Accuracy	0	8	12	15	18	20
Appearance	0	8	12	15	18	20
Integration	0	8	12	15	18	20
Completeness	0	8	12	15	18	20

Total Map Points:

Seismologist _____   Paleogeographer _____

Volcanologist _____   Geologist _____

Rock-n-Roll
Investigating the Weathering of Rocks

OBJECTIVE

Students will model mechanical and chemical weathering processes and record, analyze and compare data from the different models.

LEVEL

Middle Grades: Earth Science

NATIONAL STANDARDS

UCP.1, UCP.2, UCP.3, A.1, A.2, B.3, B.4, D.1, D.2, D.3

TEKS

6.1(A), 6.2(A), 6.2(B), 6.2(C), 6.2(D), 6.2(E), 6.3(A), 6.3(C), 6.6(C), 6.14(A)
7.1(A), 7.2(A), 7.2(B), 7.2(C), 7.2(D), 7.2(E), 7.7(A), 7.8(A), 7.14(B)
8.1(A), 8.2(A), 8.2(B), 8.2(C), 8.2(D), 8.2(E), 8.12(A), 8.14(A)
IPC: 1(A), 2(A), 2(C), 2(D)

CONNECTIONS TO AP

AP Environmental Science:
   I. Interdependence of Earth's Systems: Fundamental Principles and Concepts C. The Solid Earth
   2. Earth dynamics
   III. Renewable and Nonrenewable Resources: Distribution, Ownership, Use, Degradation C. Soils

TIME FRAME

120 minutes

MATERIALS

(For a class of 28 working in groups of 4)

1.5 liters limestone chips (small rocks)	7 small screen sieves
1.5 liters granite chips (small rocks)	7 triple beam balances or electric balances
14-28 plastic canisters	1 liter of vinegar
28 400 mL beakers	1.5 liters of marble

TEACHER NOTES

The *Rock-n-Roll* lab has students simulate both chemical and physical weathering of rocks. It can be taught during a unit covering rocks and minerals and the rock cycle, or during a unit about the atmosphere and weather. The lab can be divided so that Part I and Part II can be performed on different days.

There are several types of plastic canisters or bottles that can be used successfully in this lab. Nalgene© wide-mouth liter bottles are perfect, but you can substitute empty black plastic Cheetos©, Funyuns© or pretzel containers that are now sold in grocery and convenience stores. Be sure that whatever container/canister you choose if fit with a well-attached lid.

Both the limestone and the granite gravel can usually be purchased from a local garden or home supply store. It is sold as a decorative rock. If limestone is not available you can substitute marble. You may also purchase both types of gravel from most scientific supply companies. In both cases be sure your gravel is small enough to fit inside your wide mouth plastic bottles.

Part I of the lab has students simulate physical weathering by placing rocks and water in the canisters and vigorously shaking for ten minutes. Play the song "Shake, Rattle and Roll" on a CD or cassette player while the students work! Students should predict which type of rock will weather the most before starting the experiment. The rocks are massed before and after shaking and students record any change in mass. From their data the students can then draw conclusions and address their hypothesis.

Part II of the lab has students simulate chemical weathering by placing rocks, vinegar and water in the canisters and shaking. Again students predict which rock will weather the most prior to starting the experiment. The rocks are massed before and after shaking and students record any change in mass. Changes are recorded into student data tables.

In the data tables you will notice that variable symbols are given for some of the data recorded. For instance, initial mass has the symbol m_i. It is important for students to begin to see these symbols used in labs and equations so they will become comfortable with them prior to entering an AP course.

POSSIBLE ANSWERS TO THE CONCLUSION QUESTIONS AND SAMPLE DATA:

DATA AND OBSERVATIONS

Part I: Physical Weathering

Data Table 1			
	Initial Mass (grams)	Final Mass (grams)	Change in Mass (grams)
bottle	50.25	53.60	3.45
bottle and limestone gravel	120.13	99.21	31.12
limestone gravel	$m_i = 69.88$	$m_f = 45.61$	$\Delta m = 24.27$
bottle	50.60	50.55	.05
bottle and granite gravel	146.64	138.84	7.84
granite gravel	$m_i = 96.04$	$m_f = 88.29$	$\Delta m = 7.75$

Part II: Chemical Weathering

Data Table 2				
	Mass of Beaker (grams)	Initial Mass of Beaker and Gravel (grams)	Mass of Gravel (m_i) (grams)	Final Mass of Beaker and Gravel (grams)	Change in Mass (Δm) (grams)
Sample C	90.25	293.60	203.35	280.91	12.69
Sample W	90.26	301.12	210.86	266.33	34.79
C Control	90.20	286.35	196.15	280.39	5.96
W Control	90.22	298.26	208.04	297.40	.86

ANALYSIS

Part I: Physical Weathering

1. What factors do you think will affect the rate of weathering in your model?
   - Answers will vary. How vigorously the container is shaken and the length of time it is shaken will affect the amount of weathering in our model.

2. What procedures could your group implement in order to ensure a consistent rate of weathering?
   - The amount of time the container is shaken needs to be consistent between samples and the way the container is shaken needs to be the same for all samples.

3. How could you change the experiment to test for these variables?
   - Keeping the shaking method constant, the test could be done for different amounts of time to see if different results are produced. Keeping the amount of time constant, different shaking methods (rolling, jumping, throwing) could be performed to see how the results are affected.

4. Using the following formula determine the percent change of material in the weathering model above: $\left(\dfrac{\Delta m}{m_i}\right) \times 100$ where Δm = change in mass and m_i = initial mass.
   - 24.27g/69.88g = 34.8%= limestone
   - 7.75g/96.04g = 8.3% = granite

Part II: Chemical Weathering

1. Using the following formula determine the percent change of material in the weathering model above: $\left(\dfrac{\Delta m}{m_i}\right) \times 100$ where Δm = change in mass and m_i = initial mass.
   - 12.69g/203.35g = 6.2% = Sample C
   - 34.76g/210.86g = 16.5% = Sample W
   - 5.96g/196.15g = 3.1% = Control C
   - .86/208.4= .4% = Control W

CONCLUSION QUESTIONS

1. Based on your calculations which model (chemical or physical) had the highest rate of weathering?
   - Physical weathering showed the greatest rate of change in our model.

2. Based on your observations, what factors affect rates of weathering?
   - Time is an important factor in weathering, but so are energy and chemical reactions.

3. Describe the difference between mechanical and chemical weathering.
   - In mechanical weathering there is not a chemical reaction. Mechanical weathering simply involves the breaking down of rocks due to exposure to energy like wind or moving water. In this model we simulated moving water by shaking the rocks in a container with water. Chemical weathering involves a chemical reaction. Carbon dioxide in the atmosphere mixes with rain to make a weak acid known as carbonic acid. Over time, the acid base neutralization reaction that occurs between the carbonic acid in the rain and the basic rocks (such as limestone) will cause the rocks to weather. Plant roots also provide acid for this reaction. We simulated this in our experiment by using weak vinegar.

Rock-n-Roll
Investigating the Weathering of Rocks

Weathering is the break-up or decomposition of rock resulting from exposure to the earth's atmosphere. Weathering occurs due to either mechanical or chemical processes. Mechanical weathering takes place when rocks are broken apart by physical activities such as ice wedging or biological actions of plants and animals. Chemical weathering is generally caused by hydrolysis, oxidation, or dissolving-chemical reactions that alter and break down rocks. In this lab, you will conduct two separate weathering experiments. Part I will model mechanical weathering using limestone gravel. Part II will model chemical weathering of the same material under both warm and cold conditions.

Erosion plays an important role in the weathering process. By transporting weathered material away, erosion exposes new surface area to weathering agents. In these experiments, the erosion process will be modeled in the rinsing procedures. In Part I, using a plastic screen while rinsing establishes a protocol for what size material would be eroded away in your model.

PURPOSE
In this activity you will examine and compare the physical and chemical weathering of limestone and granite rocks.

MATERIALS
    200 mL limestone chips (small gravel)   small screen sieves
    200 mL granite chips (small gravel)   triple beam balance
    2 plastic canisters   50 mL of vinegar
    4 beakers   200 mL marble chips

PROCEDURE
Part I: Physical Weathering
1. Predict whether limestone or gravel will show the greatest change in mass after being shaken for 10 minutes. Write your hypothesis under Part I on the student answer page.

2. Using a 250 mL beaker add limestone gravel to about the 200mL mark.

3. Rinse the gravel with water and drain while covering the top of the beaker with a plastic screen. Use a paper towel to absorb excess moisture.

4. Mass an empty plastic bottle (without the cap) using triple beam balance. Record the results in data table 1 on the student answer page.

5. Add the rinsed gravel to the plastic bottle. Re-mass the bottle with gravel and record the mass in data table 1 on the student answer page. Subtract the mass of the empty bottle from the total mass of the bottle and gravel to obtain the mass of gravel. Record this mass in the appropriate space in data table 1 on the student answer page.

6. Cap the bottle and shake vigorously with a consistent rhythm for 5 minutes.

7. Rinse the gravel in the plastic bottle again. Make sure to rinse any material that is stuck to the cap back into the bottle. Again, use the screen to sort which material will be rinsed away.

8. Re-measure the mass of the bottle with gravel after the rinse and record this value in data table 1.

9. Empty, rinse and dry the bottle. Re-mass the bottle to determine whether there has been any change in mass for the bottle. Record in data table 1.

10. Subtract the final gravel mass from the initial gravel mass and record. This is the mass of the material weathered away in this experiment.

11. Repeat steps 1 through 10 using granite gravel.

Part II: Chemical Weathering

1. In this part of the lab you will be using a weak vinegar to weather rocks and varying temperature to see if it acts as a catalyst. Predict which of the four beakers containing marble chips (warm vinegar, cold vinegar, warm water, cold water) will show the most change in mass. Write your hypothesis under Part II on the student answer page.

2. Obtain four, clean, dry beakers. Label with masking tape one beaker "C," one beaker "C Control," one beaker "W" and the last "W Control."

3. Measure the mass of each beaker and record the value for each in data table 2 on the student answer page.

4. Fill each beaker up to the 50mL level with rinsed and dried marble or limestone gravel.

5. Mass each beaker with contents and record the values in data table 2 on the student answer page in the space provided for each sample.

6. Fill beaker C with enough cold vinegar to completely cover the gravel. Repeat this procedure for beaker W using warm vinegar.

7. Fill beaker "C Control" with enough cold water to completely cover the gravel. Repeat this procedure for "W Control" using warm water.

8. Allow each sample to react for 20 minutes. Gently swirl each beaker every minute.

9. Rinse and dry each sample using the screen. Take care not to lose any gravel.

10. Measure and record the mass of each sample in data table 2.

11. Subtract the final mass from the initial mass for each sample. Record in data table 2.

12. While the experiment progresses, **observe and record** any differences you notice in the beakers.

Name _____

Period _____

Rock-n-Roll
Investigating the Weathering of Rocks

HYPOTHESIS

Part I

Part II

DATA AND OBSERVATIONS

Part I Data

Data Table 1			
	Initial Mass (grams)	Final Mass (grams)	Change in Mass (grams)
bottle			
bottle and limestone gravel			
limestone gravel	$m_i =$	$m_f =$	$\Delta m =$
bottle			
bottle and granite gravel			
granite gravel	$m_i =$	$m_f =$	$\Delta m =$

Part I Observations

Part II Data

	Mass of Beaker (grams)	Initial Mass of Beaker and Gravel (grams)	Mass of Gravel (m_i) (grams)	Final Mass of Beaker and Gravel (grams)	Change in Mass (Δm) (grams)
			Data Table 2		
Sample C					
Sample W					
C Control					
W Control					

Part II Observations

ANALYSIS

Part I: Physical Weathering

1. What factors do you think will affect the rate of weathering in your model?

2. What procedures could your group implement in order to ensure a consistent rate of weathering?

3. How could you change the experiment to test for these variables?

4. Using the following formula determine the percent change of material in the weathering model above: $\left(\dfrac{\Delta m}{m_i}\right) \times 100$ where Δm = change in mass and m_i = initial mass.

Part II: Chemical Weathering
1. Using the following formula determine the percent change of material in the weathering model above: $\left(\dfrac{\Delta m}{m_i}\right) \times 100$ where Δm = change in mass and m_i = initial mass.

CONCLUSION QUESTIONS

1. Based on your calculations which model (chemical or physical) had the highest rate of weathering?

2. Based on your observations, what factors affect rates of weathering?

3. Describe the difference between mechanical and chemical weathering.

Write Your Notes and Ideas Here!

Reasons for the Seasons
Exploring What Causes the Seasons

OBJECTIVE
Students will simulate the relationship between the sun and the Earth to determine why the Earth experiences seasons.

LEVEL
Middle Grades: Earth Science

NATIONAL STANDARDS
UCP.2, UCP.3, A.1, A.2, D.1, D.3

TEKS
6.2(A), 6.2(B), 6.2(C), 6.2(D), 6.2(E), 6.3(C), 6.4(A)
7.2(A), 7.2(B), 7.2(C), 7.2(D), 7.3(C), 7.4(A), 7.13(A)
8.2(A), 8.2(B), 8.2(C), 8.2(D), 8.3(C), 8.4(A)
IPC 1(A), 2(A), 2(B), 2(C), 2(D), 3(A)

CONNECTIONS TO AP
AP Environmental Science:
  I. Interdependence of Earth's Systems: Fundamental Principles and Concepts A. The Flow of Energy
  3. sources and sinks, conversions

TIME FRAME
60 minutes

MATERIALS
(For a class of 28 working in pairs)

7 globes or large spheres	1 roll masking tape
7 thermometers	7 100 watt bulbs
7 ring stands	7 metric rulers
7 utility clamps	

TEACHER NOTES
The Reasons for the Seasons allows students to manipulate the globe and light bulb to learn about why parts of the earth have winter, spring, summer and fall. In this lab stress the fact that many people believe the misconception that the seasons are caused when the earth orbits closer to the sun or further from the sun. The actual orbit the earth follows around the sun is nearly a perfect circle! By measuring the temperature differences on their tilted model earth, students will not only learn the real reason for the seasons, but students will be more likely to remember and apply what they've learned

Students should use a globe of the Earth and a heat lamp for this activity but if globes are in short supply, then a large sphere such as a basketball may be substituted. Small thermometers are preferable to large, long thermometers that may fall off the globe. Use plenty of tape to attach the thermometer safely to the globe being careful not to cover the bulb of the thermometer with the tape.

In the first experiment students should not see a change in temperature for the two distances while in the second experiment the smaller angle to the heat lamp should have the warmer temperature. This should lead the students to conclude that the Earth's seasons are not related to the distance from the sun but rather the angle of the sun's rays created by the tilt of the axis.

If the student's data does not reflect this then discuss possible experimental errors that may have occurred, such as incorrect distance measurements or incorrect thermometer readings.

POSSIBLE ANSWERS TO THE CONCLUSION QUESTION AND SAMPLE DATA
Experiment 1: Relationship between the Temperature and Distance

	75 cm distance	100 cm distance
Beginning Temperature (°C)	25.2°C	25.5°C
Ending Temperature (°C)	26.6°C	26.7°C
Change in Temperature (°C)	1.4°C	1.2°C

Experiment 2: Relationship between the Temperature and the Angle of the Sun's Rays

	Summer Angle	Winter Angle
Beginning Temperature (°C)	25.1°C	25.0°C
Ending Temperature (°C)	35.5°C	30.2°C
Change in Temperature (°C)	10.4°C	5.2°C

1. In Experiment 1, was there a temperature difference due to the distance of the "Earth" from the "sun"?
   - No, there was not a difference in temperature for the two distances.

2. Were the results of Experiment 1 what you expected? Why or why not?
   - Answers will vary. Genererally students expect the heating of the Earth to be due to the distance from the sun, the closer the warmer and vice versa.

3. In Experiment 2, which angle corresponds to our summer season and which to our winter season, the smaller or the larger angle? Which angle showed a higher temperature?
   - The smaller angle, when the city was closer to the equator, corresponds to our summer season while the larger angle, when the city was farther from the equator corresponds to our winter season. The smaller angle had a higher temperature.

4. In Experiment 2, did you find a difference in temperature between the two seasons? From your collected data, which season had a warmer temperature?
   - Yes, there was a temperature difference between the two seasons. The summer season had a warmer temperature.

5. The latitude in Experiment 2 did not change. The amount of heat from the lamp did not change. So why are there different temperatures for summer and winter?
   - The angle of the sun's rays caused the different temperatures during summer and winter. The city closer to the equator receives direct rays from the heat lamp while the city farther from the equator receives indirect rays from the heat lamp.

6. Using what you have learned from this activity, explain why the Earth experience seasons?
   - The Earth does not experience seasons because of the distance from the sun but rather because of the tilt of the Earth's axis. The tilt causes the sun's rays to hit the Earth differently. Rays near the equator are more direct with concentrated heat while rays near the poles are indirect with heat more spread out.

Reasons for the Seasons
Exploring What Causes the Seasons

The Distance of the Earth from the Sun

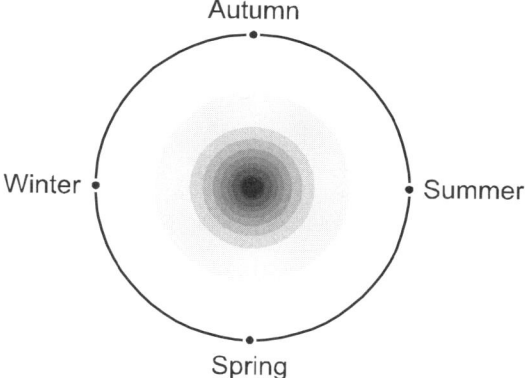

If you lived in the middle latitudes of the hemisphere, you would experience four distinct seasons — four different time periods of different temperatures and weather during the year. If you lived at the equator or the poles you would experience practically the same season all year. Many people mistakenly believe that the seasons are caused by the earth's orbit being closer or further from the sun, but the actual path of the earth, shown above, is almost circular. In fact, if you look closely you will see that the earth is slightly closer to the sun during the winter!

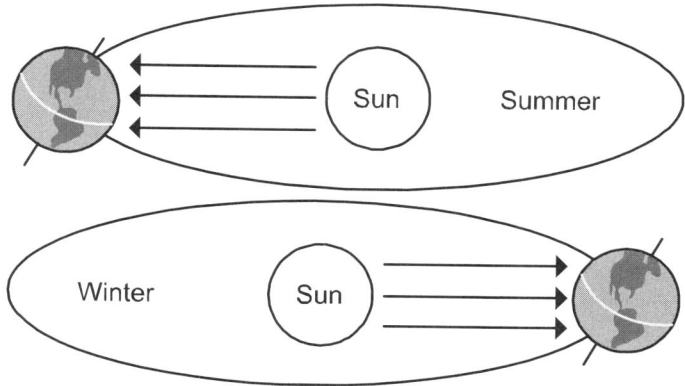

So what does cause the seasons? The seasons depend on the amount of heat received from the sun. The sun emits constant amounts of light and heat as well as various forms of energy. All parts of the Earth receive the heat from the sun, but because the Earth has a curved surface, the rays either hit a flat surface (area near the equator) or an angled surface (area near the poles). The changing angle of striking sun rays causes differential heating. So why do cities in different latitudes have different temperatures during the same season? Why are some seasons warmer than other seasons for the same city? Why does the Earth experience seasons?

PURPOSE

In this investigation, you will simulate the relationship between the sun and the Earth to determine why the Earth experiences seasons.

MATERIALS

globe or large sphere of some kind   masking tape
one thermometer   100 watt bulb
ring stand   metric ruler
utility clamp

Safety Alert
1. Be careful with the heat lamps since they can become quite hot.

PROCEDURE

Experiment 1: Relationship between the Temperature and Distance

1. Set up a ring stand and utility clamp. Attach the heat lamp so that it is held in a horizontal position. Point the heat lamp at the globe.

2. Place the globe with the North Pole facing away from the lamp and position the globe 75 cm from the heat lamp.

3. Position the heat lamp to point directly at the Equator.

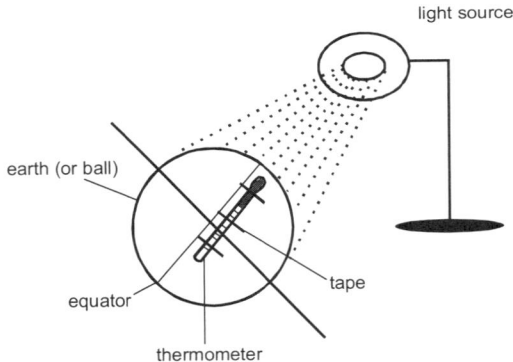

4. Tape the thermometer to the globe with the bulb at the equator.

5. Read the temperature on the thermometer and record this value in the data table on the student answer page.

6. After the first temperature has been taken and recorded, turn on the heat lamp. Begin timing and after six minutes, turn off the heat lamp and record the final temperature in the data table on the student answer page.

7. Calculate the change in temperature for the thermometer and record it in the data table.

8. Allow the thermometer to return to room temperature. Move the globe 100 cm away from the lamp and repeat steps 5 - 7 for the new distance.

Experiment 2: Relationship between the Temperature and the Angle of the Sun's Rays
1. Allow the thermometer to return to room temperature. Place the globe 30 cm away from the heat lamp and point the lamp directly at the Equator.

2. Place the thermometer at the correct latitude of your city with the bulb end closest to the lamp.

3. Position the globe so that the North Pole of the globe is facing away from the heat lamp to simulate the winter season. Be sure to tilt the earth on its' axis.

4. Before turning on the lamp record the initial temperature of the thermometer in the data table on the student answer page. Turn the lamp on and begin timing for six minutes.

5. Turn the lamp off after the six minute time period. Record the final temperature in the data table and then calculate the change in temperature.

6. Allow the thermometer to return to room temperature and change the position of the globe so that the North Pole is facing the heat lamp. Be sure to keep the globe the same 30 cm away from the heat lamp.

7. Repeat steps 4 and 5 for the new position.

Name _____

Period _____

Reasons for the Seasons
Exploring What Causes the Seasons

DATA AND OBSERVATIONS

Experiment 1: Relationship between the Temperature and Distance

	75 cm distance	100 cm distance
Beginning Temperature (^{o}C)		
Ending Temperature (^{o}C)		
Change in Temperature (^{o}C)		

Experiment 2: Relationship between the Temperature and the Angle of the Sun's Rays

	Summer Angle	Winter Angle
Beginning Temperature (^{o}C)		
Ending Temperature (^{o}C)		
Change in Temperature (^{o}C)		

CONCLUSION QUESTIONS

1. In Experiment 1, was there a temperature difference due to the distance of the "Earth" from the "sun"?

2. Were the results of Experiment 1 what you expected? Why or why not?

3. In Experiment 2, which angle corresponds to our summer season and which to our winter season, the smaller or the larger angle? Which angle showed a higher temperature?

4. In Experiment 2, did you find a difference in temperature between the two seasons? Which season had a warmer temperature?

5. The latitude in Experiment 2 did not change. The amount of heat from the lamp did not change. So why are there different temperatures for summer and winter?

6. Using what you have learned from this activity, explain why the Earth experience seasons?

Not So Lost in Space
Building a Model of the Solar System

OBJECTIVE
Students will examine planetary distances to create a scale model of each planet from the sun.

LEVEL
Middle Grades: Earth Science

NATIONAL STANDARDS
UCP.1, UCP.2, UCP.3, A.1, D.4, G.3

TEKS
6.2(A), 6.2(B), 6.2(C), 6.2(D), 6.2(E), 6.3(C), 6.4(A), 6.4(B), 6.5(A), 6.5(B), 6.13(A), 6.13(B)
7.2(A), 7.2(B), 7.2(C), 7.2(D), 7.3(C), 7.4(A), 7.4(B)
8.2(A), 8.2(B), 8.2(C), 8.2(D), 8.3(C), 8.4(A), 8.4(B), 8.5(C)
IPC 1(A), 2(A), 2(B), 2(C), 2(D), 3(A)

CONNECTIONS TO AP
AP Physics:
   I. Newtonian Mechanics F. Oscillations and gravitation

TIME FRAME
60 minutes

MATERIALS
(For a class of 28 working in groups of 3 or 4)

7 calculators	7 meter sticks
7 scissors	1 pound modeling clay or Playdoh©
butcher paper for sun (less than a meter)	7 toothpicks

TEACHER NOTES
Not So Lost In Space is an activity where students build a model of the solar system and then use their model to gain an understanding of actual distances between planets. This lab is especially important in light of NASA's recent plan for manned travel to Mars because it helps students conceptualize distances between planets.

The lab is written with two procedures. The first procedure has the students measure distances from the sun using a scale that will work in your school's hallway. You need an indoor hallway at least 35 meters long. As an alternative, the second procedure gives the scale model of the distances for use on a football field. You may wish to have students perform the lab both ways if you have time. The planets sizes are scaled for the outdoor model. In both procedures you will have to designate the sun's location.

At the beginning of the period, divide the class into nine groups and assign one planet to each group.

When the students are forming the planets out of clay, remind them that the chart contains diameter and they should measure their sphere's diameter using a meter stick or ruler. The sun is drawn on paper because the amount of clay needed to make the sun would be huge! Ask the students during the lab closure why they were not asked to make the sun out of clay and have them come up with that fact.

The students may need help with the calculations. The numbers are large and the students will have to use scientific notation. You may have the students work through the foundation lessons called "Numbers in Science" prior to this lab.

An extension to this lesson is to have the students look up the speed of different probes sent into space and calculate how long it really takes for the probes to reach a certain planet, especially Mars. You may suggest to students that some of them might be on that trip to Mars!

POSSIBLE ANSWERS TO THE CONCLUSION QUESTION AND SAMPLE DATA

Planet	Diameter (km)	Scale Diameter (mm)	Distance from Sun (AU)	Scale Distance for indoor model (cm)	Scale Distance for outdoor model (meters)
Mercury	4,878	.88	0.387	38.70	10.41
Venus	12,104	2.17	0.723	72.30	19.42
Earth	12,756	2.29	1.0	100.00	26.86
Mars	6,787	1.22	1.523	152.30	40.92
Jupiter	142,800	25.65	5.203	520.30	139.76
Saturn	120,000	21.55	9.539	953.90	256.30
Uranus	51,200	9.20	19.184	1918.40	515.54
Neptune	48,600	8.73	30.107	3011.10	808.23
Pluto	2,300	.41	39.5	395.00	1062.0
Sun (star)	1,392,000	250	NA	NA	NA

Sample calculations: Scale Diameter Venus = 12,104/5568 = 2.17 millimeters
Scale Distance indoor Venus = .723 x 100 = 72.30 centimeters
Scale Distance outdoor Venus = .723 x 26.86 = 19.42 meters

1. What do you notice about the distances between each of the terrestrial planets versus the distances between each of the Jovian planets?
   - The terrestrial planets are closer together than the Jovian planets.

2. Which planet is the closest to the Earth?
   - Venus is the closest planet to Earth, 0.3 AUs away.

3. Which planets are similar in size?
   • Venus and Earth are similar in size; Uranus and Neptune are similar in size.

4. How many times larger is the Sun than Jupiter? How many times larger is the Sun than the Earth?
   • The Sun is roughly 10 times (9.7) larger than Jupiter and roughly 109 times (109.1) larger than the Earth.

5. Let's assume that a space craft can travel at the speed of 100,000 km per hour. How many days will it take for the space craft to travel from the Earth to Mars?
   • It would take 31.25 days to travel from the Earth to the Mars.

$$0.523\,\text{AU} \times \frac{150,000,000\,\text{km}}{1\,\text{AU}} \times \frac{1\,\text{hr}}{100,000\,\text{km}} \times \frac{1\,\text{day}}{24\,\text{hr}} = 33.1\,\text{days}$$

6. Let's assume the same space craft is traveling to Pluto. How long will it take for the space craft to travel from Earth to Pluto in days?
   • It would take 2,406.25 days to travel from the Earth to Pluto, or roughly 6.6 years

$$38.5\,\text{AU} \times \frac{150,000,000\,\text{km}}{1\,\text{AU}} \times \frac{1\,\text{hr}}{100,000\,\text{km}} \times \frac{1\,\text{day}}{24\,\text{hr}} \times \frac{1\,\text{year}}{365\,\text{days}} = 6.6\,\text{years}$$

7. Are the two models, the paper cut outs and the spacing in the hallway, an accurate model of the solar system? Explain your answer.
   • No, because the scale values are not the same. The distances from the sun are in feet while the sizes of the planets are in centimeters.

Not So Lost in Space
Building a Model of the Solar System

Space is an incredibly large area. To better understand distances from the Earth to our sun, or the Earth to other planets, or even other stars, astronomers have developed a new measurement unit for large distances. The standard unit of distance between planets is the astronomical unit, the AU. One AU is equivalent to the average distance of the Earth to the Sun or 150 million kilometers or 93 million miles. Stars are a much greater distance from the Earth. To measure distances to stars astronomers use the light year. One light year is equivalent to the distance light travels in one year, which is 10 trillion kilometers, or 63,000 AU.

Our solar system consists of nine planets, the sun (our one star), meteoroids, asteroids, comets, moons, gas, dust, and numerous man made satellites, probes, and space junk.

The nine bodies conventionally referred to as planets are often further classified in several ways. Mercury, Venus, Earth and Mars are called the terrestrial or rocky planets because they are composed mostly of rock and have relative high densities. Jupiter, Saturn, Uranus and Neptune are called the Jovian or gas planets because they are primarily composed of hydrogen and helium and have relatively low densities. If the planets are divided by size, Mercury, Venus, Earth, Mars and Pluto are all considered small, while Jupiter, Saturn, Uranus and Neptune are all considered giant, and are sometimes referred to as the gas giants.

Some people refer to the planets as inner or outer, using the asteroid belt between Mars and Jupiter as the boundary between inner and outer. And finally, some people describe the planets from a historical perspective. The classical planets include Mercury, Venus, Mars, Jupiter and Saturn. All of these planets have been known about since prehistoric times. The modern planets are Neptune, Uranus and Pluto because they were not discovered until the creation of high powered telescopes in modern times.

PURPOSE
In this activity you will examine planetary distances to create a scale model of each planet from the sun. You will need to rely on your math skills to translate the immensity of our solar system to a scale model and give you some idea of the relative distances between planets.

MATERIALS

calculators
scissors
butcher paper for sun

meter sticks
pound modeling clay or Playdoh©
toothpicks

PROCEDURE I: HALLWAY

1. Using the diameters of the planets in kilometers, calculate the scale sizes of the planets in centimeters. The scale ratio is 5568 km = 1 mm, so you will need to divide the given diameters by 5568 to get the size of the model in millimeters. Show your work on the student answer page.

2. Write your results in the data table on the student answer page.

3. Obtain your assigned planet from your teacher.

4. Get enough clay to form a sphere the size of your planet.

5. On a small strip of paper, write the name of your planet and then attach it to the toothpick. Stick your toothpick in your planet to label it. If your planet is very small you may have to place the toothpick flag next to your planet.

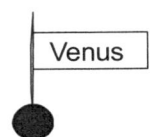

6. Convert the distances of the planets from the sun given in the data table on the student answer page into the scale distances in centimeters. The scale ratio is 1 AU = 100 centimeter. Show your math calculations on the student answer page.

7. When your teacher instructs, carry your planet into the hallway. The teacher will indicate the sun's position. Measure your planet's distance from the sun and tape your planet to the wall.

8. After all groups have taped their planets in place, describe the position of the planets on the student answer page under analysis. Be sure to include which are close together, which are far apart, which are large, which are small and the basic order of the planets.

PROCEDURE II: FOOTBALL FIELD

1. Using the diameters of the planets in kilometers, calculate the scale sizes of the planets in centimeters. The scale ratio is 5568 km = 1 mm, so you will need to divide the given diameters by 5568 to get the size of the model in millimeters. Show your work on the student answer page. Record your results in the data table on the student answer page.

2. Obtain your assigned planet from your teacher

3. Get enough clay to form a sphere the size of your planet.

4. Cut out your paper planet and label it with the planet's name. You may color your planet if time permits.

5. Convert the distances of the planets from the sun given in the data table on the student answer page into the scale distances in meters. The scale ratio is 1 AU = 26.86 meters. When your teacher instructs, carry your planet to the football field. The teacher will indicate the sun's position. Measure the scaled distance of your planet by walking away from the sun the number of yards that your planet is from the sun. Place your planet on the field.

6. After all groups have placed their planets on the field, describe the position of the planets on the student answer page under analysis. Be sure to include which are close together, which are far apart, which are large, which are small and the basic order of the planets.

Name _____

Period _____

Not So Lost in Space
Building a Model of the Solar System

DATA AND OBSERVATIONS

Show calculations here:

Planet	Diameter (km)	Scale Diameter (cm)	Distance from Sun (AU)	Scale Distance for indoor model (cm)	Scale Distance for outdoor model (meters)
Mercury	4,878		0.387		
Venus	12,104		0.723		
Earth	12,756		1.0		
Mars	6,787		1.523		
Jupiter	142,800		5.203		
Saturn	120,000		9.539		
Uranus	51,200		19.184		
Neptune	48,600		30.107		
Pluto	2,300		39.5		
Sun (star)	1,392,000		NA		NA

ANALYISIS

Part I

Part II

CONCLUSION QUESTIONS

1. What do you notice about the distances between each of the terrestrial planets versus the distances between each of the Jovian planets?

2. Which planet is the closest to the Earth?

3. Which planets are similar in size?

4. How many times larger is the Sun than Jupiter? How many times larger is the Sun than the Earth?

5. Let's assume that a spacecraft can travel at the speed of 100,000 km per hour. How many days will it take for the space craft to travel from the Earth to Mars?

6. Let's assume the same spacecraft is traveling to Pluto. How long will it take for the spacecraft to travel from Earth to Pluto in days?

7. Are the two models, the paper cut outs and the spacing in the hallway, an accurate model of the solar system? Explain your answer.

Black Holes and Beyond
Modeling a Black Hole

OBJECTIVE
Students will model a black hole to investigate the properties and formation of a black hole.

LEVEL
Middle Grades: Earth Science

NATIONAL STANDARDS
UCP.1, UCP.2, UCP.3, UCP.5, D.4

TEKS
6.2(A), 6.2(B), 6.2(C), 6.2(D), 6.2(E), 6.3(C)
7.2(A), 7.2(B), 7.2(C), 7.2(D), 7.3(C)
8.2(A), 8.2(B), 8.2(C), 8.2(D), 8.3(C), 8.13(A), 8.13(C)
IPC 1(A), 2(A), 2(C), 2(D)

CONNECTIONS TO AP
AP Physics:
    I. Newtonian Mechanics F. Oscillations and gravitation

TIME FRAME
60 minutes

MATERIALS
(For a class of 28 working in pairs)

14 small round balloons	14 basic calculators
14 pieces of string in 1 meter lengths	1 roll of aluminum foil
14 meter sticks	14 balances
14 push pins	

TEACHER NOTES

Black Holes and Beyond is designed to familiarize students with the concepts of volume, mass and density and then relate these concepts to stars and black holes. Prior to this activity introduce the stages of the life cycle of a star to your students, either through lecture, reading assignment or internet search. In addition, there are several words used to discuss the ideas surrounding black holes that your students may not be familiar with. Spacetime, for instance, can be explained as continuum of four dimensions (3 space and 1 time) in which any object or event can be located. In a way, spacetime is an area so large that you have to include the concept of time to understand it. For example, when you see the star Alpha Centauri in the night sky, you are seeing it as it was four years ago because it has taken four years for the light to reach Earth. Therefore when you look into space you are looking back in time, which means time and space (spacetime) are inseparable. Be sure to read the intro to the lab with your students and help them understand the text.

Depending on your students' math ability they may need some assistance with the formulas.

Watch the students to be sure that they do not crush the foil entirely after the star is popped. It is important that the balloons have four different sizes and that the students make the foil ball smaller and smaller each time.

Students may observe a very small change in mass, although theoretically there should not be one. Since the amount of foil is not changed by its crushing any mass change should be minimal at best. If there is a difference in mass it is probably due to experimental error. This is an important point to discuss with students during a post lab discussion.

POSSIBLE ANSWERS TO THE CONCLUSION QUESTIONS AND SAMPLE DATA

1. What stage of a star's life cycle is represented when you popped the "star" balloon?
   * The popping of the "star" balloon represents a supernova.

2. What kind of stars become black holes: low mass, medium mass, or high mass? Explain your answer.
   * High mass stars become black holes because it takes a huge mass to collapse to the form the density of a black hole

3. Describe any observed changes in the circumference of your "black hole" as you completed each trial?
   * After each trial the "black hole" decreased in size or circumference.

4. Describe any observed changes in the mass of your "black hole" as you completed each trial?
   * The mass of the black hole remained essentially constant throughout each trial.

5. Describe any observed changes in the density of your "black hole" as you completed each trial?
   * The density of the "black hole" increased with each trial.

6. How do the masses of your "star" and "black hole" compare? How do the densities of your "star" and "black hole" compare?

- The mass of the star and the black hole are relatively the same while the density of the black hole is much greater than the star.

Trial	Circumference	Radius	Volume	Mass	Density
1	45 cm	7.17 cm	1543.2 cm^3	12 g	0.0078 g/cm^3
2	40 cm	6.37 cm	1082.1 cm^3	11 g	0.0102 g/cm^3
3	35 cm	5.57 cm	723.5 cm^3	11 g	0.0152 g/cm^3
4	20 cm	3.18 cm	134.6 cm^3	11 g	0.0817 g/cm^3

Black Holes and Beyond
Modeling a Black Hole

What is a black hole? A black hole is a region of spacetime from which nothing can escape, even light. But what does that mean? Well, as the matter, or the mass of a star, is crushed into a smaller and smaller volume, the gravitational attraction increases, and hence the escape velocity also increases. Things have to move faster or be propelled harder to escape this gravitational pull. Eventually a point is reached when even light, which travels at 3×10^8 meters per second, is not traveling fast enough to escape. If light cannot get out, nothing else can either and we call it a black hole.

Einstein's general theory of relativity describes gravity as a curvature of spacetime (caused by the presence of matter. If the curvature is fairly weak, Newton's laws of gravity can explain most of what is observed. For example, the regular motions of the planets can be explained by Newton's laws. But, very massive or dense objects generate much stronger gravity. The most compact objects imaginable are predicted by general relativity to have such strong gravity that nothing, not even light, can escape their grip.

Scientists today call such an object a *black hole.* Why black? Though the history of the term is interesting, the main reason is that no light can escape from inside a black hole; it has, in effect, disappeared from the visible universe.

Do black holes actually exist? Most physicists believe they do, basing their views on a growing body of observations. In fact, present theories of how the cosmos began rest in part on Einstein's work that predicts the existence of black holes. Yet Einstein himself denied their existence, believing that black holes were a mere mathematical curiosity. He died in 1955, before the term "black hole" was coined or understood and observational evidence for black holes began to mount.

PURPOSE
To simulate the formation of a black hole.

MATERIALS

small round balloon
string
meter stick
push pin

basic calculator
aluminum foil
balance

Safety Alert
Be careful when using the push-pin to pop the balloon.

PROCEDURE

1. Inflate the balloon until the circumference is roughly 45 cm. To measure the circumference, wrap a string around the middle of the balloon. Use your finger to mark the length of the string. Measure the length of the string using a meter stick. When you have inflated your balloon to the proper circumference, tie off the end of the balloon.

2. Carefully cover the inflated balloon with aluminum foil. Try and cover the entire surface of the balloon.

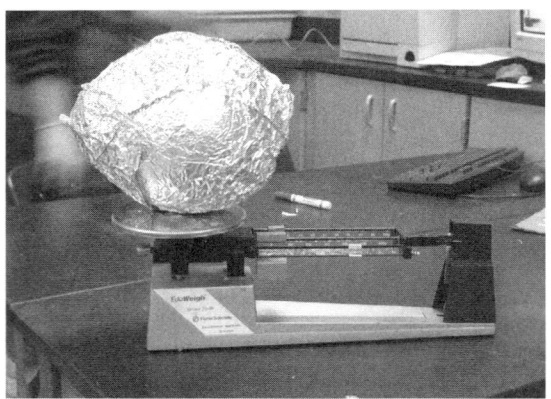

3. Using the string, measure the circumference of the "star" three times at three different angles. Take the average of the circumferences and record the average in the data table under Trial 1.

4. Place the "star" on the triple beam balance to determine the mass. Record the mass in the data table under Trial 1.

5. Using a push-pin carefully pop the "star". This represents the stage of a star becoming a supernova. DO NOT CRUSH THE FOIL. You will crush the foil in stages. Carefully crush the foil so that it is **slightly** smaller than its original volume.

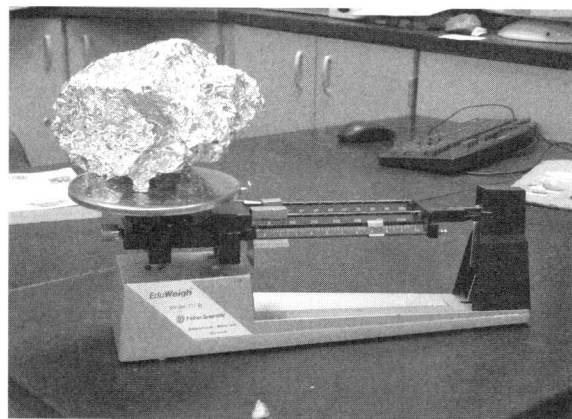

1. Measure and record the circumference of the "star" three times at three different angles. Take the average and record this circumference in the data table under Trial 2. Be careful not to further crush the "star". Measure the mass and record it under Trial 2.

2. Measure the mass of the "star" on a balance. Record the mass under Trial 2 in the data table.

3. Crush the "star" slightly once again so that it is not quite completely crushed. Measure and record the new circumference of the "star" three times at three different angles. Take the average and record this circumference in the data table under Trial 3 in the data table.

4. Crush the "star" once again so that it is now as compact as possible. Measure and record the new circumference and mass under Trial 4 in the data table.

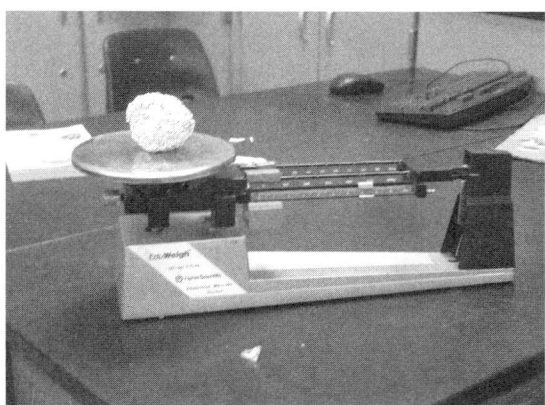

Black Holes and Beyond
Modeling a Black Hole

DATA AND OBSERVATIONS

Trial	Circumference	Radius	Volume	Mass	Density
1					
2					
3					
4					

ANALYSIS

Use the values from your data table and the formulas below to calculate the radius of the "star", the volume of the "star", and the density of the "star". Record these values in your data table.

$$\text{Circumference} = 2\pi r$$

$$\text{Volume} = \frac{4}{3}\pi r^3$$

$$\text{Density} = \frac{m}{V}$$

$$\pi = 3.14$$

CONCLUSION QUESTIONS

1. What stage of a star's life cycle is represented when you popped the "star" balloon?

2. What kind of stars become black holes: low mass, medium mass, or high mass?

3. Describe any observed changes in the circumference of your "black hole" as you completed each trial?

4. Describe any observed changes in the mass of your "black hole" as you completed each trial?

5. Describe any observed changes in the density of your "black hole" as you completed each trial?

6. Compare the mass and density of your "star" to the mass and density of your "black hole"?

Assessment

Assessment Foreword

Assessment in the foundational science courses is a critical component in the development of a successful AP program. Assessment can be used as a tool for measuring student understanding, measuring program success, and building students' test taking skills for use in a variety of settings. Well written, properly implemented assessment items play a valuable role in preparing students for success in subsequent AP courses and, ultimately, successful performance on AP exams.

Effective science programs expose students to a variety of assessment measures well in advance of the AP experience. For example, in foundational science courses, student performance can be measured through the use of laboratory practicals, laboratory reports, multiple-choice questions, open-ended questions, free response items and other assessment tools. Formal and informal assessment should occur regularly and vary in complexity. Students should be exposed to questions of similar format and rigor as those that are used on the AP exams. Specific attention needs to be given to the development of test taking skills that will allow students to confidently approach a variety of multiple choice question styles and produce thorough and logical free responses.

Collaboration with members of a vertical team allows for a progression in the development and sophistication of test-taking skills. A cohesive science vertical team can implement strategies to ensure that students have sufficient exposure and opportunities to master these skills.

The assessment section of the *Laying the Foundation* series is designed to provide background information on the AP exams in science, offer samples of appropriate questions for use in the foundational courses, and suggest strategies that can be used to build student skills necessary for different question types. The types of multiple choice and free response items typically found on the exams are identified and described. Similar questions are presented at a level appropriate for use in foundational first year science courses. Additionally, this section contains a sample unit assessment and sets of multiple choice questions that can be used to assess student understanding of selected activities found within this book.

Hopefully, this section provides sufficient insight to allow you to confidently select or construct assessment items of the appropriate level of rigor and format to empower students.

Overview of the AP Exams

AP SCIENCE EXAM INFORMATION

Course	Section	# of Questions	Relative Points	Time Allowed	Formula Sheet Provided?	Calculator Allowed
AP Biology	Multiple Choice	120 * will be 100 in 2004	60%	90 minutes	No	No
	Free Response (Problem/Essay)	4 (1 on molecules and cells, 1 on heredity and evolution, 2 on organisms and populations) At least one is lab-based.	40%	90 minutes	No	No
AP Chemistry	Multiple Choice	75	45%	90 minutes	No	No
	Free Response	2 mathematical problems (1 required; choose 1 of 2 others)	55%	40 minutes	Yes	Yes
		4 1 set of net ionic equations, 3 essay (2 mandatory, at least one of which is lab-based; choose 1 of 2 others)		50 minutes	Yes	No
AP Environmental Science	Multiple Choice	100	60%	90 minutes	No	No
	Free Response	4 (1 data, 1 document-based, 2 synthesis)	40%	90 minutes	No	No
AP Physics B	Multiple Choice	70	50%	90 minutes	Yes	Yes
	Free Response	7 (At least one lab-based)	50%	90 minutes	No	
AP Physics C (Electricity and Magnetism)	Multiple Choice	35	50%	45 minutes	No	No
	Free Response	3 (At least one lab-based)	50%	45 minutes	Yes	Yes
AP Physics C (Mechanics)	Multiple Choice	35	50%	45 minutes	No	No
	Free Response	3 (At least one lab-based)	50%	45 minutes	Yes	Yes

The multiple choice portion of the exam is graded by machine while the free response portion is graded by a group of trained "readers" consisting of college professors and high school AP science teachers. Each free-response question is scored based on a standard that is developed from input provided by not only the developers of the question but also the chief reader, question leaders, table leaders and graders. For a complete description of the grading process go to AP Central at http://apcentral.collegeboard.com and follow the exam links to "exam scoring".

The scoring of the free response section is based on a positive point system in which the student starts out with zero points and collects up to 10 points by writing correct statements or correct calculations in response to the prompt. Points are not deducted for incorrect or inaccurate statements. Students need acclimation to this positive point grading process. This process has a liberating effect in comparison to the process of deducting points for mistakes. The positive point system encourages students to write more generously and share the details of their understanding of a concept.

Strategies for Developing Test Taking Skills

Success on the Advanced Placement exam in either chemistry or physics is dependent not only on the student's understanding of the content but also upon the student's test taking skills. This section specifically addresses developing multiple choice testing skills and describes the attributes of the free response segment of the test.

Students who know what types of questions to expect on the exam and how to approach these questions will perform more successfully. You can help your students develop these skills during the first year course by exposing them to a variety of types of multiple choice questions and periodically giving attention to test taking strategies. The skills and confidence your students develop under your guidance will serve them throughout their lives. The following strategies are offered to assist you in this skill development process.

Strategies you can implement to develop multiple choice test taking skills:

1. Teach students how to approach difficult multiple choice questions by encouraging them to
    a. First read the entire question rather than simply scanning the words.
    b. Think about what the question is asking and then think of an answer before reading the choices. This will make it easier to recognize the correct answer when it is encountered.
    c. If it is a computational question, instruct students to round the quantities and constants given and estimate the correct answer.
    d. Eliminate obvious wrong answers.
    e. Make an educated guess if one or two of the choices can be eliminated.

2. Provide frequent situations in which students are timed as they answer multiple choice questions. This is particularly important for students who have grown accustom to state level tests that have no time limitations.

3. Allow students the opportunity to bubble answers on an answer document during a timed test. Students should be taught to answer a question and immediately bubble the answer on the answer sheet rather than waiting to bubble at the end of the test.

4. Use multiple choice questions as warm-ups to provide skill development opportunities in situations outside the normal testing environment.

5. Allow the students opportunities to witness your approach to difficult multiple choice questions by modeling your thought processes aloud. This metacognition process allows students the opportunity to witness effective, logical, and mature thinking and is particularly important for computational questions. It also helps them to see that answers can be derived rather than memorized.

6. Include multiple choice items on major tests that are based on or specifically related to the lab activities performed in your course. This will encourage students to pay attention to both content and process during the lab experiences.

7. Inform students about the penalty for guessing that is often used on standardized tests, including the AP exams. Have students calculate what their test score would be on any given test if you were to deduct ¼ point penalty for each wrong answer.

8. Include material and multiple choice questions from previous "units" or topics on your tests. This helps motivate students to retain the information rather than simply learn it "for the test".

9. Give students feedback regarding their performance on multiple choice questions as quickly as possible. Prompt feedback allows students the opportunity to rethink the question and learn from their mistakes.

10. Expose your students to a variety of multiple choice question styles. Several types of multiple choice questions are described in the following section.

AP Tests: Types of Questions

What do the AP tests look like? As a middle school teacher you may not be familiar with the AP exams, but it is important for you and your students for you to know about the style of the tests. The following section of material explores the tests, explains the different kinds of test questions and how they work.

Examples of the Various Types of Multiple Choice Questions Found on the AP Biology Exam

EXCEPT Questions

Multiple choice questions that contain "EXCEPT" are sometimes difficult for students because they have to look for the answer that contains the **wrong** information. Since they have been trained to look for correct information in most questions, searching for the **wrong** information can be challenging. Help your students overcome this by including "EXCEPT" questions on your quizzes and tests.

Example from the 1999 AP Biology Exam:

Nuclear divisions in which the chromosome number is reduced from $2n$ to n is part of the life cycle of all of the following organisms EXCEPT
(A) molds
(B) ferns
(C) insects
(D) bacteria
(E) protozoans

Answer: D

Example suitable for use in a first year course:

The process of meiosis occurs in all of the following organisms EXCEPT
(A) fungi
(B) plants
(C) animals
(D) bacteria
(E) protists

Answer: D

Classification Questions

In classification type questions students are given a set of answers that are to be used to answer several questions. Students need practice with this type of question because frequently the answers can be used more than once or not at all. Students must be taught to **avoid** the tendency to think that if there are five questions following the answer choices that all five answers must be used. Frequently on the AP Biology exam one or more of the answer choices in an answer set will not be used at all.

Example from the 1999 AP Biology Exam:

Questions 87-91

(A) Annelida
(B) Mollusca
(C) Arthropoda
(D) Echinodermata
(E) Chordata

87. Bilaterally symmetrical; deuterostome; dorsal hollow nerve cord
88. Coelomate; exoskeleton; jointed appendages
89. Pharyngeal slits; endoskeleton derived from mesoderm; ventral heart
90. Internal calcareous skeleton; deuterostome; water-vascular system
91. Closed circulatory system; protostome; many body segments

Answers: 87. E, 88. C, 89. E, 90. D, 91. A

Example suitable for use in a first year course:

Questions 1-5

(A) Annelida
(B) Mollusca
(C) Arthropoda
(D) Echinodermata
(E) Chordata

1. Segmented worms such as an earthworm
2. Includes classes such as Insecta and Crustacea
3. Members of this phylum have jointed appendages and exoskeletons
4. Includes organisms with a dorsal notochord such as lamprey and perch
5. Phylum containing starfish and sea urchin

Answers: 1. A, 2. C, 3. C, 4. E, 5. D

Tiered-Stem Questions with Multiple Answers

Students may struggle with tiered-stem, multiple choice questions because they require more involved thought processes than do standard multiple choice questions. In these questions students are required to think about several items simultaneously. Exposure and practice with tiered-stem questions will help students develop confidence.

Example from the 1999 AP Biology Exam:

18. In a mesophyll cell of a leaf, the synthesis of ATP occurs in which of the following?
    I.   Ribosomes
    II. Mitochondria
    III. Chloroplasts

    (A) I only
    (B) II only
    (C) III only
    (D) II and III only
    (E) I, II, and III

Answer: D

Example suitable for use in a first year course:

1. Which of the following cell organelles typically contain DNA?
   I.   Nucleus
   II. Ribosomes
   III. Chloroplasts
   IV. Mitochondria

   (A) I only
   (B) I and II
   (C) III only
   (D) II and III
   (E) I, III, and IV

Answer: E

Lab Set Questions

Students need extensive practice answering multiple choice questions that are based on experimental data or scenarios describing experimental procedures. This type of question can be used to assess science process skills as well as understanding of the scientific basis for collected data. Students will be expected to interpret diagrams, data tables, and graphs. Additionally, they will be expected to apply their understanding of the biological concept to the data recorded and recognize acceptable explanations of the trends observed.

Example from the 1999 AP Biology Exam:

Question 114 refers to an experiment in which a dialysis-tubing bag is filled with a mixture of 3% starch and 3% glucose and placed in a beaker of distilled water, as shown below. After 3 hours, glucose can be detected in the water outside the dialysis-tubing bag, but starch cannot.

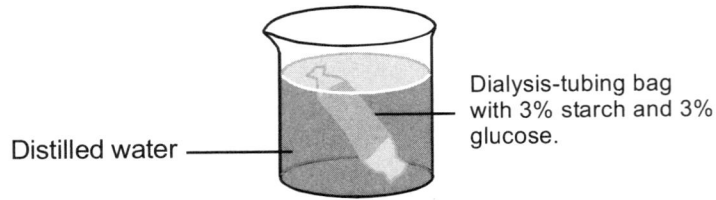

Dialysis-tubing bag with 3% starch and 3% glucose.

Distilled water

114. From the initial conditions and results described, which of the following is a logical conclusion?
    (A) The initial concentration of glucose in the bag is higher than the initial concentration of starch in the bag.
    (B) The pores of the bag are larger than the glucose molecules but smaller than the starch molecules.
    (C) The bag is not selectively permeable.
    (D) A net movement of water into the beaker has occurred.
    (E) The molarity of the solution in the bag and the molarity of the solution in the surrounding beaker are the same.

Answer: B

Example suitable for use in a first year course:

Question 1 refers to an experiment in which a dialysis-tubing bag is filled with a mixture of 3% starch and 3% glucose and placed in a beaker of distilled water, as shown below. After 3 hours, glucose can be detected in the water outside the dialysis-tubing bag, but starch cannot.

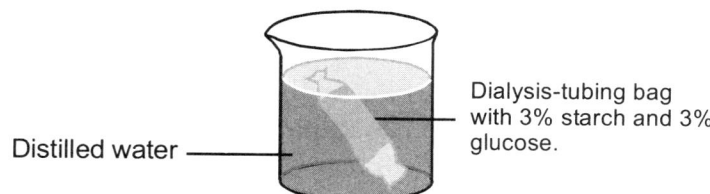

Distilled water

Dialysis-tubing bag with 3% starch and 3% glucose.

1. Which of the following statements is true based on the results described?
    (A) The bag is permeable to starch.
    (B) The distilled water caused the bag to rupture.
    (C) The pores in the bag are larger than the starch molecules.
    (D) The bag is permeable to glucose.
    (E) The bag does not allow water to move through it.

Answer: D

Free Response

In order to develop strong free response skills, students will need exposure to this type of question in the first year biology course as well as the AP Biology course. The foundation lesson IX, "Essay Writing Skills, Developing a Free Response" provides classroom strategies for developing skills in this area. The free response prompts are frequently broad and sweeping and thus require the students to consider an expansive scope of course content. The breadth of the questions makes them challenging for students. In this section we will look at examples of two types of free response questions encountered on the AP Biology Exam and explore how those types of questions may look in the first year course.

I. Thematic Content Based Questions

The College Board suggests the use of unifying themes to tie together the expansive content of the course. The themes used in the AP Biology course outline are frequent components in the free response prompts. The themes include science as a process, evolution, energy transfer, continuity and change, relationship of structure to function, regulation, interdependence in nature, and science, technology, and society. The following example shows how the themes are used in the prompts for the free response items:

Example from the 2002 AP Biology Exam:

3. The complexity of structure and function varies widely across the animal kingdom. Despite this variation, animals exhibit common processes. These include the following.
   - transport of materials
   - response to stimuli
   - gas exchange
   - locomotion
   a. Choose <u>two</u> of the processes above and for each, <u>describe</u> the relevant structures and how they function to accomplish the process in the following phyla.
      Cnidaria (e.g., hydra, jellyfish)
      Annelida (e.g., earthworm)
      Chordata (e.g., mouse)
   b. Explain the adaptive (evolutionary) value(s) of the structural examples you described in part a.

Example suitable for use in a first year course:

As animals became more complex, there was a selection pressure for a more efficient circulatory system. Compare and contrast the circulatory system for any two animals listed below and explain which animal has the more efficient system. Be sure to include a comparison of their hearts, effectiveness of gas exchange and any adaptations the system has to make it more efficient.

    a. fish
    b. grasshopper
    c. mouse
    d. snake

II. Laboratory and Experiment Based Questions

Typically, at least one of the free response questions on the AP Biology exam is based on a laboratory or experimental activity. Often the lab based prompt has connections to one of the twelve labs in the AP Biology Lab Manual. In these questions the students are frequently asked to demonstrate science process skills such as graphing, making inferences, predicting experimental outcomes, and critiquing experimental design.

Example from the 2002 AP Biology Exam:

4. The following experiment was designed to test whether different concentration gradients affect the rate of diffusion. In this experiment, four solutions (0%NaCl, 1%NaCl, 5%NaCl, and 10%NaCl) were tested under identical conditions. Fifteen milliliters (mL) of 5%NaCl were put into a bag formed of dialysis tubing that is permeable to Na^+, Cl^-, and water. The same was done for each NaCl solution. Each bag was submerged in a separate beaker containing 300 mL of distilled water. The concentration of NaCl in mg/L in the water outside each bag was measured at 40-second intervals. The results from the 5% bag are shown in the table below.

CONCENTRATION IN mg/L of NaCl OUTSIDE THE 5% NaCl BAG

Time (seconds)	NaCl (mg/L)
0	0
40	130
80	220
120	320
160	400

a. On the axes provided, graph the data for the 5% NaCl solution.
b. Using the same set of axes, draw and label three additional lines representing the results that you would predict for the 0%NaCl, 1%NaCl, and 10%NaCl solutions. Explain your predictions.
c. Farmlands located near coastal regions are being threatened by encroaching seawater seeping into the soil. In terms of water movement into or out of plant cells, explain why sea water could decrease crop production. Include a discussion of water potential in your answer.

Example suitable for use in a first year course:

A student performs a laboratory experiment with eggs, by first removing the calcium shells with vinegar. This leaves the eggs with a tough, rubbery but porous membrane. The eggs are then weighed. The student makes a number of different corn syrup solutions with varying percentages of corn syrup. She submerges a different egg into each of the different percent corn syrup solutions and allows them to sit for 24 hours. After 24 hours, the student removes the eggs and reweighs them. Analyze the student's data by completing the data table and graphing the % change in mass versus the percentage of corn syrup used to submerge the egg.

Percent of syrup	0%	10%	20%	30%	40%	50%
Original mass (g)	89	81	82	76	62	55
Final Mass (g)	78	77	86	89	82	81
Change in mass						
Percent change in mass						

Using the principles of osmosis, explain the results the student obtained. Determine from the graph what percent of corn syrup is considered isotonic with the egg and explain how you would confirm that predicted data point.

Finally, the student places her egg in distilled water for 24 hours and noticed no change in mass. Explain why "water-proof" eggs would be an important adaptation for birds.

Labs and Hands-on Activities: Connection to the AP Exams

One benefit of a strong vertical team in science is that much of the groundwork for establishing success on the AP exams can begin in middle school science. The following material shows how labs and activities specifically connect to an AP exam question.

Environmental Science – AP Question– pH levels and Indicator Organisms

The AP Environmental Science exam from 2001 contained a question that asked students to interpret a chart that represented pH ranges in which selected organisms (indicator species) could survive. The chart showed a range of both favorable pH conditions where the indicator organisms thrived, as well as less favorable conditions where they did not do as well.

In order for students to successfully answer this question they would need background knowledge in pH levels, water quality in relation to pollution levels, and possible ways to remediate the problems associated with water pollution. The following series of labs at fundamental, intermediate and advanced levels would enable students to build a knowledge base to answer the question completely.

Approaching the concept of pH on the fundamental level it would be appropriate for students to begin discovery labs that deal with pH. One simple lab involves making a pH indicator solution by boiling red cabbage and collecting the solution formed. When added to different substances students will see a color change depending on pH: acids will turn the liquid pink and bases will turn the solution blue. At this level, students would only be comparing the idea of acids and bases and might not even be dealing with the pH scale directly.

At the intermediate level students could actually calculate the pH level of different things using pH indicator solutions or pH paper. Students would then be expected to use and understand the ph scale and what specific pH numbers mean.

At the advanced level, students would be expected to use the pH scale to actually calculate the ionic concentrations.

Several other labs could be included to make a unit at each level for students to build on these concepts. The following is a series of labs that could support a unit on water quality at the fundamental, intermediate and advanced levels.

Foundation Activities

- pH lab where students discover the pH levels of everyday solutions and ends with a discussion of what pH means.
- Water quality introductory lab
- Oxygen cycle – bromothymol blue lab
- Microscope lab with water samples and microscopic organism

Intermediate Activities

- pH lab where students discover acid rain levels
- Logarithmic scale of pH measurements
- pH lab, using everyday solutions where students determine the pH
- Lab using simple water quality testing kits to test for pH levels (La Motte) – these kits use test strips.
- Point and Non-point pollution lab

Advanced Activities

- pH Lab with everyday solutions where students calculate hydroxide ion concentrations
- Lab using water quality testing kits to measure pH. These test kits require more manipulation and are more accurate. (La Motte)
- Eco-column Project (10 week project, modified from Bottle Biology) tracing the relationship between changes in pH and the living organisms (fish, Elodea) inside the column.
- Collecting rain water, testing the pH and observing the effect of acid rain on living organisms.
- Research causes of acid rain and possible ways to remediate the problem.

Lab Assessment

Assessment is an important cornerstone in preparing students for advanced science courses. At the fundamental level, students need to be comfortable with tests focused on the evaluation of lab skills as well as content learned in class.

At the intermediate level, assessment should continue to evaluate lab skills, but begin to emphasize application of learned things to new situations. At this level students should begin to produce for\mal write-ups based on templates given in class. In a test situation, students could be given data similar to what they tested in lab and asked to analyze what the data means.

Advanced level assessment must include formal write-ups, including tables, graphs and qualitative observations. Traditional tests should include multiple choice, short answers, problems and essays modeled after the AP Exam. Alternative forms of assessment could include debates, oral presentations and demonstrations.

Foundation Assessment

- Lab practical where students are given unknown solutions and asked to determine whether it is an acid or a base
- Students could be asked to draw, label and give correct magnification for microscopic organisms

Intermediate Assessment

- Students could be given water quality data and asked to interpret it
- Students could be asked to identify local point and non-point pollution sources and develop a technique for remediation

Advanced Assessment

- Students could be given pH data over a specified time period and will be asked interpret the trends observed
- Students could be given problem sets dealing with pH and asked to calculate the hydroxide ion concentrations
- Students could be asked to write a paper on the importance of indicator species, with particular emphasis on water quality

Writing Tests

Writing challenging tests in middle school is another important step for students towards AP success. The following list includes common verbs from the AP exams. Keep this list handy while you are developing your test questions.

Account for the fact…
Calculate….
Choose three of the following and...(from one free-response question)
Compare …
Describe….
Describe an experiment…
Describe and discuss…
Describe how to change the conditions of the experiment to….
Describe the…and explain….
Design a (an) (controlled) experiment…
Determine…as a function of …
Determine the value of….
Discuss….
Draw and label…
Estimate…
Explain….
Explain and defend or refute this statement.
Explain how to calculate….
Explain how you could experimentally determine…
Explain the implications of….

Explain the meaning of….
Explain your reasoning.
Express your answer in terms of…
Graph the data.
Identify a reasonable source of experimental error and the effect it would have on your measured value.
Identify another example of….
Indicate how…
Indicate whether….
Justify your answer.
Justify your prediction.
List and describe…
List measurements that must be made in order to….
Make an argument for….
Outline the procedures…
Plot these data.
Propose….
Propose a hypothesis…
Provide one argument for….
Select…. and describe (or explain)
Show calculations to support your answer.
Show the setup for the calculations…
Show your work.
Sketch…
State clearly….
Summarize the pattern…
Suppose…..What effect would this have on the calculated value of…? Justify your answer.
Trace…
Using principles of…
What effect….? Justify…

SAMPLE UNIT TEST: LIFE
Plantae/Photosynthesis Test

I: MULTIPLE CHOICE. Select the SINGLE, BEST answer to complete each statement, and write down the letter of your choice in the space provided using capital letters.

1. A fern spore grows into a
   A. haploid plant.
   B. diploid plant.
   C. pollen grain.
   D. zygote.

2. A zygote (or seed) is the result of the uniting of
   A. a spore and an ovule.
   B. a seed and a cone.
   C. a pollen grain with a seed.
   D. pollen and ovule.

3. The development of a vascular system allowed plants to
   A. grow taller.
   B. produce seeds.
   C. grow in water.
   D. reproduce faster.

4. Nonvascular plants require an environment with
   A. fairly constant winds.
   B. anaerobic bacteria.
   C. large amounts of phosphorus.
   D. high moisture levels.

5. Angiosperm reproduction centers around the
   A. fruit.
   B. seed.
   C. flower.
   D. anther.

6. Angiosperms and gymnosperms are the only groups of plants that produce
   A. zygotes.
   B. seeds.
   C. flowers.
   D. spores.

7. A fruit's primary function is to
   A. feed mankind.
   B. float to a new environment.
   C. aid in dispersing the seeds.
   D. provide water for the seeds.

8. For plants, the unneeded byproduct of photosynthesis is
   A. carbon dioxide.
   B. water.
   C. ozone.
   D. oxygen.

9. Climbing vines originally evolved in
   A. sunny grasslands.
   B. forests with a lot of shade.
   C. biomes with little rainfall.
   D. deserts with thorny plants.

10. The "raw materials" (needed to make it happen) for photosynthesis are
    A. water and glucose.
    B. carbon dioxide and glucose.
    C. water and oxygen.
    D. carbon dioxide and water.

11. In the light and photosynthesis lab why did you add bromothymol blue to the test tubes of the *Elodea* and water?
    A. To test for the presence of oxygen
    B. To test for the presence of dissolved carbon dioxide
    C. To give the plant nutrients
    D. To test for the lack of oxygen

12. In the light and photosynthesis lab, why did you put a test tube with *Elodea* and bromothymol blue and water in the dark?
    A. To serve as a control in the experiment
    B. Plants photosynthesize the most effectively in the dark
    C. Bromothymol blue works best in the dark
    D. To increase the amount of oxygen produced.

II: SHORT ANSWER and ESSAY: In the space provided, answer the following questions as briefly and accurately as possible. Use complete sentences. Answers should fit in the space provided.

13. Write the common names of each compound for the following chemical equation.
$$6CO_2 + 6H_2O + Energy \longrightarrow C_6H_{12}O_6 + 6O_2$$

14. Justify by providing two reasons why seed dispersal is crucial to the survival of a plant species.

15. Name the two types of root systems found in plants; give examples for each type.

16. What are two primary functions of roots, and discuss how the structure of a root is suited to its function?

17. What is the primary function of flowers and cones?

18. Your friend finds a small low-growing plant in the desert. He identifies the plant as a moss. You believe that your friend's identification is incorrect and this plant is probably not a moss. Justify why you think this is so.

19. You have been chosen to work on a new space probe that will carry humans deep into space. A colleague suggests that it would be important to have a population of algae on this trip. Describe two ways that algae could help human travelers on a long space voyage.

20. Even though ferns survive under many of the same environmental conditions as mosses, ferns are able to grow much larger and taller than mosses. Why is this so?

21. Discuss how a seed plant has an advantage over plants that do not produce seeds?

22. The seeds of a gymnosperm are probably not likely to be dispersed by animals, whereas the seeds of angiosperms are likely to be dispersed by animals. Explain why.

III. LAB PRACTICAL: Answer the questions after you have observed the specimens at the lab stations.

Station 1: for each of the following seeds, describe the method of seed dispersal
A. (pine seed)
B. (apple cut in half with seeds displayed)
C. (dandelion)
D. (pecan)
E. (coconut in husk)
F. (grass burr)

Station 2: View the plant sample and determine if it is a monocot or dicot and explain why by providing 2 pieces of evidence. (rose bush)
A. Circle one: monocot/dicot
B. Reason 1:
C. Reason 2:

Plantae/Photosynthesis Test

Answer Section

I: Multiple Choice
1. A
2. D
3. A
4. D
5. C
6. B
7. C
8. D
9. B
10. D
11. B
12. A

II. Short Answer and Essay
13. carbon dioxide + water ⟶ (in the presence of the sun's energy and chlorophyll) glucose + water
14. Seeds that are dispersed may grow in a new area, thus increasing the range of the plant. Plants that grow from dispersed seeds may have a better chance of surviving without competition from the parent plant.
15. taproot: carrot, radish, fibrous root: grasses
16. Roots anchor the plant.
    Roots absorb water and minerals.
    The branching, filamentous structure of roots and the large numbers of root hairs create a large surface are for water absorption and for anchoring the plant to the soil.
17. Reproduction in angiosperms and gymnosperms, respectively.
18. Mosses grow in damp places and are not well adapted to life in a desert — theyhave no structures to prevent water loss.
19. Their ability to reproduce also depends upon a supply of standing water.
    With a source of light energy, the algae could become a food source.
    The algae would produce oxygen used by astronauts and use the carbon dioxide they produce.
    Some of the wastes produced by astronauts could be used by algae as fertilizer.
20. Ferns have vascular tissue to move materials to and from their cells. Vascular tissue provides the stems with some rigidity, and leaves can grow taller above the ground and receive more light. Ferns also have a cuticle layer that prevents the plants from losing too much water to the environment.
21. A seed is a structure that protects the zygote of a seed plant.
    Seeds contain a supply of food for the developing plant.
    Seeds do not require water or a damp environment, so seed plants are able to thrive all over the globe.
22. The seeds of gymnosperms are contained within a cone.

The seeds of angiosperms are contained within a fruit. The fruit would be more likely to appeal to an animal and thus would more likely be eaten. Animals are attracted to the bright colors, smell and sweet taste of fruit. Thus, the seeds of an angiosperm might have a better chance of being dispersed by animals.

III. LAB PRACTICAL

Station 1: For each of the following seeds, describe the method of seed dispersal
- G. wind
- H. animal (eats the fruit and excretes the seeds)
- I. wind
- J. drop and bounce
- K. drop and bounce and floating
- L. picked up by animal and carried in fur

Station 2: View the plant sample and determine if it is a monocot or dicot and explain why it is by providing evidence. (rose bush)
- D. dicot
- E. Reason 1: netlike veins in leaves
- F. Reason 2: flower petals = 5

SAMPLE UNIT TEST: EARTH
Water Test

Choose the *best* answer and mark it clearly on the answer sheet.

Multiple Choice:

1. If the level of pollution in the water is 700 ppb, there are _____ pollution molecules for every _____ water molecules.
   a. 700 & 1,000,000,000
   b. 7 & 1,000,000,000
   c. 70 & 10,000,000
   d. 700 & 1,000

2. The land area in which a runoff drains into a river is called the
   a. watershed.
   b. drainage ditch.
   c. saturation zone.
   d. permeability zone.

3. Land subsidence is a problem in areas with water pumped from
   a. consolidated rocks.
   b. very deep wells.
   c. unconsolidated rocks.
   d. very shallow wells.

4. Groundwater is recharged by
   a. the filtering of precipitation through permeable rock.
   b. rivers flowing underground.
   c. sewage dripping underground slowly.
   d. glacier movement underground.

5. Porosity is
   a. the total space taken up by a substance.
   b. the available air space between soil and rock particles.
   c. how well water can travel through a substance.
   d. the density of a substance.

6. Permeability is
   a. the space inside of a substance.
   b. the available air space between soil and rock particles.
   c. how inter-connected the pore space is in a substance.
   d. the density of a substance.

7. Porosity is measured in units of
   a. time.
   b. depth.
   c. volume.
   d. volume per time.

8. Permeability is measured in units of
   a. time.
   b. depth.
   c. volume.
   d. volume per time.

9. Most of the fresh water on our planet is
   a. contained in aquifers and groundwater
   b. found in lakes streams and rivers
   c. caught up in the atmosphere
   d. frozen in icecaps and glaciers

10. Which aquifer supplies water for San Antonio?
    a. the Trinity
    b. the Ogalalla
    c. the Colorado
    d. the Edwards

11. Permeability of an aquifer is dependent on
    a. the density of a substance.
    b. the amount of water used.
    c. the amount of interconnection between the pores of a substance.
    d. the degree to which a substance can be compressed.

12. All of the following are freshwater sources except
    a. atmosphere.
    b. inland seas.
    c. groundwater.
    d. rivers.

13. An aquifer is
    a. a layer of rock or sediment that allows water to pass through it easily.
    b. an empty layer within the earth, like a cavern, that is filled with water.
    c. a type of water source in which the water always flows freely to the surface
    d. what the Romans used to transfer water.

14. About what percentage of the earth's water is fresh?
    a. 3%
    b. 10%
    c. 98%
    d. 1%

15. The rock, granite, of volcanic origin is not permeable because
    a. granite is dried clay.
    b. there is no water in granite.
    c. granite is not exposed to precipitation.
    d. there is no interconnected pore space in granite.

16. What is the porosity of clay?
    a. high
    b. low
    c. none at all
    d. impossible to measure

17. What material would not make a good aquifer?
    a. clay
    b. sandstone
    c. limestone
    d. gravel

18. Which material would make the best aquifer?
    a. clay
    b. sand
    c. gravel
    d. potatoes

19. Which of the following is NOT a potential threat to our water supply?
    a. pollution
    b. overuse
    c. the water cycle
    d. contamination

20. All of the following are a point source pollution EXCEPT
    a. a leaky gas tank in your garage.
    b. oil and antifreeze flowing down a sewer drain on your street.
    c. a three acre plot of land sprayed with pesticides.
    d. a huge chemical plant with smokestacks emitting unfiltered pollutants.

21. The City of Austin gets its water from
    a. Barton Springs.
    b. the Edwards Aquifer.
    c. the Colorado River.
    d. mostly rain water.

22. A good confining layer for an aquifer would be
    a. clay.
    b. gravel.
    c. peat moss.
    d. limestone.

23. The best filter would be
    a. clay.
    b. sand.
    c. gravel.
    d. peat moss.

24. The amount of one substance within another substance is loosely called
    a. solution.
    b. dilution.
    c. pollution.
    d. concentration.

25. Land subsidence can be slowed by
    a. decreasing recharge amounts.
    b. increasing municipal water use.
    c. population increases.
    d. decreasing water use.

26. Water in Austin creeks tends to have a pH of 8 or higher because
    a. water in Austin moves extremely fast.
    b. the limestone bedrock in the watershed.
    c. the high temperatures of lakes and rivers.
    d. of acid rain.

27. In a glass of Nestle Quik, you mix together chocolate powder and milk to make a tasty drink. To have a 25% chocolately drink you would need ___ parts chocolate and ___ parts milk.
    a. 25 & 100
    b. 100 & 25
    c. 25 & 75
    d. 75 & 25

28. Primary treatment of sewage water means
    a. settling out the chunky bits.
    b. using microbes to eat effluent.
    c. adding chemicals to kill bacteria.
    d. desalinization.

29. The water cycle is
    a. water-like tricks perfomed by fleas.
    b. exactly like the oxygen cycle (produced by plants).
    c. precipitation-saturation-evaporation-condensation.
    d. amount specific.

30. Ms. Kirby is always happy during the holiday season because
    a. she has an unnatural fondness for reindeer.
    b. she finally gets to end her fast of Ramadan.
    c. she always gets to light the candles for the menorah.
    d. her students fill her with the joy of the season and a love for all mankind.

Essays – be sure to give complete answers for full credit

31. NASA has trained a group of monkeys to perform the same lab you did in class. As they accomplished the task with ease and did not leave such a big mess, they are now in high school preparing for college. They used four samples (listed below) in a paper cup, added water, and adjusted the water level so that each had 100 mL, poked a hole in the cup, and drained the water. Using the data the magnet monkeys collected, graph the porosity in a bar graph for each sample then answer the questions that follow.

Sample	Amount held	Amount Drained	Drain Time
Fine sand	78 mL	55 mL	165 sec
Clay	2 mL	0 mL	1 hour
Gravel	66 mL	84 mL	42 sec
Coarse sand	45 mL	30 mL	300 sec

   a. Which sample is the most porous?
   b. Which sample is the most permeable?
   c. Which sample would make a poor aquifer? Why?
   d. Which sample would make a good aquifer? Why?
   e. Why is it important for water to flow through an aquifer at a good rate, not too fast, not too slow?

32. You are given 100 mL of a solution of dye that contains one part powdered pigment to four parts water. You mix it with 100 mL to make a new, more dilute dye solution.
    a. What is the original concentration of the solution?
    b. What is the dilution factor?
    c. What is the final concentration of the dye?

33. In the 1998 flood, the city of Seguin had flood stage waters well above the Extreme Peak Flow predicted by scientists. List and explain the reasons why this occurred.

34. The following is water data from an average family in Austin. Graph the data to show water use and evaluate the family according to your results.

Activity	Amount of water used in 1 week
showering (regular shower head)	4020 gallons
brushing teeth (water on)	58 gallons
washing dishes	1200 gallons
eating/cooking	600 gallons
washing clothes	3000 gallons
watering the yard	9000 gallons
flushing toilet	6800 gallons

35. You and your lab group test water from several places and get the following results. Discuss the results, including where the samples might be taken from, what the numbers mean and why they might have gotten the results they did.

	Site 1	Site 2
pH	8	7
nitrates (ppm)	0	40
phosphates (ppm)	1	2
temperature °C	15	25
dissolved oxygen	91%	71%
turbidity (JTU)	0	55

Water Test

Answer Section

Multiple Choice

1. a	16. c
2. a	17. a
3. c	18. b, c
4. a	19. c
5. b	20. d
6. c	21. c
7. c	22. a
8. d	23. b
9. d	24. b
10. d	25. d
11. c	26. b
12. b	27. c
13. a	28. a
14. a	29. c
15. d	30. a, b, c, d

Essays

31. a. Fine sand was the most porous and it held 78 mL.

    b. Gravel was the most permeable, with 84mL/42 seconds.

    c. Clay would make a poor aquifer because it is neither porous nor permeable.

    d. Gravel would make a great aquifer because it is both porous and permeable.

    e. Water that flows too fast may not have time to settle out impurities and other possible problems like sediment.

32. a. The original concentration of the solution was 20%. [20mL dye + 80 mL water = 100mL solution; 20mL is 20% of 100mL]

    b. The dilution factor is 100%.

    c. The final concentration of the solution is 10%. [20mL dye + 180mL water = 200mL solution; 20mL is 10% of 200mL]

33. The Extreme Peak Flow predicted by scientists took the flood data recorded for the last 100 years and graphed the results. The flood of 1998 was well outside this historical data set. The possible reasons are that there has been a lot of development in the Guadalupe River water shed and the increase in paved surfaces caused an increase in runoff to streams and eventually the river. Other reasons include a bizarre weather pattern that had storm clouds training over the region and most of the rain fell below the Canyon dam.

34.

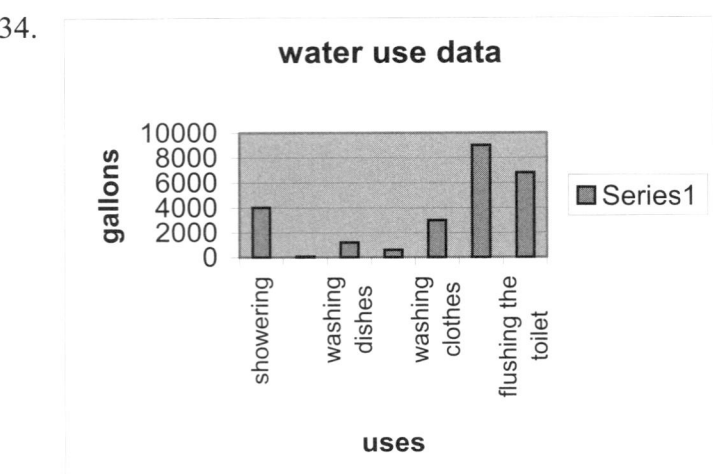

It is clear from the data that this family spends much of its' water budget watering their yard and flushing their toilets. If I were hired by the family to evaluate their usage and make suggestions for improvement I would suggest that they xeriscape their yard, catch rainwater for watering plants and water only one day a week. I would also suggest that they convert to low flow toilets and try and reduce their water flushage.

35. The results of the chemical testing show Site 1 to be better quality water than Site 2. The increase in nitrates and phosphates mean that the site might be close to a site that uses chemical fertilizers, like a golf course or farm. Site 2 also had less dissolved oxygen and high turbidity, which makes it hard for aquatic organisms and plants to survive.

Punnett Square Exercises
Solving Monohybrid Punnett Squares

Multiple Choice:

1. Two mice with black fur were mated and produced offspring with brown fur and offspring with black fur. If "B" represents the dominant allele for black fur and "b" represents the recessive allele for brown fur, which would represent the most probable genotypes of the parental mice.
   a. BB x Bb   b. BB x BB   c. Bb x Bb   d. BB x Bb

2. In pea plants, purple seed coat color is dominant over the white seed coat color. In a cross between two plants both heterozygous for the purple seed coat color, what % of offspring are expected to have white seed coats.
   a. 0%   b. 25%   c. 50%   d. 100%

3. In cats, the gene for short hair "A" is dominant over the gene for long hair "a". A short-haired male cat is mated with a long-haired female and four kittens are produced, two shot-haired and two long-haired. The genotypes of the parent cats are most likely:
   a. Aa x aa   b. AA x Aa   c. Aa x Aa   d. AA x aa

4. When a white guinea pig is crossed with a black guinea pig, all four of the resulting offspring are black. If the black color is dominant and white is recessive, what would represent the genotypes of the offspring.
   a. BB only   b. Bb only   c. bb only   d. none of the above

5. Achondroplasic dwarfism is a DOMINANT genetic disease, where D = Achondroplasiac dwarfism and d = normal non-dwarf condition. All living dwarfs are heterozygous, Dd. Any human born with DD, is sometimes referred to as a "double dominant" and the child will die tragically soon after birth 100% of the time. Two Achondroplasiac dwarfs want to have a child. What % chance do any two Achondroplasic dwarfs have in having a "double dominant" child?
   a. 0%   b. 25%   c. 50%   d. 100%

Answer Key:
1. C
2. B
3. A
4. B
5. B

Baby Dice Island
Modeling Exponential Growth

Multiple Choice:
Use the graph below to answer the questions that follow.

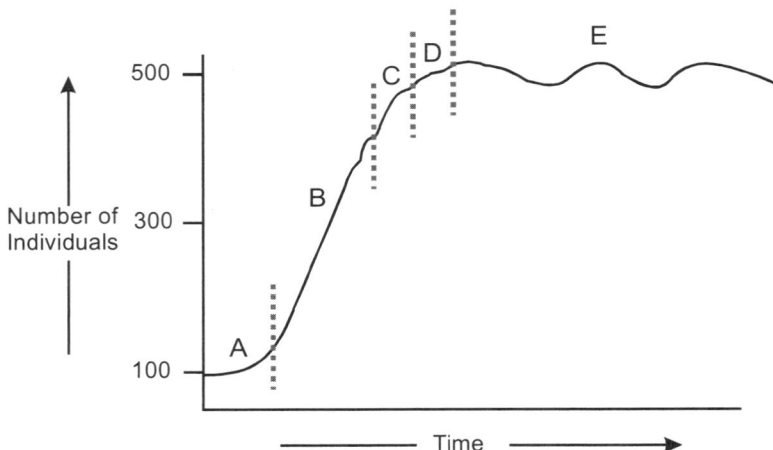

1. What is the carrying capacity for this population?
   A. ~100   B. ~200   C. ~300   D. ~500

2. Which segment of the graph — A, B, C, D or E — shows the steady state condition?
   A. A   B. B   C. C   D. E

3. What is the average population growth rate in section E?
   A. ~0%   B. ~50%   C. ~100%   D. ~200%

4. Explain why a graph of the world human population does not look like the graph above.
   A. Humans are much more complex and the human population increase is very complex.
   B. There is no carrying capacity for humans.
   C. Humans have, so far, not reached carrying capacity.
   D. Human population has been dramatically decreasing in recent years.

5. In the Baby Dice Island lab, what sort of growth were we modeling?
   A. arithmetic growth
   B. logistic growth
   C. restricted exponential growth
   D. unrestricted exponential growth

Answer Key:
1. D
2. D
3. A
4. C
5. D

Online Biome Exploration
Researching, Creating, and Presenting a Biome PowerPoint

Multiple Choice:

1. An animal in the desert that slows its metabolic rate in order to survive the heat of the day is said to be in
   a. hibernation.
   b. estivation.
   c. mimicry.
   d. camouflage.

2. Which of following shows the correct zones of vegetation as they change with increasing altitude?
   a. ice → tundra → deciduous forest → savanna
   b. ice → tundra → seasonal forest → savanna → semidesert → desert
   c. tropical rain forest → deciduous forest → coniferous forest → tundra
   d. desert → grassland → tundra → taiga

3. A maple tree is not well suited to a desert climate because
   a. it would be knocked down by the high desert winds.
   b. the wood would decay rapidly in arid regions.
   c. its broad leaves would lose too much water.
   d. it would get too much sunlight in the winter months.

4. The most important factors determining the climate of a place are temperature and
   a. soil type.
   b. amount of moisture.
   c. latitude.
   d. altitude.

5. Free-drifting microscopic organisms that are able to photosynthesize their own food are called
   a. phytoplankton.
   b. saprophytes.
   c. zooplankton.
   d. nekton.

Answer Key:
1. B
2. C
3. C
4. B
5. A

Molecular Motion: Are You Current on Convection?
Experimenting with Heat and Molecules

Multiple Choice:

1. _____ is the transfer of energy that occurs due to density differences in air.
   a. Radiation
   b. Conduction
   c. Convection
   d. Temperature gradient

2. As a substance is heated, the speed of its molecules
   a. increases.
   b. decreases.
   c. stays the same.

3. As a substance is heated, its volume
   a. increases.
   b. decreases.
   c. stays the same.

4. As a substance is heated, its density
   a. increases.
   b. decreases.
   c. stays the same.

5. As a substance is heated, its pressure
   a. increases.
   b. decreases.
   c. stays the same.

Answer Key:
   1. C
   2. A
   3. A
   4. B
   5. B

Investigating Porosity and Permeability
Aquifer Study

Multiple Choice:

1. What is the porosity of clay?
    e. high
    f. low
    g. none at all
    h. impossible to measure

2. What material would not make a good aquifer?
    e. clay
    f. sandstone
    g. limestone
    h. gravel

3. Porosity is
    e. the total space taken up by a substance.
    f. the available air space between soil and rock particles.
    g. how well water can travel through a substance.
    h. the density of a substance.

4. Permeability is
    e. the space inside of a substance.
    f. the available air space between soil and rock particles.
    g. how inter-connected the pore space is in a substance.
    h. the density of a substance.

5. Porosity is measured in units of
    e. time.
    f. depth.
    g. volume.
    h. volume per time.

6. Permeability is measured in units of
    d. time.
    e. depth.
    f. volume.
    g. volume per time.

Answer Key:
1. A
2. A
3. B
4. C
5. C
6. D

Is Dilution the Solution to Pollution?
Washing Away Our Mess

Multiple Choice:

1. If the level of pollution in the water is 700 ppb, there are _____ pollution molecules for every _____ water molecules.
    e. 700 & 1,000,000,000
    f. 7 & 1,000,000,000
    g. 70 & 10,000,000
    h. 700 & 1,000

2. In a glass of Nestle Quik®, you mix together chocolate powder and milk to make a tasty drink. To have a 25% chocolately drink you would need ___ parts chocolate and ___ parts milk.
    e. 25 & 100
    f. 100 & 25
    g. 25 & 75
    h. 75 & 25

3. All of the following are a point source pollution except
    e. a leaky gas tank in your garage.
    f. oil and antifreeze flowing down a sewer drain on your street.
    g. a three acre plot of land sprayed with pesticides.
    h. a huge chemical plant with smokestacks emitting unfiltered pollutants.

4. The amount of one substance dissolved within another substance is loosely called the:
    e. solution.
    f. dilution.
    g. pollution.
    h. concentration.

5. Mercury can be toxic to humans at a concentration of 1 ppb. What advantages does industry get from dumping waste into the ocean?
    I.   Releasing waste from the plant into the ocean requires very little technology.
    II. Releasing waste from the plant into the ocean is inexpensive.
    III. Releasing waste from the plant into the ocean provides enough dilution to make the pollution safe.

    a. I. and II.
    b. II.
    c. III.
    d. II. and III.
    e. I. and III.

Answer Key:
1. A
2. C
3. B
4. D
5. A

Appendixes

A Brief History of Science

LEVEL
All levels of science

NATIONAL STANDARDS
Science in Personal and Social Perspectives; History and Nature of Science

OBJECTIVE AND INTRODUCTION
Science is a human endeavor, and the discoveries in science, whether accidental or deliberately pursued, always occur within the contexts of culture and social environment, curiosity and necessity, and the availability of technology. The following chart which chronicles some of the major developments in science and technology over thousands of years is not meant to be complete, but is intended to convey how some of the scientific concepts were discovered and how they developed from one generation to the next. The discoveries which are highlighted were chosen because of their direct relevance to the content typically covered in middle school and high school science courses.

It should also be noted that even though modern science is often partitioned into narrow, specific areas of concentration, as science continues to progress we continue to see how one science is dependent on another. For example, the discovery of the structure of the DNA molecule is certainly one of the most important discoveries of the modern era, and could not have been possible without the cooperation and contributions of physicists, chemists, and biologists. If one of the sciences doesn't have any events highlighted during a particular time period in the chart, it should not be assumed that there were no significant developments during that time period. Often progress in one area of science ultimately reveals itself in a major development in another area of science.

It is important to help the students realize that the discoveries listed on this chart did not happen overnight. Scientists struggled for years, decades, centuries, and sometimes millennia to grasp an understanding of the difficult riddles the natural world presents us. Science teachers should continually remind students that because these concepts took considerable time and effort to discover and understand, it will likely take time and effort for the students to understand them.

Perhaps this chart can help teachers put the science they are exploring with students into perspective as it relates to time, culture, and available technology. If you like to have the students write reports on scientists and their discoveries, you might choose a particular time period or topic and have the students do further research and present their findings in a creative way, such as posters, timelines, oral and written reports, or a Power Point presentation.

The primary sources for this chart can be found at http://www.sciencetimeline.net.

	10000 BC - 800 BC	800 BC - 400 BC	400 BC - 200 BC
Biology	10,000 BC - 6500 BC Animals were probably first domesticated. 2000 BC Egyptians considere the souring of wine comparable to the souring of milk. 1600 BC Egyptian papyrus list many diagnoses of head and neck injuries and their treatment. 1000 BC Horse breeders experiment with cross-breeding of horses and donkeys.	580 BC Thales of Miletus suggests that water is the fundamental component to all life. 510 BC Almaeon of Crotona locates the seat of perception in the brain by dissection. 500 BC Xenophanes examines fossils and speculates on the evolution of the earth. 400 BC Hippocrates of Cos, maintains that diseases have natural causes.	250 BC Erasistratus of Alexandria dissects the brain and distinguishes between the cerebrum and the cerebellum.
Physics	4800 BC Astronomical calendar stones used in Egypt. 3300 BC Numerals first used in Sumerian and Egyptian hieroglyphics. 3200 BC First evidence of wheeled vehicles in Uruk. 1600 BC Documents maintain the Earth was a globe and the Earth circled the Sun. 1500 BC Babylonians understand right-triangle relationships.	From 747 BC, a continuous record of solar and lunar eclipses was kept in Mesopotamia. 585 BC Thales predicts and eclipse 530 BC Pythagoras discovers musical intervals in strings depends on length and tension. 425 BC Herodotus writes the first scientific history. 400 BC Arrow-shooting catapult developed at Syracuse.	400 BC Babylonian astronomers can predict the occurence of lunar eclipses. 370 BC Eudoxus of Cnidus invents a model of concentric spheres by which he was able to predict the motions of the moon, sin, and planets. 335 BC Aristotle writes *Physics*, which becomes the standard for science for 2000 years. 300 BC Euclid writes *Elements*, providing the basis for geometry. 260 BC Aristarchus of Samos suggests the sun-centered model of the universe, and calculates the Earth-Sun distance to the Earth-moon distance; Archimedes formulates buoyancy principles.
Chemistry	4000 BC Copper smelting is introduced in Mesopotamia. 2500 BC Smelting of bronze in Sumeria. 1200 BC Smelting of iron in Armenia.	450 BC Empedocles of Agrigento divides matter into the four elements: earth, water, air, fire. 440 BC Leucippus of Miletus suggests the existence of atoms.	250 BC 'Zero' appears in the Babylonian place-value system.

Laying the Foundation in Middle Grades Life and Earth Science

	200 BC - 200 AD	200 - 1000	1000 - 1300
Biology	1st Century AD Pedanius Dioscorides publishes recommendations as to the medicinal use of specific plant extracts. 170 Claudius Galen uses pulse-taking as a diagnostic, performs numerous animal dissections, and writes treatises on anatomy.	900 Abu Bakr al-Razi, distinguishes smallpox from measles in the course of writing several medical books in Arabic.	1200 Medical doctors, especially in Italy, begin writing case-histories, describing the symptoms and courses of numerous diseases. 1266 Hugh and Theodoric Borgogoni advocate putting surgical subjects to sleep with narcotic-soaked sponges.
Physics	134 BC Hipparchus of Rhodes measures the year with great accuracy and builds the first comprehensive star chart with 850 stars and a luminosity scale. 45 BC Sosigenes designs the calendar adopted by Julius Caesar. 100 AD Hero of Alexandria explains that the four elements consist of atoms. 141 AD Claudius Ptolemy publishes *Almagest*, the standard book for astronomy for 1500 years.	517 John Philoponus determines that falling objects fall with the same acceleration. 530 Simplicius of Cilicia writes a commentary in Greek on Aristotle's writings on 'gravity'. 1000 Ibn al-Haitam, or al-Hazen, in *Opticae Thesaurus*, introduces the idea that light rays emanate in straight lines in all directions from every point on a luminous surface.	1054 Chinese astronomers at the Sung national observatory at K'ai-feng observe the explosion of a supernova in the Crab Nebulae, visible in daylight for twenty-three days. 1268 Roger Bacon publishes proposals for educational reform, arguing for the study of nature, using observation and exact measurement, and asserting that the only basis for certainty is experience, or verification.
Chemistry	1st century AD Titus Lucretius Carus maintains that the universe came into being through the working of natural laws in the combining of atoms.	800 Jabir ibn Hayyan, later known as Geber, bases his chemical system on sulfur and mercury. 850 Moors in Spain prepare pure copper by reacting its salts with iron, a forerunner of electroplating.	1100 Alchemists develop the art of distillation to the stage at which distillates could be captured by cooling in a flask. 1260 Albertus writes a book in which he geology into a coherent theory. He was the first to produce arsenic in a free form.

Laying the Foundation in Middle Grades Life and Earth Science

	1300 - 1400	*1400 - 1500*	*1500 - 1550*
Biology	1316 Mondino of Luzzi publishes *Anatomia*, introducing the practice of public dissections for teaching. 1360, Guy de Chauliac, recommends extending fractured limbs with pulleys and weights, and replacing lost teeth with bone fastened to the sound teeth with gold wire.	1410 Benedetto Rinio publishes an herbal which contains 450 paintings of plants, botanical notes, citations of authorities used, and the names of the plants in various languages. 1482 Leonardo da Vinci begins his notebooks on dissections of the human body, the impossibility of perpetual motion, dynamics, statics, and numerous machines.	1541 Giambattista Canano publishes illustrations of each muscle and its relation with the bones. 1546 Fracastoro publishes the idea that diseases were caused by disease-specific seeds which could be contagious.
Physics	1304 Theodoric of Freiberg shows that rainbows could be explained through experiments with hexagonal crystals and spherical crystal balls. 1323 William Ockham introduces the distinction between 'being in motion' and 'being moved,' that is, as it is now called, between dynamic motion and kinematic motion. 1364 Giovanni di Dondi builds a complex clock which kept track of calendar cycles and computed the date of Easter by using various lengths of chain.	1420 Felipe Brunelleschi draws panels in scientifically-accurate perspective. 1437 Johann Gutenberg becomes the first in Europe to print with movable type cast in molds.	1543 Copernicus publishes *De revolutionibus orbium coelestium*, detailing the sun-centered model of the solar system. 1572 Tycho Brahe observes a supernova in the constellation Cassiopeia, now known as Tycho's star. 1583 Galileo Galilei discovers by experiment that the oscillations of a swinging pendulum take the same amount of time regardless of their amplitude. In 1590 Zacharias and Hans Janssen combines double convex lenses in a tube, producing the first telescope.
Chemistry	1300 Giles of Rome puts forward an atomic theory based on Avicebron's theory of matter.		Theophrastus Bombastus von Hohenheim (Paracelsus) suggests the chemical properties are combustibility, fluidity, and changeability, solidity, and permanence.

Laying the Foundation in Middle Grades Life and Earth Science

	1600 - 1625	1625 - 1650	1650 - 1700
Biology	1627 William Harvey confirms his observation that the blood circulates throughout the body.	1645 Marc Aurelio Severino discovers the heart of the higher crustacea., recognizes the respiratory function of fish gills, and recognizes the unity of vertebrates. 1650 Francis Glisson publishes an account of infantile ricketts.	1651 Harvey publishes the concept that all living things originate from eggs. 1652 Thomas Bartholin discovers the lymphatic system and determines its relation to the circulatory system. 1655 Thomas Sydenham promotes the idea that diseases are organisms inside a host. 1665 Robert Hooke names and gives the first description of cells. 1674 Anton van Leeuwenhoek reports his discovery of protozoa.
Physics	1600 William Gilbert, in *De Magnete*, holds that the earth behaves like a giant magnet with its poles near the geographic poles. 1604 Johannes Kepler and many other astronomers witness the outburst of a supernova in the constellation Serpens. 1605 Francis Bacon, with the *Advancement of Learning*, begins the publication of his philosophical works, in which he urges collaboration between the inductive and experimental methods of proof. 1609 Kepler publishes his 1st and 2nd laws of planetary motion. 1610 Galileo observes the moons of Jupiter, phases of Venus, craters on the moon. 1621 Willibrord Snell discovers the law of refraction.	1633 Galileo is placed under house arrest for his heliocentric views published in his book *Dialogue on the Two Chief World Systems*. 1638 Galileo publishes *Discourses on Two New Sciences*, outlining his theory of motion. 1644 Blaise Pascal builds a five digit adding machine. 1644 Evangelista Torricelli devises the mercury barometer and creates an artificial vacuum. 1648 Pascal shows that barometric pressure results from atmospheric pressure and that pressure applied to a confined fluid is transmitted equally to all areas and at right angles to the surface of the container.	1666 Isaac Newton discovers the essentials of calculus, the law of universal gravitation, and that white light is composed of all the colors of the spectrum. 1669 Newton circulates a manuscript containing the first notice of his calculus. 1676 Ole Roemer proves that light travels at a finite speed by repeated observations of eclipses of Jupiter's moon, Io. 1684 Gottfried Wilhelm von Leibniz publishes his system of calculus, developed independently of Newton. 1687 Newton publishes the *Principia*, a summary of his discoveries in motion, gravitation, and calculus. 1693 Edmund Halley discovers the formula for the focus of a lens. 1694 Rudolph Jakob Camerarious reports the existence of sex in flowering plants.

	1600 - 1625	1625 - 1650	1650 - 1700
Chemistry		1630 Jean Rey states that the slight increase in weight of lead and tin during their calcination could only have come from the air. 1644 Evangelista Torricelli devises the mercury barometer and creates an artificial vacuum. 1648 Jean Baptiste van Helmont concludes that plants derive their sustenance from water, demonstrates that physiological changes have chemical causes, coined the name 'gas' from the Greek *chaos*, distinguishes gases as a class with liquids and solids.	1661 Robert Boyle gives the first precise definitions of a chemical element, a chemical reaction, chemical analysis, made studies of acids and bases, and shows that pressure and volume of a gas are inversely proportional. 1670 Boyle produces hydrogen by reacting metals with acid. 1679 Denis Papin demonstrates the influence of atmospheric pressure on boiling points.

Biology	*1700 - 1750*	*1750 - 1800*	*1800 - 1850*
	1715 Thomas Fairchild produces the first artificial hybrid plant.	1752 James Lind calls attention to the value of fresh fruit in the prevention of scurvy.	1800 Karl Friedrich Burdach introduces the term 'biology,' which replaces 'natural history,' which traditionally had three components, zoology, botany, and mineralogy.
	1745 Maupertius proposes the notion of descent from a common ancestor.	1753, Carl Linné publishes *Species plantarum*, in which he distinguished plants in terms of genera and species, and later applying the system to animals.	1809 Jean-Baptiste Monet de Lamarck states that heritable changes in 'habits,' or behavior, could be brought about by the environment.
	1749 Buffon begins the publication of the 44 volumes of *Histoire Naturelle*, in which he draws attention to vestigial organs and asserted that species are mutable.	1762 Marcus Antonius Plenciz says that living agents are the cause of infectious diseases.	1820 Lamarck describes the origin of living things as a process of gradual development from matter.
	1788 Jean Senebier demonstrates that it is light, not heat, from the sun that is effective in photosynthesis.		1827 Robert Brown notices random movement of microscopic particles contained in the pollen from plants when suspended in fluid (Brownian movement).
	1791 Luigi Galvani shows that it is possible to control the motor nerves of frogs using electrical currents, i.e., that the nerves transmitted electricity.		1831 Brown discovers the cell nucleus in the course of a microscopic examination of orchids.
			1833 Marshall Hall describes the mechanism by which a stimulus can produce a response independent of both sensation and volition, and coins the term 'reflex.'
			1837 Heinrich Gustav Magnus determines that carbon dioxide released in the lungs had been carried there by blood and that more oxygen and less carbon dioxide was contained in arterial than in venous blood; René Dutrochet observes that chlorophyll is necessary for photosynthesis; Hugo von Mohl describes 'chloroplasts' as discrete bodies within the cells of green plants.
			1838 Mattias Jakob Schleiden puts forward the theory that plant tissues are composed of cells, and recognizes the significance of the nucleus.
			1839 Mohl describes the appearance of the cell plate between the daughter cells during cell division, or 'mitosis.'
			1846 William Morton demonstrates the effective use of ether as an anesthesia.
			1848 Louis Pasteur discovers molecular dissymmetry, or chirality, and coins the distinction between users and non-users of oxygen, 'aerobic' and 'anaerobic.'
			1801 Thomas Young observes that light passing through a double-slit recombines to create light and dark areas, and measures the wavelength of light using this pattern.

1700 - 1750	*1750 - 1800*	*1800 - 1850*
1704 Newton, in *Opticks*, presents his discoveries using light and elaborates his theory that it is composed of particles.	1751 Benjamin Franklin publishes *Experiments and Observations on Electricity* after several years of experiments.	1807 Young coins the word 'energy' for the fundamental quantity created by the heat which moved particles in Bernoulli's kinetic theory.
1705 Halley recognizes the orbit of the comet that bears his name and predicts its reappearance in 1758.	1752 Thomas Melvill notices that the spectra of flames into which metals or salts have been introduced show bright lines characteristic of the metal or salt.	1814 Joseph von Fraunhofer devises a primitive spectroscope by allowing light to pass through a narrow slit and then a prism.
1718 Halley states that stars move, since they had changed position since Ptolemy's *Almagest*.	1756 Franz Ulrich Theodosius Aepinus, realizes that the causes of magnetic and electrical phenomena were extremely similar.	1816 Augustin Jean Fresnel shows that diffraction, interference, and polarization can be explained in terms of the transverse wave theory of light.
1738 Daniel Bernoulli asserts the principle that as the speed of a moving fluid increases, the pressure within the fluid decreases, inventing the kinetic theory of gases.	1759 The return of Halley's comet confirms Newton''s mechanics.	1820 Hans Christian Øersted initiates the study of electromagnetism by placing a needle parallel to a wire conducting electric current and discovering that this produces a magnetic field that curls around the wire.
1746 Andreas Cunaeus invents the 'Leyden jar,' a form of capacitor.	1759 Aepinus fathers the action-at-a-distance/localization of charge theory of electricity and magnetism.	1824 Sadi Carnot shows that even under ideal conditions a steam engine cannot convert into mechanical energy all the heat energy supplied to it.
1798 Cavendish constructs a torsion balance by which he measured the mean density of the Earth.	1768 Euler proposes that the wavelength of light determines its color.	1827 Georg Simon Ohm discovers that the ratio of the potential difference between the ends of a conductor and the current flowing through it is constant, and is the resistance of the conductor.
	1783 Carnot specifies the optimal and abstract conditions for the operation for all sorts of actual machines.	1831 Faraday discovers the means of producing electricity from magnetism, i.e., electromagnetic induction.
	1785 Charles Augustin de Coulomb formulates the inverse square law for the force between electric charges.	1841 Julius Robert Mayer, working with established experimental results, derives the general relationship between heat and work, which is the first law of thermodynamics, a form of the law of conservation of energy.
		1842 Christian Doppler develops the theory that the frequency of energy in the form of the form of waves changes depending on the motion of either the sender or the receiver.
		1843 James Prescott Joule demonstrates experimentally the equivalence of the heat produced and the mechanical work spent in the operation.
		1846 Johann Gottfried Galle discovers the planet Neptune where Urbain Jean Joseph Le Verrier and, independently, John Couch Adams had predicted that a planet would be found.

Physics

	1700 - 1750	1750 - 1800	1800 - 1850
Physics (Continued)		1792 Volta discovers he could arrange metals in a series in such a way that chemical energy is converted into electrical energy.	1847 Hermann von Helmholtz formulates the law of the conservation of energy in an equation which expresses the most general form of the principle. 1850 Jean Foucault, using a rotating mirror, determines the speed of light in the air as 298,000 km/s.
Chemistry	1709 Gabriel Daniel Fahrenheit constructs an alcohol thermometer and, five years later, a mercury thermometer. 1742 Celsius develops the centigrade temperature scale which carries his name.	1754 Joseph Black heats calcium carbonate which separates into calcium oxide and carbon dioxide and then recombines back into calcium carbonate. 1757 Black discovers latent heat, distinguishing between heat and temperature. 1774 Priestly discovers sulphur dioxide, ammonia, and 'dephlogisticated air,' later named oxygen by Lavoisier. 1780 Lavoisier and Laplace develop a theory of chemical and thermal phenomena based on the assumption that heat is a substance, which they called 'caloric' and deduced the notion of 'specific heat.' 1787 Charles determines that the volume of a fixed mass of gas at constant pressure is proportional to its temperature. This was published by Joseph Louis Gay-Lussac in 1802. 1789 Lavoisier proves that mass is conserved in chemical reactions and created the first list of chemical elements.	1803 Dalton applies atomic theory to a table of atomic weights. 1808 Dalton publishes *A New System of Chemical Philosophy*, launching chemical atomic theory; Gay-Lussac enunciates the 'Law of combining volumes,' which states that when gases combine they do so in small whole number ratios. 1811 Berzelius simplifies chemistry through his suggestion that they be represented by the first letter of each element's Latin name, with the addition of the second letter when necessary. Proportions in a compound were indicated with appropriate number as subscript. 1811 Amedeo Avogadro proposes that equal volumes of gases at the same temperature and pressure contain the same number of molecules. 1815 William Prout proposes that the atomic weights of elements are multiples of that for hydrogen. 1825 Faraday discovers benzene. 1834 Faraday states that the amount of chemical change produced is proportional to the quantity of electricity passed and the amount of chemical change produced in different substances by a fixed quantity of electricity is proportional to the electrochemical equivalent of the substance. 1835 Berzelius suggests the name 'catalysis' for reactions which occurred only in the presence of some third substance. 1839 Christian Swann discovers the existence of ozone. 1848 W. Thomson proposes what became known as the 'Kelvin scale,' after the title bestowed on him by the British government. 1850 Runge demonstrates the separation of inorganic chemicals by their differential adsorption to paper. This is forerunner of chromatographic separations.

	1850 - 1875	1875 - 1900	1900 - 1925
Biology	1852 Georges Newport observes the penetration of the vitelline membrane of a frog egg by sperm. 1857 Pasteur demonstrates that lactic acid fermentation is carried out by living bacteria; Albert von Kolliker describes what were later named 'mitochondria' in the nucleus of muscle cells. 1858 Darwin''s friends, arrange for the simultaneous announcement of Wallace's and Darwin's idea of natural selection. 1859, Darwin, in *Origin of Species* asserts all life had a common ancestor. 1862 Pasteur publishes the 'germ theory': Infection is caused by self-replicating microorganisms, and that attenuated viral cultures granted immunity. These beneficent antigens he named 'vaccines' in honor of Jenner and his vaccinia virus. 1865 O.F.C. Deiters proposes the image of the nerve cell which is accepted today: cell body with its nucleus, multiple, branching dendrites, and a single axon; Lister, using carbolic acid as antiseptic and sterilizing his instrument, proved the efficacy of antiseptic surgery. 1866 Gregor Mendel interprets heredity in terms of a pairing of dominant and/or recessive unit characters. 1871 Darwin, in *The Descent of Man*, suggests that there is no sharp discontinuity between the evolution of humans and animals.	1876 Robert Koch devises the method of employing aniline dyes to stain microorganisms, isolating pure cultures of bacteria and showing the bacterial origin of many infectious diseases. 1879 Walther Flemming names 'chromatin' and 'mitosis,' made the first accurate counts of chromosome numbers,and discerned the longitudinal splitting of chromosomes. 1883 Wilhelm Roux suggests that the filaments within the cell's nucleus carry the hereditary factors. 1885 Hertwig and Strasburger develop the conception that the nucleus is the basis of heredity. 1890 Hans Driesch separates two cells of a fertilized sea urchin egg by shaking with very different results than Roux: From a single cell arose an entire sea urchin; Richard Altmann reports the presence within cells of organisms which live as intracellular symbionts, later named mitochondria. 1894 H.J.H. Fenton discovers a reaction now considered to be one of the most important mechanisms of oxidative damage in living cell.	1900 Mikhail Tsvet discerns three green pigments, chlorophyll a, b, and c, differing in color, fluorescence, and spectral absorption. 1902 Karl Landsteiner found that human blood was one of four types, A, B, A-B, and O, thus making transfusions safe; Fischer proposes that proteins consist of chains of amino acids; Ivan Pavlov combines associative learning with reflex acts, postulating the existence of associated stimuli, or 'conditioned responses.' 1903 Tsvet develops methods in chromatography. 1905 Edmund Beecher Wilson discovers that the X chromosome is linked to the sex of the bearer. 1908 Godfrey Harold Hardy works out the equilibrium formula for a population heterogenous for a single pair of alleles. 1910 Konstantin S. Mereschovsky publishes an essentially modern view of the bacterial origin of what later came to be called eukaryotic cells. 1911 Alfred Henry Sturtevant, an undergraduate student of Morgan's, constructs the first rudimentary map of the fruit fly chromosome, establishing that genes are real. 1913 Lawrence Joseph Henderson proposes that the concept of fitness be extended to the environment. This has ramifications for the origin of life.

	1850 - 1875	1875 - 1900	1900 - 1925
Biology (Continued)	1873 Anton Schneider describes chromosomes during the process of mitosis during cell division.		1921 Victor Jollos hypothesizes that the disappearance of environmentally-induced acquired traits, even after hundreds of generations, indicates that their acquisition should be assigned to the cytoplasm rather than the nucleus; Muller raises the question of the relationship of genes to viruses, or 'naked genes'. 1922 Walter Garstang shows that phylogeny is not the cause but the product of different ontogenies. 1923 Robert Feulgen discovers a selective staining technique for DNA localization, which is still in use; Jean Piaget maintains that child development proceeds in the same sequence of genetically determined stages.
Physics	1851 Foucault demonstrates that a pendulum's swing, seen relative to the Earth, would gradually precess, evidence of the Earth's rotation. 1859 Kirchhoff proves a theorem about blackbody radiation, namely, the energy emitted E depends only on the temperature and the frequency of the emitted energy. 1861, Maxwell announces his discovery that some of the properties of the vibrations in the magnetic medium are identical with those of light, and predicts the speed of light theoretically; Anders Jonas Ångström, using a spectroscope, confirms the presence of hydrogen in the Sun.	1876 Alexander Graham Bell invents the telephone. 1879 Crookes attempts to determine the paths of the 'lines of molecular pressure,' or cathode rays, in an evacuated glass tube through which two electrodes are passed. 1879 Albert Michelson determines the speed of light to be 186,350 miles per second. 1881 Venn represents logical propositions diagrammatically.	1900 Planck introduces 'quantum theory' to explain a formula, $E=hf$, where E is energy, f is frequency, and h is a new constant; Rutherford identifies a third type of radiation, which he calls 'gamma radiation.' 1903 Orville and Wilbur Wright achieve flight in a manned, gasoline power-driven, heavier-than-air flying machine. 1904 Lorentz formulates the so-called 'Lorentz transformation,' which describes the increase in mass, the shortening of length, and the time dilation of a body moving at speeds close to that of light. 1904 Hantaro Nagaoka proposes a 'Saturn model' of the atom with a nucleus and many electrons in a ring around it.

Physics (Continued)	1850 - 1875	1875 - 1900	1900 - 1925
	1865 Maxwell publishes his four equations of electromagnetism based on the work of Coulomb, Gauss, Ampere, and Faraday. 1871 Crookes creates a vacuum of about one millionth of an atmosphere which made possible the discovery of X-rays and the electron.	1887 Michelson and Edward W. Morley, using an interferometer to investigate whether the speed of light depends on the direction the light beam moves, fail to detect the motion of the Earth with respect to the aether, thereby refuting the hypothesis that the aether exists; Heinrich Hertz produces electromagnetic radio waves. 1888 Nicola Tesla patents his invention of alternating electric current. 1892 Lorentz proposes a theory in which a body carries a charge if it has an excess of positive or negative particles, and an electric current in a conductor is a flow of particulate particles. 1895 Wilhelm Conrad Röntgen, using a Crookes' tube, observes a new form of penetrating radiation, which he named X-rays. 1897 Joseph John Thomson, using a Crookes' tube, demonstrates that cathode rays consisted of units of electrical current made up of negatively charged particles of subatomic size (electrons). 1899 Ernest Rutherford characterizes 'alpha rays' and 'beta rays'; Becquerel shows that radioactivity in uranium consists of charged particles that are deflected by a magnetic field.	1905 Albert Einstein publishes three papers describing his explanation of the photoelectric effect, Brownian movement, and his theory of special relativity. 1907 Einstein deduces the expression for the equivalence of mass and energy, $E=mc^2$. 1908 Robert Andrews Millikan determines the probable minimum unit of an electrical charge, that is, of an electron. 1909 Hans Geiger and E. Marsden, under Rutherford's direction, scatter alpha particles with thin films of heavy metals, providing evidence that atoms possess a discrete nucleus. 1911 Heike Kamerlingh Onnes discovers 'superconductivity,' the ability of certain materials at low temperatures to carry electric current without resistance; Einstein postulates that light is bent by gravity. 1911-13 Hertzsprung and Russell publish graphs plotting color or spectral class against the absolute magnitude of stars. These are now called HR diagrams and are the basis of the theory of stellar evolution. 1913 Niels Bohr, applying the Planck quantum hypothesis to Rutherford's atomic model, places electrons in discrete energy levels, and postulating the quantum model of the atom; Einstein and Marcel Grossman investigate curved space and time as it relates to a theory of gravity. Einstein contributed the physics and Grossman the mathematics.

	1850 - 1875	1875 - 1900	1900 - 1925
Physics (Continued)			1919 E. Rutherford discovers the proton, which contains the positive charge within the nucleus of an atom, and publishes the first evidence of artificially-produced splitting of atomic nuclei; Eddington and Frank W. Dyson measure the bending of starlight by the gravitational pull of the sun, thus confirming Einstein's general theory of relativity. 1920 E. Rutherford postulates the existence of the neutron, required in order to keep the positively-charged protons in the nucleus from repelling each other. 1922 Arthur Compton demonstrates an increase in the wavelengths of X-rays and gamma rays when they collide with loosely bound electrons, verifying the quantum theory since the effect requires the rays be treated as particles, not waves. 1923 Louis de Broglie hypothesizes that a moving electron particle has wave-like properties.
Chemistry	1855 David Alter described the spectra of hydrogen and other gases. 1858 Friedrich August Kekulé von Stradonitz suggests that carbon atoms are formed in chains. 1859 Robert Wilhelm Bunsen discovers that each element produces its own characteristic set of lines in the spectrum. 1866 Alfred Nobel patents dynamite in Sweden.	1879 Stefan conjectures that that the radiant energy emitted by an enclosure equivalent to a black body is proportional to the fourth power of the body's temperature. 1884 Jacobus van't Hoff explains the principle of equilibrium in chemical dynamics and osmotic electrical conductivity. 1894 Strutt and William Ramsay discover and isolate argon in the process of explaining the discrepancy between the weight of nitrogen obtained from the air and from ammonia.	1905 Arrhenius expresses concern about global warming as a result of burning fossil fuels. 1913 Frederick Soddy discovers that different forms of the same element were, in fact, groups of elements with the same chemical character, but varying in their masses (isotopes), and that radioactive decay is accompanied by the transmutation of one element to another. 1916 Gilbert Newton Lewis states that the chemical bond consists of two electrons held jointly by two atoms.

Chemistry (Continued)	1850 - 1875	1875 - 1900	1900 - 1925
	1869 Dmitri Mendeléev and, independently, Julius Lother Meyer formulate the 'Periodic law.' Mendeléev placed the chemical elements in seven rows in an order where those elements having similar chemical properties were aligned vertically.	1896 Eduard Buchner discovers a chemical in yeast, which he called zymase. He noted that the crushed yeast, that is, cell-free yeast, fermented sugar. This observation opened the era of modern biochemistry.	1925 Wolfgang Pauli puts forth the principle that no two electrons in the atom can be in the same quantum state.
	1869 John Hyatt produces 'celluloid,' the first synthetic plastic to be put into wide use.	1896 Antoine Henri Becquerel discovers radioactivity in uranium.	
		1897 Felix Hoffman synthesizes a form of acetysalicylic acid that enabled the mass production of aspirin two years later.	
		1898 Marie Sklodowska Curie and P. Curie discover and isolate radium and polonium, and clarify that radiation is an atomic property. M. Curie coins the term 'radioactive.'	
		1898 J. Thomson shows that neon gas consists of two types of charged electrons, or ions, each with a different charge, or mass, or both. This raised the possibility that varieties of a single element might exist with the same atomic number but differ in mass; Wien identifies a positive particle equal in mass to the hydrogen atom, which later was named the 'proton'; Ramsey and Morris Travers discover neon, krypton, and xenon; James Dewar liquefies hydrogen.	

	1925 – 1950	1950 – 1975	1975 – 2000
Biology	1928 Alexander Fleming discovers penicillin, a relatively innocuous antibiotic. 1929 It was found that deoxyribonucleic acid (DNA) is located exclusively in the chromosomes, whereas ribonucleic acid (RNA) is located mainly outside the nucleus; Fisher provides a mathematical analysis of how the distribution of genes in a population will change as a result of natural selection, and maintained that once a species' fitness is at a maximum, any mutation will lower it. 1930 Phoebus Aaron Levene elucidate the structure of mononucleotides and showed them to be the building blocks of nucleic acids. 1931 Harriet B. Creighton and Barbara McClintock, working with maize, and Curt Stern, working with Drosophila, provide the first visual confirmation of genetic 'crossing-over.' 1935, William Cumming Rose recognizes the essential amino acid 'threonine.' 1937, Krebs discovers the citrus acid cycle, also known as the tricarboxylic acid cycle and the Krebs cycle. 1938 Hans Spemann proposes the concept of cloning and insists that cell differentiation is the outcome of an orderly sequence of specific stimuli, namely, chemical inductive agents, which were predominantly cyto-plasmic in operation; Warren Weaver coins the term 'molecular biology.' 1940 Ernst Boris Chain and Howard Walter Florey extract and purify penicillin and demonstrate its therapeutic utility.	1953 James Watson and Francis Crick build a model of DNA showing that the structure is two paired, complementary strands, helical and anti-parallel, associated by secondary, noncovalent bonds. Maurice H. F. Wilkens' and Rosalind Franklin's X-ray crystallographs of DNA supported the discovery of the structure. 1954 Salk develops an injectable killed-virus vaccine against poliomyelytis, the incidence of which began to decline after mass immunization began the following year. 1968 Norman Geschwind and Walter Levitsky show that in male and female humans there are characteristic anatomical differences, e.g., the size of the planum temporale in the hemispheres of the brain. 1973 Mertz, Davis, Lobban, Berg, Boyer, Cohen, and Morrow, animal genes were spliced into the small rings of DNA, thus beginning recombinant cloning and launching of the biotechnology industry. 1975 E. M. Southern devises an extension of gel electrophoresis, known as 'Southern blotting,' which greatly aids cloning by enabling the identification and sizing of DNA fragments.	1977 Jack Corliss, in a diving bell 2600 meters below the surface of the Pacific Ocean, observes boiling, lightless deep-sea thermal vents with hundreds of species, including a nine-foot tube worm, most of them new to science. 1978 Mary Leaky announces the discovery of fossilized human footprints from about 3.5 million years ago. 1984 Richard Leaky and Alan Walker excavate a Homo erectus skeleton, dated 1.6 million years ago; Alec John Jeffreys discovers 'genetic fingerprinting,' the pattern of nonfunctional repetitions unique to each individual's DNA. 1990 Teams led by Robin Lovell-Badge and Robin Goodfellow isolate the testis-determining factor gene SRY, the master switch for mammalian sex determination.

	1925 - 1950	1950 - 1975	1975 - 2000
Biology (continued)	1941 Astbury establishes that DNA has a crystalline structure. 1943 Thomas Francis and Jonas Edward Salk develop a formalin-killed-virus vaccine against type A and B influenzas. 1944 Oswald T. Avery, Colin MacLeod, and Maclyn McCarty establish that the material of heredity is deoxyribonucleic acid; Archer John Porter Martin and Richard Synge devise 'paper partition chromatography.' 1948 William Howard Stein and Stanford Moore isolate amino acids by passing a solution through through a chromatographic column filled with potato starch. 1949 Sven Furberg draws a model of DNA, setting sugar at right angles to base, with the correct three-dimensional configuration of the individual nucleotide; Frederick Sanger claims that proteins are uniquely specified, the implication being that, as there is no general law for their assembly, a code was necessary. 1950 Ernst L. Wynder and Evarts A. Graham publish a survey indicating a strong correlation between contracting lung cancer and smoking tobacco.		

	1925 - 1950	1950 - 1975	1975 - 2000
Physics	1926 Erwin Schrödinger initiates the development of the final quantum theory by describing wave mechanics, which predicted the positions of the electrons, vibrating as Bohr's standing waves.

1927 Heisenberg states that electrons do not possess both a well-defined position and a well-defined momentum simultaneously.

1927 George P. Thomson diffracts electrons by passing them in a vacuum through a thin foil, thus verifying de Broglie's wave hypothesis; Davisson and Germer measure the length of a de Broglie wave by observing the diffraction of electrons by single crystals of nickel.

1929 Robert van de Graaf develops an electrostatic particle accelerator; Hubble observes that all galaxies are moving away from each other.

1930 Ernest O. Lawrence publishes the principle of the cyclotron which uses a magnetic field to curl the particle trajectory of a linear accelerator into into a spiral.

1931 Pauli, in order to solve the question of where the energy went in beta decay, predicts the existence of a 'little neutral thing,' the 'neutrino.'

1932 Irène Curie and Frédéric Joliot bombard nonradioactive beryllium with alpha particles, transmuting it briefly into a radioactive element; James Chadwick isolates the neutron, the first particle discovered with zero electrical charge.

1934 I. Curie and Joliot announce the discovery of "artificial radiation obtained by bombarding certain nuclei with alpha particles." | 1950 Hoyle claims to have coined 'big-bang,' for the primal fireball, disparaging the notion that such ever occurred.

1956 Leon Cooper shows that in superconductivity the current is carried in bound pairs of electrons, or 'Cooper pairs.' This led to the BCS theory of superconductivity the following year.

1957 John Backus leads the team which creates 'Fortran,' the Formula Translation language for the IBM 704 computer.

1957 The United States government forms the Advanced Research Agency, or ARPA, in response to the Soviet Union's Sputnik, the first artificial satellite.

1958, Jack Kilby builds the first integrated circuit.

1959, James A. Van Allen, Carl E. McIlwain, and George H. Ludwig establish the existence of geometrically trapped electrons and protons in two belts above the Earth, later called the Van Allen Belts.

1960 Theodore H. Maiman describes the first laser, which used a synthetic ruby rod as the lasing medium. | 1978 Elementary particle physicists begin speaking of the 'Standard Model' as the basic theory of matter.

1979 The spacecraft Voyager 1 photographs Jupiter's rings, and subsequently visits Saturn, Uranus, and Neptune.

1983 Carlo Rubbia and Simon van der Meer, using the CERN particle accelerator, confirmed the existence of the Z and Ws particles.

1986 Johannes Georg Bednorz and Karl Alexander Müller find a new class of layered materials which superconduct at much higher temperatures than any which had been found previously.

1987 A supernova, SN 1987A, explodes in the Large Magellanic Cloud, and was the nearest supernova to have been observed since the invention of the astronomical telescope.

1990 NASA and the European Space Agency (ESA) launch the Hubble Space Telescope, or HST. Servicing missions were carried out in 1993, 1997, and 2002; Tim Berners-Lee and CERN, The European Organization for Nuclear Research, implemented a hypertext system for information access for physicists. |

	1925 - 1950	1950 - 1975	1975 - 2000
Physics (Continued)	1935 IBM introduced a punch card machine with an arithmetic unit based on relays which could do multiplication. 1938, Otto Hahn and Lise Meitner, with their colleague Fritz Strassman, bombard uranium nuclei with slow neutrons. Meitner, interprets the results to be 'nuclear fission,' the term fission being borrowed from biology. 1942 Fermi creates the first controlled, self-sustaining nuclear chain reaction. 1945 The first atomic bombs are exploded over Hiroshima, Japan, then, three days later, over Nagasaki. 1946 John Mauchly and John Presper Eckert demonstrate ENIAC, or Electronic Numerical Integrator and Computer. Its components were entirely electronic.	1963 Murray Gell-Mann and, independently, George Zweig, invent the notion of a more fundamental particle than neutrons and protons which Gell-Mann named the 'quark.' 1965 Arno Allan Penzias and Robert Woodrow Wilson discover cosmic background radiation. The implication is that intergalactic space is above absolute zero, or about 3 degrees K, leading to a drastic shift of the consensus to favor acceptance of the big-bang cosmology. 1967 Steven Weinberg, and Abdus Salam complete the observation of Glashow that the weak and electromagnetic forces result from the same fundamental force. 1968 ARPA , under Lawrence G. Roberts, contracts with BBN to build ARPANET, the prototype of the computer internet. 1970 Stephen Hawking and Penrose prove that the universe must have had a beginning in time, on the basis of Einstein's theory of General Relativity, i.e., mathematically, the big-bang must have arisen from a singularity. 1972 Ray Tomlinson creates the first electronic mail program.	1992 the United States' COBE, or 'Cosmic Background Explorer,' astronomical satellite detects very small variations, or ripples or lumps, in the background cosmic radiation which are thought to be imprints of quantum fluctuations from the early universe, or, in other words, the seeds of later giant structures; CERN releases to the public their hypertext for physicists, naming it the World Wide Web. 1995 Michel Mayor and Didier Queloz detect the first extra-solar planet using the 'wobble technique.' 1997 Ian Wilmut and Keith Campbell clone a sheep, 'Dolly,' from adult cells.

	1925 - 1950	1950 - 1975	1975 - 2000
Chemistry	1927 Walter Heitler and Fritz London show that chemical bonding, the force which holds atoms together, is electrical, and a consequence of quantum mechanics. 1931 Pauling details the rules of covalent bonding. 1944 Seaborg proposed a second 'lanthanide group' as an addition to the periodic table of the elements, as well as existence of a similar series, 90 through 103, or 'actinide group.' 1950 Leo Rainwater combines the liquid drop and shell models of the atomic nucleus.	1950 Leo Rainwater combines the liquid drop and shell models of the atomic nucleus. 1963 Stephanie Louise Kwolek synthesizes polybenzamide, or PBA, a liquid crystalline polymer, used in lightweight body armor. 1969 Calvin publishes *Chemical Evolution* in which he gives several autocatalytic scenarios for the origin of life. 1970 Woodward and Hoffman, in *The Conservation of Orbital Symmetry*, design a set of rules for postulating the areas around atoms where it is most probable that electrons will be found.	1977 Mandelbrot publishes *The Fractal Geometry of Nature* in which complex curves are reduced to straight lines, or fractals, and undergo invariant scaling.

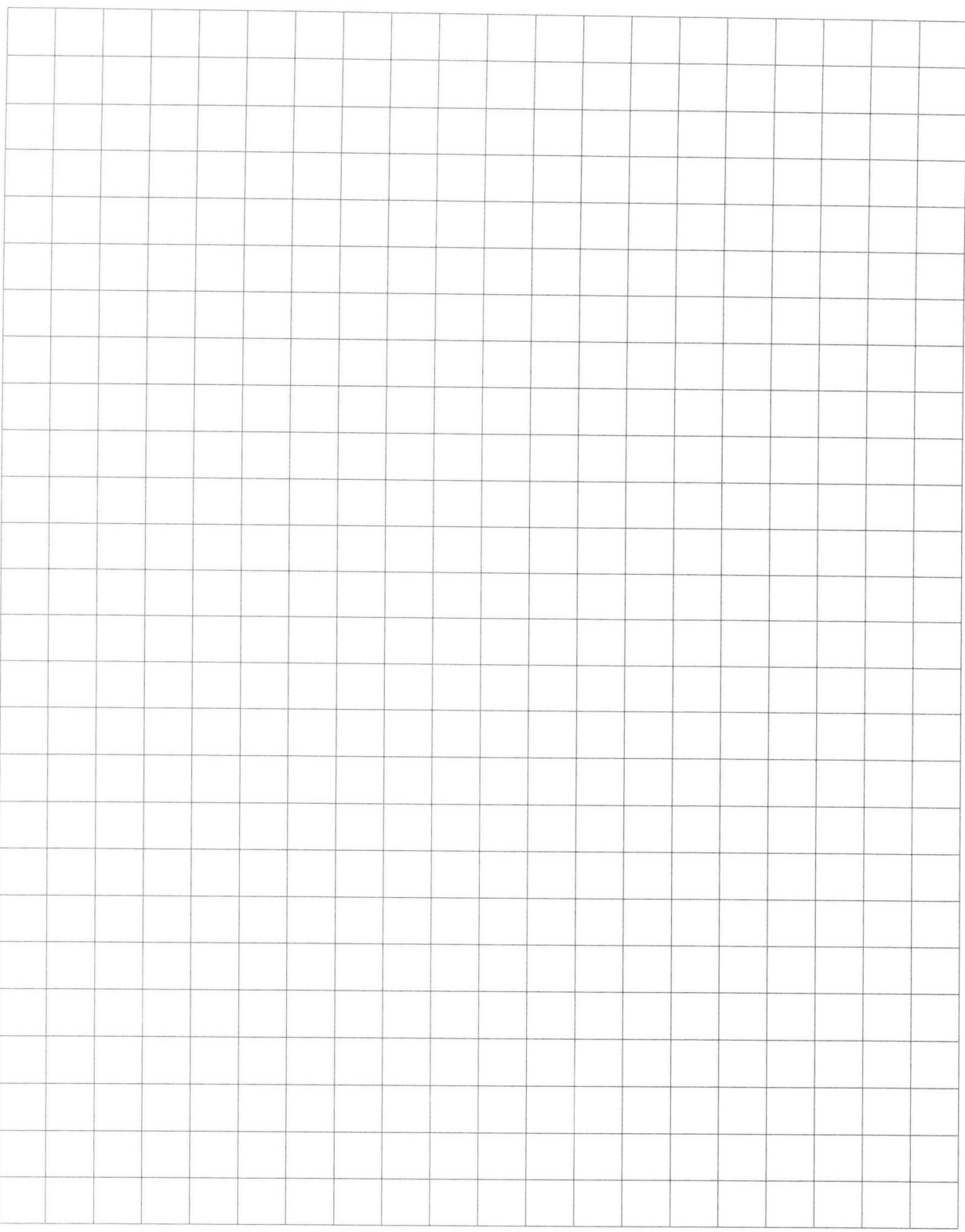

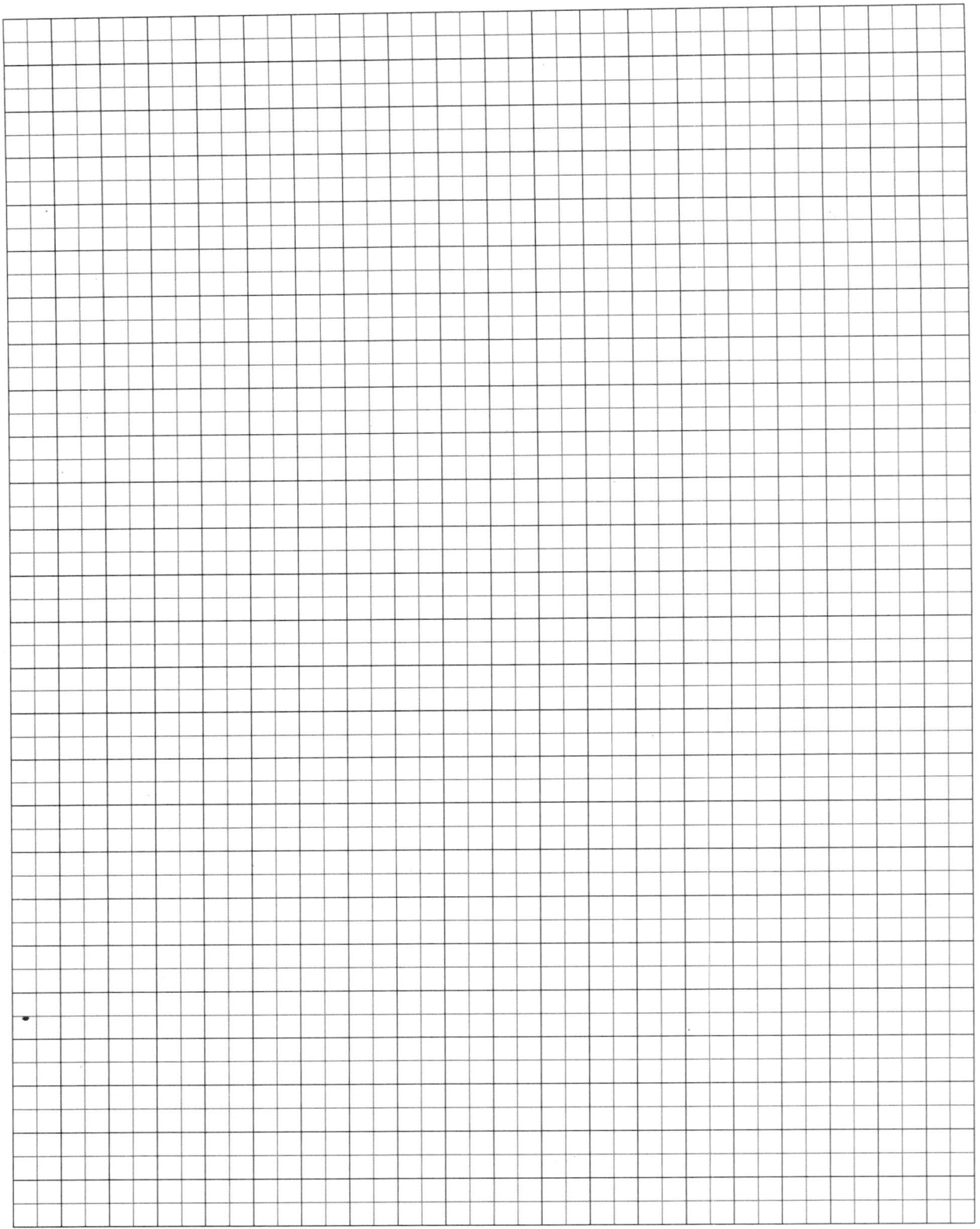